	Multiply	By	To obtain
PRESSURE	millimeters mercury (mm Hg) at 0 °C	133.32	pascals (Pa; N per m^2)
	pounds per square inch (psi)	6.895	kilopascals (kPa; 1000 pascals)
	pascals	0.0075	millimeters mercury at 0 °C
	kilopascals	0.1450	pounds per square inch
	bars	1000	millibars (mb)
	bars	100000	pascals
	bars	0.9869	atmospheres (atm)
ENERGY*	joules (J)	0.2389	calories (cal)
	kilocalories (kcal)	1000	calories
	joules	1.0	watt-seconds (W-sec)
	kilojoules (kJ)	1000	joules
	calories	0.00397	Btu (British thermal units)
	Btu	252.0	calories
POWER*	joules per second	1.0	watts (W)
	kilowatts (kW)	1000	watts
	megawatts (MW)	1000	kilowatts
	kilocalories per minute	69.93	watts
	watts	0.0013	horsepower (hp)
	kilowatts	5.7	Btu per minute
	Btu per minute	0.235	horsepower
TEMPERATURE	degrees Fahrenheit (°F) plus 459.67	0.5555	kelvins (K)
	degrees Celsius (°C) plus 273.15	1.0	kelvins
	kelvins	1.80	degrees Fahrenheit minus 459.67
	kelvins	1.0	degrees Celsius minus 273.15
	degrees Fahrenheit minus 32	0.5555	degrees Celsius
	degrees Celsius	1.80	degrees Fahrenheit plus 32

* It is important to understand the distinction between *energy* and *power* as well as the corresponding metric and English units.

 Energy in its various forms (for example, heat, chemical energy, radiant energy) represents the ability to perform work. Energy is therefore expressed in units of *work*, which is defined as a force applied over some distance. In the metric system, units of energy are the joule and the erg. By definition, 1 *joule* is the energy required when 1 newton acts through 1 meter. (One *newton* is the force that accelerates a 1-kilogram mass by 1 meter per second per second.) Alternately, 1 *erg* is the energy required when 1 dyne acts through 1 centimeter. (One *dyne* is the force that accelerates a 1-gram mass by 1 centimeter per second per second.) One joule is the equivalent of 10 ergs, and 1 calorie of heat (defined in Chapter 3) equals 4.186 joules.

 Power is the rate at which energy is used, that is, released or converted. One *watt* is defined as 1 joule per second, and 1 *kilowatt* is 1000 joules per second. One *megawatt* equals 1000 kilowatts.

 When solar radiation is absorbed by something (air, water, or land, for example), radiant energy is converted to heat. We can describe the amount of heat in terms of calories or joules. We may also be concerned about the rate at which solar radiation travels through some cross-sectional area or the rate at which it is absorbed by some surface. The flux of energy is then described in terms of watts per square meter, joules per square meter per minute, or calories per square centimeter per minute.

Meteorology

*The Atmosphere and the
Science of Weather*

THIRD EDITION

Meteorology

The Atmosphere and the Science of Weather

JOSEPH M. MORAN
MICHAEL D. MORGAN

UNIVERSITY OF WISCONSIN–GREEN BAY

Chapter 11 written by
PATRICIA M. PAULEY

UNIVERSITY OF WISCONSIN–MADISON

MACMILLAN PUBLISHING COMPANY
NEW YORK
COLLIER MACMILLAN CANADA
TORONTO

EDITOR: Robert McConnin
PRODUCTION SUPERVISOR: Elisabeth Belfer
PRODUCTION MANAGER: Nicholas Sklitsis
TEXT DESIGNERS: Rosalie Herion/Robert Freese
COVER DESIGNER: Robert Freese
PAGE LAYOUT: Rosalie Herion
COVER AND TITLE PAGE PHOTOGRAPHS: Mike Brisson

This book was set in Meridien, Torino, & Cheltenham Type
by Ruttle, Shaw & Wetherill, Inc.,
printed and bound by R. R. Donelley.
The cover was printed by Lehigh Press, Inc.

Macmillan Publishing Company
866 Third Avenue, New York, New York 10022

Collier Macmillan Canada, Inc.
Suite 200
1200 Eglinton Avenue, E.
Don Mills, Ontario, M3C 3N1

LIBRARY OF CONGRESS CATALOGING-IN-PUBLICATION DATA

Moran, Joseph M.
    Meteorology : the atmosphere and the science of weather / Joseph
M. Moran, Michael D. Morgan ; chapter 11 prepared by Patricia M.
Pauley. — 3rd ed.
    p.   cm.
    Includes bibliographical references.
    ISBN 0-02-383841-8
    1. Meteorology.   2. Weather.   3. Atmospheric physics. I. Morgan,
Michael D. II. Title.
QC861.2.M625   1991
551.5—dc20   90-5806
                                                        CIP

Printing: 1 2 3 4 5 6 7 8   Year: 1 2 3 4 5 6 7 8 9 0

Dedicated to the memory of
Joseph P. Moran
Lewis W. Morgan

PREFACE

This book is intended to introduce the college nonscience major to the fundamentals of atmospheric science. It is appropriate for an introductory course on the atmosphere, weather, or weather and climate. Our primary goal is to demonstrate how scientific principles govern the circulation of the atmosphere and the day-to-day sequence of weather events. In so doing, we also introduce the nonscience student to the nature of scientific inquiry and the methodology of science. Atmospheric science is especially well suited to achieving these goals because it is an applied science and readily lends itself to familiar illustrations.

Our approach is based on our combined 43 years of teaching science to nonscience students with little or no background in the sciences or mathematics. Our aim in writing *Meteorology* has been to present the principles of meteorology as simply as possible in the context of everyday examples but without sacrificing scientific integrity. Clear and logical explanations of the principles underlying meteorological phenomena and observations make it unnecessary to write around or avoid even the more sophisticated ideas. We break down basic concepts into elementary components and arrange them so that one concept builds logically on another. The geostrophic wind, for example, is introduced only after a detailed examination of the separate forces contributing to that wind. The atmosphere gradually emerges as a highly complex and interactive system that is governed by physical laws.

Our emphasis on scientific methodology provides the student a perspective on the accomplishments of atmospheric scientists and the challenges still facing them. The reader soon understands that weather is not an arbitrary act of nature, and yet weather forecasting has limits and the climatic future is uncertain. We have integrated topics of special contemporary interest including acid rain, potential climatic effects of rising levels of carbon dioxide, and threats to the stratospheric ozone shield.

Organization

Chapters are arranged in a traditional sequence. Chapter 1 introduces the basic properties and structure of the atmosphere. Chapters 2, 3, and 4 cover radiation and energy conversions within the atmosphere and focus on the global radiation balance. The main theme in these chapters is that weather is a response to heat imbalances within the Earth-atmosphere system. Air pressure is discussed in Chapter 5 and is related to other atmospheric properties through the gas law. Water in the atmosphere is the subject of Chapters 6, 7, and 8, with special emphasis on saturation and precipitation processes. Having established the basis for atmospheric circulation, we then examine the forces governing weather systems (Chapter 9). There follow descriptions of global-scale circulation (Chapter 10), synoptic-scale weather systems (Chapter 11), and local and regional weather systems (Chapter 12). Chapters 13 and 14 deal with the genesis and characteristics of thunderstorms and severe weather phenomena such as tornadoes and hurricanes. Our treatment of weather closes with Chapter 15 on weather forecasting, which is structured to integrate and apply the key concepts developed earlier in the book. A special feature of the book is the inclusion of three final chapters on topics that are of particular contemporary interest: Chapter 16 on air pollution meteorology, Chapter 17 on world climates, and Chapter 18 on the climatic record and the variability of climate.

Although it is desirable to cover Chapters 1 through 15 in sequence, certain sections of most chapters may be dropped without loss of continuity. For example, sections on weather instruments or atmospheric optics may be deleted. Chapters 16, 17, and 18 may be covered in any order, and any one or all may be omitted to fit the available time.

Special sections, set apart from the main flow of the chapters, provide considerable flexibility with regard to depth and breadth of topic coverage. Each chapter contains one or two supplementary *Special Topics*, such as "Why Is the Sky Blue?" and "Solar Power," which are related to a major theme of the chapter. *Mathematical Notes* at the ends of seven chapters provide quantitative discussions of concepts in the chapters, with the basic meteorological equations for those who are interested.

In addition, each chapter features the following elements, designed to guide student understanding:

- chapter outline
- summary statements
- key word list
- review questions
- points to ponder
- projects
- selected readings with annotations

Key words are boldfaced and defined at their first use in the chapter and in the Glossary at the back of the book.

Metric units are used throughout the book, with the English equivalents in parentheses. Unit conversions are given inside the front cover. Psychrometric tables appear inside the back cover. The appendixes feature the history of meteorology, the standard atmosphere, weather map symbols, weather extremes, and climatic data for the United States and Canada.

The Revision

In this third edition, we have significantly revised and/or updated our treatment of the ozone shield, the greenhouse effect, atmospheric stability, the North American drought of 1988, precipitation processes, El Niño and La Niña, charge separation in thunderstorms, the hurricane threat to the Southeast, long-range weather forecasting, climate controls, and global climate models. We are very pleased that Patricia M. Pauley of the University of Wisconsin–Madison took responsibility for revising and updating Chapter 11, Synoptic-Scale Weather. In addition, we have added a hurricane tracking chart to Appendix III and one new Special Topic: Clouds by Mixing (Chapter 6).

We have retained the same basic organization as in the second edition. However, in response to suggestions by many reviewers, we have introduced a second color to improve the physical appearance and utility of line drawings. In addition, we have upgraded the quality of photographs.

Supplements

Supplements available to adopters of this book include an *Instructor's Manual* (with a set of transparency masters of selected line drawings from the text), a Computerized Test Bank, and a set of 50 35-mm color slides of atmospheric phenomena. New to this edition is a companion *Student Study Guide*. Also available is *Meteorology Exercise Manual and Study Guide*, by Robert A. Paul. Both are published by Macmillan.

Acknowledgments

In preparing the three editions of this book, we have profited greatly from the creative ideas and constructive criticisms of many reviewers. We are especially grateful to Gaylen L. Ashcroft, Utah State Climatologist; David D. Houghton, University of Wisconsin–Madison; John E. Oliver, Indiana University–Terre Haute; Patricia M. Pauley, University of Wisconsin–Madison; Robert A. Ragotzkie, University of Wisconsin–Madison; and Clayton H. Reitan, Northern Illinois University.

We also thank L. Dean Bark, Kansas State University; Bradley Bramer,

University of Wisconsin–Madison; Arnold Court, California State University, Northridge; William A. Dando, University of North Dakota; Russell L. De-Souza, Millersville State University; Lee Guernsey, Indiana State University–Terre Haute; B. Ross Guest, Northern Illinois University; Edward J. Hopkins, University of Wisconsin–Madison; William C. Kaufman, University of Wisconsin–Green Bay; Bruce E. Kopplin, University of Nebraska–Lincoln; Garrick B. Lee, Butte College; William H. Long, Florida State University, Tallahassee; Thomas H. McIntosh, University of Wisconsin–Green Bay; David W. Marczely, Southern Connecticut State University, New Haven; Shamin Naim, Illinois State University; T. R. Oke, University of British Columbia; Robert A. Paul, Northern Essex Community College; Peter J. Robinson, University of North Carolina at Chapel Hill; Charles C. Ryerson, University of Vermont; Russell Schneider, University of Wisconsin–Madison; William M. Smith, University of Wisconsin–Green Bay; Gregory E. Taylor, Creighton University; Charles L. Wax, Mississippi State University; Wayne M. Wendland, Illinois State Water Survey; James H. Wiersma, University of Wisconsin–Green Bay; Thomas B. Williams, Indiana University at Indianapolis; George Wooten, Hillsborough Community College.

We give special thanks to Professor Reid A. Bryson of the University of Wisconsin–Madison for his insight, guidance, and example.

We also have been fortunate in working with some of the most talented and enthusiastic professionals in college publishing. We thank John H. Staples, Nancy Flight, Margaret Mason, Gary Brahms, and Richard Abel for their encouragement in the early days. We owe much to the tireless efforts of our development editor on the first edition, Janilyn M. Richardson: her dedication, fine attention to detail, and objectivity were essential to the successful evolution of this book and made possible subsequent editions. We thank Robert A. McConnin of Macmillan for guiding this project to completion. And we extend a special note of appreciation to Elisabeth Belfer whose editorial skills and commitment to quality contributed much to the second and third editions.

We thank our skillful typists, Jeanne Broeren and Joy Phillips, for their patience. We thank our wives, Jennifer and Gloria, for their understanding and encouragement. Finally, we acknowledge with gratitude the contributions of our students and colleagues at the University of Wisconsin–Green Bay.

J. M. M.
M. D. M.

CONTENTS

6   Humidity and Stability   125

7   Dew, Frost, Fog, and Clouds   157

11 Synoptic-Scale Weather   *by Patricia M. Pauley*   277

12 Local and Regional Circulations   313

16 Air Pollution Meteorology 439

17 World Climates 467

18 Climatic Record and Climatic Variability   499

Appendixes   539

Glossary   557

Index   571

Meteorology

The Atmosphere and the
Science of Weather

*A generation goes, and a
    generation comes. But the earth
    remains forever.*
*The sun rises and the sun goes
    down, and hastens to the place
    where it rises.*
*The wind blows to the south, and
    goes round to the north; round
    and round goes the wind, and
    on its circuits the wind returns.*

ECCLESIASTES 1:4–6

1

Atmosphere: Origin, Composition, and Structure

The Earth's atmosphere is a thin envelope of gases, clouds, and tiny particles that surrounds the globe. Weather in its myriad forms occurs primarily in the lowest 10 km (6.2 mi) of the atmosphere. [NASA photograph]

FIGURE 1.1
The basic variability of weather makes possible a wide variety of recreational activities
[Photographs by J. M. Moran (left) and Mike Brisson (right)]

EVERYONE SEEMS to be interested in the weather, probably because it affects virtually every aspect of daily life—our clothing, our outdoor activities, the prices of oranges and coffee in the grocery store, and even the outcome of a football game. Tranquil, pleasant weather allows us to enjoy a variety of outdoor recreational activities (Figure 1.1). A turn to stormy weather may bring mixed blessings: heavy rains wash out picnics but also benefit crops wilting under the searing summer sun. Occasionally, the weather is extreme, and the impact may range from mere inconvenience to a disaster that is costly in human lives and property (Table 1.1). Heavy snowfall snarls commuter traffic, thick fog causes flight delays and cancellations, and a night of subfreezing cold takes its toll on Florida citrus. But these impacts are minor when compared to the death, injury, and property damage that may attend a hurricane or tornado (Figure 1.2).

Regardless of where we live in the world, each of us is well aware from personal experience that weather is variable. This variability prompted Mark Twain—not one to shy from exaggeration—to quip of spring weather in New England: "I have counted one hundred and thirty-six different kinds of weather inside of four-and-twenty hours."* Of course, weather's variability is not the same everywhere. For example, the temperature contrast between

---

* Excerpted from "Address of Mr. Samuel L. Clemens: The Weather in New England," *New England Society in the City of New York, Annual Report,* 1876, p. 59.

Table 1.1
Approximate Annual Losses to Selected Weather Hazards in the
United States

Hazard	Fatalities	Cost (millions of dollars)
Flood	163	3,175[a]
Hurricane	33	796[b]
Tornado	98	300
Lightning	97	200
Hail		750
Drought		800

[a] In 1985 dollars.
[b] In 1982 dollars.
Source: W. E. Riebsame et al. "The Social Burden of Weather and Climate Hazards." *Bulletin of the American Meteorological Society* 67 (1986):1379.

winter and summer is much more pronounced on Canada's prairies, where summers are very warm and winters are bitter cold, than in south Florida, where the weather is usually tropical year-round.

Any reasonable description of weather must reflect its geographical and temporal variability. Hence, we define **weather** as the state of the atmosphere at some place and time, described in terms of such variables as temperature, cloudiness, precipitation, and wind. **Meteorology** is the study of the atmosphere and the processes that cause weather.

We define **climate** as weather conditions at some locality averaged over a specified time period, but climate encompasses more than this. Departures from averages and extremes in weather are also important aspects of climate. For example, farmers are interested in knowing not only the long-term average

FIGURE 1.2
Occasionally, the weather turns violent and may take lives and cause considerable property damage. Hurricane Camille caused this damage in Biloxi, Mississippi, in August 1969. [NOAA photograph]

rainfall for July, but also the frequency of extremely dry Julys. Climate can be considered the ultimate environmental control in that climate determines, for example, what crops can be cultivated, the long-term water supply, and the average heating and cooling requirements for homes.

The **atmosphere,** where weather takes place, encircles the globe as a relatively thin envelope of gases and tiny, suspended particles. In fact, 99 percent of the atmosphere's mass is confined to a layer having a thickness that is only about 0.25 percent of the Earth's diameter. Hence, in thickness, the Earth's atmosphere is comparable to the thin skin on an apple. Yet, the thin atmospheric skin is essential for life and the orderly functioning of physical and biological processes on Earth. The atmosphere shields organisms from exposure to hazardous levels of ultraviolet radiation;* it contains the gases required for the life-sustaining processes of cellular respiration and photosynthesis; and it supplies the water needed by all life.

Understanding the Atmosphere

Our present understanding of the atmosphere, weather, and climate is the culmination of centuries of painstaking inquiry by scientists from many disciplines. Physicists, chemists, astronomers, and others applied basic principles in unlocking the mysteries of the atmosphere. The roots of modern meteorology, in fact, extend back to the fourth century B.C. and Aristotle's *Meteorologica,* the first treatise on atmospheric science. Other milestones in the early history of meteorology are listed in Appendix 1.

Although much progress has been made, many questions remain regarding the workings of the atmosphere, weather, and climate. Modern atmospheric scientists (meteorologists and climatologists) continue the efforts of their predecessors, and, although armed with more sophisticated tools such as satellites and electronic computers, they still rely on the scientific method of investigation.

THE SCIENTIFIC METHOD

The **scientific method** is a systematic form of inquiry involving observation, speculation, and reasoning. An investigation of an acid rain problem provides an illustration of how scientists apply the scientific method.

In the 1970s, some ponds and lakes in the Adirondack Mountains of New York State showed declines in fish populations. In the past, these same lakes had supported abundant aquatic life. Some local businesses were concerned because barren lakes meant no fishing, and no fishing meant fewer tourist dollars. Environmentalists speculated that the lakes were becoming yet another casualty of pollution.

---

* Radiation in its various forms is discussed in detail in Chapter 2.

Initially, some biologists proposed that toxic materials (poisons) were entering the lakes and were creating conditions intolerable to aquatic life. If this were true, what were the toxins and where were they coming from? Testing of lake water samples revealed that the waters did not contain hazardous concentrations of toxins, but the waters were abnormally acidic. Laboratory studies have shown that excessively acidic waters are lethal to young fish. On the basis of this observation, a new hypothesis was formulated: Acidic rainwater was making the lake waters excessively acidic.

Normal rainwater is slightly acidic because rainwater dissolves some of the carbon dioxide (CO_2) in the air to form weak and harmless carbonic acid (H_2CO_3). Rainwater samples collected in the vicinity of the Adirondack lakes and tested in the laboratory, however, were more than 100 times more acidic than expected. Why was the rainwater so acidic? Further laboratory study identified sulfuric acid (H_2SO_4) and nitric acid (HNO_3) in the rainwater. Such acids form when rainwater dissolves oxides of sulfur and nitrogen, common industrial air pollutants. The Adirondacks are downwind of some major industrial sources of these air pollutants, such as coal-fired electric power plants.

By this reasoning, the loss of fish in Adirondack Mountain lakes was at least circumstantially linked to industrial air pollution. The hypothesis that abnormally acidic rainwater was responsible for fish kills in the ponds and lakes is consistent with what we know about the tolerance of fish for acidic waters, the chemical reactions involving rainwater and air pollutants, and the type of air pollutants that are transported to the Adirondack region. After repeated field testing of rainwater samples and laboratory studies in which fish were exposed to varying levels of acidic water, the hypothesis was accepted as fact.

From our acid rain example, it is evident that the scientific method involves a sequence of steps in which scientists (1) identify a specific question, (2) propose an answer to the question in the form of an ''educated guess,'' (3) state the educated guess in such a way that it can be tested, that is, formulate a hypothesis, (4) predict what the consequences would be if the hypothesis were correct, (5) test the hypothesis by checking to see if the prediction is correct, and (6) revise or restate the hypothesis if the prediction is wrong.

These steps are not always followed in this order, and some steps, such as 3 and 4, may be combined. Indeed, in actual practice, a scientist does not follow this scheme cookbook style, and discrete steps are usually thoroughly integrated as a single avenue of inquiry. Note that the scientific method is not in itself a recipe for creativity. The method does not provide the key idea, the ''hunch,'' the educated guess that becomes the hypothesis. Rather, the method is a technique to test the validity, or worth, of that creative key idea—however or wherever the idea originates.

As in the acid rain example, a hypothesis is a tool that is used to suggest new experiments or observations or open new avenues of inquiry. Hence, in this way, even an erroneous hypothesis is sometimes fruitful. Above all else, the scientist must remember that a hypothesis is merely a working assumption that eventually may be accepted, modified, or rejected. He or she must be objective in evaluating a hypothesis and not allow personal biases or expectations to cloud that evaluation. And, in fact, inquiry, creative thinking, and imagination are stifled when hypotheses are considered to be immutable.

A new hypothesis (or an old, resurrected one) may be hotly debated within the scientific community. History tells us that there is a natural human resistance to new ideas that threaten to displace long-held notions. And sometimes, on a particularly controversial issue, disagreement among scientists receives wide publicity, which may confuse the general public. Thus, in some cases, the prevailing public reaction may be, "Well, if the so-called experts can't agree among themselves, who am I to believe? Is there really a problem after all? However, debate and disagreement are essential steps in the process of reaching scientific understanding in that they generate useful suggestions, stimulate new thinking, and uncover errors. In fact, such debate and skepticism buffer the scientific community from too hastily accepting new ideas. If a hypothesis survives the scrutiny and skepticism of scientists and the public, chances are that it is on target.

ATMOSPHERIC MODELS

In applying the scientific method, scientists may find that the use of models will aid their investigation. This is certainly the case in the atmospheric sciences. Because we will be using models throughout this book, it is a good idea to consider at the outset some of the general objectives and limitations of scientific models, especially as they apply to meteorology and climatology.

A **scientific model** is defined as an approximate representation or simulation of a real system. A model eliminates all but the essential variables, or characteristics that can change. For example, to learn how to improve the fuel efficiency of automobiles, we might examine a model automobile in a wind tunnel. An automobile can be designed to reduce its air resistance and thereby increase its miles per gallon. The shape of the model automobile is the critical variable and is the focus of study. Other variables, such as the color of the model automobile or whether it is equipped with whitewall tires, are irrelevant to the experiment and can be ignored. Often what is relevant and what is not are determined by trial and error.

Sometimes models are used to organize information. Because they are not cluttered with extraneous and distracting details, they may provide important insights into how things interact, or they may trigger creative thinking about complex phenomena. Models can also be used to make predictions. For example, in a model composed of many variables, one variable may be perturbed in order to assess its effect on other variables.

Depending on their particular functions, scientific models are classified as conceptual, graphical, physical, or numerical. A **conceptual model** describes the general relationships among components of a system. For example, the geostrophic wind (described in detail in Chapter 9) is a conceptual model that

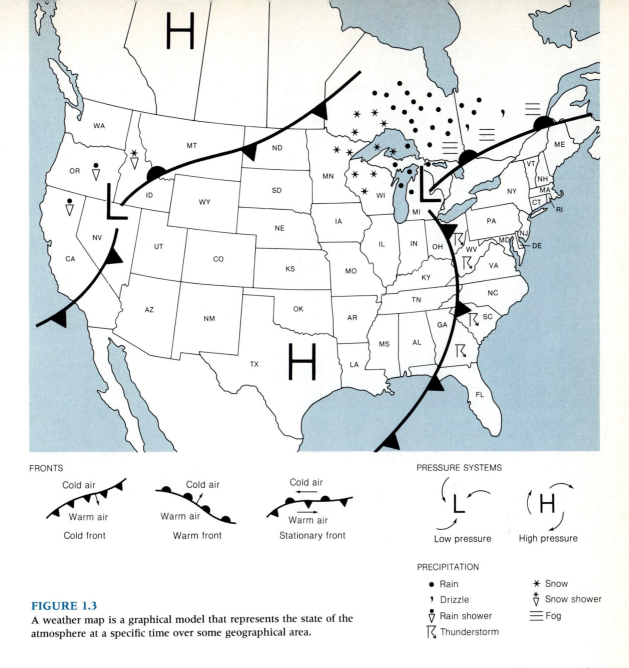

FRONTS

Cold front — Cold air / Warm air

Warm front — Cold air / Warm air

Stationary front — Cold air / Warm air

PRESSURE SYSTEMS

Low pressure — L

High pressure — H

PRECIPITATION

- Rain
, Drizzle
Rain shower
Thunderstorm

* Snow
Snow shower
Fog

FIGURE 1.3
A weather map is a graphical model that represents the state of the atmosphere at a specific time over some geographical area.

relates the interaction of certain forces operating in the atmosphere to straight, horizontal air movement above an altitude of about 1.0 kilometer (km), or 0.6 mi.* A **graphical model** assembles and displays data in an organized format that can be readily interpreted. For example, a weather map (Figure 1.3) integrates weather observations taken simultaneously at hundreds of

---

* Metric units, followed by the English-unit equivalents, are used throughout this book. For unit conversions, see the inside front cover.

FIGURE 1.4

A dishpan model of the atmosphere. The outer rim is heated, and the center is cooled, to simulate the temperature contrast between the Earth's equator and poles. As the dishpan rotates, the fluid flows in a manner that approximates the large-scale circulation of the atmosphere. [Courtesy of D. Fultz, Hydrodynamics Laboratory, The University of Chicago (Fultz et al., *Meteorological Monographs* 4, No. 21 (1959):64)]

locations into a coherent representation of the state of the atmosphere. In recent years, scientists have also employed the techniques of computer graphics to simulate complex weather systems; an example is Plate 24. A **physical model** is usually a miniaturized version of some system. For example, scientists have tried to duplicate the circulation of the atmosphere using dishpan models (Figure 1.4).

In recent decades scientific modeling has been greatly assisted by electronic computers that can accommodate enormous quantities of data and perform calculations very rapidly. Typically, a computer is programmed with a **numerical model** consisting of one or more mathematical equations that portray, for example, the behavior of a particular physical system such as the atmosphere. Variables in the numerical model (such as temperature or humidity) may be manipulated, individually or in groups, in order to assess the impact on the system.

Computerized numerical models of the atmosphere have been used to forecast the weather since the 1950s. More recently, they have also been used to predict the potential impact on climate of rising levels of carbon dioxide in the atmosphere. As discussed in more detail in Chapters 2 and 18, the atmospheric carbon dioxide concentration has been increasing for many decades, primarily as a byproduct of fossil fuel (coal and oil) burning. Warmer air temperatures may be the consequence because carbon dioxide slows the loss of the Earth's heat to space.

Atmospheric scientists have employed numerical models of the atmosphere in an effort to determine the temperature rise that might accompany a continued increase in atmospheric carbon dioxide concentration. Three steps are followed. First, they design a numerical model of the atmosphere that accurately depicts the present worldwide air temperature pattern, given the current level of atmospheric carbon dixide. Second, holding all other variables (except temperature) constant, they elevate the carbon dioxide concentration (typically, doubled), and the numerical model computes a new worldwide air temperature pattern. Third, they subtract the initial temperature pattern from the final, predicted temperature pattern. Presumably, the net temperature change can be attributed to the elevated carbon dioxide concentration.

It is important to emphasize that models are only simulations of reality, and hence, they are subject to error. For example, one potential difficulty with numerical models concerns the accuracy of their component equations. Typically, the equations are only approximations of the way a system really works in nature, and they may not account for all the relevant variables. This is one reason why long-range weather forecasting, based on numerical models of the atmosphere, declines in accuracy as the forecast period lengthens.

Evolution of the Atmosphere

Now that we have a general idea of how meteorologists and climatologists (and other scientists) go about the job of studying the atmosphere, it is time to begin our investigation of the basic properties of the atmosphere. In this section we review what is understood about the origins of the atmosphere.

The Earth's atmosphere is the product of a lengthy evolutionary process that began at the Earth's birth about 4.6 billion years ago. Astronomers scanning the solar system and geologists analyzing evidence obtained from meteorites, rocks, and fossils have given us a reasonable, albeit as yet incomplete, scenario of the origins of the atmosphere.

PRIMEVAL PHASE

Planet Earth as well as the sun and the entire solar system is believed to have developed out of an immense cloud of dust and gases within the Milky Way galaxy. In the beginning, Earth was an aggregate of dust and meteorites surrounded by a gaseous envelope of mostly hydrogen and helium. There was no gaseous oxygen. For millions of years, the Earth's mass grew by accretion as the planet swept up cosmic dust in its path. Bombardment by meteorites heated up the Earth and its atmosphere, driving off most of the original atmospheric gases.

Earth became geologically active as volcanoes spewed forth huge quantities of lava, ash, and a variety of gases. Then as now the principal gaseous emissions of volcanoes were probably water vapor (H_2O), carbon dioxide

(CO_2), and some nitrogen (N_2). Free oxygen (O_2) was notably absent, although oxygen was combined with other elements in various chemical compounds such as carbon dioxide. Millions of years of volcanic activity produced a dense atmosphere rich in carbon dioxide and some nitrogen. Intense solar radiation in the upper atmosphere caused some water vapor to break down into its constituent atoms, thereby contributing minor amounts of oxygen to the early atmosphere. Radioactive decay of an isotope of potassium (potassium-40) in the Earth's crustal rock added the inert gas argon to the evolving atmosphere.

This early atmosphere was perhaps 10 to 20 times denser than the present atmosphere. The abundance of carbon dioxide meant higher surface temperatures than now because, as mentioned previously, carbon dioxide slows the loss of the Earth's heat to space. Numerical models that simulate this early atmosphere predict an average surface temperature in the range of 85 to 110 °C (185 to 230 °F).* This phase of the Earth's primeval atmosphere probably persisted for several hundred million years.

With the subsequent formation of seas and the coming of life, the atmosphere continued to evolve. Perhaps 3.8 billion years ago, volcanic activity subsided somewhat, and the Earth and its atmosphere gradually cooled. Cooling caused some of the water vapor to condense into clouds, and rains gave rise to the first rivers, lakes, and seas. The global water cycle began. In those seas, the first primitive forms of life appeared about 3.5 billion years ago, and photosynthesis began with the coming of the first marine plants. **Photosynthesis** is the process whereby plants use sunlight, water, and carbon dioxide to manufacture their food. A byproduct of this process is oxygen, which is released to the atmosphere. Through subsequent millions of years, photosynthesis added oxygen to the atmosphere, and oxygen eventually became the second most abundant atmospheric gas after nitrogen.

We cannot assume, however, that the concentration of atmospheric oxygen climbed steadily after the arrival of photosynthetic organisms. For reasons not completely understood, atmospheric oxygen apparently underwent significant fluctuations at least over the past 100 million years. For example, scientists reporting on their analysis of ancient air bubbles trapped in 85-million-year-old amber (hardened tree resin) concluded that at that time the atmosphere contained up to 50 percent more oxygen than it does today.

While all of this was taking place, the concentration of carbon dioxide in the atmosphere dropped significantly. Photosynthesis removed some, but much of the carbon dioxide dissolved in ocean waters. Marine organisms used some of the dissolved carbon dioxide in building their shells, and when they died their remains accumulated as sediment on the ocean floor. In time these sediments converted to marine carbonate rocks such as limestone ($CaCO_3$). The carbon cycle was complete, and carbon dioxide declined to only a small fraction of atmospheric gases.

---

* The various temperature scales are described in detail in Chapter 3.

MODERN PHASE

Ultimately, these slow, evolutionary processes produced the modern atmosphere, which is a mixture of many different gases. Because the lower atmosphere is mixed continually, the principal atmospheric gases occur almost everywhere in about the same relative proportions up to an altitude of about 80 km (50 mi). That portion of the atmosphere is called the **homosphere.** Above 80 km, gases are stratified. This means that concentrations of the heavier gases decrease more rapidly with altitude than do concentrations of the lighter gases. The region of the atmosphere above 80 km is known as the **heterosphere.**

Nitrogen and oxygen are the chief gases composing the homosphere. Not counting water vapor (which has a highly variable concentration), nitrogen occupies 78.08 percent by volume of the homosphere, and oxygen is 20.95 percent by volume. The next most abundant gases are argon (0.93 percent) and carbon dioxide (0.03 percent). As shown in Table 1.2, the atmosphere also contains small quantities of helium, methane, hydrogen, ozone, and several other gases. Unlike the atmosphere's principal gases, the percent volumes of some of these trace gases vary within the homosphere and with time.

Within the heterosphere, above about 150 km (93 mi), oxygen is the chief atmospheric gas and occurs primarily in the atomic (O) rather than diatomic (O_2) form. Ultraviolet radiation from the sun causes photodissociation of O_2 into its constituent atoms. (**Photodissociation** is the breakdown of molecules by radiation.) Two oxygen atoms can recombine into a molecule only by colliding with a third body (another atom or molecule). However, air

Table 1.2

Relative Proportions of Gases Composing Dry Air in the Lower Atmosphere (below 80 km)

Gas	Percent by volume	Parts per million
Nitrogen	78.08	780,840.0
Oxygen	20.95	209,460.0
Argon	0.93	9,340.0
Carbon dioxide	0.035	350.0
Neon	0.0018	18.0
Helium	0.00052	5.2
Methane	0.00014	1.4
Krypton	0.00010	1.0
Nitrous oxide	0.00005	0.5
Hydrogen	0.00005	0.5
Ozone	0.000007	0.07
Xenon	0.000009	0.09

density (mass per unit volume) at these altitudes is so low that such collisions are infrequent. At lower altitudes the intensity of incoming solar ultraviolet radiation weakens, thereby reducing the rate of photodissociation of O_2. With greater air density at lower altitudes, molecular collisions are more frequent. Hence, below about 100 km (62 mi), the rate of recombination of oxygen atoms exceeds the rate of photodissociation of oxygen molecules and oxygen occurs mostly as O_2.

The Earth's nitrogen/oxygen-dominated atmosphere is in striking contrast to the carbon dioxide-rich atmospheres of neighboring planets, Venus and Mars. The atmosphere of Venus is almost 100 times denser than the Earth's atmosphere and features an average surface temperature of about 460 °C (860 °F). The Martian atmosphere, on the other hand, is much thinner than the Earth's atmosphere and has an average surface temperature of about −53 °C (−63 °F). This is in spite of the probability that all three planets started out with atmospheres having similar compositions. The atmospheres of these planets evidently followed different evolutionary paths. In the Special Topic, "The Martian Atmosphere," we contrast the evolutionary paths of the atmospheres of Earth and Mars.

In addition to gases, the Earth's atmosphere contains minute liquid and solid particles, collectively called **aerosols.** Some aerosols—water droplets and ice crystals—are visible as clouds. Most aerosols are found in the lower atmosphere near their source, the Earth's surface. They originate through forest fires, from wind erosion of soil, as tiny sea-salt crystals from ocean spray, in volcanic emissions, and from industrial and agricultural activities. Some aerosols, such as meteoric dust, also enter the atmosphere from above.

It may be tempting to dismiss as unimportant those substances that make up only a small fraction of the atmosphere, but the significance of an atmospheric gas or aerosol is not necessarily related to its relative abundance. For example, water vapor, carbon dioxide, and ozone (O_3) occur in minute concentrations, yet they are essential for life. By volume, no more than about 4 percent of the lowest kilometer of the atmosphere is water vapor—even in the warm, humid air over tropical oceans and rainforests. Without water vapor, however, there would be no clouds, and no rain or snow to replenish soil moisture, rivers, lakes, and seas. Although composing only about 0.03 percent of the atmosphere, carbon dioxide is essential for photosynthesis. Together, water vapor and carbon dioxide act as a blanket over the Earth's surface, causing the lower atmosphere to retain heat and making the planet more amenable to life. Although the volume percentage of ozone is minute, this vital gas shields living things from exposure to potentially lethal intensities of ultraviolet radiation.

The aerosol concentration of the atmosphere is also relatively small, yet these suspended particles participate in important processes. Some aerosols act as nuclei for the development of clouds and precipitation, and some influence air temperature by interacting with solar radiation.

The Martian Atmosphere

Sensors onboard NASA's unmanned Mariner and Viking spacecraft discovered that the Martian atmosphere differs considerably from the Earth's atmosphere. The Martian atmosphere is much thinner (producing a surface pressure that is only 0.6 percent of the Earth's mean sea-level air pressure) and 95.3 percent carbon dioxide (CO_2). In the beginning, more than 4 billion years ago, however, the atmospheres of Earth and Mars probably were quite similar, but, for several reasons, the two atmospheres followed distinctly different evolutionary paths.

Outgassing produced the primeval atmospheres of both Earth and Mars. Gases were released from ancient planetary rock through volcanic eruptions and through impact when meteorites collided with the rocky surfaces of the planets. The gases were probably the same because the source rocks on both planets were chemically the same. Rock chemistry depends on the temperature at which crystallization takes place, which in turn depends on the planet's distance from the sun. Mars's greater distance from the sun (about 50 percent farther than Earth) would not produce significantly different crystallization (rock forming) conditions. Hence, through outgassing, the primeval atmospheres of both planets were composed of mostly carbon dioxide along with some nitrogen and water vapor—the principal gaseous emissions of volcanoes, both ancient and modern.

As noted elsewhere in this chapter, formation of oceans, establishment of the carbon cycle, and the arrival of photosynthetic plants gradually altered the Earth's primeval atmosphere so that eventually nitrogen and oxygen became the primary ingredients and carbon dioxide and water vapor became minor components. On Mars, however, carbon dioxide remained the principal gas even though the density of the Martian atmosphere declined considerably.

Within the scientific community currently there is no consensus of opinion on why the atmospheres of Earth and Mars evolved differently. One hypothesis attributes the difference to contrasts in the volcanic histories of the two planets. On Mars, the bulk of volcanic activity apparently took place during the planet's first 2 billion years, whereas, on Earth, volcanism has been more or less continuous throughout the planet's history. A decline in Martian volcanism meant less

CO_2 released to the atmosphere. At the same time, other geologic processes began to play important roles in removing CO_2 from the Martian atmosphere. Some carbon dioxide adhered to the fine sediment that blankets the planet's surface, and some carbon dioxide was locked up in carbonate rocks. Consequently, CO_2 in the Martian atmosphere eventually fell to a very small fraction of its original value.

As the amount of atmospheric CO_2 on Mars declined, so too did the planet's surface temperature. Carbon dioxide slows the loss of heat to space so that as CO_2 thins, more heat escapes and air temperatures drop. Three to four billion years ago Martian surface temperatures were so high that water occurred in liquid form. Seas, lakes, and rivers formed; in fact, channels cut by these ancient rivers were photographed by Mariner spacecraft. Today the mean temperature on the Martian surface is about -53 °C (-63 °F), much too low for running water. Water in unknown quantities likely remains on the planet mixed with solid carbon dioxide (dry ice) in the polar ice caps and as scattered patches of permafrost. Severe cold also meant no life on Mars, no photosynthesis, and hence, no free oxygen in the atmosphere.

The decline in Martian volcanism also cut the supply of nitrogen, and much of the original nitrogen gradually escaped the planet's relatively weak gravitational field. Gravity, the force that holds the atmosphere to a planet, is about 38 percent weaker on Mars than on Earth. This is because Mars is smaller (53 percent of the Earth's diameter) and less massive (10 percent of the Earth's mass).

In summary, the difference in evolutionary paths taken by the atmospheres of Earth and Mars may be due largely to physical contrasts between the two planets. The Earth has been more volcanically active (greater outgassing) and is more massive (stronger gravitational field) than Mars. The story on the Martian atmosphere is far from complete, however. Future spacecraft missions to Mars promise to provide more data that will add to our understanding of the evolution of the Martian atmosphere. The United States next planned mission to Mars is the Mars Observer satellite, which is designed to orbit the planet for one Martian year (687 Earth days) and is scheduled for launch in 1992.

AIR POLLUTANTS

Human activity also plays a role in the evolution of the atmosphere primarily through our contribution to air pollution. **Air pollutants** are gases or aerosols that occur in concentrations that threaten the well-being of living organisms (especially humans) or that disrupt the orderly functioning of the environment. Many of these substances occur naturally in the atmosphere. Sulfur dioxide (SO_2) and carbon monoxide (CO), for example, are normal minor components of the atmosphere that are considered pollutants only when their concentrations approach or exceed the tolerance limits of organisms. In sufficiently high concentrations, sulfur dioxide irritates the human respiratory system and carbon monoxide is an asphyxiating agent (reduces the blood's oxygen-carrying ability). Certain air pollutants, however, do not occur naturally in the atmosphere, and some of these are hazardous in any concentration. An example is asbestos fibers, which are known to cause cancer.

Air pollutants are products of both natural events and human activities. Natural sources of air pollutants include forest fires, dispersal of pollen, wind erosion of soil, decay of dead plants and animals, and volcanic eruptions. The single most important human-related source of air pollutants is the internal combustion engine that propels most motor vehicles. According to the U.S. Environmental Protection Agency (EPA), transportation vehicles yearly emit almost 60 million metric tons (63 million tons) of the major air pollutants. Many industrial sources also contribute to air quality problems. Unless emissions are controlled, pulp and paper mills, zinc and lead smelters, oil refineries, and chemical plants can be prodigious producers of air pollutants. Additional pollutants come from fuel combustion for space heating and generation of electricity, from refuse burning, and from various agricultural activities such as crop dusting. In the United States, almost 130 million metric tons (145

Table 1.3
Estimated Emissions of Principal Air Pollutants in the United States for 1986 (in millions of metric tons)

Source	Carbon monoxide	Sulfur oxides	Volatile organics	Particulates	Nitrogen oxides	Total
Transportation	42.6	0.9	6.5	1.4	8.5	59.9
Fuel combustion	7.2	17.2	2.3	1.8	10.0	38.5
Industrial processes	4.5	3.1	7.9	2.5	0.6	18.6
Solid waste disposal	1.7	—	0.6	0.3	0.1	2.7
Others	5.0	—	2.2	0.8	0.1	8.1
Totals	61.0	21.2	19.5	6.8	19.3	127.8

Source: U.S. Environmental Protection Agency, *National Air Pollutant Emission Estimates, 1940–1986,* 1988.

million tons) of the chief air contaminants are emitted to the atmosphere each year as the result of human activity—almost 0.6 metric ton per person (Table 1.3).

Some substances are harmful immediately upon emission into the atmosphere and are designated **primary air pollutants.** Carbon monoxide in automobile exhaust is an example. In addition, within the atmosphere, chemical reactions involving primary air pollutants, both gases and aerosols, produce **secondary air pollutants.** An example is smog, produced by the action of sunlight on automobile exhaust. We have much more to say about air pollution in Chapter 16.

Probing the Atmosphere

Much of what we know about the properties of the atmosphere is derived from direct sampling and measurement. At first, scientists explored the atmosphere from the ground, scaling rugged mountain peaks to sample the rarefied air. Through the years, exploration of the atmosphere has progressed from tentative probing using primitive instruments to sophisticated remote sensing that employs space-age technology.

Early efforts at remote sensing of the atmosphere employed kites. In 1749, in Glasgow, Scotland, a thermometer carried aloft by a kite provided the first temperature profile of the lower atmosphere. Benjamin Franklin, an inventive genius, is credited with designing an experiment utilizing a kite to demonstrate the electrical nature of lightning. He proposed flying a kite during a thunderstorm in an effort to attract an electrical discharge to a brass key that was attached to the kite string. Contrary to popular belief, Franklin probably never conducted the experiment himself—a wise decision since a single bolt would have been fatal. But such experiments were conducted in France during the summer of 1752, and the findings prompted Franklin the next year to extol the effectiveness of lightning rods in his *Poor Richard's Almanac.*

In 1804, the French scientist J. L. Gay-Lussac ushered in the age of manned balloon exploration of the atmosphere. He took air samples and measured temperature and humidity, and on one ascent he reached an altitude of 7000 meters (m), or 23,000 ft. In 1862, the British scientist James Glaisher and his fellow aeronaut Henry Coxwell took weather instruments aloft in a series of balloon ascents from Wolverhampton, England. They almost perished from severe cold and oxygen deprivation when they set a manned balloon altitude record of 9000 m (29,000 ft). By the close of the nineteenth century, the invention of recording instruments made possible unmanned balloon monitoring of the atmosphere. As a balloon ascended, its instrument package profiled air temperature, pressure, and humidity. Eventually, the balloon burst, and the instrument package parachuted back to the Earth's surface where the recordings were read.

Through the early part of the twentieth century, weather instruments borne by kites, balloons, and aircraft provided information chiefly on the lowest 5 km (3 mi) of the atmosphere. A leap forward in monitoring higher altitudes came in the late 1920s when the first **radiosonde,** a small instrument package equipped with a radio transmitter, was carried aloft by a helium-filled balloon. These devices transmit to ground stations continuous altitude measurements, called **soundings,** of temperature, air pressure, and relative humidity. With radiosondes, the data are received immediately; no recovery of a recording instrument is needed. By World War II, radiosonde movements were being tracked from the ground by a radio direction-finding antenna, thus giving an indication of variations in wind direction and wind speed with altitude. A radiosonde used in this way is called a **rawinsonde.**

Today, radiosondes are launched simultaneously at 12-hour intervals from hundreds of ground stations around the world (Figure 1.5). Information is monitored up to the altitude at which the balloon bursts (typically about 30 km, or 20 mi), and then the instrument package descends to the surface under a parachute. Some radiosondes (about 20 percent) are recovered, refurbished, and reused. Each radiosonde contains a prepaid mailbag and instructions to the finder for its return to the National Weather Service (NWS).

A **dropwindsonde** is very similar to a rawinsonde except that instead of being launched by a balloon from a surface station, the instrument package is dropped from an aircraft on a parachute. The dropwindsonde was developed at the National Center for Atmospheric Research (NCAR) in Boulder, Colorado, to obtain soundings over oceans where conventional rawinsonde stations are virtually absent. Dropwindsondes provide vertical profiles of air temperature, pressure, humidity, and wind.

Robert H. Goddard is credited with conducting the first rocket probe of the atmosphere in 1929. The payload of Goddard's primitive rocket included a thermometer and a barometer (for measuring air pressure). World War II spurred advances in rocketry, and, by the late 1940s, rockets were used to investigate the middle and upper atmosphere. In 1947, a vertically fired V2

A

B

FIGURE 1.5
(A) Launch of a radiosonde, a balloon-borne instrument package (B) that measures vertical profiles of air temperature, pressure, and relative humidity. [Photographs by Mike Brisson]

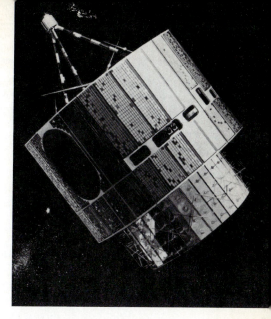

FIGURE 1.6

Artist's view of a geosynchronous meteorological satellite positioned over the equator at about 37,000 km (23,000 mi) altitude. This satellite orbits at the same rate as the Earth rotates, so the satellite always surveys the same portion of the planet. From the satellite's relatively high orbit, the on-board sensors "see" almost one third of the Earth's surface. [NASA photograph]

rocket took the first successful photographs of the Earth's cloud cover from altitudes of 110 to 165 km (65 to 100 mi). This and subsequent rocket probes of the atmosphere convinced scientists of the value of cloud pattern photographs in monitoring weather systems and inspired the first serious proposals for orbiting a weather satellite.

In the mid- to late 1950s, the United States' fledgling space program was directed at developing a launch vehicle (rocket) capable of putting a satellite in orbit. Those attempts to be the first in space were thwarted by the Soviet Union's successful orbiting of Sputnik I on 4 October 1957. The age of remote sensing by satellite had begun. On 1 April 1960 the United States orbited the world's first weather satellite, TIROS-I (Television Infra-Red Observation Satellite). Since then, a series of increasingly sophisticated weather satellites have been orbited (Figure 1.6).

Weather satellites have proved to be invaluable tools in weather observation. These "eyes in the sky" offer distinct advantages over the network of surface weather stations in providing a broad and continuous field of view; surface weather stations are discrete and often widely spaced data sources. In fact, weather observations are sparse or absent over vast areas of the Earth's surface—especially the oceans. Today's weather satellites carry sophisticated sensors capable of determining patterns of temperature and water vapor concentration, upper-air winds, and the life cycles of severe storms. By the early 1990s, vertical profiling of atmospheric temperature and humidity by satellite is expected to be a routine practice of the National Weather Service. We have more to say about meteorology by satellite in Chapter 15.

Also since World War II, weather radar has become an increasingly important tool for surveillance of severe weather systems such as hurricanes and intense thunderstorms. Radar signals locate and track the movement of areas of rainfall and can distinguish the rate of rainfall. As we will see in Chapter

14, a new generation of weather radar can also detect the detailed circulation of air within a severe thunderstorm and thereby provide advance warning of tornado development.

These various methods of probing the Earth's atmosphere have provided us with detailed information on the properties of the atmosphere. In the remainder of this chapter we consider some of the important results of that probing: the vertical profile of temperature within the atmosphere and the electrical characteristics of the upper atmosphere.

Temperature Profile of the Atmosphere

For convenience of study, the atmosphere is usually subdivided into concentric layers, according to the vertical profile of the average air temperature, as shown in Figure 1.7. Most weather occurs within the lowest layer, the **troposphere,** which extends from the Earth's surface to an average altitude

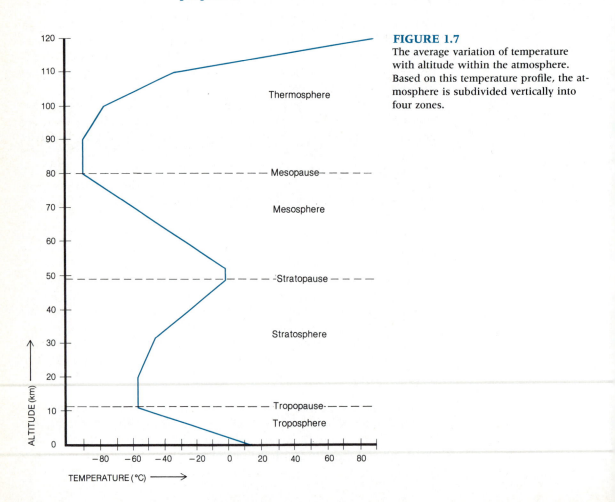

FIGURE 1.7
The average variation of temperature with altitude within the atmosphere. Based on this temperature profile, the atmosphere is subdivided vertically into four zones.

ranging from about 18 km (11 mi) at the equator down to 8 km (5 mi) at the poles. Normally, but not always, the temperature within the troposphere decreases with altitude. Hence, air temperatures on mountaintops are usually lower than those in surrounding lowlands. The upper boundary of the troposphere, called the **tropopause,** is a transition zone between the troposphere and the next higher layer, the stratosphere.

The **stratosphere** extends from the tropopause up to about 50 km (31 mi). On average, in the lower portion of the stratosphere, the temperature does not change with altitude. When temperature is constant, the condition is described as **isothermal.** Above about 20 km (13 mi), the temperature increases with altitude up to the top of the stratosphere, the **stratopause.**

The stratosphere is ideal for jet aircraft travel because it is above the weather. Therefore, it offers excellent visibility and features generally smooth flying conditions. Since the early 1970s, however, scientists have been concerned about possible detrimental effects of pollutants entering the stratosphere. Because little exchange of air takes place between the troposphere and the stratosphere, pollutants that enter the lower stratosphere may stay there for long periods. Gases thrown into the stratosphere during violent volcanic eruptions, for example, can persist there for many months to years and perhaps trigger changes in climate. Other pollutants threaten to erode the protective ozone layer within the stratosphere. We have more to say about these problems in Chapters 2, 16, and 18.

The stratopause is the transition zone between the stratosphere and the next higher layer, the **mesosphere.** Within this layer, the temperature once again decreases with increasing altitude. The mesosphere extends up to the **mesopause,** which is about 80 km (50 mi) above the Earth's surface, and features the lowest average temperature in the atmosphere ($-90\,°C$, or $-130\,°F$). Above this is the **thermosphere,** where temperatures at first are isothermal and then increase rapidly with altitude. Within the thermosphere, temperature is more variable with time than in any other region of the atmosphere.

The Ionosphere

The **ionosphere** is located primarily within the thermosphere, at altitudes from about 80 to 900 km (50 to 600 mi). The region is named for its relatively high concentration of ions. An **ion** is an atomic-scale particle possessing an electrical charge. High-energy solar radiation entering the upper atmosphere strips electrons from oxygen and nitrogen atoms and molecules, leaving them as positively charged ions. The highest concentration of ions occurs within the lower portion of the thermosphere.

Although conditions in the upper atmosphere do not greatly influence day-to-day weather, the ionosphere is important for long-distance radio transmissions. Ions reflect radio waves and thereby extend the range of radio

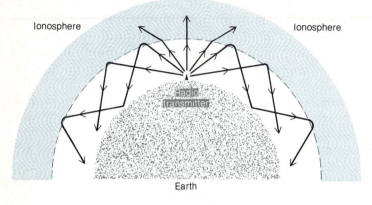

FIGURE 1.8
Within the ionosphere, at altitudes above 80 km (50 mi), regions of charged subatomic particles reflect outgoing radio waves. Multiple reflections involving both the ionosphere and the Earth's surface greatly extend the range of radio transmissions.

transmissions (Figure 1.8). Radio signals travel in straight lines and bounce back and forth between the Earth's surface and the ionosphere. By repeated reflections, a radio signal may travel completely around the globe. We have more to say on radio signals in the Special Topic "The Ionosphere and Radio Transmission."

The ionosphere is also the site of the spectacular **aurora borealis** (northern lights) in the Northern Hemisphere and the **aurora australis** (southern lights) in the Southern Hemisphere. Auroras appear in the polar night sky as overlapping ribbons or curtains of blue-green light, occasionally fringed with red and pink (Figure 1.9).

An aurora is triggered by the **solar wind,** a stream of electrically charged subatomic particles (protons and electrons) that continually emanates from the sun and travels into space at speeds of several hundred kilometers per second. The solar wind deflects the Earth's magnetic field into a teardrop-shaped cavity surrounding the Earth known as the **magnetosphere** (Figure 1.10). A complex interaction between the solar wind and the Earth's magnetosphere generates beams of electrons that collide with atoms and molecules within the ionosphere. Collisions rip apart molecules, excite atoms, and increase ion and electron densities. As atoms shift down from their excited

FIGURE 1.9
The aurora borealis (northern lights) viewed in the Alaskan night sky. [NOAA photograph]

The Ionosphere and Radio Transmission

Reception of distant radio signals at night is not at all unusual. Late-night radio listeners in Illinois and Wisconsin, for example, routinely pick up WBZ, a Boston radio station (1030 on the AM dial) even though the station's transmitter is more than 1500 km (950 mi) away.

The arrival of distant radio signals at night and their subsequent disappearance during sunlit hours are due to the interaction of radio signals with the ionosphere, the portion of the atmosphere above the mesosphere. A radio signal is a form of radiation that travels in straight paths in all directions away from the source transmitter. The curvature of the Earth, however, limits the distance from the transmitter that direct radio waves can be picked up. That is, the Earth's surface gradually curves under and away from those direct radio signals. About 100 km (62 mi) from the transmitter begins a quiet zone where direct radio waves are not received. At night, beyond the quiet zone, reception resumes because radio waves are reflected back to the Earth's surface from the upper region of the ionosphere.

Recall that the ionosphere is a region of ions and free electrons. Highly energetic radiation from the sun causes the atmosphere's molecular nitrogen (N_2) and molecular oxygen (O_2) to split into atoms, positively charged ions, and free electrons. The production rate of ions and electrons depends on two factors, both of which vary with altitude: (1) the density of atoms and molecules available for ionization, which decreases rapidly with altitude, and (2) the intensity of solar radiation, which increases with altitude. Combined, these two factors maximize the concentration of ions and free electrons in the ionosphere.

By convention, the ionosphere is subdivided vertically into several layers. From lowest to highest, layers are designated as D (60 to 90 km), E (90 to 140 km), and F (above 140 km). The original basis for this subdivision was the belief that each layer is a distinct zone of maximum electron density. Measurements by rockets and satellites, however, show that the ionosphere is not made up of discrete layers; rather, electron density increases nearly continuously with altitude to a maximum at an average level close to 300 km (190 mi). Hence, the D, E, and F labels used in this discussion refer to specific subregions within the ionosphere.

Radio waves that enter the ionosphere interact with free electrons in the D, E, and F regions and are either absorbed or reflected toward the Earth's surface depending on the amount of solar radiation. At night, when there is no ionizing radiation, the D region virtually disappears as ions and electrons recombine into neutral particles. The recombination rate depends on air density; that is, the denser the air, the greater is the likelihood of collision of particles and the capture of electrons by positive ions. In the E and F regions, the air is so rarefied that collisions are infrequent, and although the E region weakens, both regions persist through the night. Radio waves that reach the F region are reflected back toward the Earth's surface. (Exceptions are radio waves that enter the F region at nearly a right angle; these waves pass through the F region and on into space.) At night, because of F-region reflection, radio signals propagate many hundreds of kilometers from their point of origin.

With the return of the sun's ionizing radiation during the day, the D region redevelops. Most radio waves that reach the D region are absorbed rather than reflected. Waves that do penetrate the D region are reflected by the overlying F region back to the D region where they are absorbed. Consequently, during sunlit hours, radio-wave propagation is not aided by F-region reflection. In summary, then, radio signals travel greater distances at night because direct radio waves are reflected by the upper ionosphere back to the Earth's surface.

Because the ionosphere is generated by ionizing radiation from the sun, any solar activity that disturbs the flow of this radiation may affect ion density and, consequently, radio communication on earth. *Sudden ionospheric disturbances (SIDs),* typically lasting 15 to 30 minutes, are caused by bursts of ultraviolet radiation from the sun. Ionization temporarily increases, D-region absorption strengthens, and radio transmissions fade. The same solar activity responsible for auroral displays also increases ionization and causes radio fade-out.

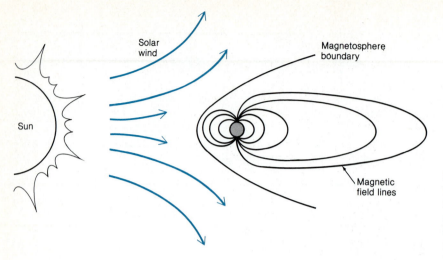

Solar
wind

Magnetosphere
boundary

Sun

Magnetic
field lines

FIGURE 1.10

The solar wind deflects the
Earth's magnetic field into
the magnetosphere. Solar
wind particles trigger auroral displays.

(energized) states and as ions combine with free electrons they emit radiation,
part of which is visible as the aurora. Excited nitrogen emits pink light whereas
excited oxygen emits whitish green light.

The Earth's magnetic field channels some solar wind particles into two
doughnut-shaped belts, each of which is centered on one of the Earth's north
and south geomagnetic poles.* These belts of more or less continuous auroral
activity, known as **auroral zones,** are situated between 20 and 30 degrees of
latitude from the geomagnetic poles. The Northern Hemisphere auroral zone
is centered on the northwest tip of Greenland at latitude 78.5 degrees N and
longitude 69 degrees W. As a result, auroral displays are usually visible only
at high latitudes.

Auroral activity varies with the sun's activity. When the sun is quiet, the
auroral zone shrinks, but when the sun is active, the auroral zone expands
equatorward, and the aurora may be seen across southern Canada and the
northern United States or, rarely, even further south. An active sun features
gigantic explosions called solar flares (Figure 1.11). A **solar flare** is a brief
event (lasting perhaps an hour) that produces a shock wave that propagates
rapidly (500 to 1000 km per second) through the solar wind. Collision of
the shock wave with the magnetosphere causes the auroral zone to expand
equatorward.

_____

* Latitude is distance on the Earth's surface measured in degrees and minutes north or
south of the equator. The latitude of the equator is 0 degrees, and the geographical poles are
at 90 degrees N and 90 degrees S; 1 degree equals 60 minutes. Longitude is measured in
degrees east and west of the meridian that passes through Greenwich, England. The Greenwich meridian is called the "prime meridian" and is assigned a longitude of 0 degrees.
Longitude is measured east and west of the prime meridian to 180 degrees E(ast) and 180
degrees W(est). The Earth's geomagnetic poles do not coincide with its geographical or
compass poles (the compass north is at 76 degrees N, 102 degrees W).

FIGURE 1.11
Solar flares emit highly energetic streams of electrically charged subatomic particles off into space. [National Center for Atmospheric Research/National Science Foundation]

Conclusions

In this chapter we covered the origin, evolution, and basic compositional and structural characteristics of the atmosphere. We emphasized the importance of minor components in the functioning of the atmosphere, and we surveyed the various techniques that are used to monitor the properties of the atmosphere. We also saw how the vertical variation of average air temperature enables us to subdivide the atmosphere into four layers. Because the primary focus of this book is weather and climate, in our study of the atmosphere we will be concerned primarily with processes occurring within the troposphere.

Our next major objective is to examine the driving force behind weather. To do so, we require an understanding of energy input and energy conversions within the atmosphere. In the next chapter we learn how the sun supplies the energy that drives the atmosphere's circulation. As we will see, the circulation of the atmosphere ultimately is responsible for the variation of weather from one place to another and with time.

Summary Statements

Geographic and temporal variability is a basic characteristic of weather.

Weather is defined as the state of the atmosphere at a specified place and time; climate refers to both average weather conditions and the variability (extremes) in weather.

The scientific method is a systematic form of inquiry that involves observation, speculation, and reasoning.

A scientific model is an approximation or simulation of a real system. Examples in the atmospheric sciences include weather maps and numerical models of the atmosphere.

The Earth's modern atmosphere is the product of a lengthy evolutionary process that began billions of years ago and involved volcanic activity, formation of seas, and the coming of life.

The Earth's atmosphere is a mixture of gases and suspended aerosols. The proportions of the major gases are fixed within the lower 80 km (50 mi). Some trace gases, such as water vapor, exhibit important variations in concentration within the lower atmosphere.

The significance of an atmospheric gas or aerosol is not necessarily related to its relative concentration within the atmosphere. Water vapor, carbon dioxide, and ozone, for example, are "minor" in concentration but extremely important in the roles they play within the atmosphere.

Air pollutants are produced by both human activities and natural processes and include gases or aerosols whose concentrations threaten the well-being of organisms or disrupt the orderly functioning of the environment.

Through the years, tools for investigating the atmosphere have progressed from instrumented kites and manned balloons to rockets, radiosondes, satellites, and radar.

The atmosphere may be subdivided into concentric layers based on the average vertical temperature profile. The troposphere, the lowest subdivision of the atmosphere, is the site of most weather.

The ionosphere is located within the thermosphere and contains a relatively high concentration of charged particles (ions and electrons). It is also the site of the aurora.

Key Words

weather	photosynthesis	soundings	thermosphere
meteorology	homosphere	rawinsonde	ionosphere
climate	heterosphere	dropwindsonde	ion
atmosphere	photodissociation	troposphere	aurora borealis
scientific method	aerosols	tropopause	aurora australis
scientific model	air pollutants	stratosphere	solar wind
conceptual model	primary air pollutants	isothermal	magnetosphere
graphical model	secondary air	stratopause	auroral zones
physical model	pollutants	mesosphere	solar flare
numerical model	radiosonde	mesopause	

Review Questions

1. In your own words, define the scientific method.

2. What is the basic purpose of a scientific model? Provide some examples of models that are useful in studying weather and climate.

3. What are some of the potential problems with atmospheric numerical models used for weather prediction?

4. Distinguish between weather and climate, and explain why a description of climate only in terms of average weather is incomplete and perhaps misleading.

5. Provide several examples of how the atmosphere sustains life on earth.

6. What role has volcanic activity played in the evolution of the Earth's atmosphere?

7. In the distant past carbon dioxide was the chief component of the Earth's atmosphere. Explain why its concentration declined.

8. Explain how the evolution of the Earth's atmosphere is linked to the formation of oceans and the coming of life on Earth.

9. Distinguish between the homosphere and the heterosphere.

10. Most atmospheric aerosols are products of processes that occur at the Earth's surface. Identify several of these sources.

11. Present several examples of how some "minor" constituents of the atmosphere are important in the functioning of the environment.

12. Under what conditions is a gas or aerosol considered an air pollutant?

13. Identify several natural sources of air pollution and several sources related to human activity.

14. Describe the various approaches that scientists have used in their investigations of the properties of the atmosphere.

15. Explain why the invention of the radiosonde marked a major step forward in the exploration of the atmosphere.

16. Identify some of the advantages of satellite observations of the atmosphere as compared with other techniques of atmospheric monitoring.

17. What advantages does the stratosphere offer over the troposphere for jet aircraft travel? Speculate on some disadvantages.

18. Why do pollutants tend to persist for long periods once they enter the stratosphere?

19. Within the thermosphere above 150 km (93 mi), oxygen occurs chiefly in the atomic (O) rather than the diatomic (O_2) form. Explain why.

20. What is the source of ions within the ionosphere?

21. Why is the aurora visible only at high latitudes?

22. How and why does auroral activity vary with solar activity?

Points to Ponder

1. Photosynthesis occurs chiefly during the growing season. Speculate on how variations in the rate of photosynthesis through the course of a year might influence the concentration of carbon dioxide in the atmosphere.

2. Why does a radiosonde balloon eventually burst as it ascends to altitudes above about 30 km (19 mi)?

3. The sun is the ultimate source of heat for the Earth. Mountaintops are closer to the sun than lowlands, and yet mountaintops are colder than lowlands. Why?

Projects

1. Keep a daily log of weather conditions in your area. Use your own (or the school's) weather instruments or rely on television or radio weather reports.

2. Keep track of the changing patterns of weather on a national scale by watching televised weather summaries or by listening to the NOAA (National Oceanic and Atmospheric Administration) weather radio reports (described in Chapter 15) on a daily basis.

Selected Readings

Beveridge, W.I.B. *The Art of Scientific Investigation.* New York: Vintage Books, 1957, 239 pp. Considers the nature of scientific inquiry as a creative art.

Conover, J. H. "The Blue Hill Observatory." *Weatherwise* 37 (1984):296–303. Presents a historical account of a long-term weather station located near Boston, site of the world's first continuous recording of air temperature variation with altitude.

Haberle, R. M. "The Climate of Mars." *Scientific American* 254, No. 5 (1986):54–62. Reviews what is known about the Martian atmosphere's origin and circulation.

Kasting, J. F., et al. "How Climate Evolved on the Terrestrial Planets." *Scientific American* 258, No. 2 (1988):90–97. Proposes that differences in the carbon cycle operating on Venus, Earth, and Mars resulted in drastically different climates on these planets.

Riebsame, W. E., et al. "The Social Burden of Weather and Climate Hazards." *Bulletin of the American Meteorological Society* 67 (1986):1378–1388. Includes a summary of trends in mortality and property damage caused by weather extremes.

Radiation

The sun supplies the energy that drives the atmosphere's circulation. The most intense portion of the solar energy that reaches Earth is visible as sunlight. [Photograph by Arjen and Jerrine Verkaik/SKYART]

27

THE SUN drives the atmosphere; that is, the sun is the source of energy that drives the circulation of the atmosphere and powers winds and storms. The circulation of the atmosphere ultimately is responsible for weather and the temporal and spatial variability that characterizes weather.

The sun ceaselessly emits energy to space in the form of electromagnetic radiation. A very small portion of that energy is intercepted by the Earth-atmosphere system* and is converted into other forms of energy including, for example, heat and the kinetic energy of the atmosphere's circulation. In this regard, it is important to note that, although energy can be converted from one form to another, it cannot be created nor destroyed. This is the **law of energy conservation.**

In this chapter, we examine the basic properties of electromagnetic radiation, how solar radiation interacts with the components of the Earth-atmosphere system, and its conversion to heat. We also learn how the Earth-atmosphere system responds to solar heating. First we consider the nature of electromagnetic radiation in general, and then we describe some of the specific properties of its various forms.

Electromagnetic Radiation

Planet Earth is bathed continuously in **electromagnetic radiation,** so named because this form of energy exhibits both electrical and magnetic properties. All known objects emit electromagnetic radiation.† Forms of electromagnetic radiation include radio waves, microwaves, visible light, infrared radiation, ultraviolet radiation, X rays, and gamma radiation. Together, the many forms of electromagnetic radiation make up the **electromagnetic spectrum,** illustrated in Figure 2.1.

Electromagnetic radiation travels in the form of waves, which are usually described in terms of wavelength or frequency. **Wavelength** is the distance between successive wave crests or, equivalently, from wave trough to wave trough, as shown in Figure 2.2. **Wave frequency** is defined as the number of crests (or troughs) that passes a given point in a specified period of time, usually 1 second. Passage of one complete wave is called a cycle, and a frequency of one cycle per second equals 1.0 hertz (Hz). A wave's frequency

---

* The Earth-atmosphere system is the Earth's surface and its atmosphere considered together.

† As we will see in Chapter 3, objects emit no electromagnetic radiation at a temperature of absolute zero.

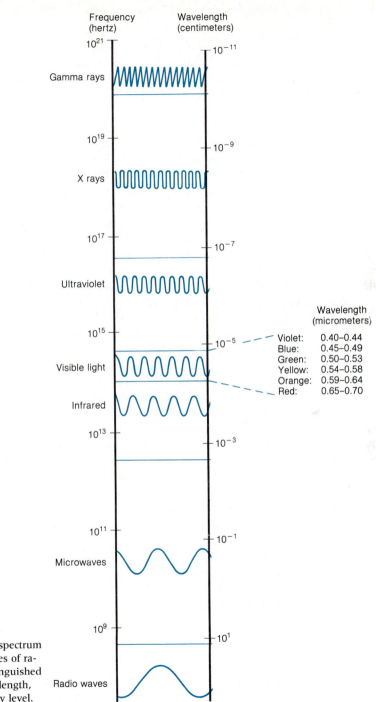

Frequency
(hertz)

Wavelength
(centimeters)

10^{21} — 10^{-11}

Gamma rays

10^{19} — 10^{-9}

X rays

10^{17} — 10^{-7}

Ultraviolet

10^{15} — 10^{-5}

Visible light

Infrared

10^{13} — 10^{-3}

10^{11} — 10^{-1}

Microwaves

10^{9} — 10^{1}

Radio waves

Wavelength
(micrometers)

Violet:	0.40–0.44
Blue:	0.45–0.49
Green:	0.50–0.53
Yellow:	0.54–0.58
Orange:	0.59–0.64
Red:	0.65–0.70

FIGURE 2.1
The electromagnetic spectrum
consists of many types of radiation that are distinguished
on the basis of wavelength,
frequency, and energy level.

29

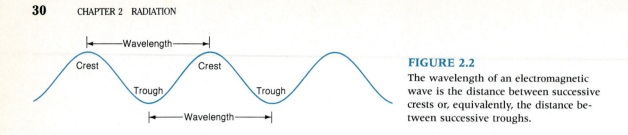

FIGURE 2.2
The wavelength of an electromagnetic wave is the distance between successive crests or, equivalently, the distance between successive troughs.

is inversely proportional to its wavelength; that is, the higher the frequency, the shorter the wavelength. As shown in Figure 2.1, radio waves have frequencies of millions of hertz and wavelengths up to hundreds of kilometers. At the other end of the electromagnetic spectrum, in contrast, gamma rays have frequencies as high as 10^{24} (a trillion trillion) Hz and wavelengths as short as 10^{-14} (a hundred trillionth) m.

Electromagnetic radiation travels through space as well as through gases, liquids, and solids. In a vacuum, all electromagnetic waves travel at maximum speed, 300,000 km (186,000 mi) per second. All forms of electromagnetic radiation slow down when passing through materials, the speed varying with wavelength and type of material. As electromagnetic radiation passes from one medium to another, it may be reflected or refracted (that is, bent) at the boundary between the two media. This happens, for example, when solar radiation strikes the ocean surface: some is reflected and some is bent upon penetrating the water. Electromagnetic radiation can also be absorbed, as when solar radiation is absorbed at the Earth's surface and converted to heat.

Although the electromagnetic spectrum is continuous, different names are assigned to different segments because we detect, measure, generate, and use those segments in different ways. Note that the different types of electromagnetic radiation do not begin or end at precise points along the spectrum. For example, red light shades into invisible infrared radiation (infrared, meaning *below red*) on the frequency scale. At the other end of the visible portion of the electromagnetic spectrum, violet light shades into invisible ultraviolet radiation (ultraviolet, meaning *beyond violet*).

At the low-energy (low-frequency, long-wavelength) end of the electromagnetic spectrum are **radio waves.** Their wavelengths range from a small fraction of a centimeter to hundreds of kilometers, and their frequencies range up to a billion hertz. FM (frequency modulation) radio waves, for example, span 88 million Hz to 108 million Hz, hence the familiar 88 and 108 at opposite ends of the FM radio dial.

Next comes the **microwave** portion of the electromagnetic spectrum, which has wavelengths ranging from 300 to 0.1 millimeter (mm). (A millimeter is a thousandth of a meter.) Some microwave frequencies are used for long-distance telephone communications, Earth-to-satellite television systems, tracking weather systems (radar), and in microwave ovens.

Infrared radiation lies between microwaves and visible light. We cannot see infrared radiation, but we can feel the heat generated if the infrared radiation is intense enough, as it is, for example, when emitted by a hot stove. Some naturally occurring infrared radiation comes directly from the sun, but most is converted by the Earth-atmosphere system from visible solar radiation.

At its uppermost frequencies, infrared radiation shades into the lowest frequency of visible radiation, red light. Wavelengths of **visible light** range from about 0.70 micrometer (μm) at the red end down to about 0.40 μm at the violet end of the spectrum. (A micrometer is a millionth of a meter.) Visible light is essential for photosynthesis and for many activities of plants and animals. For plants, the duration of sunlight coordinates flowering, the opening of buds, and the dropping of leaves. In some animals, the duration of sunlight regulates reproductive activity, hibernation, and migration.

Beyond visible light on the electromagnetic spectrum and in order of increasing frequency, increasing energy level, and decreasing wavelength, are **ultraviolet radiation (UV), X rays,** and **gamma radiation.** All three occur naturally, and all can be produced artificially. All have medical uses: ultraviolet radiation is a potent germicide; X rays are a powerful diagnostic tool; and both X rays and gamma radiation are used to treat cancer patients. These three highly energetic types of radiation can be dangerous as well as useful. Ultraviolet rays can cause blindness by irreparably damaging the retinal cells of the eye, and prolonged exposure to UV can cause cancer on exposed skin. Overexposure to X rays or gamma rays can cause sterilization, cancer, genetic mutations, and tissue damage to a fetus. Fortunately for us, the Earth's atmosphere blocks out most incoming ultraviolet radiation and virtually all X rays and gamma radiation. Without this protective atmospheric shield, all life on Earth would be quickly destroyed.

Radiation Laws

Several physical laws describe the properties of electromagnetic radiation that is emitted by a perfect radiator, a so-called blackbody. (These laws are summarized in the Mathematical Note, "Blackbody Radiation Laws," at the end of this chapter.) By definition, at a given temperature, a **blackbody*** absorbs all radiation incident on it at every wavelength and emits all radiation at every wavelength; no radiation is reflected. A blackbody is therefore a perfect absorber and a perfect emitter. Although neither the sun nor the Earth is a precise blackbody, they so closely approximate perfect radiators that we can apply blackbody radiation laws to them with very useful results. Here we apply two blackbody radiation laws: Wien's displacement law and the Stefan–Boltzmann law.

---

* The term *blackbody* can be misleading because the concept does not refer to color. Objects that do not appear black nonetheless may be blackbodies, that is, perfect radiators. For example, bright white snow is very nearly a blackbody for infrared radiation.

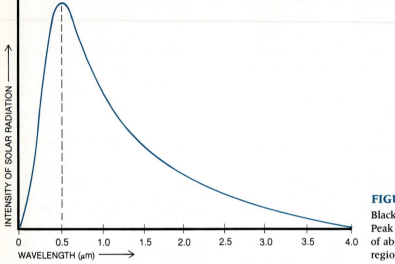

FIGURE 2.3

Blackbody radiation emission for the sun. Peak energy intensity is emitted by waves of about 0.50 μm in length (in the green region of the visible spectrum).

Although all known objects emit all forms of electromagnetic radiation, the wavelength of most intense radiation is inversely proportional to the temperature of the object. This is a statement of **Wien's displacement law.** Hence, relatively warm objects (such as the sun) emit peak radiation at relatively short wavelengths, whereas colder objects (such as the Earth-atmosphere system) emit peak radiation at longer wavelengths. Figure 2.3 illustrates the variation of radiation intensity with wavelength for a blackbody at the radiating temperature of the sun, about 6000 °C (10,000 °F). Thus, the sun emits a band of radiation (mostly between 0.25 and 2.5 micrometers) that is most intense at a wavelength of about 0.5 micrometer (in the green of visible

FIGURE 2.4

Blackbody radiation emission for the Earth's surface. Peak energy intensity is emitted by waves of about 10 μm in length (in the infrared region of the electromagnetic spectrum).

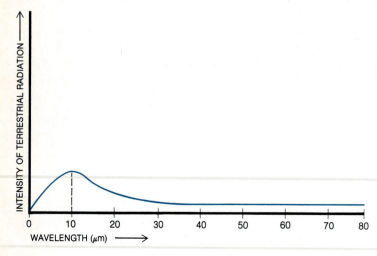

light). Figure 2.4 shows how the intensity of radiation varies with wavelength for a blackbody at the average radiating temperature of the Earth's surface, about 15 °C (59 °F). The Earth's surface therefore emits a band of infrared radiation (mostly between 4 and 24 micrometers) that has peak intensity at a wavelength of about 10 micrometers.

The curves in Figures 2.3 (for the sun) and 2.4 (for the Earth's surface) represent the total radiational energy emitted per unit surface area at all wavelengths. Note, however, that the vertical scales in these two figures are not the same because the sun emits considerably more total radiational energy than does the Earth-atmosphere system. This contrast in energy emission is described by the **Stefan–Boltzmann law:** The total energy radiated by an object is proportional to the fourth power of its absolute temperature (T^4 K).* The sun radiates at a much higher temperature than does the Earth-atmosphere system, so that the Stefan–Boltzmann law predicts that the sun's energy output per square meter is about 160,000 times that of the Earth-atmosphere system. As solar radiation diverges outward from the sun into space, its intensity diminishes rapidly (as the inverse square of the distance traveled). Hence, when solar radiation reaches Earth, its intensity is reduced considerably.

The total energy (in the form of solar radiation) absorbed by planet Earth is equal to the total energy (in the form of infrared radiation) emitted by the Earth-atmosphere system back to space. This balance between energy input and energy output is known as **global radiative equilibrium** and is an example of the law of energy conservation.

Input of Solar Radiation

The sun, our closest star, is a huge gaseous body composed almost entirely of hydrogen (about 80 percent by mass) and helium and featuring internal temperatures that may exceed 20 million °C. The ultimate source of solar energy is a continuous nuclear fusion reaction in the sun's interior. Simply put, in this reaction four hydrogen nuclei fuse to produce one helium nucleus. However, the mass of the four hydrogen nuclei is about 0.7 percent greater than the mass of one helium nucleus. This excess mass is converted to energy as described by Albert Einstein's equation

$$E = mc^2$$

whereby mass, m, is related to energy, E, and c is the speed of light (300,000 km per second). Note that c^2 is such a huge number that a very small mass is converted into an enormous quantity of energy. Some of the energy produced

---

* The absolute or Kelvin (K) temperature scale is the number of degrees above absolute zero. The various temperature scales are described in detail in Chapter 3.

by nuclear fusion is used to bind the helium nucleus together. The rest of the energy is radiated and convected to the sun's surface, and from there, energy is radiated off to space.

At temperatures near 6000 °C, the visible surface of the sun, known as the **photosphere,** is much cooler than the sun's interior. The photosphere's honeycomb appearance is due to a network of huge, irregularly shaped convective cells, called **granules.** One granule is typically the size of Texas and consists of a broad central area of rising hot gas surrounded by a thin layer of cooler gas sinking back into the sun. Relatively hot spots, called **faculae,** and relatively cool spots, called **sunspots,** dot the photosphere.

Outward from the photosphere is the **chromosphere** consisting of ions of hydrogen and helium at 4000 to 40,000 °C. Beyond this is the outermost portion of the sun's atmosphere, the **corona** (Plate 1), a region of hot (1 to 4 million °C) and highly rarefied gases that extends millions of kilometers into space, to the outer limits of the solar system. Solar flares erupt from the photosphere into the corona and intensify the solar wind as described in Chapter 1.

Of the enormous quantity of energy radiated by the sun to space, planet Earth intercepts only about one two-billionth of the total amount. Solar radiation that reaches Earth is known as **insolation** (for *in*coming *sol*ar ra*di*ation). About 45 percent is visible as sunlight; the remainder consists of infrared (46 percent) and ultraviolet (9 percent).

SOLAR ALTITUDE

We know from experience that the intensity of solar radiation varies significantly over the course of a year. In winter, the sun is lower in the sky, and at the Earth's surface the sun's rays are weaker than in summer. In winter, the days are shorter than they are in summer. Even in the course of a single day, noticeable changes in insolation occur: the noon sun is more intense than the rising or setting sun.

It is evident, then, that the angle of the sun above the horizon, called the **solar altitude,** influences the intensity of solar radiation. By intensity we mean the amount of energy striking or passing through a unit area in a unit time.* The sun is so far away from Earth, about 150 million km (93 million mi) on average, that solar radiation reaches the planet as a beam that in cross section is uniform in intensity. However, the intensity of solar radiation received at the Earth's surface changes in response to changes in the solar altitude. As shown in Figure 2.5, when the noon sun is directly overhead (solar altitude of 90 degrees), the sun's rays reaching the Earth's surface are most concentrated and therefore most intense. As the sun moves lower in the

---

* This is also called the *flux density.*

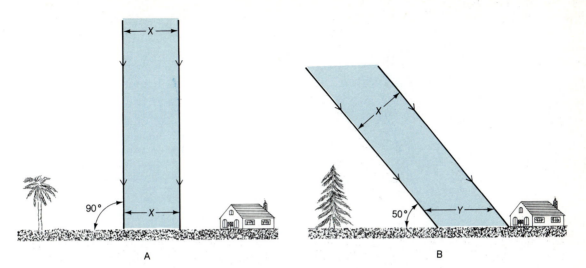

FIGURE 2.5

The intensity of solar radiation that strikes the Earth's surface varies with changes in solar altitude (the sun's angle above the horizon). (A) Solar radiation is most intense when the sun is directly overhead (solar altitude of 90 degrees). (B) With decreasing solar altitude, the solar radiation received at the Earth's surface spreads over an increasingly larger area (Y is greater than X) so that the radiation becomes less intense.

sky, the solar altitude decreases and solar radiation is spread over a greater area of the Earth's surface, thus becoming less intense at the surface. Because the Earth presents a curved surface to incoming solar radiation (Figure 2.6), solar altitude and solar radiation intensity always vary with latitude.

Solar altitude also influences the interaction between insolation and the atmosphere. With decreasing solar altitude, the path of the sun's rays through the atmosphere lengthens (Figure 2.7). As the path lengthens, solar radiation

FIGURE 2.6

On any day of the year, the solar altitude varies with latitude because the Earth's surface is curved. In this example, the solar altitude is 90 degrees at the equator and decreases with latitude (toward the poles). Hence, solar radiation is most intense at the equator and least intense at the poles.

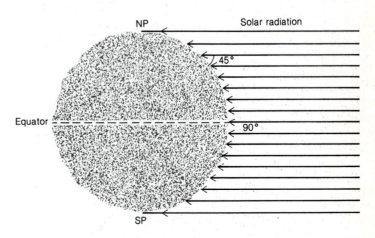

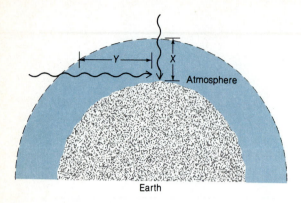

FIGURE 2.7
The path of solar radiation through the atmosphere lengthens as the solar altitude decreases, that is, as the sun moves lower in the sky. X is the path length at high solar altitude, and Y is the path length at low solar altitude.

interacts more with the component gases and aerosols of the atmosphere.* Solar radiation weakens as a consequence of this interaction. Thus, even with clear skies, the longer the path of the solar radiation through the atmosphere, the less intense is the radiation that strikes the Earth's surface.

Although solar altitude has an important influence on the intensity of solar radiation striking the Earth's surface, the length of day affects the total amount of radiational energy that is received. For some specified intensity of solar radiation, those areas of the Earth with more daylight hours will receive more total energy. Variations both in solar altitude and in day length accompany the annual march of the seasons. Before examining these relationships, let us first consider the fundamental motions of Earth in space: rotation of the planet on its axis and the planet's orbit about the sun.

EARTH'S MOTIONS IN SPACE

Rotation of the Earth on its axis accounts for day and night. Approximately once every 24 hours, the Earth makes one complete rotation. Consequently, at any point in time, half the planet is in darkness (night) and the other half is illuminated by solar radiation (day).

In one year, which is actually 365.25 days, the Earth makes one complete revolution about the sun in a slightly elliptical orbit (Figure 2.8). The Earth's orbital eccentricity, that is, its departure from a circular orbit, is so slight that the Earth-to-sun distance varies by only about 3.3 percent through the year. Earth is closest to the sun (147 million km, or 91 million mi) on about 3 January and farthest from the sun (152 million km, or 94 million mi) on 4 July. These are the dates of **perihelion** and **aphelion,** respectively. In the

---

* The nature of this interaction (absorption, reflection, and scattering) is discussed later in this chapter.

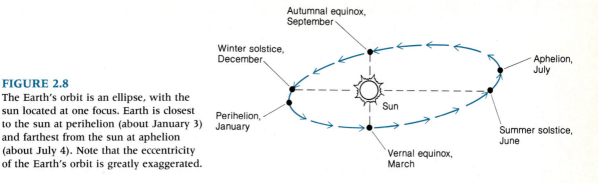

FIGURE 2.8
The Earth's orbit is an ellipse, with the sun located at one focus. Earth is closest to the sun at perihelion (about January 3) and farthest from the sun at aphelion (about July 4). Note that the eccentricity of the Earth's orbit is greatly exaggerated.

Northern Hemisphere, Earth is therefore closest to the sun in winter and farthest from the sun in summer. The eccentricity of the Earth's orbit does not, therefore, explain the seasons. What, then, does account for them?

THE SEASONS

Seasons on Earth are attributed to the 23 degree 27 minute tilt of the Earth's rotational axis to the plane defined by the Earth's orbit (Figure 2.9). This tilt causes the Earth's orientation to the sun to change continually as the planet revolves about the sun. The Northern Hemisphere thus leans away from the sun during winter and toward the sun in summer. At the same time that the Northern Hemisphere is leaning away from the sun, the Southern Hemisphere is leaning toward the sun. When it is winter in the Northern Hemisphere, it is therefore summer in the Southern Hemisphere, and vice versa.

As Earth's orientation to the sun changes, so too do the solar altitude and length of day. Hence, the intensity and total amount of solar radiation received at the Earth's surface vary seasonally. In the winter hemisphere, solar altitudes

FIGURE 2.9
The seasons change because the Earth's equatorial plane is inclined (at 23 degrees, 27 minutes) to its orbital plane. The seasons given are for the Northern Hemisphere. Note that the eccentricity of the Earth's orbit is greatly exaggerated.

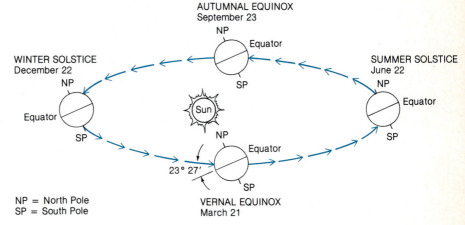

are lower, days are shorter, and there is less solar radiation. In the summer hemisphere, solar altitudes are higher, days are longer, and there is more solar radiation. Less solar radiation in winter than in summer means that winters are colder than summers.

If the Earth's rotational axis were perpendicular to its orbital plane (no tilt), Earth would always have the same orientation to the sun. Without an axial tilt, changes in Earth-to-sun distance between aphelion and perihelion would cause only a slight seasonal contrast.

How does the Earth's orientation with respect to the sun change over the course of a year? Viewed from the Earth's surface, the location of the sun's most intense radiation (solar altitude of 90 degrees) shifts from 23 degrees 27 minutes south of the equator to 23 degrees 27 minutes north of the equator, and then back to 23 degrees 27 minutes south. On 21 March, and again on 23 September, the sun's noon position is directly over the equator. Day and night are of equal length (12 hours) everywhere (Figure 2.10). For this reason, these dates* are labeled **equinoxes,** from the Latin for equal nights.

Following the equinoxes, the sun continues its apparent journey toward maximum poleward locations, its **solstice** latitudes. On 22 June, the sun's noon rays are vertical at 23 degrees 27 minutes N, the **Tropic of Cancer.** As shown in Figure 2.11, daylight is continuous north of the **Arctic Circle** (66 degrees 33 minutes N), and absent south of the **Antarctic Circle** (66 degrees 33 minutes S). Elsewhere, days are longer than nights in the Northern Hemisphere, where it is summer, and days are shorter than nights in the Southern Hemisphere, where it is winter.

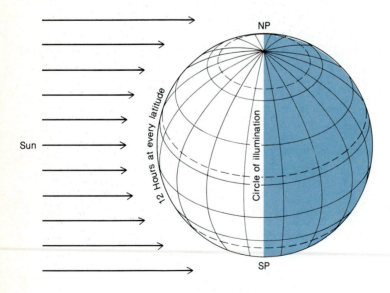

FIGURE 2.10
At the autumnal and vernal equinoxes, insolation is maximum at the equator, and day and night are of equal length everywhere.

---

* Actual dates of equinoxes and solstices vary because of leap years.

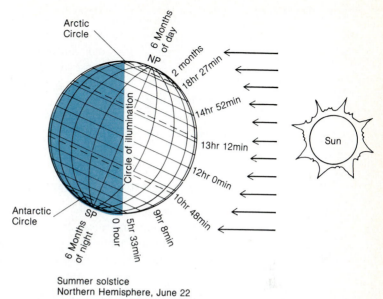

FIGURE 2.11
At the Northern Hemisphere summer solstice (June 22), maximum insolation is at 23 degrees, 27 minutes N, and days are longer than nights everywhere north of the equator. Duration of daylight is given for every 20 degrees of latitude.

Summer solstice
Northern Hemisphere, June 22

On 22 December, the noon sun is directly over 23 degrees 27 minutes S latitude, the **Tropic of Capricorn,** and the situation is reversed (Figure 2.12). Daylight is continuous south of the Antarctic Circle and absent north of the Arctic Circle. Elsewhere, nights are longer than days in the Northern Hemisphere, where it is winter, and days are longer than nights in the Southern Hemisphere, where it is summer.

FIGURE 2.12
At the Northern Hemisphere winter solstice (December 22), maximum insolation is at 23 degrees, 27 minutes S, and days are shorter than nights everywhere north of the equator. Duration of daylight is given for every 20 degrees of latitude.

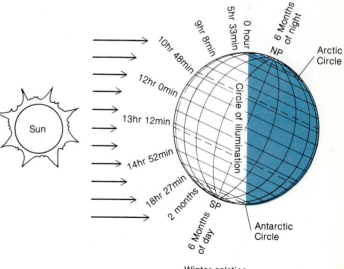

Winter solstice
Northern Hemisphere, Dec. 22

Solar radiation is at its maximum intensity where the noon sun is directly overhead. North and south of this latitude, the intensity of solar radiation diminishes as the solar altitude decreases. For example, at the equinoxes, insolation is most intense at the equator at noon and decreases with latitude to zero at the poles. At the Northern Hemisphere summer solstice, insolation is most intense along the Tropic of Cancer and decreases to zero at the Antarctic Circle. At the Northern Hemisphere winter solstice, insolation is most intense along the Tropic of Capricorn and decreases to zero at the Arctic Circle.

THE SOLAR CONSTANT

For convenience of study, the solar energy input into the Earth-atmosphere system is often expressed as the solar constant. The **solar constant** is defined as the rate at which solar radiation falls on a surface located at the top of the atmosphere and positioned perpendicular to the sun's rays when Earth is at its mean distance from the sun. The "constant" designation is actually misleading because solar energy input fluctuates by a very small fraction of a percent over a year and exhibits some longer-term variations (discussed in Chapter 18).* Nonetheless, we can approximate the solar constant in energy units as 1.97 calories per square centimeter (cal per cm^2) per minute, or in power units as 1372 watts per square meter (W per m^2). For the distinction between energy units and power units, see the inside front cover.

The rate of solar energy input varies through the course of a year from a maximum when Earth is closest to the sun (perihelion) to a minimum when Earth is farthest from the sun (aphelion). At perihelion, Earth is about 3.3 percent closer to the sun than at aphelion. The intensity of solar radiation traveling through space diminishes as the inverse square of the distance traveled, so that the planet receives about 6.7 percent more radiation at perihelion (2.04 cal per cm^2 per minute) than at aphelion (1.91 cal per cm^2 per minute).

The perihelion/aphelion contrast in solar energy coupled with the seasonal variation in radiation has some implications for global climate. The Southern Hemisphere receives more radiation during summer and less radiation during winter than does the Northern Hemisphere. All other factors being equal, this would cause a greater winter-to-summer temperature contrast in the Southern Hemisphere. As we will see in Chapter 17, this effect is offset by the greater percentage of ocean surface area in the Southern Hemisphere and the great thermal stability of ocean water.

---

* Because the solar constant is not constant, *solar parameter* has been suggested as an alternate name for the flux of solar energy input.

Solar Radiation and the Atmosphere

As solar radiation travels through the atmosphere, it interacts with the gases and aerosols that compose the atmosphere. These interactions involve reflection, scattering, and absorption. Solar radiation that is not reflected or scattered back to space, or absorbed by gases and aerosols, reaches the Earth's surface, where further interactions take place.

REFLECTION, SCATTERING, AND ABSORPTION

Reflection occurs at the interface between two different media such as air and cloud when some of the radiation striking that interface is thrown back. At such an interface (Figure 2.13), the angle of incident radiation (i) equals the angle of reflected radiation (r); this is known as the **law of reflection.** The fraction of incident radiation that is reflected by some surface (or interface) is the **albedo** of that surface, that is,

$$\text{albedo} = \frac{\text{reflected radiation}}{\text{incident radiation}}$$

where albedo is expressed either as a fraction or a percentage. As a rule, surfaces that have a high albedo (reflect a large fraction of incident solar radiation) appear light colored and surfaces with a low albedo are perceived as dark colored.

FIGURE 2.13
An illustration of the law of reflection whereby the angle of incident radiation (i) equals the angle of reflected radiation (r).

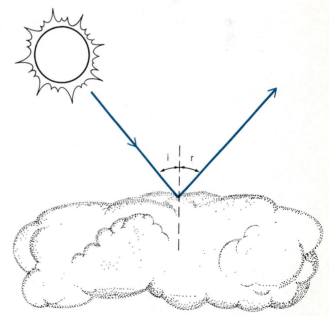

Within the atmosphere, the tops of clouds are the most important reflectors of insolation (Figure 2.14). The albedo of cloud tops depends primarily on cloud thickness and varies from under 40 percent for thin clouds (less than 50 m, or 150 ft, thick) to 80 percent or more for thick clouds (more than 5000 m, or 15,000 ft, thick). The average albedo for all cloud types and thicknesses is about 55 percent, and at any given time, clouds cover about 60 percent of the planet.

Special Topic

Why Is the Sky Blue?

Visible light is composed of the entire spectrum of colors—red, orange, yellow, green, blue, and violet. If a beam of sunlight is passed through a glass prism and onto a screen (Plate 2), we see a spectrum of colors because the light beam is refracted (bent) by the glass. The more energetic violet end of the spectrum is bent the most, and the less energetic red end is bent the least.

Refraction is not the only way that visible light is dispersed into its component colors. Scattering of sunlight can produce a similar effect and is responsible for the color of the daytime sky. Scattering occurs when tiny particles in the atmosphere interact with light waves and send those light waves in random directions. If the radii of the scattering particles are much smaller than the wavelength of the scattered light, then the amount of scattering varies with wavelength. This is the case, for example, when visible light is scattered by the gas molecules composing the atmosphere. In a now classic experiment performed in 1881, Lord Rayleigh demonstrated that this scattering of light is inversely proportional to the fourth power of the wavelength. This means that violet light, at the short-wavelength end of the visible spectrum, is scattered much more than red light, which is at the long-wavelength end. As sunlight travels through the atmosphere, the various colors are therefore scattered selectively out of the solar beam: violet is scattered more than blue, blue more than green, green more than yellow, and so forth.

The dependence of scattering on wavelength would suggest that scattered sunlight is mostly violet in color. Why, then, does the sky appear blue rather

Scattering is the dispersal of radiation in all directions—up, down, and sideways. Actually, reflection is a special case of scattering. Both gas molecules and aerosols in the atmosphere scatter solar radiation but with some important differences. Scattering by molecules is wavelength dependent. As described in the Special Topic, "Why Is the Sky Blue?," the preferential scattering of blue-violet light by oxygen and nitrogen molecules is the principal reason for the color of the daytime sky. On the other hand, water droplets and ice crystals that compose clouds scatter visible solar radiation equally at all wavelengths so that clouds appear white.

Reflection and scattering within the atmosphere merely change the direction of insolation. However, through **absorption,** radiation is converted to heat. Most absorption within the atmosphere is by oxygen, ozone, water vapor, and various aerosols. For example, warming in the upper stratosphere (refer back to Figure 1.7) is due to absorption of solar ultraviolet radiation by ozone and oxygen. Clouds, though excellent reflectors, are poor absorbers of solar radiation. Typically, clouds absorb less than 10 percent of the radiation that strikes the cloud top, although exceptionally thick clouds such as thunderclouds (cumulonimbus) may absorb somewhat more. Within the atmosphere, the sum of the percentage of insolation that is absorbed (the **absorptivity**) plus the percentage reflected or scattered (the albedo) plus the percentage transmitted to the Earth's surface (the **transmissivity**) must equal 100 percent. This is another example of the law of energy conservation.

than violet? The principal reason is that the human eye is more sensitive to blue light than to violet light, so that the sky appears bluer to the human eye than it really is. Another factor contributing to the sky's blueness may be the dilution of violet light by all the other scattered colors. Although the other colors are scattered less than violet, they tend to wash violet into blue.

We can now understand why on a clear evening the setting sun is red (Plate 3). At very low solar altitudes, the sun's rays must traverse the maximum thickness of the atmosphere. Hence, at sunset (or sunrise) there is considerable interaction between incoming solar radiation and the atmosphere's component gas molecules. Consequently, nearly all the blue-violet light is scattered out of the solar beam, leaving it rich in the red end of the visible spectrum.

If we were on the moon, which has a highly rarefied atmosphere, the sun would appear as a white disk in a black sky and stars would be visible even in bright sunshine; within a rarefied atmosphere, scattering is inconsequential.

A different effect occurs when radiation is scattered by particles that have radii approaching or exceeding the wavelength of the radiation being scattered. In these instances, scattering is not wavelength dependent; instead, radiation is scattered equally at all wavelengths. This is known as *Mie scattering*. The particles composing clouds (tiny ice crystals or water droplets or both) and most other atmospheric aerosols are sufficiently large to scatter sunlight in this way. For this reason, clouds appear white, and when the atmosphere contains a considerable concentration of aerosols, the entire sky is a hazy white.

THE OZONE SHIELD

As noted above, absorption of solar ultraviolet radiation (UV) warms the upper stratosphere. Much more importantly, UV absorption in the stratosphere prevents potentially lethal levels of UV from reaching the Earth's surface. UV is absorbed during both the formation and destruction of ozone within the stratosphere. Without this **ozone shield,** life as we know it would not exist on Earth.

At altitudes between about 10 and 50 km (6 to 31 mi), UV dissociates

---

Special Topic

The Hazards of Sunbathing

Many people believe that a deep, dark tan is attractive and a sign of good health (Figure 1). But mounting evidence indicates that too much sun can age the skin and cause wrinkles and dark spots to appear prematurely. The sun is also the primary cause of skin cancer, the most common form of cancer in the United States. More than 400,000 new cases of skin cancer are diagnosed each year so that about one in seven Americans will contract this disease during his/her lifetime. The culprit is overexposure to the ultraviolet portion of solar radiation, specifically the long-wavelength end of ultraviolet, known as *UVB*. How, then, can you protect yourself from this dangerous radiation?

The body does have some natural defenses against UVB radiation. The outermost layer of skin is composed of tightly packed dead cells, usually about 20 cells thick, that shield the underlying layers of skin from potentially damaging radiation. Dead dells are continually shed and replaced by new cells that form below the dead skin layer. The sequence of cell formation and shedding typically takes three to four weeks. UVB radiation speeds up the process so that a thicker layer of dead skin forms that better protects the underlying layers of skin. The consequent increased rate of shedding that occurs several days after a sunburn is commonly described as peeling.

The body's second line of defense consists of specialized cells called *melanocytes,* which lie beneath the dead skin layer. Melanocytes respond to UVB by producing *melanin,* a dark pigment that absorbs UVB and

is responsible for a tan. Melanin spreads to surrounding cells, which then migrate upward and become part of the dead cell layer. The time required for this migration means that a tan takes several days to develop. Cell migration also explains why a tan gradually fades since the melanin is lost during normal shedding.

FIGURE 1
Sunbathers at the beach run the risk of excessive exposure to solar ultraviolet radiation that could cause health problems. [Photograph by J. M. Moran]

oxygen molecules (O_2) into oxygen atoms (O) that subsequently combine with other oxygen molecules to form ozone molecules (O_3). While all this is going on, other photochemical reactions—some quite complex—destroy ozone. The consequence of these competing chemical reactions is a peak ozone concentration of only about 10 ppm (parts per million) near an altitude of 25 km (40 mi). Some UV penetrates the ozone shield and reaches the Earth's surface where overexposure may cause serious health problems for humans. For more information on this subject, refer to the Special Topic, "The Hazards of Sunbathing."

Melanin does not provide total protection against UVB. Typically, several days elapse between the skin's initial exposure to UVB and the development of a protective tan. In the meantime, considerable damage can occur, especially if the initial exposure causes sunburn. Moreover, melanin does not absorb all the UVB so that no one is immune to the hazards of sunbathing. Although the risk of contracting skin cancer is greater for fair-skinned individuals (whose bodies produce less melanin), people with dark complexions (whose bodies produce more melanin) are not immune.

Often, people are exposed to much more UVB than they realize. Ultraviolet radiation penetrates

clouds more readily than does visible solar radiation so that even when the sky is completely overcast, some protection from UV is necessary. At the beach a person may believe that the best way to avoid too much sun is to sit under a beach umbrella, jump in the water, or wear a T-shirt. Such strategies provide only minimal protection. Sand reflects up to 50 percent of the incident UVB, thereby exposing a person to dangerous radiation even in the shade of a beach umbrella. Water transmits UVB to a depth of a meter or so, and a wet T-shirt allows 20 to 30 percent of incident UVB to reach the skin.

The beach is not the only place where a person is likely to be exposed to high levels of UVB. Skiing at high mountain elevations, where UVB is more intense than at sea level, results in significant exposure. In addition, snow has an even higher reflectivity for UVB than does sand.

Fortunately, many of the adverse effects of ultraviolet exposure can be prevented. Obviously, a person can try to avoid the direct sun or wear protective clothing. For most people, however, a more practical solution is to apply a sunscreen. The most popular sunscreen agents have ingredients that selectively block UVB wavelengths. When purchasing a sunscreen, the buyer should be aware of the degree of protection offered: the sunscreen's *sun protection factor (SPF)*. The higher the SPF value, the greater the protection. Experts recommend a SPF of at least 15 to gain maximum protection. Note that suntan lotions are not sunscreens; suntan lotions merely help to keep the skin moist and do not provide any protection from UVB.

One of today's major environmental concerns is directed at a group of chemicals, collectively known as CFCs (for chlorofluorocarbons), which threaten to erode the ozone shield. CFCs have been used for decades as refrigerants, propellants in aerosol sprays, and blowing agents for foam insulation. Within the troposphere, CFCs are virtually inert (chemically nonreactive), but eventually some CFCs drift into the stratosphere where intense UV breaks them down, liberating chlorine atoms. Chlorine acts as a catalyst* in chemical reactions that convert ozone to oxygen. In this way, each chlorine atom can destroy tens of thousands of ozone molecules.

A thinner ozone shield would likely mean more intense UV radiation at the Earth's surface and, for humans, a greater incidence of skin cancer and eye damage including cataracts. We have much more to say on threats to the ozone shield and what is being done about this problem in Chapter 16.

Solar Radiation and the Earth's Surface

Solar radiation that is transmitted directly through the atmosphere to the Earth's surface constitutes **direct insolation.** This is augmented by **diffuse insolation,** solar radiation that is scattered or reflected to the surface or both. The direct and diffuse insolation that strikes the Earth's surface is either reflected or absorbed by it depending on the surface albedo. The fraction that is not reflected is absorbed—that is, converted to heat—and the fraction that is not absorbed is reflected.

As noted earlier, light surfaces are more reflective of solar radiation than are dark surfaces. For example, skiers who have been sunburned on the slopes know full well that a snow cover is very reflective of sunlight. Fresh-fallen snow typically has an albedo of between 75 and 95 percent; that is, 75 to 95 percent of the solar radiation striking the surface of a snow cover is reflected, and the rest (5 to 25 percent) is absorbed. At the other extreme, the albedo of a dark surface, such as a blacktopped road or a green forest, may be as low as 5 percent. From this it is obvious why light-colored clothing is usually a more comfortable choice than dark-colored clothing during hot weather. The albedos of some common surfaces are listed in Table 2.1.

The albedo of some but not all surfaces also varies with the angle of incidence of solar radiation, that is, the angle of the sun above the horizon (solar altitude). The variation of albedo with solar altitude is especially pronounced for the surfaces of water bodies—oceans and lakes. As shown in Figure 2.15, under clear skies, the albedo of a water surface increases with

---

* A catalyst accelerates a chemical reaction without itself being altered by the reaction.

Table 2.1
Reflectivity (Albedo) of Some Common Surface Types for Visible Solar Radiation

Surface	Albedo (percent reflected)	Surface	Albedo (percent reflected)
Grass	16–26	Sea ice	30–40
Deciduous forest	15–20	Fresh snow	75–95
Coniferous forest	5–15	Old snow	40–70
Crops	15–25	Glacier ice	20–40
Tundra	15–20	Water (high sun)	3–10
Desert	25–30	Water (low sun)	10–100
Blacktopped road	5–10		

decreasing solar altitude. The increase in albedo is especially sharp for solar altitudes less than 30 degrees and approaches a mirrorlike 100 percent near sunrise and sunset. On the other hand, when the sky is completely cloud covered, only diffuse radiation is incident on a water surface and the albedo changes little with solar altitude and is uniformly very low (less than 10 percent). On a global basis, the average albedo of the ocean surface is only about 8 percent; that is, the ocean is a strong absorber of solar radiation.

FIGURE 2.15
Variation of albedo with solar altitude for a flat and undisturbed water surface. The albedo increases sharply for solar altitudes less than 30 degrees. An agitated, wave-covered surface has a slightly higher albedo at high solar altitudes and slightly lower albedo at low solar altitudes.

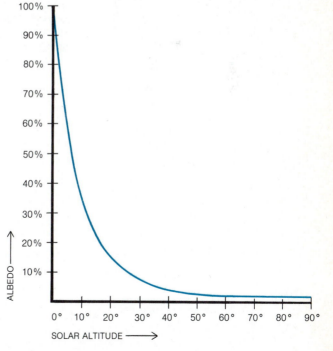

Solar radiation is selectively absorbed by wavelength as it penetrates ocean or lake waters. Absorption increases for increasing wavelength so that within clear, clean water, red light is totally absorbed within about 15 m (45 ft) of the surface, whereas blue-violet light may penetrate to depths of 250 m (800 ft). Usually, however, suspended sediments significantly boost the rate of absorption so that sunlight is completely absorbed within shallower depths. In fact, most water bodies are sufficiently turbid that little if any sunlight reaches below 10 m (30 ft).

Important changes in surface albedo can occur seasonally. Over land, a heavy winter snow cover increases the surface albedo considerably. In autumn in forested areas, the loss of leaves from deciduous trees raises the surface albedo. In middle and high latitudes, significant increases in surface albedo accompany the winter freeze-over of lakes and the formation of sea ice.

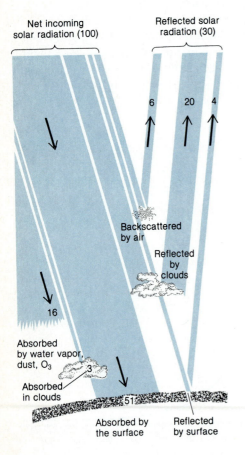

Net incoming solar radiation (100)

Reflected solar radiation (30)

6   20   4

Backscattered by air

Reflected by clouds

16

Absorbed by water vapor, dust, O₃

Absorbed in clouds   3

51

Absorbed by the surface

Reflected by surface

FIGURE 2.16
The disposition of solar radiation as it interacts with the atmosphere and the Earth's surface. Note that of the 100 units of solar energy entering the atmosphere, 16 units are absorbed by water vapor, dust, and ozone, 3 units are absorbed by clouds, 51 units are absorbed at the Earth's surface, 6 units are scattered back into space by the atmosphere, 20 units are reflected back into space by clouds, and 4 units are reflected into space by the Earth's surface. Hence, $16 + 3 + 51 + 6 + 20 + 4 = 100$. [Adapted from A. P. Ingersoll, "The Atmosphere," *Scientific American* 249, No. 3 (1983):164. Copyright © 1983 by Scientific American, Inc. All rights reserved.]

Solar Radiation Budget

Satellite measurements indicate that about 30 percent of the incoming solar radiation is reflected or scattered off to space by the Earth-atmosphere system. This reflected and scattered light, which has been seen as "Earthshine" by United States astronauts on the moon, is known as the Earth's **planetary albedo.** By contrast, the moon's albedo is only about 7 percent primarily because of the absence of clouds in the highly rarefied lunar atmosphere. This means that Earth viewed from the moon is more than four times brighter than the moon viewed from Earth on a clear night.

Of the total solar radiation that is intercepted by the Earth-atmosphere system, only about 19 percent is absorbed by constituents of the atmosphere. In other words, the atmosphere is relatively transparent to solar radiation. The remaining 51 percent of solar radiation (30 percent was reflected to space) is absorbed by the Earth's surface, chiefly because of the low average albedo of ocean waters covering nearly three quarters of the globe. The distribution of solar radiation within the Earth-atmosphere system on an average annual basis is summarized in Figure 2.16.

The Earth's surface is the principal recipient of solar heating, and the Earth's surface, in turn, continuously radiates heat back to the atmosphere, which eventually radiates it back into space. The Earth's surface is thus the main source of heat for the atmosphere. This is evident in the vertical temperature profile of the troposphere. Air is normally warmest close to the Earth's surface, and air temperature drops with increasing altitude, that is, away from the main source of heat.

Infrared Response and the Greenhouse Effect

If solar radiation were continually absorbed by the Earth-atmosphere system without any compensating flow of heat out of the system, the air temperature would rise steadily. In reality, the average global air temperature changes little from year to year. This is because an equal amount of heat leaves the Earth-atmosphere system, mainly via infrared (IR) radiation, as enters the system from the sun. Although solar radiation is supplied only to the illuminated portion of the planet, infrared radiation is emitted ceaselessly, both day and night, by the entire Earth-atmosphere system.

Because solar radiation and terrestrial infrared radiation peak in different portions of the electromagnetic spectrum, they have different properties, and they interact differently with the atmosphere. As noted earlier, atmospheric gases and aerosols absorb only about 19 percent of the incoming solar radia-

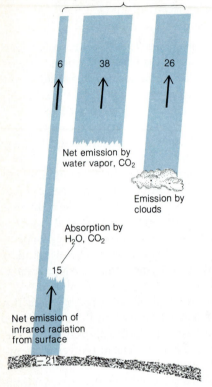

Outgoing infrared radiation (70)

6   38   26

Net emission by
water vapor, CO_2

Emission by
clouds

Absorption by
H_2O, CO_2

15

Net emission of
infrared radiation
from surface

21

FIGURE 2.17

Infrared radiation interacts with components of the atmosphere. The
numbers are based on 100 units of incident solar radiation in FIGURE
2.16. Note absorption by CO_2 and H_2O. The 70 units of solar energy absorbed by the Earth-atmosphere system are ultimately emitted to space as
infrared radiation. [Adapted from A. P. Ingersoll, "The Atmosphere," *Scientific American* 249, No. 3 (1983):164. Copyright © 1983 by Scientific
American, Inc. All rights reserved.]

tion. In contrast, the atmosphere absorbs a proportionately greater amount of
infrared radiation emitted by the Earth's surface. The atmosphere, in turn,
reradiates some infrared radiation back to the Earth's surface (Figure 2.17).
This reradiation by the atmosphere slows the escape of heat to space and
causes the lower atmosphere to have a higher average temperature than the
upper atmosphere. The lower atmosphere is thus hospitable to life. Thanks to
this reradiation, the Earth's average surface temperature is more than 30
Celsius degrees (54 Fahrenheit degrees) higher than it would otherwise be.
That is, viewed from space, the planet radiates at about −18 °C (0 °F), whereas
the average surface temperature is about 15 °C (59 °F).

Atmospheric gases, primarily water vapor and to a lesser extent carbon
dioxide, ozone, methane (CH_4), and nitrous oxide (N_2O), absorb terrestrial
infrared radiation and thus impede its loss to space. The percentage of infrared
radiation absorbed varies with wavelength, as shown in Figure 2.18. Interestingly, the percentage absorbed is very low at wavelengths near the peak
infrared intensity. It is through these so-called **atmospheric windows,** wavelength bands in which there is little or no absorption of radiation, that most
heat from the Earth's surface eventually escapes to space as infrared radiation.

Like the Earth's atmosphere, window glass is relatively transparent to visible solar radiation but slows the transmission of infrared radiation. Greenhouses, where plants are stored or grown, are designed to take advantage of this property of glass by being constructed almost entirely of glass panes. Sunlight readily enters a greenhouse and is absorbed (that is, converted to heat). Some of the heat energy is radiated as IR which is absorbed and not transmitted by the glass. Because of similarities in radiative properties, the behavior of IR-absorbing atmospheric gases is often referred to as the **greenhouse effect.** The analogy is not always correct, however. Trapping of IR radiation by glass is only part of the reason why some greenhouses retain internal heat. Those greenhouses cut heat loss principally by acting as a shelter from the wind, thereby reducing conductive and convective heat loss (discussed in Chapter 3). As a rule, the thinner the greenhouse glass and the greater the external wind speed, the more important is the shelter effect compared to radiation trapping. Nonetheless, because reference to the greenhouse effect is so common in discussions of the Earth-atmosphere radiation balance, we use the term here.

To illustrate the greenhouse effect in the atmosphere, compare the typical summer weather in the American Southwest with that of the Gulf of Mexico coast. Both areas are at about the same latitude and hence receive about the

FIGURE 2.18

Absorption of radiation by selected components of the atmosphere as a function of wavelength. Absorptivity is the fraction of the radiation absorbed and ranges from 0 to 1 (0 to 100 percent absorption). Absorptivity is very low or near zero in atmospheric windows. Note the infrared windows near 8 and 10 μm. [From R. G. Fleagle and J. Businger, *An Introduction to Atmospheric Physics.* New York: Academic Press, 1963, p. 153.]

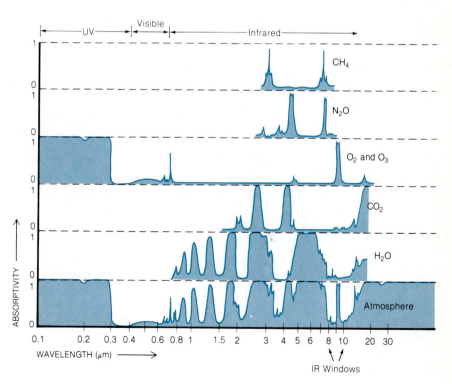

same intensity of solar radiation. In both areas, afternoon temperatures typically top 30 °C (86 °F). At night, however, air temperatures in the two areas often differ markedly. In the Southwest, there is usually less water vapor in the air to impede the escape of infrared radiation, and heat is readily lost to space. Surface air temperatures may fall below 15 °C (59 °F) by sunrise. On the other hand, along the Gulf Coast, the air often is more humid and thus absorbs more infrared radiation. Because a portion of this heat is reradiated back toward the Earth's surface, nighttime temperatures may fall only into the 20s Celsius (the 70s Fahrenheit).

Clouds, which are composed of water droplets and/or ice crystals, also produce a greenhouse effect. Hence, nights usually are colder when the sky is clear than when the sky is cloud covered. Even a high, thin overcast through which the moon is visible can elevate temperatures at the Earth's surface by 5 or more Celsius degrees. Recall, however, that clouds also have a high albedo for solar radiation. As a result, clouds affect climate in opposing ways. By absorbing and reradiating IR, they warm the Earth's surface and by reflecting solar radiation, they cool the Earth's surface. On a global scale, which of these effects is more important? Recent analysis of satellite measurements of radiation indicates that clouds have a net cooling effect on global climate.

The greenhouse effect is also the reason why mountaintops, though closer to the sun, are usually colder than lowlands, Water vapor, the principal contributor to the greenhouse effect, is concentrated near the Earth's surface, declines in concentration with altitude, and is virtually absent above 10 km (6.2 mi). On mountaintops, less water vapor means that IR radiation more readily escapes to space, and air temperatures are relatively low in spite of intense sunshine.

Today, considerable concern is directed at the potential global climatic consequences of increasing concentrations of atmospheric IR-absorbing trace gases, especially carbon dioxide. Atmospheric CO_2 has probably been on the rise since the Industrial Revolution because of the burning of coal and oil and, to a lesser extent, because of the clearing of forests. More CO_2 in the atmosphere enhances the greenhouse effect and, unless compensated, may trigger significant global warming.*

As noted in Chapter 1, atmospheric scientists have employed numerical models of the atmosphere in an effort to predict the magnitude of global warming that could attend a continued increase in CO_2. These experiments predict that the Earth's average surface temperature could rise between 2 and 5 C° (4 and 10 F°) with a doubling of CO_2—possibly by the middle of the next century. As Stephen Schneider of the National Center for Atmospheric

---

* In the popular media, the term *greenhouse effect* often refers to CO_2-induced global warming when, actually, an *enhanced* greenhouse effect is the intended meaning.

Research (NCAR) recently pointed out, the warming could match the total post–Ice Age temperature rise—albeit at a rate 10 to 100 times faster. The agricultural, socioeconomic, and political implications of such a climate change are likely to be far reaching and disruptive. Already some atmospheric scientists are attributing the global warming trend of the 1980s to an enhanced greenhouse effect.

In addition to CO_2, upward trends in other atmospheric IR-absorbing trace gases may also enhance the natural greenhouse effect. Specifically, concentrations of methane (CH_4), nitrous oxide (N_2O), and CFCs (chlorofluorocarbons) are rising. Although atmospheric concentrations of those three gases are considerably less than carbon dioxide, they are more efficient absorbers of IR because they strongly absorb within the atmospheric windows. In fact, trace gas absorption is directly proportional to concentration, so that doubling the concentration of a gas doubles the absorption. The combined climatic impact of rising levels of these gases could equal CO_2-induced global warming. We have more to say about prospects for an enhanced greenhouse effect in Chapter 18.

Radiation Measurement

The **pyranometer** is the standard instrument for measuring the intensity of solar radiation striking a horizontal surface. The instrument consists of a sensor enclosed in a transparent hemisphere (Figure 2.19) that transmits total (direct plus diffuse) short-wave (less than 3.5 μm wavelength) insolation. The sensor is a disk consisting of alternating black and white wedge-shaped segments

FIGURE 2.19
The pyranometer is the standard instrument used to measure solar radiation. [Courtesy of Qualimetrics, Inc.]

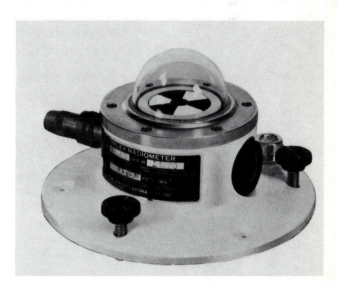

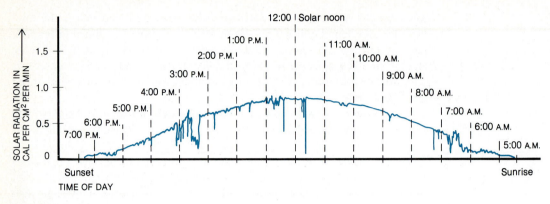

FIGURE 2.20
A recording pyranometer produces a continuous record of the incoming solar radiation. Scattered clouds cast shadows on the radiation sensor and produce spikes in the curve. This record was obtained at Green Bay, Wisconsin, in early July.

that form a starlike pattern. The black wedges are highly absorptive, and the white wedges are highly reflective of solar radiation. Differences in absorptivity and albedo mean that the temperatures of the black and white portions of the sensor respond differently to the same intensity of solar radiation. The temperature contrast between the black and white segments is calibrated in terms of radiation flux (calories per square cm per minute, for example). A pyranometer may be linked electronically to a pen recorder that traces a continuous record of insolation (Figure 2.20), or the instrument's output signal may be recorded on a magnetic tape cassette for processing and storage.

Special care must be taken in mounting and maintaining a pyranometer. The instrument should be situated where it will not be affected by shadows, by any highly reflective surfaces nearby, or by other sources of radiation. The glass bulb must also be kept clean and dry.

Conclusions

Net incoming solar radiation is balanced by the infrared radiation emitted to space by the Earth-atmosphere system. Absorption of solar radiation causes warming, and emission of infrared radiation to space causes cooling. Within the Earth-atmosphere system, however, the rates of radiational heating and radiational cooling are not the same everywhere. Before we examine the reasons for these energy imbalances and their implications for atmospheric circulation, we first need to distinguish between heat and temperature and describe heat transfer processes. This we do in the next chapter.

Mathematical Note

Blackbody Radiation Laws

Several laws describe properties of the electromagnetic radiation emitted by a *perfect radiator*. A perfect radiator, usually called a *blackbody*, is a hypothetical object that absorbs all the radiation that strikes it; that is, a perfect radiator neither reflects nor transmits any radiation. In reality, no perfect radiators exist, but the sun and the Earth's surface radiate *approximately* as blackbodies. We can therefore apply blackbody radiation laws to solar and terrestrial radiation, with some qualifications. Here, in brief, are the basic blackbody radiation laws.

Kirchhoff's law holds that a perfect absorber of radiation of a given wavelength is also a perfect emitter of radiation at that same wavelength. A blackbody thus emits, as well as absorbs, all incident radiation at all wavelengths. In general, for all objects the efficiency of radiation absorption, called *absorptivity,* equals the efficiency of radiation emission, called *emissivity.* A good absorber is therefore a good emitter, and a poor absorber is a poor emitter. The emissivity and absorptivity of a blackbody are both 100 percent.

As noted elsewhere in this chapter, in a transparent substance such as the atmosphere, the law of energy conservation requires the sum of the absorptivity plus albedo plus transmissivity to equal 100 percent. From Kirchhoff's law, since absorptivity equals emissivity, it follows that the sum of emissivity plus albedo plus transmissivity equals 100 percent. For an opaque substance, such as the ground, transmissivity equals zero, and the sum of emissivity plus albedo equals 100 percent.

The *Stefan–Boltzmann law* states that the rate at which a blackbody radiates energy across all wavelengths (called *emittance, E*) is directly proportional to the fourth power of the absolute temperature, T (measured in kelvins), of the radiating body. The mathematical statement of this law is

$$E = \sigma T^4$$

where σ is the Stefan–Boltzmann constant equal to 5.67×10^{-8} W m^{-2} K^{-4}.

Planck's law states that the rate at which radiation is emitted by a blackbody depends on the absolute temperature of the body and the specific wavelength of the radiation. This law enables one to compute the amount of radiation emitted at some wavelength and at a specified temperature.

Wien's displacement law holds that the wavelength at which a blackbody emits the maximum intensity of radiation, λ_{max}, is inversely proportional to the absolute temperature, T, of the blackbody, that is,

$$\lambda_{max} = C/T$$

where C is the constant of proportionality (equal to 2897 if λ_{max} is expressed in micrometers).

Summary Statements

The many forms of electromagnetic radiation make up the electromagnetic spectrum and are distinguished on the basis of wavelength, frequency, and energy level.

Wien's displacement law predicts that relatively warm bodies emit shorter wavelength radiation than do relatively cold bodies. Hence, the sun emits shorter wavelength radiation than does the Earth-atmosphere system.

The total energy (in the form of solar radiation) that is absorbed by the Earth-atmosphere system is equal to the total energy (in the form of infrared radiation) emitted by the Earth-atmosphere system back to space.

Only a minute fraction of the radiation emitted by the sun to space is intercepted by planet Earth. As a consequence of the Earth's elliptical orbit, spherical shape, and tilted rotational axis, insolation is unevenly distributed over the Earth's surface and changes through the course of a year.

Solar radiation that is not absorbed by the atmosphere or returned to space via reflection and scattering reaches the Earth's surface, where it is either reflected or absorbed, depending on the surface albedo.

The Earth's surface and atmosphere emit infrared radiation; the surface is the primary heat source for the lower atmosphere.

Water vapor, carbon dioxide, and other gases absorb and reradiate infrared energy back toward the Earth's surface, thereby moderating the temperature of the lower atmosphere. Clouds also contribute to this so-called greenhouse effect.

Key Words

law of energy conservation	X rays	corona	law of reflection
electromagnetic radiation	gamma radiation	insolation	albedo
electromagnetic spectrum	blackbody	solar altitude	scattering
wavelength	Wien's displacement law	perihelion	absorption
wave frequency	Stefan–Boltzmann law	aphelion	absorptivity
radio waves	global radiative equilibrium	equinoxes	transmissivity
microwave	photosphere	solstice	ozone shield
infrared radiation	granules	Tropic of Cancer	direct insolation
visible light	faculae	Arctic Circle	diffuse insolation
ultraviolet radiation	sunspots	Antarctic Circle	planetary albedo
	chromosphere	Tropic of Capricorn	atmospheric windows
		solar constant	greenhouse effect
		reflection	pyranometer

Review Questions

1. What is the relationship between the wavelength and frequency of electromagnetic radiation?

2. Describe how energy level varies within the electromagnetic spectrum.

3. In the Northern Hemisphere, we are closer to the sun during the winter than during the summer. Why, then, is winter colder than summer?

4. In middle latitudes, days are longer than nights between the spring and fall equinoxes. Why?

5. What is the significance of the Tropic of Cancer and the Tropic of Capricorn?

6. What is the significance of the Arctic Circle and the Antarctic Circle?

7. How and why does the solar altitude affect the intensity of solar radiation received at the Earth's surface?

8. Define in detail the solar constant.

9. Absorption of solar radiation by constituents of the atmosphere is an energy conversion process. Explain what is meant by this statement.

10. Why is the sky blue, and why are clouds white?

11. In your locality, are there seasonal changes in surface albedos? What are the implications of these changes for air temperatures?

12. Although insolation reaches its maximum intensity at noon, surface air temperatures typically do not reach a maximum until several hours later. Explain why.

13. How does the relatively high albedo of a snow cover influence air temperature?

14. The atmosphere is relatively transparent to solar radiation. Elaborate on this statement.

15. For the entire globe, why must the net incoming solar radiation balance outgoing infrared radiation? What would be the implications for global climate if this energy balance did not prevail?

16. What is the relationship between the temperature of a radiating object and the wavelength of most intense radiation?

17. What is meant by an atmospheric window for infrared radiation?

18. What is the significance of the greenhouse effect for temperatures at the Earth's surface?

19. Are air temperatures likely to be lower on a clear night or a cloudy night? Explain your choice.

20. Explain why the day-to-night temperature difference is typically much greater in a warm and dry locality (such as the Nevada desert) than in a warm and humid locality (such as New Orleans).

Points to Ponder

1. Radiation intensity decreases as the inverse square of the distance traversed. If the mean Earth-to-sun distance were *three times* what it is today, the solar constant would be reduced to what fraction of its present value?

2. If there were no atmosphere, the total solar radiation received per day at the Earth's surface would reach a maximum in summer at the South Pole. Explain why the maximum is at the pole and not at the solstice latitude.

3. Explain why the albedo of the moon is considerably less than the Earth's planetary albedo.

4. What natural or human-related activities might cause a change in the Earth's planetary albedo? What are the implications of such changes for global climate?

5. What basic assumptions are made when blackbody radiation laws are applied to the sun or the Earth-atmosphere system?

Projects

1. Determine whether solar radiation data are available for your locality. If available, find out how variations in cloud cover influence the radiation.

2. For your area, determine whether there is a

relationship between cloud cover at night and the early morning low temperature. For reasons that are discussed in Chapter 4, limit your observations to days when the wind is light or calm.

Selected Readings

Brune, W. H. "Ozone Crisis: The Case Against Chlorofluorocarbons." *Weatherwise* 43, No. 3 (1990):136–143. Reviews threats to the delicate stratospheric ozone layer.

Foster, K. R., and A. W. Guy. "The Microwave Problem." *Scientific American* 255, No. 3 (1986):32–39. Reviews the possible hazards of exposure to low levels of microwave radiation.

Foukal, P. V. "The Variable Sun." *Scientific American* 262, No. 2 (1990):34–41. Describes how the sun's output of radiation and particles varies with time.

Lindzen, R. S. "Some Coolness Concerning Global Warming." *Bulletin of the American Meteorological Society* 71 (1990):288–299. Questions the widely held view that increasing levels of atmospheric CO_2 will trigger global warming.

Walsh, J. E. "Snow Cover and Atmospheric Variability." *American Scientist* 72, No. 1 (1984):50–57. Discusses the influence of a regional snow cover on the radiation balance.

Heat and Temperature

Heat and temperature, the subjects of this chapter, are two distinct concepts. The temperatures of different substances respond differently to the addition (or loss) of the same quantity of heat. This is one reason why, at the beach in summer, the sand feels warmer than the water. [Photograph by J. M. Moran]

"TEMPERATURE" is one of the most important and common variables used to describe the state of the atmosphere; it is a usual component of weather reports and forecasts. From everyday experience, we know that air temperature varies with time: from one season to another, between day and night, and even from one hour to the next. Air temperature also varies from one place to another: highlands and higher latitudes are usually colder than lowlands and lower latitudes.

We also know that temperature and heat are related concepts. When we heat a pan of soup on the stove, the soup's temperature rises. When we drop an ice cube into a drink, the temperature of the drink drops. Granted that the two concepts are closely linked, what is the distinction between heat and temperature?

Distinguishing Heat and Temperature

All substances are made up of a multitude of minute particles (atoms or molecules) that are continually in rapid, random motion. Hence, atoms and molecules possess energy of motion, called **kinetic energy. Heat** is defined as the *total* kinetic energy of the atoms or molecules composing a substance.* The atoms or molecules in a substance do not all move at the same velocity so that there's actually a range of kinetic energy among the atoms or molecules. **Temperature** is a measure of the *average* kinetic energy of the individual atoms or molecules.

The distinction between heat and temperature is made clearer by the following illustration. A cup of water at 80 °C (176 °F) is much hotter than a bathtub of water at 30 °C (86 °F). That is, the average kinetic energy of individual water molecules at 80 °C is greater than at 30 °C. However, the greater volume of the bathtub water means that it contains more total kinetic molecular energy (heat) than does a cup of water. Consequently, the cup of water will cool down to room temperature much more rapidly than will the bathtub of water. Much more heat energy must be removed from the bathtub water than from the cup of water in order for both to cool to the same temperature.

From the above discussion it would appear that air temperature always changes whenever the air gains or loses heat. However, this is not necessarily

---

* Note that heat is a characteristic of a substance and is not a separate entity. Strictly speaking, therefore, it is imprecise to refer to heat apart from the particular substance that possesses the heat. For example, heat does not rise, but heated air does. In our discussions of heat, we refer either explicitly or implicitly to the heat energy of some substance.

the case. Water is a component of air and occurs in all three phases (as ice crystals, droplets, and vapor), and, as we will see in Chapter 4, heat is either required or released when water changes phase. Furthermore, air is a compressible mixture of gases; that is, an air sample can change volume. As we will discuss in Chapter 6, heat energy is required for the work of expansion or compression of air. Hence, as a sample of air gains or loses heat, that heat may be involved in some combination of temperature change, phase change of water, or volume change.

Temperature Scales

For most scientific purposes, temperature is described in terms of the Celsius scale. Established by the Swedish astronomer Anders Celsius in 1742, the Celsius temperature scale has the numerical convenience of a 100-degree interval between the melting point of ice and the boiling point of pure water. The United States is virtually the only nation that still uses the numerically more cumbersome Fahrenheit temperature scale for everyday measurements, including weather reports. This scale was introduced in 1714 by a German scientist, Gabriel Fahrenheit. If a thermometer graduated in both scales is immersed in a glass containing a mixture of ice and water, the Fahrenheit scale will read 32 °F and the Celsius scale will read 0 °C. In boiling water at sea level, the readings will be 212 °F and 100 °C.

The average kinetic energy of individual molecules is less in cold substances than in hot substances. There is, theoretically at least, a temperature at which all molecular motion ceases. It is called **absolute zero** and corresponds to −273.15 °C (−459.67 °F). Actually, some atomic-level activity

FIGURE 3.1
A comparison of the three temperature scales: Kelvin, Celsius, and Fahrenheit.

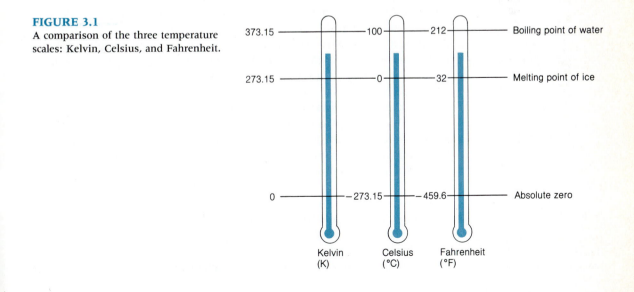

occurs at absolute zero, but, as noted in Chapter 2, an object at that temperature does not emit electromagnetic radiation.

On the Kelvin scale, temperature is the number of degrees above absolute zero; hence, the Kelvin scale is a more direct measure of average kinetic molecular activity than the Fahrenheit and Celsius temperature scales. Whereas units of temperature are expressed as "degrees Celsius" on the Celsius scale and "degrees Fahrenheit" on the Fahrenheit scale, on the Kelvin scale they are expressed simply as "kelvins." Since nothing can be colder than absolute zero, there are no negative temperatures on the Kelvin scale. A 1-degree interval on the Kelvin scale corresponds precisely to a 1-degree increment on the Celsius scale. The three scales are contrasted in Figure 3.1, and conversion formulas are given inside the front cover.

Temperature Measurement

A **thermometer** is the usual instrument for monitoring variations in air temperature. Perhaps the most common type of thermometer was invented in the midseventeenth century and consists of a liquid-in-glass tube attached to a graduated scale (Figure 3.2). Typically, the liquid is either mercury (which freezes at −39 °C, or −38 °F) or alcohol (which freezes at −117 °C, or −179 °F). As the air warms, the liquid expands and rises in the glass tube; as the air cools, the liquid contracts and drops in the tube.

Some liquid-in-glass thermometers are designed to record the maximum and minimum temperatures over a specified period. In one common design, maximum and minimum thermometers are mounted side by side (Figure 3.3). The maximum thermometer has a small constriction in the tube just above the bulb. As heat is supplied to the bulb, the fluid (usually mercury) expands upward and beyond the constriction. Then, when the temperature falls, the fluid thread breaks at the constriction so that the end of the mercury column is positioned at the highest (maximum) temperature. The maximum thermometer is reset by whirling the tube which drives the fluid past the constriction and back into the bulb. In the minimum thermometer, a metal index floats at the surface of the fluid column (usually alcohol). As the temperature falls, the index is drawn downward; as the temperature rises again, the fluid expands and the index is left behind at the lowest (minimum) temperature. The minimum thermometer is reset by tilting the thermometer, bulb end upward. Usually, maximum and minimum thermometers are reset once every 24 hours.

A second type of thermometer uses a bimetallic sensing element to take advantage of the expansion and contraction that accompany the warming and cooling of metals. A bimetallic sensing element consists of two different metal strips that are welded together side by side. The two metals have different rates of thermal expansion; that is, one metal expands more than the other

FIGURE 3.2
A liquid-in-glass thermometer graduated in both the Celsius (°C) and Fahrenheit (°F) scales. In this case, the liquid is alcohol. [Photograph by Laura Carlson]

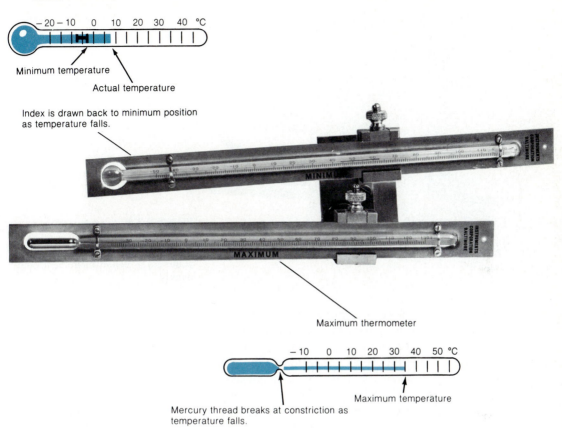

FIGURE 3.3
Liquid-in-glass thermometers mounted side by side and designed to register maximum and minimum temperatures, usually over a 24-hour period. [Courtesy of Belfort Instrument Company]

in response to the same heating. Because the two metals are bonded together, heating causes the bimetallic strip to bend, and the greater the heating, the greater is the bending. For example, the rate of thermal expansion of brass is about twice that of iron so that a bimetallic strip composed of those two metals will bend in the direction of the iron when heated. A series of gears or levers translates the response of the bimetallic strip to a pointer and a dial calibrated to read in °C or °F. Alternatively, this device may be rigged to a pen and a clock-driven drum to give a continuous trace of temperature with time (Figure 3.4). This instrument is called a **thermograph.**

Another type of thermometer employs a thermistor, a type of electrical conductor whose resistance changes as the temperature fluctuates. Variations in electrical resistance are calibrated in terms of temperature. Radiosondes are

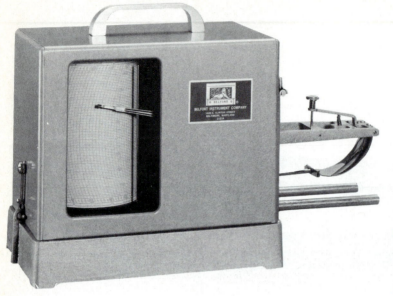

FIGURE 3.4

A thermograph provides a continuous trace of fluctuations in air temperature [Courtesy of Belfort Instrument Company]

equipped with this type of thermometer. In addition, a thermistor thermometer may be designed to give remote temperature readings by mounting the sensor at the end of a long cable joined to the instrument. This type of system is currently being installed at National Weather Service facilities in place of the standard liquid-in-glass thermometers shown in Figure 3.3.

Regardless of the type of thermometer used, two important considerations in selecting an instrument are accuracy and response time. For most meteorological purposes, a thermometer that is reliable to within 0.3 C° (0.5 F°) is sufficient. Response time refers to the instrument's capability of resolving oscillations in temperature. Most liquid-in-glass and electrical resistance thermometers have rapid response times, whereas bimetallic thermometers tend to be more sluggish.

For accurate measurements of air temperature, ideally a thermometer should be adequately ventilated and shielded from precipitation, direct sunlight, and the night sky. Enclosing thermometers (and other weather instruments) in a louvered wooden shelter painted white (such as shown in Figure 3.5A) has been standard practice for official temperature measurements.* The sensor for the new National Weather Service thermistor system is mounted

---

* Widespread use of instrument shelters dates back only to the 1870s even at official weather stations. In North America the earliest report of a sheltered thermometer was at the Toronto Magnetic and Meteorological Observatory in 1841. Previously, official thermometers were usually suspended unprotected just outside a window—and not always a north-facing window. An earlier practice of mounting a thermometer indoors in an unheated room apparently had been largely abandoned by the mideighteenth century.

A   B

FIGURE 3.5

National Weather Service (NWS) thermometers are
mounted inside a shelter that provides ventilation and protection from precipitation and direct sunlight. (A) Standard
white wooden louvered instrument shelter. Cylindrical instruments to the right of the shelter are rain and snow
gauges. (B) Shelter for the new NWS thermistor system.
[Photographs by J. M. Moran]

inside a shield made of louvered plastic rings (Figure 3.5B), and the digital
read-out box is indoors. So that temperature readings are representative, an
instrument shelter should be located in an open grassy area well away from
trees, buildings, or other obstacles. As a rule of thumb, the shelter should be
no closer than four times the height of the nearest obstacle. Where a shelter
is not available, mounting a thermometer outside a window on the shady
northside of a building is usually sufficient for general purposes.

In addition to using standard thermometers, air temperature can sometimes be deduced in unconventional ways. One interesting and surprisingly
accurate approach is to count cricket chirps. For temperatures above about 12
°C (54 °F), the number of cricket chirps heard in an 8-second period plus 4
approximates the air temperature in degrees Celsius.

Heat Units

Although temperature is a convenient way to describe the degree of hotness or coldness, we can quantify heat energy directly. Until recently, meteorologists commonly measured heat energy in units called calories. A **calorie (cal)** is defined as the quantity of heat needed to raise the temperature of 1 gram (g) of water 1 Celsius degree (technically, from 14.5 to 15.5 °C). (The "calorie" used to measure the energy content of food is actually 1000 heat calories, or 1.0 kcal.) Today, the more usual unit for energy of any form, including heat, is the joule (J). One calorie equals 4.1868 J. In this book we use both units of heat measurement.

In the English system, heat is quantified as **British thermal units (Btu).** A Btu is defined as the amount of heat required to raise the temperature of 1 pound (lb) of water 1 Fahrenheit degree (technically, from 62 to 63 °F). One Btu is equivalent to 252 cal and to 1055 J.

Transport of Heat

In response to unequal rates of radiational heating and radiational cooling within the Earth-atmosphere system, air temperature varies from one place to another. (In Chapter 4, we examine the nature and implications of these imbalances in radiational heating and cooling rates.) A change in temperature with distance is known as a **temperature gradient.** A familiar temperature gradient is between the hot equator and the cold poles (a horizontal temperature gradient). Another is the temperature gradient between the relatively mild Earth's surface and the relatively cold tropopause (a vertical temperature gradient). According to a principle of thermodynamics,* when a temperature gradient develops within a system, heat always flows from locations of higher temperature toward locations of lower temperature. In addition, the greater the temperature difference (that is, the steeper the temperature gradient), the more rapid is the rate of heat flow. In response to temperature gradients within the Earth-atmosphere system, heat is transferred via conduction, convection, and radiation.

CONDUCTION

Conduction occurs within a substance or between substances that are in direct physical contact. In **conduction,** the kinetic energy of atoms or molecules (that is, heat) is transferred by collisions between neighboring atoms or mol-

---

* Thermodynamics is the study of the transport of heat and the relationship between heat and other forms of energy.

Table 3.1
Heat Conductivities of Some Familiar Substances

Substance	Heat conductivity[a]	Substance	Heat conductivity[a]
Copper	0.92	Concrete	0.0022
Aluminum	0.50	Water (at 10 °C)	0.0014
Iron	0.16	Dry sand	0.0013
Ice (at 0 °C)	0.0054	Air (at 20 °C)	0.000061
Limestone	0.0048	Air (at 0 °C)	0.000058

[a] Heat conductivity is defined as the quantity of heat (cal) that would flow through a unit area of a substance (cm^2) in one second in response to a temperature gradient of one Celsius degree per centimeter. Hence, heat conductivity is in units of calories per cm^2 per sec per C° per cm.

ecules. This is why a spoon heats up when placed in a steaming cup of coffee. As the more energetic molecules of the hot coffee collide with the less energetic atoms of the cooler spoon, some kinetic energy is transferred to the atoms of the spoon. These atoms then transmit some of their heat energy, via collisions, to their neighboring atoms, so heat is eventually conducted up the handle of the spoon and the handle becomes hot to the touch.

Some substances conduct heat much more readily than others. As a rule, solids are better conductors than liquids, and liquids are better conductors than gases. At one extreme, metals are excellent conductors of heat, and at the other extreme, air is a very poor conductor of heat. The heat conductivities of some common substances are listed in Table 3.1.

Differences in heat conductivity can cause one substance to feel colder than another, even though both substances have the same temperature. For example, at the same relatively low temperature, a metallic object feels colder than a wooden object. The heat conductivity of metals is much greater than that of wood so that when you grasp the two objects, your hand more rapidly conducts heat to the metallic object than to the wooden object. Consequently, you have the sensation that the metal is colder than the wood.

The relatively low heat conductivity of air makes it useful as a heat-insulating medium. Heat conductivity is lower for still air than for air in motion. Thus, to take maximum advantage of air as a heat insulator, air must be confined. For example, when a fiberglass blanket is used as attic insulation, it is primarily the still air trapped between individual fiberglass fibers that inhibits heat loss. In time, as fiberglass fibers settle, air is excluded, and the blanket loses much of its insulating value.

A fresh snow cover has an extremely low heat conductivity because of the air trapped between the individual snowflakes. A thick snow cover (20 to 30 cm, or 8 to 12 in.) can thus inhibit or prevent freezing of the underlying soil, even though air temperatures may drop well below freezing. In time, however, like the fiberglass, the snow cover loses some of its insulating property as the snow settles and the air escapes (Figure 3.6).

Heat is conducted from warm ground to cooler overlying air, but because the air has a low heat conductivity, conduction is significant only in a very thin layer of air that is in immediate contact with the Earth's surface. Convection is much more important than conduction in transporting heat vertically within the troposphere.

CONVECTION

Although conduction takes place in solids, liquids, or gases, convection occurs only in liquids or gases.* **Convection** is the transport of heat within a fluid via motions of the fluid itself.

Convection occurs within the atmosphere as a consequence of differences in air density. As heat is conducted from relatively warm ground to cooler overlying air, the air becomes warmer than the surrounding air. Warm air is less dense than cold air so that the cooler, denser air sinks downward and forces the warmer, lighter air to rise. The cooler air is then heated by the ground and the process is repeated. In this way, as shown in Figure 3.7, a convective circulation of air transports heat vertically from the Earth's surface into the troposphere. As we will see later in this book, heat is also transported horizontally within the atmosphere; such a process is known as **advection.**

Convection is readily visualized in a pan of boiling water on a stove. We actually see the circulating water that is redistributing the heat that is conducted from the bottom of the pan into the water. As in this case and in the atmosphere, conduction and convection often work together; heat transported by the combined processes of conduction and convection is referred to as **sensible heating.**

RADIATION

As we saw in Chapter 2, **radiation** consists of electromagnetic waves traveling at the speed of light. Unlike conduction and convection, radiation does not require an intervening physical medium; it can take place in a vacuum. While not precisely a vacuum, interplanetary space is so highly rarefied that conduction and convection play no role in the transport of heat from the sun to Earth. Rather, radiation is the principal means whereby the Earth-atmosphere system gains heat from the sun and loses heat to space.

---

* Geologists point out an important exception: convection currents probably occur in the Earth's solid interior under conditions of great confining pressures.

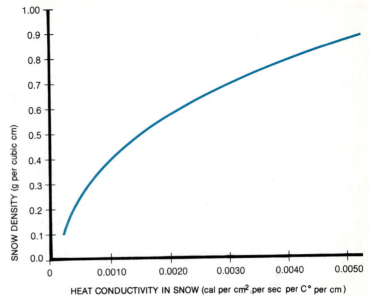

FIGURE 3.6

A thick fresh cover of snow is a good heat insulator primarily because of the air trapped between the snowflakes. (Air is a poor conductor of heat.) But, in time, the snow settles, air escapes, snow density increases, and the snow cover's insulating property decreases.

FIGURE 3.7

Convection currents transport heat from near the Earth's surface into the troposphere.

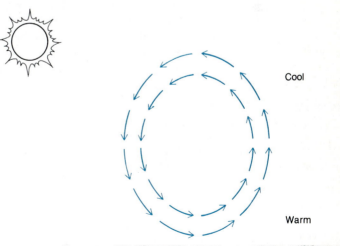

Specific Heat

Whether by conduction, convection, or radiation, the transport of heat from one place to another within the atmosphere is accompanied by changes in air temperature. But, as pointed out at the beginning of this chapter, some of that heat energy may, instead, be used to change either the phase of water or the volume of the air. In circumstances where heat gain and heat loss affect the temperature, then air that gains heat always exhibits a temperature rise, and air that loses heat always exhibits a temperature drop.

In a more general sense, the temperature response to an input (or output) of some specified quantity of heat varies from one substance to another. The amount of heat required to change the temperature of 1 g of a substance by 1 C° is defined as the **specific heat** of that substance. Joseph Black, a Scottish chemist, first proposed the concept of specific heat back in 1760. Two different materials registering the same temperature do not necessarily possess the same amount of heat energy. Different quantities of heat may also be required to raise or lower the temperature of equal amounts of two different substances by one degree. The specific heat of all substances is measured relative to that of water, which is 1 cal per g per C° (at 15 °C). Table 3.2 lists the specific heats of some familiar materials.

On exposure to the same heat source, substances with low specific heat warm up more than substances with high specific heat. Water has the greatest specific heat of any naturally occurring substance. For example, its specific heat is about five times that of dry sand. Hence, 1 cal of heat will raise the temperature of 1 g of water 1 C°, whereas 1 cal will raise the temperature of 1 g of sand about 5 C°. This is one reason why in the summer the sand of the beach feels hot relative to the water.

The difference in specific heat is one of the principal reasons why the surface temperature of land is more variable with time than that of a body of water such as a lake or the sea. Compared to an adjacent body of water, land

Table 3.2
Specific Heats of Some Familiar Substances

Substance	Specific Heat (cal per g per °C)	Substance	Specific heat (cal per g per °C)
Water	1.000	Sand	0.188
Ice (at 0 °C)	0.478	Dry air[a]	0.171
Wood	0.420	Copper	0.093
Aluminum	0.214	Silver	0.056
Brick	0.200	Gold	0.031
Granite	0.192		

[a] At constant volume.

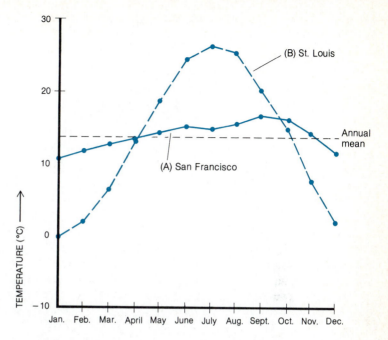

FIGURE 3.8

Variation of average monthly temperatures for (A) maritime San Francisco and (B) continental St. Louis. [NOAA data]

heats up more during the day and in summer and cools down more at night and during the winter. Hence, a body of water exhibits greater resistance to temperature change, called **thermal stability,** than does a land area.

Another factor that contributes to the greater thermal stability of water bodies compared to land is a difference in heat transport. Solar radiation penetrates water to significant depths, but not the opaque land surface. Water circulates so that heat is readily transported into great volumes of water, whereas heat is conducted only very slowly into the soil. Thus, an identical input of heat causes a land surface to warm up more than an equivalent surface area of a water body.

The contrast in thermal stability between land and water has important implications for climate because air temperatures are regulated to a considerable extent by the temperature of the surface over which the air resides or travels. Localities that are immediately downwind of an ocean (maritime localities) exhibit smaller seasonal temperature variations than do localities that are situated well inland (continental localities). An illustration of this effect is the comparison of the average monthly temperatures of San Francisco and St. Louis (Figure 3.8). Although the two cities are at about the same latitude, summers are cooler and winters are milder in maritime San Francisco than in continental St. Louis.

Climatologists use an **index of continentality** to describe the degree of maritime influence on average air temperatures. Several different indexes are available, but most are based on the difference between average winter and

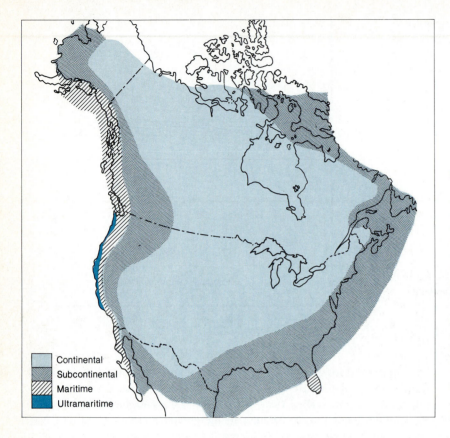

FIGURE 3.9
An index of continentality gauges the influence of oceans on air temperature over continents. In this scheme, North America is divided into zones of increasing maritime influence: continental, subcontinental, maritime, and ultramaritime. The greater the maritime influence, the less is the average temperature contrast between winter and summer. [Modified after D. R. Currey, "Continentality of Extratropical Climates," *Annals of the Association of American Geographers* 64, No. 2 (1974):274]

Legend:
- Continental
- Subcontinental
- Maritime
- Ultramaritime

summer temperatures. A generalized index of continentality for North America is shown in Figure 3.9. Note that because winds blow primarily from west to east, western North America is more maritime (less continental) than eastern North America.

Heating and Cooling Degree-Days

With the contemporary concern for energy conservation, television and newspaper weather summaries routinely report heating or cooling degree-day totals in addition to daily maximum and minimum temperature. Heating and cooling degree-days are indicators of household energy consumption for space heating and cooling, respectively.

In the United States, **heating degree-day units** are based on the Fahrenheit temperature scale and are computed only for days when the mean outdoor air temperature is lower than 65 °F (18 °C). Heating engineers who formulated this index early in this century found that when the mean outdoor temperature falls below 65 °F, space heating is required in most buildings to maintain an indoor air temperature of 70 °F (21 °C). The mean daily temperature is the simple arithmetic average of the 24-hour maximum and minimum air temperatures. Subtracting the daily mean temperature from 65 °F yields the number of heating degree-day units for that day. For example, suppose that this morning's low temperature was 36 °F (2 °C), and this afternoon's high temperature was 52 °F (11 °C). Today's mean temperature would then be 44 °F (7 °C), giving a total of 65 − 44 = 21 heating degree-day units. It is usual to keep a running total of heating degree-day units, that is, to add degree-day units for successive days through the heating season (actually from July of one year through June of the next).

Fuel distributors and power companies closely monitor cumulative degree-day unit totals. Fuel oil dealers base fuel use rates on degree-day units and schedule home deliveries accordingly. Natural gas and electrical utilities anticipate power demands on the basis of degree-day unit totals, and implement priority use policies on the same basis when capacity fails to keep pace with demand.

Figure 3.10 shows the average annual heating degree-day totals over the United States. Outside of mountainous areas, regions of equal heating degree-day totals tend to parallel latitude circles with degree-day totals increasing poleward. As an example, the annual space heating requirement in Chicago (6100 heating degree-day units) is about four times that of New Orleans (1500 heating degree-day units). If the per unit fuel costs are the same in both cities, then, in an average winter, Chicago homeowners can expect to pay four times as much for space heating as homeowners in New Orleans.

Cooling degree-day units are computed only for days when the mean outdoor air temperature is higher than 65 °F.* Supplemental air conditioning may be needed on such days. Again, a cumulative total is maintained through

_____
* Higher base temperatures are sometimes used.

FIGURE 3.10
Average annual heating degree-day totals over 48 states of the United States. [EIDS data]

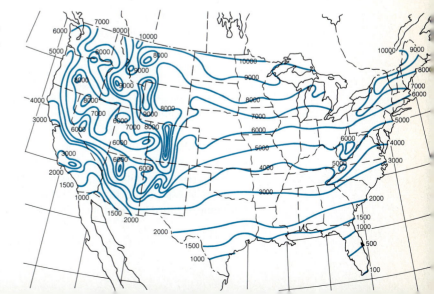

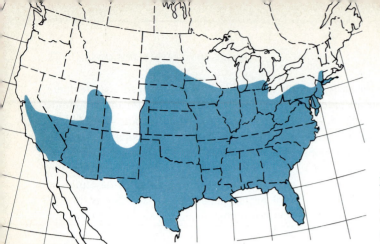

FIGURE 3.11
Shaded area of the United States is where average annual cooling degree-days total more than 700 (using a base of 65 °F, or 18 °C). Air conditioning is desirable in this area. [NOAA data]

the cooling (summer) season. Generally, however, air conditioning is needed mostly in those localities where cumulative cooling degree-day unit totals exceed 700 (Figure 3.11).

Note that indexes of heating and cooling requirements are based on outside air temperatures and do not take into account other weather elements, such as wind speed and humidity, which influence human comfort and demands for space heating and cooling. Heating and cooling degree-day units are therefore only approximations of our residential fuel demands for heating and cooling.

Windchill

At low air temperatures, the wind increases human discomfort outdoors and heightens the danger of **frostbite,** the freezing of body tissue. Air in motion increases the rate of sensible heat loss (the combined effect of conduction and convection) from the body. Immediately adjacent to the body is a very thin (measured in millimeters) layer of still air, called the **boundary layer,** that helps to insulate the body from heat loss. Within the boundary layer, heat is transferred by the very slow process of conduction. (Recall that air is a poor conductor of heat.) However, as wind speed increases, the thickness of the boundary layer diminishes, and the rate of sensible heat loss from the body increases.

Because of the danger of frostbite, weather reports during winter in northern localities and in mountainous regions include the **windchill equivalent temperature (WET),** sometimes referred to simply as the windchill index. This index, first introduced in the early 1940s by the polar scientist P. A. Siple and his colleague C. F. Passel, and later refined by A. Court, is presented in Table 3.3. As an illustration, suppose the actual air temperature is − 1 °C (30 °F) and the air is calm; then the windchill equivalent temperature is the same as the air temperature. If, however, the wind speed increases to 32 km (20 mi) per hour, then the WET drops to − 15 °C (4 °F). Contrary to popular opinion, this does not mean that skin temperature actually drops to − 15 °C.

Through sensible heat transfer, skin temperature can drop no lower than the temperature of the surrounding air, which is -1 °C (30 °F) in this example. What it does mean is that any exposed body parts lose heat at a *rate* equivalent to conditions induced by calm winds at -15 °C. Hence, air in motion (wind) is more effective in removing heat from the body than air temperature alone would imply.

Many factors besides wind speed affect human comfort during cold weather. The windchill equivalent temperature does not take into account variations in rates of body heat production due to changes in physical activity or metabolic processes. Nor does it consider the effects of humidity or radiational heating and cooling. Furthermore, we may argue that comfort is very subjective and perceived differently by different individuals. Nonetheless, the WET has proven to be a very useful guide in regions prone to harsh winter weather.

Table 3.3A
Windchill Equivalent Temperature (°C)

Wind speed (m/sec)	Air temperature (°C)															
	6	3	0	−3	−6	−9	−12	−15	−18	−21	−24	−27	−30	−33	−36	−39
3		3	−1	−4	−7	−11	−14	−18	−21	−24	−28	−31	−34	−38	−41	−45	−48
6		−2	−6	−10	−14	−18	−22	−26	−30	−34	−38	−42	−46	−50	−54	−58	−62
9		−6	−10	−14	−18	−23	−27	−31	−35	−40	−44	−48	−53	−57	−61	−65	−70
12		−8	−12	−17	−21	−26	−30	−35	−39	−44	−48	−53	−57	−62	−66	−71	−75
15		−9	−14	−18	−23	−27	−32	−37	−41	−46	−51	−55	−60	−65	−69	−74	−79
18		−10	−14	−19	−24	−29	−33	−38	−43	−48	−52	−57	−62	−67	−71	−76	−81
21		−10	−15	−20	−25	−29	−34	−39	−44	−49	−53	−58	−63	−68	−73	−77	−82
24		−10	−15	−20	−25	−30	−35	−39	−44	−49	−54	−59	−63	−68	−73	−78	−83

Table 3.3B
Windchill Equivalent Temperature (°F)

Wind speed (mph)	Air temperature (°F)																		
	45	40	35	30	25	20	15	10	5	0	−5	−10	−15	−20	−25	−30	−35	−40	−45
5	43	37	32	27	22	16	11	6	1	−5	−10	−15	−20	−26	−31	−36	−41	−47	−52
10	34	28	22	16	10	4	−3	−9	−15	−21	−27	−33	−40	−46	−52	−58	−64	−70	−76
15	29	22	16	9	2	−5	−11	−18	−25	−32	−38	−45	−52	−58	−65	−72	−79	−85	−92
20	25	18	11	4	−3	−10	−17	−25	−32	−39	−46	−53	−60	−67	−74	−82	−89	−96	−103
25	23	15	8	0	−7	−15	−22	−29	−37	−44	−52	−59	−66	−74	−81	−89	−96	−104	−111
30	21	13	5	−2	−10	−18	−25	−33	−41	−48	−56	−63	−71	−79	−86	−94	−102	−109	−117
35	19	11	3	−4	−12	−20	−28	−35	−43	−51	−59	−67	−74	−82	−90	−98	−106	−113	−121
40	18	10	2	−6	−14	−22	−29	−37	−45	−53	−61	−69	−77	−85	−93	−101	−108	−116	−124
45	17	9	1	−7	−15	−23	−31	−39	−47	−55	−62	−70	−78	−86	−94	−102	−110	−118	−126

When the WET is significantly lower than the air temperature, we are well advised to dress more warmly than the actual air temperature might suggest. Especially at low windchill equivalent temperatures, all body parts should be protected if a person expects to be exposed to the wind for more than a few minutes at a time. Low air temperatures and high winds are especially hazardous to those body parts that are usually exposed and have a high surface area-to-volume ratio, such as the ears, nose, and fingers. These

Special Topic

Temperature and Human Comfort

To appreciate how air temperature influences human comfort, we must first understand that humans are *homeothermic*. This means that we regulate our internal or core temperature within 2 C° (3.6 F°) of 37 °C (98.6 °F), despite much greater variations in the temperature of the surrounding air (the *ambient air*). The *core* refers to those regions of the body where vital organs such as the brain, heart, lungs, and digestive tract are located. If vital organs are not maintained at a nearly constant temperature, they do not function properly. Other parts of the body, such as the legs and arms, however, may undergo much greater temperature changes without ill effects.

At ambient air temperatures of 20 to 25 °C (68 to 77 °F), someone who is fully clothed and indoors will feel comfortable at rest. In this temperature range, the body readily maintains a core temperature of 37 °C without having to resort to special temperature-regulating mechanisms.

When we are exposed to ambient air temperatures above or below the 20 to 25 °C range, the body must initiate processes that maintain the core temperature at 37 °C. For example, if you stand in the sun on a day when air temperatures hit 30 °C (86 °F), your core temperature will begin to rise. In response, you perspire. The heat required to evaporate the perspiration is provided by your skin, and as a consequence, your skin cools. You experience the same cooling effects of evaporation as you step out of a shower or climb out of a swimming pool. *Evaporative cooling* reduces skin temperature, which in turn, normally leads to a drop in the body's core temperature.

If you are exposed to air temperatures below 20 °C (68 °F), the core temperature may begin to fall, and you start to shiver. This increased muscular activity produces additional heat, which helps to raise the core temperature back to 37 °C. Perspiring and shivering are examples of *thermoregulation*, natural physiological mechanisms that assist in maintaining a nearly constant core temperature regardless of the ambient air temperature.

Another thermoregulation process involves blood flow. Heat transfer from the body core to the skin occurs via the circulatory system. When you are exposed to air temperatures below 20 °C, your body can limit heat loss by restricting blood flow to the skin. By direction of the nervous system, many of the tiny blood vessels in the skin constrict. As ambient air temperatures decline, additional blood vessels constrict, further reducing blood flow to the body surface. This phenomenon can be observed by immersing your hand in a container of ice water; the skin becomes paler. The reduced blood flow produces a thicker insulating layer between the heat-producing core and the skin surface.

In contrast, ambient air temperatures above 25 °C (77 °F) trigger increased blood flow to the body surface. Blood vessels near the skin surface dilate, giving the skin a flushed or reddish appearance. Greater flow of blood to the body surface raises skin temperature. As a consequence, the body-to-air temperature gradient increases, and cooling by radiation, conduction, and convection is enhanced. Greater blood flow to the skin also increases evaporative cooling by supplying more water for perspiring.

In addition to physiological changes, behavioral responses assist in thermoregulation. For example, if we feel hot, we shed clothing, seek shelter from the sun, or turn on a fan or air conditioner.

Under some conditions, thermoregulation is insufficient to maintain a 37 °C core temperature. For

body parts are especially susceptible to frostbite. For more information on human responses to heat stress, see the Special Topic, ''Temperature and Human Comfort.''

Temperature is also a critical factor in agricultural success. Of all the climatic variables, air temperature is the single most important consideration in determining where specific types of crop can be grown. For information on this, see the Special Topic, ''Temperature and Crop Yields.''

example, if a person hiking in the woods is drenched by a cold rain, and if the soaked hiker then overexerts himself physically and becomes exhausted, thermoregulation may not be able to compensate for heat loss. Consequently, the core temperature drops, and hypothermia may ensue.

Hypothermia refers to those responses that occur when the human core temperature drops below 35 °C (95 °F). Initially, shivering becomes more violent and uncontrollable. In addition, the victim begins to have difficulty speaking and becomes apathetic and lethargic. If the core temperature falls below 32 °C (90 °F), shivering is replaced by muscular rigidity, and coordination deteriorates. Mental abilities are impaired, and the victim is generally unable to help himself. At a core temperature of 30 °C (86 °F), the person may drift into unconsciousness. Death may occur at core temperatures below about 24 °C (75 °F) because the heart rhythm becomes uncontrollably irregular (ventricular fibrillation) or uncontrollably halted (cardiac arrest).

Once hypothermia begins, the victim is in serious trouble. With only a 3 C° (5.7 F°) drop in core temperature, the ability of the body to regulate its core temperature is already greatly impaired. If the core temperature drops to 29 °C (85 °F), thermoregulation is essentially ineffective. The first signs of hypothermia should never be ignored, and action should be taken immediately. Treatment takes two forms: prevention of further heat loss and the addition of heat. Further heat loss can be prevented by replacing wet clothing with dry clothing, finding shelter, and insulating the person from the ground so that body heat is not conducted to the colder ground surface. The body can be heated by an external source, such as a space heater or other human bodies. Administering hot nonalcoholic drinks, if the victim is conscious, helps warm the core from

the inside. In any event, medical attention should be sought as soon as possible.

In some situations, thermoregulation may be unable to prevent a rise in core temperature. For instance, a person exposed to hot desert conditions with an inadequate supply of water will eventually experience an increase in core temperature. If the core temperature continues to rise, hyperthermia may ensue. *Hyperthermia* refers to those responses that take place when the core temperature tops 39 °C (102 °F). As the core temperature climbs to 41 °C (105 °F), thermoregulation breaks down, and a person may suddenly and quite unexpectedly collapse. The victim also experiences muscle cramps or spasms and slips into unconsciousness. Sweating ceases, although it is not known if this is a cause or a result of hyperthermia. With serious heat stress, the individual may die within a few hours unless the core temperature can be lowered artificially. These responses are collectively identified by various names, including heatstroke, sunstroke, and heat apoplexy.

The victim of hyperthermia must be treated promptly because, once thermoregulation fails, the core temperature rises rapidly. To save the victim, the core temperature must be lowered from outside the body. The victim should be moved to a cooler environment, and, if possible, the body should be placed in cold water. Alternatively, sponging the body with alcohol enhances cooling as the alcohol evaporates. Again, medical attention is needed as soon as possible.

The human body possesses remarkable capabilities to adjust to changing air temperatures and to provide a sense of comfort. These capabilities are limited, however. If the core temperature begins to deviate from normal, the individual or his or her companions must take corrective action promptly. Failure to do so may be fatal.

Temperature and Crop Yields

In the Midwest, there's an old saying that corn should not be planted before the leaves of an oak tree reach the size of a squirrel's ear. While this adage is quaint, farmers today use more sophisticated and reliable indexes of crop–climate relationships to help optimize crop yields. Two important indexes are (1) length of growing season and (2) growing-degree units.

The period during the year when air temperatures remain sufficiently mild to permit plant growth is a critical determinant of agricultural productivity. For many years, the *growing season* was defined as the number of days between the date of last killing frost in spring and the date of first killing frost in fall. However, "killing frost" is an ambiguous term. Whether or not a frost kills a plant depends on the plant species and the stage of its life cycle, the duration of freezing temperatures, and the rate of freezing. Thus, no single temperature reading can fully convey the actual impact of subfreezing temperatures on agriculture, particularly where a variety of crops are grown. Currently, the period most commonly used to delineate the growing season is the freeze-free period, that is, the time between the last day of 0 °C (32 °F) or lower in the spring and the first date of 0 °C (32 °F) or lower in the autumn.

As a rule, the average length of the growing season shortens with increasing latitude so that the types of crops that can be grown successfully also change with latitude. The growing season may be lengthened locally by the moderating influence of nearby large bodies of water. A good example comes from Wisconsin, where localities bordering Lake Michigan have a growing season that is 20 to 40 days longer than sites at the same latitude, but 100 km (62 mi) inland. Topography can also influence the length of the growing season. Because cold air is relatively dense, it tends to drain downhill so that the growing season typically is several weeks shorter in valleys than on surrounding hillsides. For that reason, vineyards and orchards are usually situated on hillsides rather than in valley bottoms.

One problem with relying on growing season length as a predictor of agricultural yields is that temperatures above 0 °C (32 °F) are not equally effective in promoting crop growth. For example, corn grows very little at temperatures below 10 °C (50 °F) or above 30 °C (86 °F). Hence, the actual air temperature during the growing season can either accelerate or retard plant growth and often is as significant as the length of the growing season in affecting yields.

To better gauge the influence of temperature on

Conclusions

Heat and temperature are distinct and yet closely related quantities. In response to temperature gradients, conduction, convection, and radiation redistribute heat from one place to another. Depending primarily on the specific heat of the medium (air, water, or land), temperatures change in response to this heat redistribution. Imbalances in rates of radiational heating and radiational cooling ultimately are responsible for producing temperature gradients within the Earth-atmosphere system. We examine the reasons for these imbalances in the next chapter.

plant growth during the growing season, an index called *growing-degree units (GDUs)* has been developed. Because one of its most successful applications is in growing corn, we will use corn as an example of how the index works. To determine the GDUs for a particular day, the average daily temperature is computed by adding the highest and lowest temperatures in °F for the day (24 hours) and dividing by two. Because most corn growth occurs when air temperatures are between 50 and 86 °F (10 and 30 °C), any daily high temperature above 86 °F is counted as 86 °F and any daily low temperature below 50 °F is counted as 50 °F. The number 50 (the lower threshold for growth of corn plants) is then subtracted from the daily average temperature to obtain the number of GDUs for that day. GDUs are then added from one day to the next to give a cumulative total.

There are several important applications of GDUs. One is to match the GDU requirement of a particular corn hybrid with the long-term average number of GDUs for a region. For example, a farmer living in central Illinois or Indiana typically selects a corn hybrid that requires 3000 to 3400 GDUs to mature. If the planting season is delayed (perhaps by early spring rains) or if fields have to be replanted (perhaps because of poor germination owing to low soil moisture after planting), the farmer may have to switch to a shorter-season hybrid (that is, a hybrid that requires fewer GDUs). By computing the number of GDUs that have already accumulated during the growing season and subtracting this total from the number of GDUs for an average growing season, the farmer obtains a good estimate of what hybrid to use so that the corn reaches maturity before the first killing freeze of the fall.

For grains such as corn, oats, and wheat GDUs provide a useful guide in selecting the best varieties to plant in specific areas and can be used in planning planting and harvesting times. For fruit crops such as apples, pears, and peaches, GDUs have been used to successfully predict bloom dates and other stages of development, including ripening. This aids in choosing varieties that bloom after the danger of freezing temperatures has passed but are still capable of producing fruit. GDUs have also been used to predict the time of insect infestation, which aids in planning spraying operations.

Although GDUs are valuable in modeling the growth and development of plants, often specific weather factors affect quality, color, and size of fruits. For example, day-to-night temperature shifts are critical for growing wine grapes in the Napa Valley of California.

Summary Statements

Heat is the total kinetic molecular energy contained in a substance.

Temperature is a commonly used, but incomplete, way of describing the relative heat energy of substances. Temperature is a measure of the average kinetic energy of individual atoms or molecules composing a substance. We use the Fahrenheit, Celsius, and Kelvin temperature scales.

Because the Kelvin scale is based on absolute zero, it is a better measure of the average kinetic molecular energy than either the Celsius or Fahrenheit temperature scale.

Heat energy is quantified directly as calories, joules, or British thermal units.

In response to gradients in temperature, heat is transported from one place to another via conduction, convection, and radiation. Heat always flows from areas of higher temperature toward areas of lower temperature.

Convection is much more important than conduction in transporting heat within the troposphere. However, heat is initially transported from the Earth's surface to the air immediately overlying the surface via conduction.

The temperature response to an input (or output) of heat differs from one substance to another, depending on the specific heat of the substance.

The warming and cooling rates of large bodies of water and land have important implications for climate. Water bodies, such as seas and lakes, exhibit less variability in temperature than do continents.

An index of continentality is used to represent the degree of maritime influence on air temperatures.

The greater the maritime influence, the less is the seasonal (winter-to-summer) temperature contrast.

Heating and cooling degree-day units are indicators of household energy consumption for space heating and cooling, respectively.

Windchill equivalent temperature is an index that estimates the combined effect of low air temperature and wind speed on human comfort.

Key Words

kinetic energy	British thermal units	radiation	cooling degree-day
heat	(Btu)	specific heat	units
temperature	temperature gradient	thermal stability	frostbite
absolute zero	conduction	index of continentality	boundary layer
thermometer	convection	heating degree-day	windchill equivalent
thermograph	advection	units	temperature (WET)
calorie (cal)	sensible heating		

Review Questions

1. Distinguish between heat and temperature.

2. Why is temperature an incomplete way of comparing the heat energy of different substances?

3. Why is the Celsius temperature scale considered to be more convenient than the Fahrenheit temperature scale for most scientific purposes?

4. What is the significance of "absolute zero," that is, zero kelvins?

5. In some localities, winter temperatures dip below −40 °C (−40 °F). In those areas, what types of thermometers must be used?

6. What is meant by a thermometer's response time? Why is it an important consideration in selecting a thermometer?

7. Present some examples of temperature gradients (a) within the same substance and (b) between different substances.

8. Describe three processes by which heat is transferred from one place to another. Provide a common example of each.

9. Air has a low thermal conductivity. How is this property of air used in insulating a home?

10. Why is convection much more important than conduction in the transport of heat within the troposphere?

11. Contrast the heat conductivity of solids, liquids, and gases.

12. Why does a deep snow cover inhibit the freezing of the underlying soil, even though the temperature of the air drops well below the freezing point?

13. Define specific heat.

14. Referring to Table 3.2, determine the temperature change of 1.0 gram of each of the following substances upon an addition of 1.0 calorie of heat: (a) ice, (b) dry air, (c) aluminum.

15. How and why do maritime climates differ from continental climates?

16. Heating degree-day units are computed using a base of 65 °F. Why?

17. Why do large bodies of water, such as lakes or seas, reduce the temperature fluctuations of the overlying air?

18. Convert the following.

$$0 \,°C = \text{_____} \,°F = \text{_____} \,K$$

$$20 \,°C = \text{_____} \,°F = \text{_____} \,K$$

$$65 \,°F = \text{_____} \,°C = \text{_____} \,K$$

$$300 \,K = \text{_____} \,°C = \text{_____} \,°F$$

19. Today's high temperature was 20 °F and this morning's low temperature was −5 °F. Compute today's heating degree-day units.

20. Explain how air in motion (the wind) affects the rate of heat transport from the human body. On a bitterly cold and windy day, does a windchill equivalent temperature of −40 °C (−40 °F) mean that exposed skin will actually drop to this temperature? Elaborate on your response.

Points to Ponder

1. From Figure 1.7, compute the average temperature gradient between the Earth's surface and the tropopause. Speculate on how the magnitude of this gradient compares with that of the average horizontal temperature gradient between the equator and poles.

2. In winter, double-glazed windows or storm windows are used to cut heat loss from buildings. Speculate on how these windows reduce conductive and convective heat flow.

3. Why do you suppose the specific heat differs from one substance to another?

4. How do the thermal conductivity and specific heat of dry air compare in magnitude with the same properties of materials composing the Earth's surface? Speculate on the significance of these differences for weather and climate.

5. Air is a very poor conductor of heat. Why, then, do people go to the trouble of wrapping hot water pipes with insulating materials?

Projects

1. Determine whether your location has a continental or maritime climate. You may wish to plot mean monthly temperatures in order to determine the winter-to-summer temperature contrast. These data are available from the nearest National Weather Service office, but first check the government documents section of your library.

2. Maintain a running total of heating degree-day units and cooling degree-day units for your area. Is air conditioning really necessary in your area?

3. Design an experiment to determine the temperature response of samples of dry and wet sand to the same heat input. Explain any difference in temperature response.

Selected Readings

Asimov, I. *Understanding Physics: Motion, Sound, and Heat.* New York: New American Library Mentor Book, 1969. 248 pp. Contains a well-written explanation of basic physical principles with chapters on heat and temperature.

Driscoll, D. M. "Windchill: The 'Brrr' Index." *Weatherwise* 40, No. 6 (1987):321–326. Gives a critique of the windchill concept along with its historical background.

Middleton, W.E.K. *A History of the Thermometer and Its Use in Meteorology.* Baltimore, Md.: Johns Hopkins University Press, 1966. 249 pp. Provides an authoritative summary of the history of development of an essential instrument in weather observation.

Quayle, R., and F. Doehring. "Heat Stress: A Comparison of Four Different Indices." *Weatherwise* 34, No. 3 (1981):120–124. Compares four available ways of measuring heat stress.

Snow, J. T., and S. B. Harley. "Basic Meteorological Observations for Schools: Temperature." *Bulletin of the American Meteorological Society* 68, No. 5 (1987):468–496. Describes methods of temperature measurement using economical instruments.

Then the sea
And heaven rolled as one
and from the two
Came fresh transfigurings
of freshest blue.

Wallace Stevens
SEA SURFACE FULL OF CLOUDS

4

Heat Imbalances and Weather

Imbalances in radiational heating and cooling trigger processes that redistribute heat within the Earth-atmosphere system. Evaporation of water at the Earth's surface and its subsequent condensation as clouds are an important heat-transfer process. [Photograph by Mike Brisson]

WEATHER IS not a capricious act of nature. It is a response to unequal rates of radiational heating and cooling within the Earth-atmosphere system. As we saw in Chapter 2, absorption of solar radiation causes heating and emission of infrared radiation causes cooling. Imbalances in rates of heating and cooling from one place to another within the Earth-atmosphere system produce temperature gradients. In response to temperature gradients, the atmosphere circulates and redistributes heat.

In this chapter we examine the heat imbalances that develop within the Earth-atmosphere system. We then consider the various means whereby heat is redistributed in response to temperature gradients. All this enables us to better understand atmospheric circulation, energy conversions within the Earth-atmosphere system, and the processes that govern air temperature within the troposphere.

Heat Imbalance: Atmosphere Versus Earth's Surface

If we measure the annual rate of radiational heating and the annual rate of radiational cooling for the Earth's atmosphere, and if we then do the same for just the Earth's surface, we will find an important heat imbalance. In the atmosphere, the rate of cooling due to infrared emission is greater than the rate of warming due to absorption of solar radiation. At the Earth's surface, however, it's just the opposite. The rate of warming due to absorption of solar radiation exceeds the rate of cooling due to infrared emission. This heat imbalance is summarized in Figure 4.1, which combines the distribution of the net incoming solar radiation with that of the outgoing infrared radiation.

The global distribution of radiation implies a net cooling of the atmosphere and a net warming of the Earth's surface. Yet, in reality, the atmosphere is *not* cooling relative to the Earth's surface. Hence, a compensating transfer of heat must be taking place from the Earth's surface to the atmosphere via processes other than radiation. How is this heat transfer accomplished? The flow of heat energy from the Earth's surface to the atmosphere is brought about by sensible heating (about 23 percent) and by latent heating (about 77 percent). In Figure 4.1, 30 units of heat energy are transferred from the Earth's surface to the atmosphere by these two processes: 7 by sensible heating (7/30 or 23 percent) and 23 by latent heating (23/30 or 77 percent).

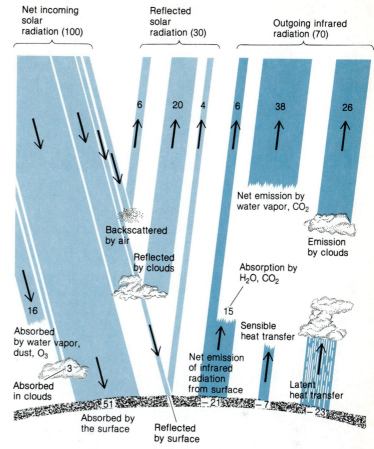

Net incoming solar radiation (100)

Reflected solar radiation (30)

Outgoing infrared radiation (70)

6   20   4   6   38   26

Net emission by water vapor, CO_2

Backscattered by air

Emission by clouds

Reflected by clouds

Absorption by H_2O, CO_2

16

Absorbed by water vapor, dust, O_3

15

Sensible heat transfer

3

Net emission of infrared radiation from surface

Latent heat transfer

Absorbed in clouds

51   –21   –7   –23

Absorbed by the surface

Reflected by surface

FIGURE 4.1

The distribution of 100 units of incoming solar radiation and outgoing infrared radiation on a global scale indicates excess heating at the Earth's surface. This excess heat is transferred to the atmosphere via sensible and latent heating. [Adapted after A. P. Ingersoll, "The Atmosphere," *Scientific American* 249, No. 3 (1983):164. Copyright © 1983 by Scientific American, Inc. All rights reserved.]

SENSIBLE HEATING

The term *sensible* is used to describe this heat-transfer mechanism because heat redistribution brought about by sensible heating can be monitored, or "sensed," as temperature changes. **Sensible heating** involves two processes: conduction and convection. Heat is conducted from the relatively warm surface of the Earth to the cooler overlying air. As the lowest portion of the troposphere is heated by contact with the Earth's surface, the surrounding air, which is cooler and denser, sinks downward and forces the warmer, lighter air to rise. The net result is a transfer of heat via conduction and convection from the surface of the Earth into the troposphere.

LATENT HEATING

Latent heating refers to the movement of heat from one place to another as a consequence of changes in the phase of water. Depending on the specific type of phase change, water either absorbs heat from its environment or releases heat to its environment. The quantity of heat that is involved in phase changes is known as latent heat. Heat is required for phase changes because of differences in molecular activity represented by the three physical phases of water. In the solid phase (ice), water molecules are relatively inactive and vibrate about fixed locations. In the liquid phase, molecules move about with greater freedom, so that liquid water takes the shape of its container. In the vapor phase, molecules exhibit maximum activity and diffuse readily throughout the entire volume of a container. A change of phase is thus linked to a change in level of molecular activity. Hence, heat must be either added to or released from the water undergoing a phase change.

During any phase change, heat is exchanged between the water and its environment. Although the temperature of the environment changes in response, the temperature of the water undergoing the phase change remains

---

Special Topic

The Unique Thermal Properties of Water

Water has some unusual thermal properties. Water's freezing and boiling temperatures are so anomalously high that within the range of temperatures observed on Earth, water occurs in all three phases—as solid, liquid, and vapor. (Interestingly, water is the only substance that exists naturally in all three phases on the Earth's surface.) Water's latent heat of fusion and latent heat of vaporization are also unusually high, and its specific heat is the highest among commonly occurring substances. All these thermal properties of water have important implications for weather and climate as is evident in this and other chapters.

Why is water so unusual? The answer is found in its molecular structure. As shown in Figure 1, each water molecule consists of two hydrogen atoms and one atom of oxygen. These atoms are bonded together by sharing electrons. Each hydrogen atom shares its electron with the six bonding electrons of the oxygen atom. However, the oxygen atom has a stronger attraction for the shared electrons than do the hydro-

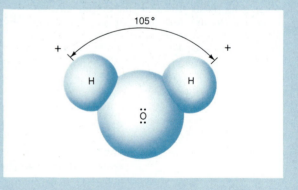

FIGURE 1
The dipolar structure of the water molecule.

gen atoms. Because the hydrogen–oxygen–hydrogen atoms describe a 105-degree angle, a charge separation develops within a water molecule: a negatively charged oxygen region and positively charged hy-

constant until the phase change is complete. That is, the available heat, the latent heat, is involved solely in changing the phase of water and not in changing its temperature. Interestingly, unusually great quantities of heat are required to bring about phase changes of water as compared to phase changes of other naturally occurring substances. The basic reasons for this effect are given in the Special Topic, "The Unique Thermal Properties of Water."

To illustrate the concept of latent heat, follow the fate of a 1-g ice cube as it is heated (Figure 4.2). The specific heat of ice is about 0.5 cal per gram per Celsius degree, which means that 0.5 cal of heat must be supplied for every 1 Celsius degree of temperature rise. Hence, warming our ice cube from −20 to 0 °C (−4 to 32 °F) requires an input of 10 cal of heat. Once the freezing (or melting) point is reached, an additional 80 cal of heat (called the **latent heat of fusion,** or the **latent heat of melting**) must be supplied per gram to break the forces that bind molecules in the ice phase. The temperature of the water and ice remains at 0 °C until all the ice is melted.

The specific heat of liquid water is 1.0 cal per gram per Celsius degree. Thus, once our ice cube is melted, 1.0 cal is needed for every 1 Celsius degree rise in water temperature. A phase change from liquid water to water vapor

drogen regions. A molecule that exhibits such a separation of positive and negative charges is said to be *dipolar*.

Opposite charges attract so that, like tiny magnets, the dipolar water molecules link together. The negative (oxygen) pole of one is attracted to a positive (hydrogen) pole of another. This special attractive force among neighboring water molecules is known as *hydrogen bonding*. A single water molecule forms hydrogen bonds in three directions because each of the water molecule's three atoms is a potential site for bonding.

Hydrogen bonding inhibits changes in the kinetic activity of individual water molecules. Hence, for example, as heat is added to water, the resulting increase in kinetic molecular activity and corresponding rise in temperature are unusually small. That is, water exhibits an anomalously high specific heat. In addition, greater amounts of heat, and thus anomalously high temperatures, are required for water to reach its melting and boiling points.

The stabilizing effect of hydrogen bonding also means that an unusually great amount of heat is required to change the phase of water. For water to change from the solid to liquid phase, heat energy must be supplied to break the hydrogen bonds that maintain water in the crystalline (solid) phase. The relative strength of the hydrogen bond dictates an unusually high latent heat of fusion for water. Not all hydrogen bonds are broken as water changes from ice to liquid; that is, numerous small clusters of bonded molecules persist into the liquid phase. Considerably greater amounts of heat are required for water to vaporize (either evaporate or boil) because all hydrogen bonds must be broken. For this reason, water's latent heat of vaporization is considerably greater than its latent heat of fusion.

In summary, in the solid and liquid phases, dipolar water molecules are linked by hydrogen bonds. Hydrogen bonding means that water is more thermally stable than most other substances. As a consequence, water temperature responds sluggishly to an addition or loss of heat, and anomalously great quantities of heat are either released or must be supplied when water changes phase.

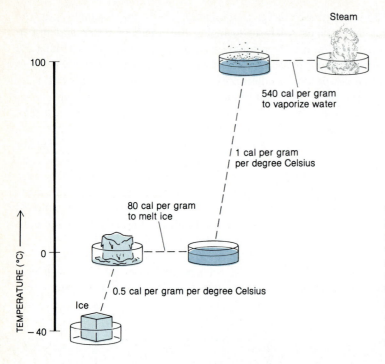

FIGURE 4.2
Heat is added to raise the temperature of ice and water and to change the phase of water. Note, however, that during a phase change (ice to water, water to vapor, for example), the temperature of the water undergoing the phase change is constant. [From J. M. Moran, M. D. Morgan, and J. H. Wiersma, *Introduction to Environmental Science*. New York: W. H. Freeman and Company, p. 273. Copyright © 1980.]

can occur at any temperature and requires the addition of much more heat than the phase change from ice to liquid water. The **latent heat of vaporization** varies from about 600 cal per gram at 0 °C to 540 cal per gram at 100 °C (212 °F). For our 1-g ice cube to vaporize directly without melting, a process called **sublimation,** the latent heat of fusion plus the latent heat of vaporization must be supplied to the ice cube. This would amount to 680 cal per gram at 0 °C.

If the sequence just described is reversed—that is, if the water vapor is cooled until it becomes liquid and then ice—the temperature drops, or phase changes take place, as equivalent amounts of heat are released to the environment. When the water vapor becomes liquid, a process called **condensation,** latent heat of vaporization is released, and when the water freezes, latent heat of fusion is released.

Applying this phase-change concept to the Earth-atmosphere system illustrates the mechanism of heat transfer by latent heating. As the Earth's surface absorbs solar radiation, some of the heat produced thereby is used to vaporize water from oceans, lakes, rivers, soil, and vegetation. Within the troposphere, some of the water vapor is converted to tiny liquid water droplets or ice crystals that are visible as clouds. During cloud formation, then, water changes phase and latent heat is released to the atmosphere. The heat required for vaporization is supplied at the Earth's surface, and that same heat is subsequently released to the atmosphere during cloud development. Through latent heating, then, heat is transferred from the Earth's surface into the troposphere.

The formation of familiar convective clouds combines sensible heating with latent heating to channel heat from the Earth's surface into the troposphere. Rising columns of relatively warm air in convective currents often produce **cumulus clouds,** which resemble puffs of cotton floating in the sky (Figure 4.3). These clouds are sometimes referred to as "fair weather" cumulus because they are seldom accompanied by rain or snow. On the other hand, when certain atmospheric conditions develop (see Chapter 13), convective

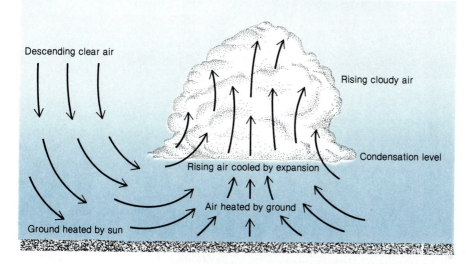

FIGURE 4.3

Cloud development transports excess heat at the Earth's surface into the troposphere via conduction, convection, and latent heat transfer. [Top: After M. Neiburger, J. G. Edinger, and W. D. Bonner, *Understanding Our Atmospheric Environment*. New York: W. H. Freeman and Company, p. 6. Copyright © 1973. All rights reserved. Bottom: Photograph by J. M. Moran]

FIGURE 4.4
When convection currents surge to great altitudes within the troposphere, cumulus clouds may billow upward and form thunderclouds. [Photograph by Mike Brisson]

currents surge to great altitudes, and cumulus clouds billow upward to form towering **cumulonimbus clouds,** also known as thunderstorm clouds (Figure 4.4). In retrospect, then, an important heat transfer process was occurring when that thunderstorm washed out your ball game or sent you scurrying for shelter at the beach last summer.

Since ocean waters cover about 71 percent of the Earth's surface, it is not surprising that latent heating is more significant on a global scale than sensible heating. As shown in Table 4.1, the ratio of sensible heating to latent heating, called the **Bowen ratio,** varies from one locality to another and depends on the amount of surface moisture. The ratio of sensible heating to latent heating is about 1:10 for oceans, but about 2:1 for a relatively dry area like the desert interior of Australia. The drier the land surface is, the less important is latent heating and the more important is sensible heating.

Table 4.1
Bowen Ratio for Various Geographical Areas

Geographical area	Bowen ratio[a]	Geographical area	Bowen ratio[a]
Europe	0.62	Atlantic Ocean	0.11
Asia	1.14	Indian Ocean	0.09
North America	0.74	Pacific Ocean	0.10
South America	0.56	All land	0.96
Africa	1.61	All oceans	0.11
Australia	2.18		

[a] The Bowen ratio is the ratio of heat used for conduction and convection (sensible heating) to heat used for vaporization of water (latent heating).
Source: W. D. Sellers, *Physical Climatology.* Chicago: The University of Chicago Press, 1965, p. 105.

In some places at some times, heat transport is directed from the troposphere to the Earth's surface, which is the reverse of the average global situation that we just described. This occurs, for example, when warm air flows over cold, snow-covered ground or when warm air blows over a relatively cool ocean or lake surface. It occurs frequently at night when the Earth's surface cools by infrared radiation and becomes colder than the overlying air.

We noted in Chapter 1 that nearly all weather is confined to the troposphere, the lowest subdivision of the atmosphere. This implies that sensible and latent heat transfer operate primarily within the troposphere. Heat and temperature distributions above the troposphere are determined primarily by radiational processes.

Heat Imbalance: Variation by Latitude

On a global scale, imbalances in radiational heating and radiational cooling occur not only vertically, between the Earth's surface and the atmosphere, but also horizontally, with latitude. Because Earth is nearly spherical, parallel beams of incoming solar radiation strike lower latitudes more directly than higher latitudes (refer back to Figure 2.6). At higher latitudes, solar radiation spreads over a greater area and is less intense per unit surface area than at lower latitudes. Emission of IR radiation by the Earth-atmosphere system also varies with latitude but less so than solar radiation. IR emission declines with increasing latitude in response to the drop in temperature with latitude. (Recall from Chapter 2 that radiation emission is temperature dependent.) Consequently, over the course of a year at higher latitudes, the rate of infrared cooling exceeds the rate of warming caused by the absorption of solar radiation. At lower latitudes the reverse is true: the rate of solar radiational warming is greater than the rate of infrared radiational cooling (Figure 4.5).

Satellite measurements* indicate that the division between regions of *net* radiational cooling and regions of *net* radiational warming is close to the 30 degree latitude circle in both hemispheres. The implication is that latitudes poleward of about 30 degrees N and 30 degrees S experience net cooling over the course of a year, whereas tropical latitudes are sites of net warming. In fact, lower latitudes do not become progressively warmer relative to higher latitudes. Heat must therefore be transported from the tropics to the middle and high latitudes. This **poleward heat transport** is brought about primarily

---

* The Earth Radiation Budget Experiment (ERBE) promises to provide more precise top-of-the-atmosphere measurements of the flux of both solar and terrestrial radiation. Underway since the mid-1980s, ERBE employs three satellites equipped with radiation sensors that provide global scenes of infrared emission and reflected sunlight. An international team of scientists is currently engaged in analyzing ERBE radiation data.

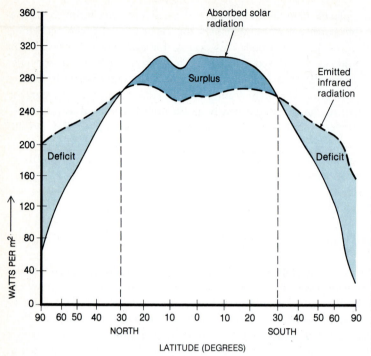

FIGURE 4.5

Variation by latitude of absorbed solar radiation (solid line) and outgoing infrared radiation (dashed line) as obtained by satellite measurements made between June 1974 and February 1978. Radiative heating and cooling rates are equal at about 30 degrees N and S. [From J. S. Winston et al., *Earth-Atmosphere Radiation Budget Analyses Derived from NOAA Satellite Data,* June 1974–February 1978. Washington, DC: NOAA Meteorological Satellite Laboratory, 1979.]

by the exchange of air masses.* Warm air masses that form in lower latitudes flow poleward and are replaced by cold air masses that flow toward the equator from higher latitudes. In this way, sensible heat is transported poleward.

Air mass exchange accounts for about half the total poleward heat transport. The remaining poleward heat transport is due to release of latent heat in storms (about 30 percent) and to ocean currents (about 20 percent). In low latitudes, water that evaporates from the warm ocean surface is drawn into the circulation of developing storms. As storms travel poleward, some of that water vapor condenses as clouds, thereby releasing latent heat. The latent heat of vaporization acquired in lower latitudes is thus transported into higher latitudes. In addition, cold ocean currents drift toward the tropics, whereas warm ocean currents drift poleward. In the tropics, a relatively cool surface ocean current is a "heat sink"; that is, heat is conducted from the warm air to the cool ocean water. In middle and high latitudes, on the other hand, a relatively warm surface ocean current is a "heat source" for the atmosphere; that is, heat is conducted from sea to air.

---

* An air mass is a huge volume of air, covering hundreds of thousands of square kilometers, that is relatively uniform horizontally in temperature and water vapor concentration.

Weather: Response to Heat Imbalances

As we have seen, imbalances in rates of radiational heating and cooling give rise to temperature gradients between (1) the Earth's surface and troposphere and (2) low and high latitudes. In response, heat is transported within the Earth-atmosphere system via conduction, convection, air mass exchange, and storms. That is, the atmosphere circulates and brings about changes in weather. A cause-and-effect chain thus operates in the Earth-atmosphere system starting with the sun as the prime energy source and resulting in weather.

ENERGY CONVERSION

We have also seen that within the Earth-atmosphere system, some solar radiation is converted to heat through absorption, and some of this heat is emitted to space as infrared radiation from the Earth's surface and atmosphere. Some solar energy is also converted to **kinetic energy,** the energy of motion, in the circulation of the atmosphere. Kinetic energy is manifested in winds, in convection currents, and in the exchange of air masses. Circulation (weather) systems do not last indefinitely, however. The kinetic energy of atmospheric circulation ultimately is dissipated as frictional heat as winds blow in contact with the Earth's surface. This heat is, in turn, emitted into space as infrared radiation. Figure 4.6 is a schematic diagram of the major energy transformations operating within the Earth-atmosphere system. Technology also taps solar energy and transforms it into heat and electricity. For more on such conversions, see the Special Topic, "Solar Power."

FIGURE 4.6

A series of energy transformations operate within the Earth-atmosphere system. [Reprinted with permission of Merrill, an imprint of Macmillan Publishing Company, from *Meteorology,* 5th ed., by Albert Miller and Richard A. Anthes, Copyright © 1985 Merrill Publishing Company, Columbus, Ohio.]

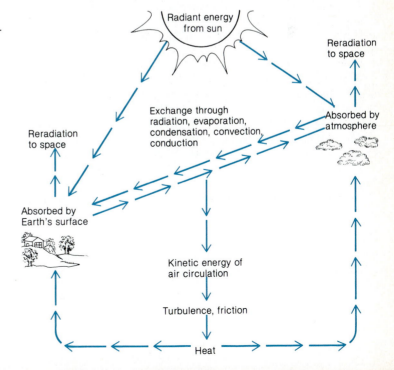

Solar Power

The total amount of solar energy that falls yearly on U.S. lands is about 600 times greater than the amount of energy the United States consumes during the same period. If it is assumed that solar energy can be collected and sold at average electricity rates, more than $5000 worth of energy shines on the roof of a small house (93 m², or 1000 ft²) in the course of a year.

Collection of solar radiation presents some problems, however, because, unlike fossil fuels (coal, natural gas, oil) or nuclear fuels, sunlight is a diffuse energy source that is reduced by cloudiness and available only during part of the day. The solar power reaching United States lands averages annually only about 190 W per m². (By comparison, recall that the solar constant is about 1372 W per m².) The area with the greatest potential for solar power in the United States is the Southwest where cloudiness is relatively minimal and insolation is intense.

To tap solar power, we use *solar collectors*, panels that collect and concentrate the sun's rays. At present, solar collectors are used for space or water heating in small buildings such as homes, apartments, schools, and small businesses. In the near future, solar-powered air conditioners may cool these same places. Solar collectors do not produce extremely high temperatures, so that these devices are not adequate for most industrial purposes.

Basically, a solar collector is a framed panel of glass. Sunlight passes through two layers of glass before it is absorbed by a blackened (low albedo) metal plate. Heat is then conducted from the absorbing plate to either air or a liquid, which is conveyed by fans or pumps to wherever the heat is needed. Figure 1 is an example of the type of solar panels currently in use. Solar collectors typically capture 30 to 50 percent of the solar energy that reaches them.

All technologies based on the collection of solar radiation are limited by the fact that such radiation is not continuous. Not only does the sun not shine at night, but its intensity varies seasonally, geographically, and with cloud cover. Seasonal changes are especially troublesome in middle and high latitudes. For example,

FIGURE 1

Solar collectors on the roof of a roadside rest area off Interstate 43 in northeastern Wisconsin. [Photograph by J. M. Moran]

in eastern Washington State, a relatively sunny locality, the average annual insolation is 194 W per m², but monthly mean values vary from a low of 50 to a high of 343 W per m², a sevenfold difference through the course of a year. This seasonal variation underscores the desirability of developing long-term (summer-to-winter) storage systems for such localities. In tropical latitudes, seasonal differences in insolation are considerably less.

To reduce the variability of insolation due to changes in solar altitude, solar collectors are tilted, and some are designed to track the sun so the collectors are always perpendicular to the solar beam. The advantage of tilted and tracking collectors over collectors that are fixed and horizontal depends on the average cloud cover and latitude of the site. An optimal situation occurs in winter at a midlatitude locality favored by clear skies. There, tilted and tracking solar collectors can double the amount of absorbed insolation.

Scientists and engineers are currently studying ways to convert solar energy to electricity on a large scale. In one conversion system, called a *power tower system*, computer-controlled mirrors, called *heliostats*, track the sun and focus its energy on a single heat-collection point at the top of a tower. Concentrated sunlight in these systems can produce temperatures up to 480 °C (900 °F), temperatures high enough to convert water into high-pressure steam for driving turbines that generate electricity. A solar-powered system of this sort requires a large land area filled with tracking mirrors. In fact, the generation of 50 megawatts of electricity (enough for 15,000 homes) would require about 2.6 km² (1 mi²) of land and would render the land unusable for other purposes.

A federally sponsored experimental power tower is located outside Barstow, California (Figure 2).

FIGURE 2

A power tower stands amid a field of mirrors. Computer-controlled mirrors (called heliostats) track the sun and focus solar radiation on a heat exchanger in the tower. This radiation is converted to heat that produces steam in a boiler. The steam is then used to drive a turbine and generate electricity. This facility near Barstow, California, is operated by Southern California Edison. [U.S. Department of Energy photograph]

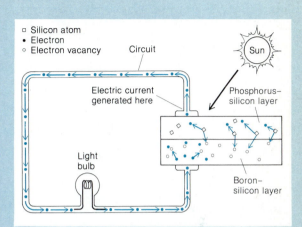

FIGURE 3

Schematic drawing of a silicon solar cell. Sunlight frees electrons from silicon atoms, and the electrons then flow through the circuit to generate an electrical current.

Southern California Edison operates the Solar One facility, which consists of an array of 1818 heliostats and can generate 10 megawatts of electricity.

An alternative to solar-driven turbines for generating electricity is the *photovoltaic cell,* also called *solar cell.* Solar cells convert insolation directly into electricity and routinely power a number of devices ranging from hand-held calculators to space vehicles. Two very thin layers of specially treated crystals (silicon is commonly used) are sandwiched together to form a solar cell. Figure 3 is a schematic drawing of a silicon solar cell. When sunlight strikes the junction between the two crystal layers, a direct electrical current flows through wires connecting the layers. In order to produce significant quantities of electricity, many solar cells are connected in a sealed panel such as shown in Figure 4. About 40 solar cells must be linked to match the power output of a single automobile battery.

One drawback of today's solar cells is their low efficiencies. In this context, *efficiency* is defined as the percentage of insolation striking a cell that is converted to electric energy. A hand-held solar-powered calculator, for example, has an efficiency of only 3 percent

FIGURE 4
Solar electricity-generating panel composed of intercon-
nected solar cells. [U.S. Department of Energy photo-
graph]

or less. Some mass-produced silicon cells have conversion efficiencies that approach 14 percent. In August
1988, researchers at Sandia National Laboratories reported that they were able to achieve an all-time high
efficiency of 31 percent by stacking two solar cells (a
gallium arsenide cell on top and a silicon cell on the
bottom). Among the many factors that contribute to

the low efficiency of solar cells are cell reflectivity and
the conversion of radiation to heat, and the sensitivity
of cells to only a portion of the solar spectrum.

The principal obstacle to greater use of solar cells
for generating electricity is cost, although future prospects are encouraging. The cost of solar-cell-generated
electricity has dropped by an order of magnitude since
1974 and is projected to drop by another order of
magnitude by the mid-1990s. This may not be achieved
by raising the efficiency of traditional solar cells because, as a general rule, costs spiral with increasing
efficiency. A more promising alternative is further development of thin-film solar cells which are less efficient but also much less expensive to manufacture than
traditional solar cells. A *thin-film solar cell* consists of a
film of silicon or other light-sensitive substance that is
deposited on a base material, whereas traditional solar
cells consist of individual crystals.

Before the end of the century, multimegawatt
solar-cell power plants are expected to be feeding electricity into regional grids at costs that are competitive
with conventional power plants. This bright outlook is
based on current technological trends in developing
more efficient thin-film solar cells (up to 11.2 percent
efficient in 1988), plus declining manufacturing costs
made possible by mass production and economy of
scale. In addition, environmental concerns and the inevitable decline in supplies of fossil fuels should spur
demand for solar power in the future.

Conversion of solar radiation into heat, kinetic energy, or electricity follows the **law of energy conservation.** That is, energy is neither created nor
destroyed, but is changed into other forms. In summary then, the sun drives
the atmosphere: imbalances in solar heating spur atmospheric circulation
(weather), which redistributes heat. Hence, solar energy is the ultimate source
of kinetic energy, which is manifest in the circulation of the atmosphere.

SEASONAL CONTRASTS

Because the need for heat redistribution varies with the season, atmospheric
circulation and weather change through the year. For example, when steep

temperature gradients prevail across North America, the weather tends to be energetic. Storm systems are large and intense, winds are strong, and the weather is changeable. Such weather is typical of winter, when it is not unusual for daily temperatures in the southern United States to be more than 30 C° (54 F°) warmer than temperatures across southern Canada.

In contrast, when temperature varies little across the continent, as in summer, the weather tends to be more tranquil, and large-scale weather systems are generally weak and not well defined. Nevertheless, summer weather is sometimes very active. Intense heating of the ground by the hot summer sun often triggers strong convection and the development of thunderstorms. Some of these weather systems spawn destructive hail, strong and gusty winds, and heavy rains. However, these systems are usually shorter lived and more localized than winter storms.

Variation of Air Temperature

Air temperature is variable, fluctuating from hour to hour, from one day to the next, with the seasons, and from one place to another. Our discussion of the basic reasons for weather provides some insight as to why air temperature varies. Air temperature is regulated by local radiation and by the movement (advection) of air masses. Although these two factors actually work in concert, for the purpose of study we first consider them separately.

RADIATIONAL CONTROLS

Conditions that influence the local radiation balance, and hence the local air temperature, include (1) the time of day and day of the year, which determine the solar altitude and the intensity and duration of incoming solar radiation, (2) cloud cover, since cloudiness affects the flux of both solar and terrestrial radiation, and (3) the nature of the surface cover because surface characteristics determine the albedo and the percentage of absorbed solar radiation used for sensible heating and latent heating. Hence, air temperature is generally higher in June than in January (in the Northern Hemisphere), during the day than at night, under clear rather than cloudy afternoon skies, when the ground is bare instead of snow covered, and when the ground is dry rather than wet.

The annual temperature cycle, also called the march of mean monthly temperature, clearly reflects the systematic variation in incoming solar radiation over the course of a year. In the latitude belt between the tropics of Cancer and Capricorn, solar radiation varies little over the course of a year. Thus, the variation in average monthly air temperature during the year exhibits

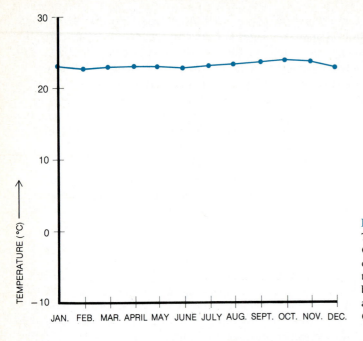

FIGURE 4.7

The march of mean monthly temperatures at Clevelandia, Amazon Basin (4 degrees N, 52 degrees W). At this near-equator location, little temperature change occurs during a year because of minimal variation in solar radiation and length of daylight. [World Meteorological Organization data]

no significant seasonal contrasts (Figure 4.7). In fact, in the tropics, the temperature difference between night and day often is greater than the winter-to-summer temperature contrast.

In middle latitudes, solar radiation features a pronounced annual maximum and minimum. In high latitudes, poleward of the Arctic and Antarctic circles, the seasonal difference in solar radiation is extreme, varying from zero in winter to a maximum in summer. This marked periodicity of incident solar radiation outside of the tropics accounts for the distinct winter-to-summer temperature contrasts observed in middle and high latitudes (Figure 4.8).

In middle and high latitudes, the march of mean monthly temperature lags behind the monthly variation in insolation, so that the warmest and coldest months of the year typically do not coincide with the times of maximum and minimum solar radiation, respectively. This is because the troposphere's temperature profile takes time to adjust to the changing solar energy input. Typically, the warmest portion of the year is about a month after the summer solstice, and the coldest part of the year usually occurs about a month after the winter solstice. In the United States, the temperature cycle lags the solar cycle by an average of 27 days. However, in coastal localities with a strong maritime influence (Florida, the shoreline of New England, and California, for example), the average lag time is up to 36 days. In addition, as we saw in Chapter 3, the maritime influence reduces the amplitude of the annual march of monthly mean temperature; that is, the winter-to-summer temperature contrast is less.

To some extent, the variation of air temperature over the course of a 24-hour day reflects the day-to-night (diurnal) variation in radiation. Typically, the day's lowest temperature occurs near sunrise as the culmination of a night of radiational cooling. The day's highest temperature is usually recorded in early or midafternoon, even though insolation peaks around noon. The time required for the temperature of the lower troposphere to adjust to the day-to-night variation in radiation accounts for the lag of several hours between radiational forcing and air temperature response.

The diurnal lag between insolation and air temperature explains why in summer the greatest risk of sunburn is about noon and not during the warmest time of day. Incoming solar ultraviolet radiation, the cause of sunburn, is most intense near noon, but the air temperature usually reaches a maximum several hours later.

As a further illustration of local radiational controls, consider the influence of ground characteristics on air temperature. All other factors being equal, in response to the same insolation, the air over a dry surface warms up more than it would if that surface were moist. When the surface is dry, the absorbed solar radiation is used primarily for sensible heating of the air (conduction

FIGURE 4.8

The march of mean monthly temperature at Regina, Saskatchewan, Canada (50 degrees 26 minutes N, 104 degrees 40 minutes W). The temperature regime at this extremely continental midlatitude locality is strongly influenced by the seasonal variation of incoming solar radiation. [Atmospheric Environment Service data]

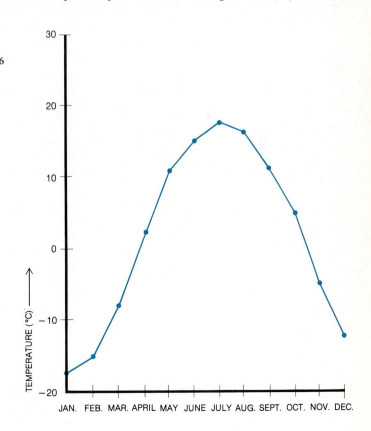

and convection of heat from the surface into the overlying air). Hence, the Bowen ratio is relatively high, and so is the air temperature. On the other hand, when the surface is moist, much of the absorbed solar radiation is used to vaporize water, the Bowen ratio is lower, and so is the air temperature. This suggests a simple means of reducing summer air conditioning needs: where water is in plentiful supply, shallow pools of water placed on rooftops reduce direct solar heating of the building.

This also helps explain why unusually high air temperatures often accompany **drought,** a lengthy period of extreme moisture deficit. Soils dry out, crops wither and die, and reservoirs shrink. Less surface moisture is available for vaporization, and more of the absorbed solar radiation is channeled into raising the air temperature through sensible heating. Consider as an example the severe drought that gripped a ten-state area of southeastern United States between December 1985 and July 1986. Rainfall was generally less than 70 percent of the long-term average, and in the hardest hit areas, portions of the Carolinas, it was less than 40 percent. By July, many weather stations in the drought-stricken region were setting new high temperature records. Columbia, South Carolina, Savannah, Georgia, and Raleigh/Durham, North Carolina, all reported the warmest July on record. Also contributing to record heat was more intense solar radiation reaching the ground, a consequence of less than the usual daytime cloud cover.

The same association between exceptionally dry surface conditions and unusually high air temperatures was observed during the severe drought that afflicted the Midwest and Great Plains during the summer of 1988. At many long-term weather stations, the summer of 1988 was one of the driest and hottest on record. The record high temperatures triggered much speculation (especially in the media) that a major climatic change was underway and that the much heralded CO_2-induced global warming had at last begun. But, as we will see in Chapter 18, one long, hot summer does not necessarily indicate a climatic change. Actually, researchers at the National Center for Atmospheric Research report that the drought of 1988 was caused by an atmospheric circulation pattern that was linked to an unusual sea-surface temperature pattern over the tropical Pacific.

Because of the heat required for vaporization or melting of snow, a snow cover reduces sensible heating of the overlying air. In addition, the relatively high albedo of snow substantially decreases the amount of available solar radiation that is absorbed at the surface of the snow cover and converted to heat. Consequently, a snow cover lowers the day's maximum air temperature. Because snow is also an excellent radiator of infrared, nocturnal radiational cooling is extreme where the ground is snow covered—especially when skies are clear and winds are very light or calm. On such nights, the air temperature near the surface may drop 10 C° (19 F°), or more, lower than it would if the ground were bare of snow. The net effect of a snow cover, then, is to reduce significantly the 24-hour average temperature.

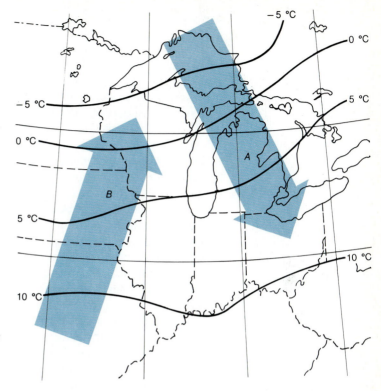

FIGURE 4.9

Cold air advection occurs when (A) the horizontal wind blows across isotherms from cold areas toward warmer areas, and warm advection occurs when (B) the horizontal wind blows across isotherms from warm areas toward colder areas. Solid lines are isotherms.

AIR MASS CONTROLS

Air mass advection refers to the movement of an air mass from one locality to another. **Cold air advection** occurs when the wind blows in a direction across regional isotherms from a colder area to a warmer area (arrow A in Figure 4.9), and **warm air advection** takes place when the wind blows in a direction across regional isotherms from a warmer area to a colder area (arrow B in Figure 4.9). **Isotherms** are lines drawn on a map through localities having the same air temperature.* Air mass advection thus occurs when one air mass replaces another air mass having different temperature characteristics. Recall that this is the most important process in poleward heat transport.

In terms of temperature variations at a given locality, the significance of air mass advection depends on the initial temperature characteristics of the

---

* Air temperatures are reported by weather stations that are often widely spaced. Hence, isotherms usually are drawn by interpolating between stations. Selection of the interval between successive isotherms hinges on the temperature range and the need to resolve the temperature field in sufficient detail.

new air mass, as well as on the degree of modification the air mass undergoes as it travels across the Earth's surface. For example, a surge of bitterly cold arctic air loses much of its punch when it travels over ground that has no snow cover because the arctic air mass is warmed from below by sensible heating (conduction and convection). In contrast, modification of an arctic air mass by sensible heating is minimized when the air mass travels over a cold, snow-covered surface. We have more to say about air masses and air mass modification in Chapter 11.

So far, we have been considering how horizontal movement of air (advection) influences air temperature at some locality. However, as we saw in our discussion of convection currents, air also moves vertically. As air moves up and down, its temperature changes: air cools as it rises and warms as it descends. We discuss the reasons for these temperature changes in Chapter 6.

Although we have considered local radiational controls and air mass advection separately, the two actually regulate air temperature together. For example, air mass advection may compensate for, or even overwhelm, local radiational influences on air temperature. As noted earlier, the local radiation balance usually causes the air temperature to rise from an early morning minimum to an early or midafternoon maximum. This typical pattern can change, however, if an influx of cold air occurs during the same period. Depending on how cold the incoming air is, air temperatures may climb more slowly than usual, may remain steady, or may even fall during daylight hours. If cold air advection is extreme, air temperatures may drop precipitously throughout the day, in spite of bright, sunny skies. In another example, air temperatures may climb through the evening hours as a consequence of strong warm air advection, so the day's high temperature occurs just before midnight.

Conclusions

Imbalances in radiational heating and radiational cooling are ultimately responsible for the circulation of the atmosphere. These imbalances occur both vertically (between the Earth's surface and the atmosphere) and horizontally (between tropical and higher latitudes). Through circulation of the atmosphere, heat is redistributed within the Earth-atmosphere system. This enables us to understand how air temperature is regulated by a combination of local radiation and air mass advection.

Another important consequence of atmospheric circulation is the formation of clouds that can produce rain and snow. Before examining cloud- and precipitation-forming processes in detail, we must first examine another variable of the atmosphere, air pressure.

Summary Statements

Heat imbalances develop between the Earth's surface and the atmosphere, and between tropical and high latitudes, because of differences in the rates of solar heating and infrared cooling.

Excess heat is transported from the Earth's surface to the troposphere by conduction and convection (sensible heating) and by vaporization and condensation of water (latent heating).

In the development of convective clouds (cumulus, cumulonimbus), sensible heating combines with latent heating in transporting heat from the Earth's surface into the troposphere.

The ratio of sensible heating to latent heating, the Bowen ratio, depends on how moist the surface is. The drier the land surface becomes, the less important is latent heating and the more important is sensible heating.

Over the course of a year, latitudes poleward of about 30 degrees N and 30 degrees S experience net radiational cooling, whereas tropical latitudes are sites of net radiational warming.

Excess heat is transported poleward from the tropics by air mass exchange, storm systems, and ocean currents. The most important of these processes is air mass exchange.

The sun drives the circulation of the atmosphere by producing heat differentials within the Earth-atmosphere system and by being the ultimate source of kinetic energy.

Air temperature is regulated by local radiational controls and by advection of air masses. In some cases, cold or warm air advection overwhelms the local radiation balance.

Key Words

sensible heating	sublimation	poleward heat	drought
latent heating	condensation	transport	air mass advection
latent heat of fusion	cumulus clouds	kinetic energy	cold air advection
latent heat of melting	cumulonimbus clouds	law of energy	warm air advection
latent heat of vaporization	Bowen ratio	conservation	isotherms

Review Questions

1. What is meant by radiational heating and radiational cooling within the Earth-atmosphere system?

2. On a global scale and over the course of a year, compare the radiational heating and cooling of the Earth's surface with the radiational heating and cooling of the troposphere.

3. How is the excess heat energy at the Earth's surface transported into the troposphere? Which of these processes is most important and why?

4. Describe the processes that are involved in sensible heating.

5. How does molecular activity vary with the three physical phases of water?

6. Explain why heat is either added or released when water undergoes a phase change.

7. What happens to the temperature of water undergoing a phase change?

8. When clouds form, heat is released to the atmosphere. Why?

9. Define the Bowen ratio.

10. The Bowen ratio for Asia is greater than that for North America. Why?

11. What is the global average Bowen ratio?

12. Under what conditions is the net flow of heat from the troposphere to the Earth's surface?

13. Describe how radiational heating and radiational cooling vary with latitude.

14. What processes are involved in poleward heat transport? Which process is most important?

15. State the law of energy conservation and provide some examples of how this law applies to atmospheric circulation.

16. Why is air temperature variable?

17. On a winter day, is the air temperature more likely to be higher if the ground is snow covered or if it is bare? Explain your answer.

18. Why is the winter solstice usually not the coldest day of the year?

19. Explain why in summer the greatest risk of sunburn is not during the warmest time of day.

20. Under what conditions might the day's high temperature be recorded at 11 P.M.?

Points to Ponder

1. In your own words, explain the following statements: (a) The sun drives the atmosphere. (b) The atmosphere is heated from below.

2. List some of the advantages of solar power over other more conventional energy sources. What might be some disadvantages?

3. To evaluate the solar power potential of a given locality, what climate records would you consult?

4. In spite of clear skies, the air temperature remains steady throughout the day. Explain this observation.

5. Speculate on whether there might be seasonal changes in the poleward transport of heat.

6. Why does the atmosphere circulate?

7. Carefully study Figure 4.1 and compare the rate of cooling of the Earth's surface via emission of IR versus latent heating. Which process is more important? Speculate on the significance of your answer for the temperature of the lower atmosphere.

Projects

1. Without utilizing solar collectors or solar cells, design a simple dwelling that takes advantage of solar heating during the winter but does not overheat in summer.

2. How extensively are solar power systems utilized in your community? You may wish to arrange a tour of a public facility.

Selected Readings

ERBE Science Team. "First Data from the Earth Radiation Budget Experiment (ERBE)." *Bulletin of the American Meteorological Society* 67 (1986):818–824. Discusses results of recent satellite monitoring of solar and terrestrial radiation.

Hamakawa, Y. "Photovoltaic Power." *Scientific American* 256, No 4 (1987):87–92. Discusses new developments in photovoltaic technology and future prospects for large-scale power plants.

Ingersoll, A. P. "The Atmosphere." *Scientific American* 249, No. 3 (1983):162–174. Presents a concise summary of atmospheric characteristics and the global radiation balance.

Rosenfeld, A. H., and D. Hafemeister. "Energy-efficient Buildings." *Scientific American* 258, No. 4 (1988):78–85. Discusses techniques and policies for energy conservation in office buildings and houses.

Trenberth, K. E. "What Are the Seasons?" *Bulletin of the American Meteorological Society* 64 (1983):1276–1282. Analyzes the distinction between the meteorological seasons and the astronomical seasons.

*"Yes, what a climb that was!
    I was scared to death, I can tell
    you.
Sixteen hundred meters—that is
    over five thousand feet, as I
    reckon it. . . ."
And Hans Castrop took in a deep,
    experimental breath of the
    strange air. It was fresh, and
    that was all. It had no
    perfume, no content, no
    humidity; it breathed in easily,
    and held for him no
    associations.*

Thomas Mann
THE MAGIC MOUNTAIN

5

Air Pressure

Air pressure and air density drop
rapidly with increasing altitude. At
high elevations, the air is so thin
that mountain climbers may require
a supplementary oxygen supply.
[Photograph by J. M. Moran]

107

I N DESCRIBING the state of the atmosphere, television and radio weather-casters usually report the latest air pressure reading along with air temperature and relative humidity. Although we are physically aware of changes in temperature and humidity, we do not sense changes in air pressure as readily. If we follow air pressure reports over a period of time, however, we quickly learn that important shifts in weather accompany relatively small variations in air pressure.

In this chapter, we examine the properties of air pressure and the reasons for spatial and temporal variations in air pressure. In later chapters, we describe how this variability of air pressure is involved in the circulation of the atmosphere.

Defining Air Pressure

Air exerts a force on the surfaces of all objects that it contacts. Air pressure is a measure of that force per unit of surface area. Molecules composing air are always in rapid, random motion, and each molecule exerts a force as it collides with the surface of a solid (the ground, for example) or liquid (the sea surface, for example). In a millionth of a second, each square centimeter of the Earth's surface is bombarded by billions upon billions of gas molecules. The total air pressure exerted, then, is the cumulative force of a multitude of molecules colliding with the surface of any object in contact with air.

The amount of pressure produced by the gas molecules composing air depends on (1) the mass of the molecules, (2) the pull of gravity,* and (3) the kinetic energy of the molecules. Usually, **air pressure** at a given location (the Earth's surface, for example) is described as the weight per unit area of the column of air above that location. Weight is the force exerted by gravity on a mass of air, that is,

$$\text{weight} = \text{mass} \times \text{gravity}$$

Pressure Balance

The average air pressure at sea level is about 14.7 lb per square in. This same pressure is produced by a column of water about 10 m high. Hence, the total weight of the atmosphere on the roof of a typical three-bedroom ranch-style house at sea level is about 2.1 million kg (4.6 million lb), equivalent to the

---

* Gravity is the force that holds you and all other objects on the Earth's surface.

combined weight of 1500 full-size autos. Why doesn't the roof collapse? It doesn't because air pressure at any point is the same in all directions—up, down, and sideways. Thus, the air pressure within the house exactly counterbalances the air pressure outside the house, and the *net* pressure acting on the roof is zero. This pressure balance (or equilibrium) is the prevailing condition in the atmosphere.

Variation with Altitude

We know from pumping air into a bicycle tire that air is compressible; that is, its volume and density are variable. The pull of gravity compresses the atmosphere so that the maximum **air density** (mass per unit volume) is at the Earth's surface. In other words, the atmosphere's gas molecules are spaced most closely at the Earth's surface, and the spacing between molecules increases with increasing altitude. The number of gas molecules per unit volume thus decreases with altitude. This "thinning" of air is so rapid that at an altitude of only 16 km (10 mi), air density is about 10 percent of its average sea-level value. Thinning of air means fewer molecular collisions and a rapid decline in air pressure with altitude (Figure 5.1). The drop of air pressure with increasing altitude was first verified by Pascal in 1658 when he directed mountain climbers to take barometer* readings as they ascended the Puy-de-Dome in France.

---

    * A barometer, the instrument used to monitor air pressure, is described later in this chapter.

FIGURE 5.1
Variation of air pressure in millibars (mb) with altitude. The average air pressure at sea level is 1013.25 mb.

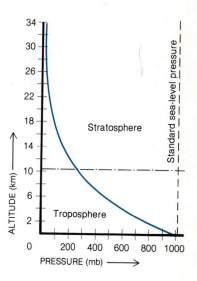

The vertical profile of air pressure in Figure 5.1, as well as the vertical profile of air temperature (refer back to Figure 1.7), is based on the **standard atmosphere,** a model of the real atmosphere. The standard atmosphere is the state of the atmosphere averaged for all latitudes and seasons. It features a fixed sea-level air temperature (15 °C, or 59 °F) and pressure (1013.25 mb) and fixed vertical profiles of temperature and pressure. Appendix II presents the standard atmosphere in tabular format.

Although air pressure and air density drop rapidly with altitude, it is not possible to specify an altitude where the Earth's atmosphere definitely ends. That is, no one altitude can be clearly identified as the beginning of interplanetary space. At best, we can describe the vertical extent of the atmosphere in terms of the relative distribution of its mass with altitude. Half the atmosphere's mass lies between the Earth's surface and an altitude of about 5.5 km (3.4 mi). About 99 percent of the atmosphere is below 32 km (20 mi). Above 80 km (50 mi), the relative proportions of atmospheric gases change markedly, and by about 1000 km (600 mi), the atmosphere has merged with the highly rarefied interplanetry gases, hydrogen and helium. Interestingly, if the Earth's atmosphere had a uniform density throughout, it would have a well-defined top. Assuming a temperature equal to the average value at sea level, we observe that the top of a uniform density atmosphere would be at an altitude of only 8 km (5 mi).

From a slightly different perspective, the atmosphere thins so rapidly with altitude that at only 32 km (20 mi), air pressure is less than 1 percent of the average air pressure at sea level. This rapid pressure drop means that appreciable changes in air pressure accompany even relatively minor changes in land elevation. For example, the average air pressure at Denver, a mile above sea level, is about 83 percent of the average air pressure at Boston, located just above sea level.

The expansion and thinning of air that accompany the fall in air pressure with altitude trigger physiological adjustments in humans. Visitors to mountainous areas may complain of dizziness, headache, and shortness of breath—symptoms brought on by reduced oxygen concentration in the relatively thin air. After several days, these distressing symptoms usually disappear as the number of red blood cells, which carry oxygen in the bloodstream, increases. The body thus adjusts gradually, or is acclimatized, to low oxygen levels. Some people are better able to acclimatize than others because of differences in individual genetic makeup. An upper altitude limit does exist, however, above which no human can adjust. At the summit of Mount Everest, the Earth's highest peak at 8850 m (29,028 ft), the oxygen concentration is so low that only those individuals who are both genetically endowed and carefully acclimatized can survive without an auxiliary oxygen supply.

Above about 5000 m (15,000 ft), an aircraft cabin must be pressurized unless a supplemental oxygen supply is available to the flight crew and passengers. In actual practice, commercial aircraft cabins are pressurized from takeoff on up. For example, the air pressure outside a transcontinental jet

Ear-Popping

On a rapidly ascending or descending airplane or elevator, we commonly feel the effects of changing air pressure by a popping sensation in our ears. Ear-popping is symptomatic of a natural response that helps to protect the eardrum from damage.

As Figure 1 indicates, the eardrum separates the outer ear from the middle ear chamber. As air pressure in the outer ear changes, the eardrum will become distorted if air pressure in the middle ear does not equal the air pressure in the outer ear. For example, as an aircraft takes off and the cabin pressure lowers, the air pressure on the outer ear declines. If the middle ear chamber does not experience an equal drop in air pressure during the ascent, the air pressure in the middle ear will be greater than the pressure in the outer ear. As a consequence, the eardrum bulges outward. Such deformation not only causes physical discomfort, but the bulging eardrum does not vibrate efficiently and sounds are muffled. If the air pressure difference between the middle ear and outer ear continues to increase, the eardrum could rupture, perhaps causing a permanent hearing loss.

Fortunately, there is a natural mechanism whereby the air pressure on the middle ear is altered. The *eustachian tube* connects the middle ear with the upper throat region, which, in turn, leads to the outside via the oral and nasal cavities. Normally, the eustachian tube is closed where it enters the throat, but it will open if a sufficient air pressure difference develops between the middle ear and the throat. When the eustachian tube opens, air pressure in the middle ear quickly equalizes with the external air pressure and the eardrum pops back to its normal shape. The vibrations of the eardrum that are associated with its rapid change in shape are what we hear as "ear-popping," the human body's way of preventing a permanent hearing loss when we experience a rapid change in air pressure.

To reduce discomfort, we can hasten the opening of the eustachian tube by yawning or swallowing. For this reason, air travelers are advised to chew gum because the subsequent swallowing helps to equalize air pressure on both sides of the eardrum.

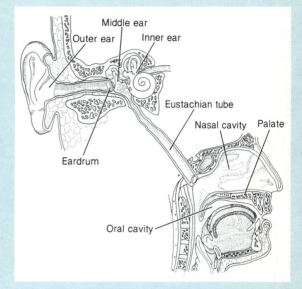

FIGURE 1

When air pressure differences develop between the middle ear and the outer ear, the eardrum is distorted and sounds are muffled. Opening of the eustachian tube equalizes the pressure and causes a popping sensation in the ears.

aircraft cruising at 10,000 m (33,000 ft) is only about 25 percent of the mean sea-level air pressure. The cabin is typically pressurized to about 75 percent of its sea-level value. Nonetheless, as an aircraft changes altitude, especially during takeoff and landing, passengers notice changes in air pressure by a "popping" sensation in their ears. The reasons for this phenomenon are discussed in the Special Topic, "Ear-Popping."

Very low air density at high altitudes also has interesting implications in terms of air temperature and heat transfer. In the thermosphere, the highest thermal subdivision of the atmosphere, individual atoms and molecules move about with an average kinetic activity that is indicative of very high temperatures (refer back to Figure 1.7). There are so few atoms and molecules per unit volume, however, that the *total* kinetic molecular energy (that is, heat) is relatively low. Hence, in spite of temperatures that approach 1200 °C (2200 °F) in the thermosphere, heat is not readily conducted to cooler bodies. Satellites orbiting at these altitudes, for example, do not acquire such temperatures.

Horizontal Variations

Air pressure differs from one place to another, and variations are not always due to differences in the elevation of the land. In fact, meteorologists are more interested in air pressure variations that arise from factors other than land elevation. Hence, weather observers usually determine an equivalent sea-level value; that is, they adjust local air pressure measurements to what the air pressure would be if the station were actually located at sea level. When this **reduction to sea level** is carried out everywhere, air pressure still varies from one place to another and fluctuates from day to day and even from hour to hour (Figure 5.2).

Although spatial and temporal fluctuations in surface air pressure (reduced to sea level) are relatively small, they can accompany some important changes in weather. In middle latitudes, weather is dominated by a continuous procession of different air masses that bring about changes in air pressure and changes in weather. An **air mass** is a huge volume of air that is relatively uniform (horizontally) in temperature and water vapor concentration. As air masses move from place to place, surface air pressures fall or rise, and the weather changes. As a general rule, weather becomes stormy when air pressures fall and becomes fair when air pressures rise.

Why do some air masses exert greater pressure than other air masses? One reason is the difference in air density that arises from differences in air temperature, or from differences in water vapor concentration, or from both. As a rule, temperature has a much more important influence on air pressure than does water vapor concentration.

TEMPERATURE/WATER VAPOR INFLUENCE

Recall from Chapter 3 that temperature is a measure of the average kinetic energy of individual molecules. As air temperature rises, the constituent molecules move about with greater activity. If air is heated within a closed con-

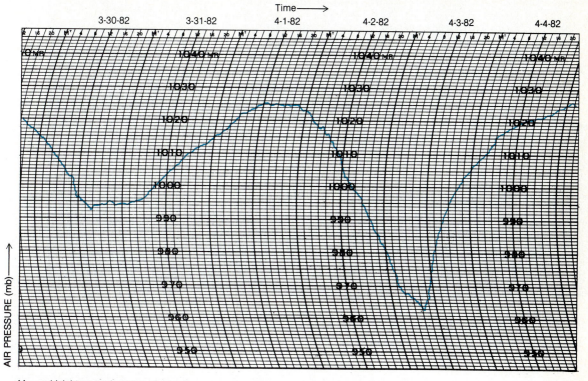

FIGURE 5.2

A trace from a barograph showing the variation in air pressure in millibars reduced to sea level at Green Bay, Wisconsin, from 30 March through 4 April 1982. Note that significant changes in air pressure occur from day to day and even from one hour to the next.

tainer, such as a rigid metal can, we would expect the air pressure on the internal walls of the container to rise as the increasingly energetic molecules bombard the walls with greater force. The air density inside the container would not change because no air is added to or removed from the container and the air volume is constant. In contrast, the atmosphere is not confined by walls (except the Earth's surface), so the air is free to expand and contract. That is, within the atmosphere, air density is variable.

When air is heated within the atmosphere by conduction, convection, or radiation, air pressure decreases with rising temperature. This is because the greater activity of the heated molecules increases the spacing between neighboring molecules and thus reduces air density. Decreasing air density lowers

FIGURE 5.3
A hot air balloon rises within the atmosphere because heated air within the balloon is less dense than the cooler air surrounding the balloon. [Photograph by Mike Brisson]

the pressure exerted by the air. Given equal volumes, warm air is thus lighter (less dense) than cold air and consequently exerts less pressure (Figure 5.3).

The greater the concentration of water vapor in air, the *less* dense is the air. This statement is contrary to the popular misconception that humid air is "heavier" than dry air. Although hot, muggy air may weigh heavily on a person's disposition, humid air is, in fact, less dense than dry air at the same temperature. Water vapor reduces the density of air because the molecular weight of water is less than the average molecular weight of dry air.*

The atmosphere is a continuous fluid. This implies that when water molecules enter the atmosphere as a gas, they take the place of other gas molecules, principally nitrogen and oxygen. The molecular weight of water vapor is less than that of either N_2 or O_2. Hence, as the water vapor concentration in air increases, the net effect is for air density to decrease. At equal temperatures and volumes, then, a humid air mass exerts less pressure than a relatively dry air mass.

Cold, dry air masses produce higher surface pressures than do warm, humid air masses. Warm, dry air, in turn, produces higher surface pressures

---

* The molecular weight of water is 18, and the mass-weighted mean molecular weight of dry air is about 29.

than an equally warm, but more humid, air mass. Hence, replacement of one air mass by another, that is, **air mass advection,** usually is accompanied by a change in surface air pressure. For example, we would expect the air pressure to rise as a surge of cold air replaces the mild air that has been with us for the past several days. Air mass modification (changes in air mass temperature and/or water vapor concentration) also produces changes in surface air pressure. These modifications may occur when an air mass travels over different surface types (from cold snow cover to mild bare ground, for example) or, if the air mass is stationary, when the air is locally heated or cooled. In Chapter 4, we examined the regulation of air temperature by local conditions (radiation balance) plus air mass advection. From the above discussion, we see that local conditions and air mass advection also influence surface air pressure.

DIVERGENCE AND CONVERGENCE

In addition to air pressure changes caused by variations of temperature and (to a lesser extent) water vapor concentration, air pressure can also be influenced by circulation patterns that cause divergence or convergence of air. Consider some examples. Suppose that at the Earth's surface, horizontal winds blow radially away from a central point, as in Figure 5.4A. This is an example of **divergence** of air. At the center, air descends from above and takes the place of the air diverging at the surface. If more air diverges at the surface than descends from aloft, then the air density and air pressure decrease. On the other hand, suppose that at the Earth's surface, horizontal winds blow radially inward toward a central point, as in Figure 5.4B. This is an example of **convergence** of air. If more air converges at the surface than ascends, then air density and air pressure increase. In later chapters, we discuss in much greater detail air pressure changes caused by the divergence and convergence of air within weather systems.

FIGURE 5.4
(A) Air descends from aloft and diverges at the Earth's surface. (B) Air converges at the Earth's surface and ascends. Such patterns of airflow can cause changes in air density and air pressure.

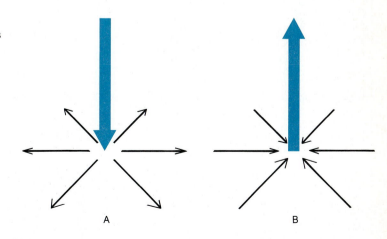

A   B

Highs and Lows

We now have some insight into the meaning of those "H" and "L" symbols on newspaper and television weather maps (Figure 5.5). After air pressure readings are reduced to sea level, lines called **isobars** are drawn on a map connecting locations with the same air pressure.* An "H" (or "High") symbol is used to designate places where sea-level air pressure is relatively high compared with the air pressure of surrounding areas, and an "L" (or "Low") symbol is used to indicate regions where sea-level air pressure is relatively low by comparison. For reasons presented in Chapter 9, a "High" usually is a fair weather system, and a "Low" is a stormy weather system.

_____

* Isobars are interpolated between weather stations.

FIGURE 5.5

A typical surface weather map showing variations in air pressure (reduced to sea level) from one place to another. Solid lines are isobars, lines joining locales having the same air pressure. Shaded areas are where rain or snow is falling. [NOAA photograph]

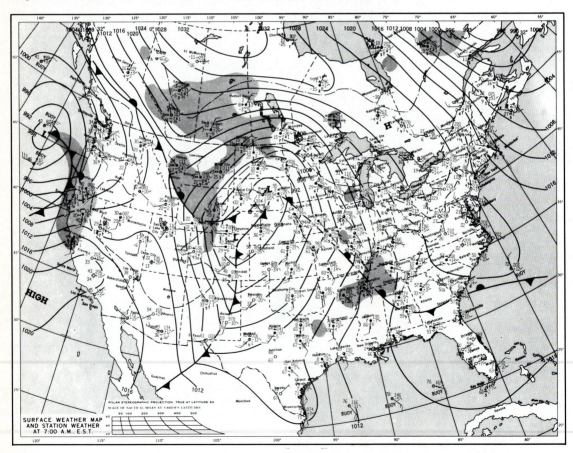

Air Pressure Measurement

A **barometer** is the instrument used to monitor changes in air pressure. There are two basic types of barometer: mercury and aneroid.

The more accurate, though cumbersome, of the two is the **mercury barometer,** invented in 1643 by Evangelista Torricelli, an Italian physicist and student of Galileo. The instrument consists of a glass tube a little less than 1 m (39 in.) long, sealed at one end, open at the other end, and filled with mercury, a very dense liquid. The open end of the tube is inverted into a small

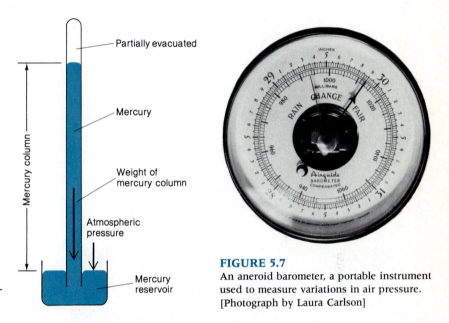

FIGURE 5.6
Schematic drawing of a mercury barometer.

FIGURE 5.7
An aneroid barometer, a portable instrument used to measure variations in air pressure.
[Photograph by Laura Carlson]

open container of mercury, as shown in Figure 5.6. Mercury settles down the tube (and into the container) until the weight of the mercury column exactly balances the weight of the atmosphere acting on the surface of the mercury in the container.

The average atmospheric pressure at sea level will support the mercury column in the tube to a height of 760 mm (29.92 in.). When air pressure changes, however, the height of the mercury column changes. Falling air pressure allows the mercury column to drop, and increasing air pressure forces the mercury column to rise. This, then, is the origin of the practice of expressing air pressure in units of length.

The **aneroid** (nonliquid) **barometer,** pictured in Figure 5.7, is less precise but more portable than the mercury barometer. It consists of a coil of tubing from which much of the air has been evacuated. A spring keeps the chamber from collapsing. As air pressure changes, the coil flexes, bending in when pressure rises and bowing out when pressure drops. A series of gears and levers transmits these movements to a pointer on a dial, which is calibrated to read in equivalent millimeters (or inches) of mercury, or to read directly in units of air pressure.

Some aneroid barometers, especially those designed for home use, have dials with legends, such as "fair," "changeable," and "stormy," corresponding to certain ranges of air pressure. These designations should not be taken too literally, since a given air pressure reading does not always correspond to a specific type of weather.

Much more useful than these legends for local weather forecasting is **air pressure tendency,** that is, the change in air pressure with time. Rising air pressure usually means fair weather, whereas falling air pressure generally signals stormy weather. For determining pressure tendency, some aneroid barometers are equipped with a second pointer that serves as a reference marker. By turning a knob on the barometer face, the user sets the second pointer to correspond to the current air pressure reading. At a later time (perhaps in 2 or 3 hours), the user can observe the new pressure reading and compare it with the earlier set reading to determine the air pressure tendency.

An aneroid barometer may also be linked to a pen that records on a clock-driven drum chart, as shown in Figure 5.8. This instrument, called a **barograph,** provides a continuous record of air pressure variations with time. Because air pressure drops with altitude, an aneroid barometer can be calibrated to read elevation. Such an instrument is called an **altimeter.** For more on altimetry, see the Special Topic.

FIGURE 5.8
A barograph provides a continuous trace of air pressure variations with time. Such a trace is shown as Figure 5.2. [Photograph by J. M. Moran]

Altimetry

Altimetry is the determination of altitude above mean sea level on the basis of air pressure. An *altimeter* is an aneroid barometer that is graduated in increments of altitude. The graduation (that is, the calibration of altitude against air pressure) is prescribed by the *standard atmosphere* which is described elsewhere in this chapter and in Appendix II.

At any time and place the real atmosphere frequently differs from the standard atmosphere so that an altimeter often does not give the true altitude. For aircraft this discrepancy can pose serious problems, especially during the crucial takeoff and landing phases of flight. Hence, aircraft altimeters are equipped with a movable scale that enables the pilot to "zero" the altimeter to the airport's elevation and its latest air pressure reading. While in flight, the altimeter is continually adjusted to elevations and surface air pressure radioed from points en route.

In-flight adjustments of altimeters to surface conditions, however, do not correct for pressure variations that arise en route principally from temperature variations within the air column beneath the aircraft. Cold air is denser than warm air so that air pressure drops more rapidly with altitude in cold air than it does in warm air. Hence, within a column of cold air a given air pressure occurs at a lower altitude than does the same air pressure in a column of warm air. This means, for example, that as an aircraft flies into a column of air that is warmer than specified by the standard atmosphere (Figure 1), the altitude indicated by the altimeter will be lower than the true altitude. Conversely, the altimeter onboard an aircraft flying into air colder than specified by the standard atmosphere will read too high.

The danger, of course, is that an erroneous altimeter reading may impede a pilot's ability to clear an obstacle such as a mountain peak. In practice this hazard can be greatly reduced by an onboard computer that measures air temperature at flight level and makes an appropriate adjustment to the altimeter reading. Note that this correction is not based on the mean

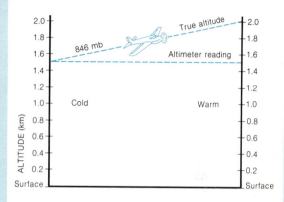

FIGURE 1

An aircraft altimeter initially is calibrated to the standard atmosphere. But as the aircraft flies into a column of warmer air, its altimeter will indicate an altitude that is lower than its true altitude. According to the standard atmosphere, 846 mb is at an altitude of 1.5 km.

temperature of the air column so that, although the error is reduced, it is not totally eliminated.

Federal Aviation Administration (FAA) regulations require aircraft flying below 6000 m (18,000 ft) to employ altimeter corrections as described above. Above 6000 m (18,000 ft), aircraft follow a constant pressure (isobaric) surface with the altimeter zeroed at the standard sea-level pressure of 1013.25 mb.

An alternative to the air pressure-based altimeter is the *radio altimeter*, which is flown aboard most commercial aircraft. This instrument emits radio waves from the plane to the ground where the signal is reflected back to the aircraft. Altitude is calibrated in terms of the time elapsed between emission and reception of the radio signal. The longer the signal takes, the higher is the indicated altitude of the aircraft. Use of this device requires a thorough knowledge of the underlying terrain.

Pressure Units

On television and radio weather reports, air pressure readings are usually given in units of length (millimeters or inches), based on the type of instrumentation used to measure the pressure. It is more appropriate, however, to express air pressure in units of pressure. Physicists use the pascal as the metric unit of pressure and have determined the average air pressure at sea level to be 101,325 Pa, or 101.325 kPa. Meteorologists, on the other hand, traditionally designate air pressure in millibar units (1 mb = 100 Pa).

The average sea-level air pressure reading is 1013.25 mb, and the usual worldwide range in sea-level air pressure is about 970 to 1050 mb. The lowest sea-level air pressure ever recorded (excluding tornadoes) was 870 mb, measured on 12 October 1979 in the eye of Typhoon Tip over the Pacific Ocean northwest of Guam. The highest sea-level air pressure ever recorded was 1083.8 mb at Agata, Siberia, USSR, on 31 December 1968 and was associated with an extremely cold air mass.

The Gas Law

To this point, we have described the state of the atmosphere in terms of variations in temperature, pressure, and density. These important properties, collectively known as **variables of state,** change in magnitude from one place to another across the Earth's surface, with altitude above the Earth's surface, and with time. The three variables of state are also interrelated through the **gas law.** Although the gas law was derived for a single "ideal" gas,* the law provides a reasonably accurate description of the behavior of the atmosphere, which, as noted earlier, is a mixture of many gases.

Simply put, the gas law states that air pressure is proportional to the product of air density and temperature. Expressed as a word equation, the gas law becomes

$$\text{air pressure} = \text{constant} \times \text{density} \times \text{temperature}$$

(The constant is an experimentally derived number that changes a proportional relationship to an equation.) The physical basis for this law is presented in the Mathematical Note at the end of this chapter.

The dependence of air pressure on two interdependent variables, density and temperature, complicates matters. Since within the atmosphere the volume of air can change, temperature variations affect the air density. That is, as we saw earlier, air temperature varies inversely with air density. In terms

---

* An ideal gas follows the kinetic molecular theory precisely. For more on this subject, see the Mathematical Note at the end of the chapter.

of the gas law, this means that raising the air temperature is not normally accompanied by an increase in air pressure, nor is declining air temperature usually associated with falling air pressure. For example, in the winter it is usual for temperatures to drop (not rise) as the surface air pressure rises. The gas law is still satisfied in this case because air density increases at the same time as the temperature drops.

Conclusions

We now have a working definition of air pressure, and we have examined the causes of spatial and temporal variations in air pressure within the Earth-atmosphere system. Air pressure is controlled by air density which, in turn, is dependent on air temperature and to a lesser extent on the concentration of water vapor in air. Divergence and convergence of air may also affect air density and surface air pressure.

The range of air temperature and pressure variations in the Earth-atmosphere system allows water to be present in all three phases. Earlier we saw how phase changes of water help to redistribute heat within the atmosphere (latent heating). Changes in air temperature also trigger the phase changes of water that cause clouds to form or to dissipate. In the next three chapters, we take a closer look at water within the atmosphere with a special emphasis on cloud- and precipitation-forming processes.

Mathematical Note

The Gas Law

Although the atmosphere is a mixture of many gases, it behaves much as if it were a single ideal gas. By definition, an *ideal gas* follows the kinetic-molecular theory precisely. That is, an ideal gas is made up of a very large number of minute particles, called *molecules,* that are in rapid and random motion. As they move about, molecules experience perfectly elastic collisions, so they lose no momentum. Molecules are so small that the attractive forces between them are negligible.

An ideal gas follows Charles's law and Boyle's law exactly, whereas a real gas behaves only approximately as these laws dictate. *Charles's law* holds that with constant presure, P, the absolute temperature, T (in kelvins), of an ideal gas is inversely proportional to the density, ρ, of the gas. That is,

$$T \propto 1/\rho$$

As a sample of gas is heated, the gas expands and its density decreases. According to Boyle's law, on the other hand, when the temperature is held constant, the pressure and density of an ideal gas are directly proportional. That is,

$$P \propto \rho$$

As the pressure on an ideal gas increases, its volume decreases and its density increases.

Charles's and Boyle's laws are combined as the *ideal gas law,* that is,

$$P \propto \rho T$$

The constant of proportionality, R, varies depending on the specific gas. The ideal gas law is thus expressed as an equation relating variables of state,

$$P = \rho R T$$

This so-called *equation of state* describes approximately the behavior of dry air (air minus water vapor) when we assign R a value of 287 J/kg-K.

We can modify the equation of state so that it relates the variables of state for humid air, the more realistic situation. One approach is to keep the same R, the gas constant for dry air, and replace T by T_v, the virtual temperature. The *virtual temperature* is defined to be the temperature of dry air having the same density and pressure as a given sample of humid air. The virtual temperature is computed from the specific humidity, q, that is,

$$T_v = T(1 + 0.61q)$$

Specific humidity is the ratio of the mass of water vapor to the mass of humid air containing the water vapor and is usually expressed as grams of water vapor per kilogram of humid air. The virtual temperature is higher than the actual temperature but rarely by more than 3 C°.

For humid air, the equation of state thus becomes

$$P = \rho R T_v$$

This equation is valid for humid air in which no condensation takes place. Condensation complicates matters because it is accompanied by release of latent heat of vaporization, which elevates air temperature and lowers density.

Summary Statements

The pressure exerted by air depends on the pull of gravity and the mass and kinetic energy of the gas molecules that compose air.

At any point within the atmosphere, air pressure has the same magnitude in all directions.

Air pressure and air density decrease very rapidly with altitude, but the atmosphere has no clearly defined upper boundary. The atmosphere gradually merges with the gases of interplanetary space, hydrogen and helium.

The thinning of air with altitude has adverse physiological effects on humans, and our ability to adjust to thin air is limited.

Variations in air mass characteristics cause the surface air pressure (reduced to sea level) to differ from one place to another. Air temperature and, to a lesser extent, water vapor concentration affect air density and hence surface air pressure.

Air pressure may change in response to divergence or convergence of air, which is produced by patterns of atmospheric circulation.

Important weather changes often accompany relatively small changes in surface air pressure.

The mercury barometer is more accurate, but also more cumbersome, than the aneroid barometer.

The variables of state of the atmosphere (temperature, pressure, and density) are interrelated through the gas law.

Key Words

air pressure	air mass advection	barometer	barograph
air density	divergence	mercury barometer	altimeter
standard atmosphere	convergence	aneroid barometer	variables of state
reduction to sea level	isobars	air pressure tendency	gas law
air mass			

Review Questions

1. Define air pressure.

2. Why are we not particularly aware of day-to-day variations in air pressure?

3. Why does the greatest density of the atmosphere occur adjacent to the surface of the Earth?

4. Based on the pressure variation with altitude, is there a clearly defined top to the Earth's atmosphere? Explain your response.

5. Why are the cabins of commercial jet aircraft pressurized?

6. Why do weather stations routinely adjust air pressure readings to sea-level values?

7. As a general rule, how does the weather change as air pressure rises and falls?

8. How does temperature influence the pressure exerted by a sample of air?

9. Explain why increasing the concentration of water vapor reduces the density of air.

10. Which air mass exerts a greater surface pressure: a warm and humid air mass or a cold and dry air mass? Explain your response.

11. On a particularly hot and muggy evening, a sportscaster comments that baseballs hit to the outfield will not carry far in the "heavy" air. How valid is this observation?

12. What is the meaning of the "H" and "L" symbols that are displayed on newspaper and television weather maps?

13. Air pressure readings are often reported in units of length (millimeters or inches) rather than units of pressure. Why?

14. Explain the principle of the mercury barometer.

15. What are the advantages of an aneroid barometer over a mercury barometer?

16. Why is a barometer also an altimeter?

17. Explain why air pressure tendency is useful for local weather forecasting.

18. As an auto is driven, its tires heat up because of friction. From the gas law, predict what (if anything) will happen to the tire pressure as the tires heat up.

19. In your own words, provide a statement of the gas law.

20. Cold air masses produce higher surface air pressures than do warm air masses. Explain how the gas law is still satisfied.

Points to Ponder

1. The atmosphere thins rapidly with altitude, and yet the proportion of oxygen and nitrogen is constant throughout the troposphere. How, then, do the concentrations of oxygen and nitrogen molecules change with altitude within the troposphere?

2. A jet aircraft is flying at the 400-mb level, that is, at the altitude where air pressure is 400 mb. What fraction of the atmosphere's mass is below the aircraft?

3. What is an ideal gas? What assumptions are made in applying the gas law to the atmosphere?

4. What is the purpose of the virtual temperature? The virtual temperature is never more than a few degrees higher than the actual air temperature. What does this imply about the relative concentration of water vapor in the atmosphere?

Projects

1. If you have access to a barometer, or better yet a barograph, determine the relationship between air pressure tendency and major weather changes.

2. Design an experiment to demonstrate that, at equal temperatures, relatively dry air exerts more pressure than relatively humid air.

Selected Readings

FAA and NOAA. *Aviation Weather.* Washington, D.C.: Federal Aviation Administration, 1975. 219 pp. Intended for pilots and flight operations personnel, a well-illustrated handbook that applies the basics of meteorology to flight.

Perlman, E. "For a Breath of Thin Air." *Science 83,* No. 4 (1983):52–58. Describes a scientific expedition to the summit of Mount Everest to study human response to severe oxygen deprivation.

6

Humidity and Stability

Water occurs in three phrases: ice, liquid, and vapor. Water changes phase as it circulates between the Earth's surface and the atmosphere. This circulation is an important part of the global water cycle that supplies us with all of our fresh water. [Photograph by Mike Brisson]

125

WITHIN THE atmosphere, water occurs in all three phases: as water vapor (an invisible gas) and as tiny ice crystals and water droplets (visible as clouds). The total amount of water within the atmosphere is very small, and most of that is in the lower portion of the troposphere. Indeed, if at any moment all water were removed from the atmosphere as rain and distributed uniformly over the globe, the rainfall would cover the Earth's surface to a depth of only about 2.5 cm (1.0 in.).

Water continually cycles into the atmosphere as vapor from reservoirs of water at the Earth's surface, and water continually leaves the atmosphere and returns to the Earth's surface as rain, snow, and other forms of precipitation. On average, the residence time of a water molecule in the atmosphere is about 10 days. This cycling is an essential component of a global-scale system known as the hydrologic cycle.

The Hydrologic Cycle

On Earth, water is distributed among oceanic, terrestrial, and atmospheric reservoirs (Table 6.1). The ocean, the largest of these reservoirs, contains 97.2 percent of the planet's water; most of the remainder is tied up as ice in Antarctic and Greenland glaciers. The ceaseless flow of water among the

Table 6.1
Water Stored in Global Reservoirs of the Hydrologic Cycle

Reservoir	Percent of Earth's total water
World oceans	97.2
Ice sheets and glaciers	2.15
Groundwater	0.62
Lakes (freshwater)	0.009
Inland seas, saline lakes	0.008
Soil water	0.005
Atmosphere	0.001
Rivers and streams	0.0001

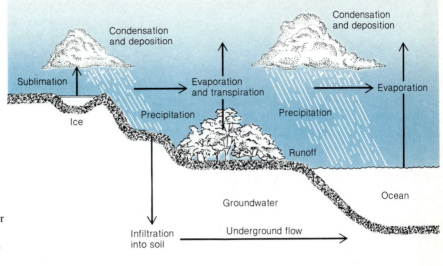

FIGURE 6.1
The hydrologic cycle is a continuous transfer of water among terrestrial, oceanic, and atmospheric reservoirs.

reservoirs, known as the **hydrologic cycle,** is illustrated in Figure 6.1. In brief, water vaporizes from the sea and land to the atmosphere where clouds form. From clouds, rain and snow fall back to the Earth's surface, thus supplying rivers, which flow back to the sea. The endlessness of the hydrolgic cycle is expressed in the verse from Ecclesiastes on the preceding page: "All the rivers run into the sea, yet the sea is never full; unto the place from whence the rivers come thither they return again." Here we focus on a critical link in the hydrologic cycle, the link that joins the atmosphere to the oceanic and terrestrial reservoirs of water.

Within the usual range of air temperature and pressure on planet Earth, all three phases of water coexist naturally. That is, water vapor is in equilibrium with water's liquid and solid phases. Under these conditions, water molecules continually change phase. At the interface between water and air (the surface of the ocean, for example), some water molecules escape from the water surface and enter the air as vapor, whereas other water molecules leave the vapor phase and return to the water surface as liquid. **Evaporation** occurs if more water molecules enter the air than return to the liquid water, and **condensation** takes place if more water molecules return to the liquid water than enter the atmosphere as vapor.

Evaporation occurs from all open bodies of water, as well as from the wetted surfaces of plant leaves and stems and from the soil. Direct evaporation of ocean water is the principal source of atmospheric water vapor. **Transpiration** is the process by which water that is taken up through plant root systems is eventually released as vapor through tiny pores on the surface of green leaves. On continents, transpiration is considerable and is often more important than direct evaporation from the surfaces of lakes, streams, and the soil. For example, a single hectare (2.5 acres) of corn typically transpires 35,000 L (8750 gal) of water per day. Measurements of direct evaporation and transpiration are usually combined as **evapotranspiration.**

A two-way exchange of water molecules also takes place at the interface between ice (or snow) and air, except that water molecules cannot escape from an ice surface as readily as they can from a liquid water surface. Water molecules are more tightly bound in the solid phase than in the liquid phase. **Sublimation** occurs when more water enters the vapor phase than returns to the ice, and **deposition*** takes place when more water returns to the ice than vaporizes.

The gradual shrinkage of snowbanks, even though the air temperature remains well below the freezing point, is a consequence of sublimation plus settling. Condensation or deposition on exposed surfaces on the ground is visible as dew or frost, respectively. These same processes operating within the atmosphere produce clouds.

Precipitation—rain, drizzle, snow, ice pellets, and hail—returns a major portion of atmospheric water from clouds to the Earth's surface, where most of it vaporizes back into the atmosphere. Evaporation and sublimation, with subsequent condensation and deposition, purify water. As water vaporizes, suspended and soluble substances like sea salts are left behind. Through these cleansing mechanisms, saline water from the sea eventually falls on land as freshwater precipitation, which replenishes terrestrial reservoirs.

Comparing the movement of water into and out of the terrestrial reservoirs with the movement of water into and out of the ocean is instructive. The balance sheet for the inputs and outputs of water to and from the various global reservoirs is called the **water budget** (Table 6.2). Each year on the continents the total precipitation exceeds evapotranspiration by about one-third. At sea, however, the annual evaporation exceeds precipitation. Over the course of a year, the water budget therefore shows a net gain of water on land and a net loss of water from the oceans, with the excess on land approximately equal to the ocean's deficit. The land, however, is not getting any soggier, nor are the world's oceans drying up. This is because the excess precipitation on land drips, seeps, and flows from the land back to the seas.

Once precipitation reaches the ground, it follows various routes (Figure 6.2). Some water vaporizes, and the remainder either seeps into the ground as groundwater or runs off as rivers to the sea. The ratio of the portion that infiltrates the Earth's surface to the portion that runs off depends on the intensity of rainfall and on the vegetation, topography, and physical properties

---

* Sometimes sublimation is applied to the change from vapor to solid as well as the change from solid to vapor.

Table 6.2
Global Water Budget[a]

Source	m^3 per year (gal per year)
Precipitation on sea	3.24×10^{14} (85.5×10^{15})
Evaporation from sea	3.60×10^{14} (95.2×10^{15})
Net loss from sea	-0.36×10^{14} (-9.7×10^{15})
Precipitation on land	0.98×10^{14} (26.1×10^{15})
Evaporation from land	0.62×10^{14} (16.4×10^{15})
Net gain on land	$+0.36 \times 10^{14}$ ($+9.7 \times 10^{15}$)

[a] Note that the excess of water on land equals the deficit of water at sea.

of the surface. For example, frozen soil is impermeable so that heavy rains falling on frozen soil will mostly run off and may cause flooding.

A river drains water from a fixed geographical region called a **drainage basin,** or **watershed.** Climate, vegetation, topography, and various activities, both natural and human, in the drainage basin affect both the quantity and the quality of river and groundwater.

It is well worth noting that the sun drives the hydrologic cycle. As we saw in Chapter 4, some of the solar radiation that strikes the Earth's surface is absorbed, that is, converted to heat. Some of this heat is used to vaporize water. If water did not vaporize, there would be no clouds, no precipitation, and no hydrologic cycle.

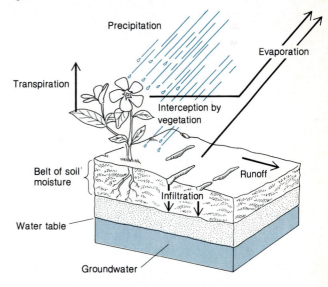

FIGURE 6.2

The various pathways taken by precipitation that falls on the land. [From J. M. Moran et al., *Introduction to Environmental Science.* New York: W. H. Freeman, p. 163. Copyright © 1980.]

How Humid Is It?

"It's not the heat, it's the humidity." This popular statement ascribes the discomfort we feel on a hot, muggy day to the water vapor content of the air. Water vapor concentration in air is an important determinant of our physical comfort, as discussed in the Special Topic, "Humidity and Human Comfort." We know from experience that the water vapor content of the air is quite variable. In this section, we examine some ways to quantify the water vapor content of air.

VAPOR PRESSURE AND MIXING RATIO

Water vapor, just like all other gaseous components of the atmosphere, obeys Dalton's law. **Dalton's law** states that the total pressure of a mixture of gases equals the sum of the pressures exerted by each constituent gas. When water enters the atmosphere as vapor, the water molecules disperse rapidly and mix with the other gases composing air and contribute to the total pressure exerted by the atmosphere. **Vapor pressure** is simply the pressure contributed by water vapor alone and is directly proportional to the concentration of water vapor in air.

From a practical perspective, the vapor pressure is very unlikely anywhere to be greater than about 40 mb. As noted in Chapter 1, no more than about 4 percent of the atmosphere's lowest kilometer is water vapor even in the warm, humid air over tropical oceans and rainforests. Hence, we can estimate the maximum possible vapor pressure to be 4 percent of the mean sea-level air pressure (1013.25 mb) or about 40 mb.

The water vapor concentration of air can also be described as the mass of water vapor per mass of dry air—usually expressed as grams of water vapor per kilogram of dry air. This measure of water vapor in air is called the **mixing ratio** because it specifies how much water vapor is mixed with the other atmospheric gases. Vapor pressure and mixing ratio are two of the more common ways of specifying the water vapor concentration of air and are used in this book.

SPECIFIC AND ABSOLUTE HUMIDITY

Other, less frequently used measures of the water vapor content of air are specific humidity and absolute humidity. **Specific humidity** is defined as the ratio of the mass of water vapor to the mass of the air containing the water vapor (that is, the combined mass of dry air plus water vapor). Specific humidity usually differs from mixing ratio by no more than about 2 percent. **Absolute humidity** is defined as the mass of water vapor (in grams) per unit volume of dry air (in cubic meters), equivalently, the vapor density of

Humidity and Human Comfort

Why do we feel uncomfortable on a hot, muggy day? Our physical comfort during such weather depends, in part, on our body's ability to lose heat via evaporative cooling. This ability depends, in turn, on the vapor pressure gradient between our skin (the evaporative surface) and the surrounding air. The greater the vapor pressure of the ambient air, the smaller is the vapor pressure gradient, and the less water evaporates from the skin surface. Hence, evaporative cooling declines as air becomes more humid.

In light of these conditions, we can better understand why people often experience greater discomfort when the weather is *both* hot and humid. At air temperatures above 25 °C (77 °F), sweating and consequent evaporative cooling promote heat loss from the body. If the air is also quite humid, however, the rate of evaporation is reduced, which hampers the body's ability to maintain a constant core temperature (37 °C, or 98.6 °F). In contrast, in arid regions, the drier air facilitates evaporative cooling. Hence, at an air temperature of say 32 °C (90 °F), we feel more comfortable when the air is dry than when the air is humid.

Scientists at the National Oceanic and Atmospheric Administration (NOAA) estimate that heat waves accompanied by high humidities kill approximately 150 people each year in the United States. Everyone is adversely affected to some extent by high temperatures and humidities. As the weather causes increasing discomfort, people become more irritable and are less able to perform physical and mental tasks. In hot, humid weather, the efficiency of factory workers declines, and students do not concentrate as well on their studies.

In recent decades, a variety of indexes have been

Table 1
Apparent Temperature Index

Relative humidity (%)	Air temperature (°F)										
	70	75	80	85	90	95	100	105	110	115	120
	Apparent temperature (°F)										
0	64	69	73	78	83	87	91	95	99	103	107
10	65	70	75	80	85	90	95	100	105	111	116
20	66	72	77	82	87	93	99	105	112	120	130
30	67	73	78	84	90	96	104	113	123	135	48
40	68	74	79	86	93	101	110	123	137	151	
50	69	75	81	88	96	107	120	135	150		
60	70	76	82	90	100	114	132	149			
70	70	77	85	93	106	124	144				
80	71	78	86	97	113	136					
90	71	79	88	102	122						
100	72	80	91	108							

Source: National Weather Service, NOAA.

Table 2
Hazards Posed by Heat Stress by Range of Apparent Temperature[a]

Category	Apparent temperature	Heat syndrome
I	54 °C or higher (130 °F or higher)	Heatstroke or sunstroke *imminent*.
II	41 to 54 °C (105 to 130 °F)	Sunstroke, heat cramps, or heat exhaustion *likely*. Heatstroke *possible* with prolonged exposure and physical activity.
III	32 to 41 °C (90 to 105 °F)	Sunstroke, heat cramps, and heat exhaustion *possible* with prolonged exposure and physical activity.
IV	27 to 32 °C (80 to 90 °F)	Fatigue *possible* with prolonged exposure and physical activity.

Source: National Weather Service, NOAA.
[a] Apparent temperature combines the effects of heat and humidity on human comfort.

developed that combine temperature and humidity to indicate their impact on human comfort and to advise people of the potential danger from heat stress. The *heat index*, or *apparent temperature index*, developed by R. G. Steadman in 1979, has been reported regularly by the National Weather Service since the summer of 1984. As Table 1 illustrates, high temperature and high humidity combine to produce an apparent temperature that is considerably greater than the actual temperature. For example, a person exposed to an air temperature of 38 °C (100 °F) and a relative humidity of 50 percent experiences the same discomfort and stress as if the air temperature were 49 °C (120 °F) because the high humidity retards evaporative cooling from the skin. The heat index can be divided into four categories based on the severity of impact on human well-being; these categories are listed in Table 2.

When apparent temperatures reach between 41 and 54 °C (105 and 130 °F), heat cramps and heat exhaustion are likely with prolonged exposure and physical activity. At apparent temperatures over 54 °C (130 °F), heatstroke (hyperthermia) is imminent. Keep in mind that the actual degree of heat stress experi-

enced will vary with individual age, health, and body characteristics.

At low humidities, we experience the apparent temperature as lower than the actual temperature, another indication of the role of evaporative cooling. At low humidities, the ambient air feels cooler than it actually is. It is therefore desirable to raise indoor humidity during the winter heating season. Increasing the humidity makes us feel comfortable at a lower room temperature, and we can turn down the thermostat and save on fuel bills.

So far in our discussion of the combined effect of air temperature and humidity on human comfort, we have ignored the effects of radiative exchange between the body and its surroundings. Although the apparent temperature suggests comfortable conditions, discomfort might result from radiative heat loss to cold walls or windows or through radiative heat gain from hot surfaces. Our discussion also neglects the role played by wind in promoting human comfort when conditions are hot and humid. Air in motion accelerates evaporative cooling and helps to transport heat away from the body.

water. However, absolute humidity is not very useful for most meteorological purposes because the volume of a unit mass of air may change. A change in volume means a change in absolute humidity even though there is no gain or loss of water vapor.

The Saturation Concept

Earlier in this chapter we noted that a two-way exchange of water molecules takes place at the interface between water and air (or between ice and air). Water molecules are in a continual state of flux between the liquid and vapor phases. During evaporation, more water molecules enter the vapor phase than return to the liquid phase, and during condensation, more water molecules return to the liquid phase than enter the vapor phase. Eventually, a state of dynamic equilibrium may develop such that liquid water becomes vapor at the same rate that water vapor becomes liquid. That is, the flux of water molecules is the same in both directions. At equilibrium, the vapor pressure is constant and is known as the **saturation vapor pressure.**

If we elevate the temperature, we upset this dynamic equilibrium at least temporarily. The higher the temperature, the greater is the average kinetic energy of individual molecules, and the more readily do water molecules escape the water surface as vapor. Initially, evaporation prevails. If there's a sufficient supply of water and if the water vapor is not continually carried away by winds, eventually a new equilibrium is established. That is, a new balance develops between the flux of water molecules becoming vapor and the flux of water molecules becoming liquid. Now, however, at a higher temperature than before, the water vapor concentration is greater so that the saturation vapor pressure is higher.

Ultimately then, the saturation vapor pressure depends on the rate of vaporization of liquid water which, in turn, is regulated by temperature. The same argument applies to the **saturation mixing ratio.** Hence, as indicate in Tables 6.3 and 6.4, the saturation vapor pressure and saturation mixing ratio increase as the temperature rises. As a rule, the saturation vapor pressure and saturation mixing ratio double for about every 11 C° (20 F°) rise in air temperature. At water's boiling point (100 °C at sea level), the saturation vapor pressure is the same as the surrounding air pressure.

The relationship between temperature and the saturation vapor pressure (or saturation mixing ratio) has been popularly interpreted to imply that warm air can "hold" more water vapor than cold air. In this misleading viewpoint, the atmosphere is likened to a sponge that can accommodate only so much water. However, as argued above, air does not "hold" water vapor; rather, air coexists with water vapor. Water vapor is just one of the many gases that compose air. Temperature changes affect the rate of vaporization of water. Hence, the saturation vapor pressure, rather than being a measure of air's "water holding capacity," is actually a measure of water's vaporization rate.

Table 6.3

Variation of Saturation Vapor Pressure with Temperature

Temperature		Saturation vapor pressure (mb)	
(°C)	(°F)	Over water	Over ice
50	122	123.40	
40	104	73.78	
30	86	42.43	
20	68	23.37	
10	50	12.27	
0	32	6.11	6.11
−10	14	2.86[a]	2.60[a]
−20	−4	1.25	1.03
−30	−22	0.51	0.38
−40	−40	0.19	0.13

[a] Note that for temperatures below freezing, two different values are given: one over super-cooled water and the other over ice.

Table 6.4

Variation of Saturation Mixing Ratio with Temperature

Temperature		Saturation mixing ratio (g water per kg dry air)	
(°C)	(°F)	Over water	Over ice
50	122	88.12	
40	104	49.81	
30	86	27.69	
20	68	14.95	
10	50	7.76	
0	32	3.84	3.84
−10	14	1.79[a]	1.63[a]
−20	−4	0.78	0.65
−30	−22	0.32	0.24
−40	−40	0.12	0.08

[a] Note that for temperatures below freezing, two different values are given: one over super-cooled water and the other over ice.

Relative Humidity

Relative humidity is perhaps the most familiar way of describing the water vapor content of air. It is the water vapor measurement most often cited by television and radio weathercasters. **Relative humidity** compares the actual concentration of water vapor in the air with the concentration of water vapor in that same air at saturation. Relative humidity (RH) is usually expressed as a percentage and can be computed approximately from either the vapor pressure or the mixing ratio. That is,

$$RH = \frac{vapor\ pressure}{saturation\ vapor\ pressure} \times 100\%$$

or

$$RH = \frac{mixing\ ratio}{saturation\ mixing\ ratio} \times 100\%$$

Note that when the actual concentration of water vapor in air is equal to the water vapor concentration at saturation, the relative humidity is 100 percent, that is, the air is saturated with respect to water vapor.

Consider an example of how relative humidity is computed. Suppose that the air temperature is 10 °C (50 °F), and the vapor pressure is 6.1 mb. From Table 6.3, we determine that at 10 °C the saturation vapor pressure of air is about 12.3 mb. Using the formula above, we compute the relative humidity to be about 50 percent [(6.1 mb/12.3 mb) × 100%].

Note that the relative humidity is directly proportional to the vapor pressure and inversely proportional to the saturation vapor pressure. Because the saturation vapor pressure is directly proportional to air temperature, the relative humidity is also inversely proportional to air temperature. The dependence of relative humidity on air temperature can be confusing because even if the actual concentration of water vapor in air does not change, the relative humidity rises or falls depending on how the air temperature changes.

On a clear and calm day, the air temperature usually rises from a minimum near sunrise to a maximum during early to midafternoon and falls off thereafter. If the vapor pressure (or mixing ratio) does not change through the day, then the relative humidity varies inversely with air temperature: the relative humidity is highest when the temperature is lowest and vice versa (Figure 6.3). After sunrise, as the air warms, the relative humidity drops, but not because the air is drying out. The relative humidity decreases because rising air temperature means a greater saturation vapor pressure.

In the previous example, the relative humidity responded only to variations in local air temperature; the wind was calm, and there was no air mass advection. The situation is complicated when air mass advection occurs. Advection can influence both air temperature and vapor pressure. For example, when a warm and humid air mass replaces a cool and dry air mass, the relative humidity is affected both by higher temperatures and increasing vapor pressure.

Perhaps surprisingly, the mean annual water vapor content of the troposphere is about the same over the desert area of the southwestern United States as it is over the Great Lakes region. There are, however, significant differences in relative humidity. Higher average air temperatures in the Southwest mean lower relative humidities and the reduced likelihood of cloudiness and precipitation.

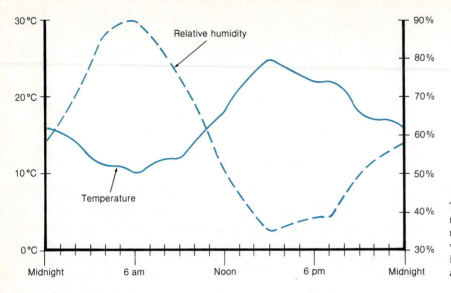

30°C

20°C

10°C

0°C

Midnight   6 am   Noon   6 pm   Midnight

90%

80%

70%

60%

50%

40%

30%

Relative humidity

Temperature

FIGURE 6.3
The variation of relative humidity on a day with no air mass advection (calm winds). The relative humidity varies inversely with the air temperature.

Humidification

In public buildings as well as in individual homes, it is sometimes desirable to alter the relative humidity in the interest of human comfort. When indoor air is extremely dry, a humidifier adds water vapor to the air, and when indoor air is excessively muggy, a dehumidifier removes water vapor.

In winter, as the air that is drawn in from outdoors is heated by a furnace, its relative humidity declines—sometimes to uncomfortably low levels. In very cold localities, indoor relative humidities may be so extremely low that artificial humidification may be necessary. For example, when outdoor air at − 20 °C (−4 °F) and relative humidity of 50 percent is drawn indoors and heated to 20 °C (68 °F), the relative humidity drops to about 3 percent. At such low relative humidities, people may experience discomfort caused by irritation of mucous membranes in the nose and throat. House plants require more frequent watering, and wood furniture may crack and become unjointed.

One remedy for excessively low relative humidities indoors, as noted above, is a humidifier. A humidifier raises the relative humidity of indoor air to more comfortable levels by vaporizing water into the air, thereby increasing the vapor pressure (or mixing ratio). Usually, a humidifier consists of a wheel that continually rotates a porous belt into and out of a reservoir of water. A fan then blows air through the wetted belt, thereby humidifying the air stream.

On the other hand, in summer when the weather is hot and humid, a dehumidifier may be desirable. A dehumidifier lowers the relative humidity of indoor air to more comfortable levels by lowering the actual vapor pressure (or mixing ratio) by inducing condensation. In a standard dehumidifier, a fan draws humid air past cold refrigerated coils. The air is cooled to saturation, water vapor condenses and collects on the cold coils, and liquid water drips from the coils into a collection reservoir.

Humidity Instruments

Leonardo da Vinci, a creative genius of the fifteenth century, was probably the first to conceive of an instrument to gauge the water vapor content of air. His design was a simple balance. A small wad of dry cotton on one side of the scale was exactly balanced by a weight on the other side of the scale. As the cotton absorbed water vapor from the air, an imbalance would develop, the amount of imbalance being a measure of humidity.

Today, an instrument commonly used to measure atmospheric humidity is the **psychrometer.** It usually consists of two identical mercury-in-glass thermometers mounted side by side, as shown in Figure 6.4. The bulb of one thermometer is wrapped in a muslin wick. Readings are taken by first soaking the wick in distilled water, and then whirling both thermometers (or aerating them by a fan) until the wet-bulb thermometer reading steadies. The dry-bulb thermometer measures the ambient air temperature. Water vaporizes from the muslin wick into air streaming past the wet-bulb thermometer, and evaporative cooling lowers the wet-bulb temperature. The drier the air, the greater the evaporation, and the lower the reading will be on the wet-bulb thermometer compared with the dry-bulb reading. The temperature difference between the two thermometers, called the **wet-bulb depression,** is calibrated in terms of percent relative humidity on a psychrometric table (see inside back cover).

For example, if the ambient air temperature (dry-bulb) is 20 °C (68 °F), and the wet-bulb depression (dry-bulb reading minus wet-bulb reading) is 5 C° (9 F°), the relative humidity is 58 percent. If there is no wet-bulb depression, the air is saturated; that is, the relative humidity is 100 percent.

An instrument that is less accurate than a psychrometer, but measures atmospheric humidity more directly and conveniently, is the **hair hygrometer.** As the name implies, this device uses hair as the humidity sensor. The hair lengthens slightly as the relative humidity increases and shrinks slightly as the relative humidity drops. Usually, a sheaf of blond human hairs is linked mechanically to a pointer on a dial that is calibrated to read in percent relative humidity. Hair typically changes length by about 2.5 percent over the full range of relative humidity from 0 to 100 percent.

FIGURE 6.4

A sling psychrometer, an instrument used to measure the humidity of the air. Together with the calibration table (inside back cover), the air temperature and the difference between the air temperature and the wet-bulb temperature can be used to determine relative humidity. [Courtesy of Belfort Instrument Company]

FIGURE 6.5
A hydrograph gives a continuous trace of relative humidity variations.
[Courtesy of Qualimetrics]

A hair hygrometer may be designed to activate a pen on a clock-driven drum, as shown in Figure 6.5. This instrument, called a **hygrograph,** traces a continuous record of fluctuations in relative humidity with time.

The type of hygrometer used in radiosondes (described in Chapter 1) is based on changes in the electrical resistance of certain chemicals as they adsorb* moisture from the air. The adsorbing element may be a thin carbon coating on a glass or plastic strip. The more humid the air, the more moisture that is adsorbed, and the greater is the change in the resistance to an electric current passing through the sensing element. Variations in electrical resistance are calibrated in terms of percent relative humidity.

---

* Adsorption refers to the adhesion of a gas or soluble substance to the *surface* of some object.

Expansional Cooling and Saturation

The relative humidity is 100 percent at saturation, but it is possible for air to become supersaturated, that is, for the relative humidity to be greater than 100 percent. As the relative humidity approaches or exceeds saturation, however, condensation or deposition of water vapor becomes more and more likely. Clouds are products of condensation or deposition within the atmosphere. Hence, the probability of cloud development increases as the relative humidity approaches saturation.

What, then, causes the relative humidity to increase? Relative humidity increases

1. When air is cooled; the saturation vapor pressure declines whereas the vapor pressure remains constant.
2. When water vapor is added to the air; the vapor pressure increases whereas the saturation vapor pressure remains constant.

In this chapter we focus on the first process, and in the next chapter we give some examples of the second process.

One way in which air cools and increases in relative humidity is by expansion. In fact, expansional cooling is the principal means of cloud formation in the atmosphere. **Expansional cooling** takes place when the pressure on a volume of air drops, as it does when air ascends in the atmosphere. (Recall from Chapter 5 that atmospheric pressure declines rapidly with altitude.) As air ascends, it expands in the same way that a helium-filled balloon expands (and eventually bursts) as it drifts skyward. Now, if you have ever released air from a tire, you know that air cools as it expands. Compressed air expands as it escapes from the tire, and it is cool to the touch. Conversely, when air is compressed, its temperature rises. A familiar illustration of this phenomenon is the warming of the cylinder wall of a bicycle tire pump as air is pumped (compressed) into a tire. **Compressional warming** thus occurs when air descends within the atmosphere.

ADIABATIC PROCESSES

Expansional cooling and compressional warming within the atmosphere are examples of **adiabatic processes.** To understand adiabatic processes, think of upward- and downward-moving currents of air as composed of streams of discrete mass units, called **air parcels.** If a rising air parcel is unsaturated and is not heated or cooled from outside, the expansional cooling is dry adiabatic. In a dry adiabatic process, radiation and conduction are not effective in adding heat to or removing heat from the air parcel. As the parcel rises and expands, it pushes against the surrounding air and cools at a rate equivalent to the

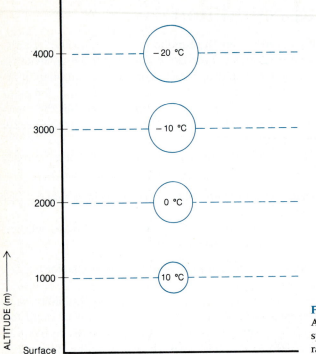

FIGURE 6.6

As an unsaturated parcel of air ascends in the atmosphere, it expands and cools at the dry adiabatic lapse rate (10 C° per 1000 m of ascent).

energy* expended. This cooling rate, termed the **dry adiabatic lapse rate,** amounts to 10 C° for every 1000 m of ascent (or 5.5 F° for every 1000 ft). The dry adiabatic cooling rate is illustrated in Figure 6.6.

An air parcel descending within the atmosphere undergoes compressional warming. Whether the parcel is unsaturated or initially saturated, the compressional warming is adiabatic if (1) warming occurs without any radiative or conductive heat exchange between the air parcel and the surrounding air, and (2) the parcel contains no liquid water or ice available to evaporate or sublimate. The atmosphere compresses the air parcel, and the parcel temperature rises 10 C° for every 1000 m of descent. (For more on energy conservation and the adiabatic process, refer to the Mathematical Note at the end of this chapter.)

Interestingly, the ascent and descent of unsaturated air parcels within the real atmosphere are usually very close to adiabatic. Hence, the dry adiabatic

---

*In expanding, the air parcel does work as it "pushes aside" the air that formerly occupied the volume into which it expanded, and work requires energy. The energy for the work of expansion is drawn from the internal molecular energy (heat) of the air parcel; hence the parcel temperature decreases.

lapse rate enables us to accurately predict the temperature change of unsaturated air as it moves upward and downward within the atmosphere.

Should rising air parcels cool to the point that the relative humidity approaches 100 percent and condensation or deposition takes place, the ascending air then no longer cools at the dry adiabatic rate (Figure 6.7). This is because latent heat that is liberated during condensation or deposition *partially* counters the expansional cooling. Consequently, an ascending saturated air parcel cools more slowly than an ascending unsaturated air parcel. Rising saturated air parcels cool at the **moist adiabatic lapse rate.**

Although the dry adiabatic lapse rate is fixed in value, the moist adiabatic lapse rate varies with temperature. As noted earlier, temperature governs the rate of vaporization of water so that the saturation vapor pressure (or saturation mixing ratio) is directly proportional to temperature. That is, warm saturated air has a higher vapor pressure (or mixing ratio) than cool saturated air. Thus, condensation or deposition in warm saturated air releases more latent heat than condensation or deposition in cool saturated air. The greater the quantity of latent heat released, the more the expansional cooling is offset, and the smaller is the moist adiabatic lapse rate. The moist adiabatic lapse rate ranges from about 4 C° per 1000 m (2.2 F° per 1000 ft) for very warm

FIGURE 6.7

Rising parcels of unsaturated air cool at the dry adiabatic lapse rate, but as the rising parcels cool, their relative humidity increases. At the level of condensation, the relative humidity is 100 percent, and the parcels are saturated. Continued lifting of the saturated parcels triggers condensation (or deposition) and the release of latent heat. The latent heat release partially offsets the adiabatic cooling so that rising saturated air parcels do not cool as rapidly as rising unsaturated air parcels.

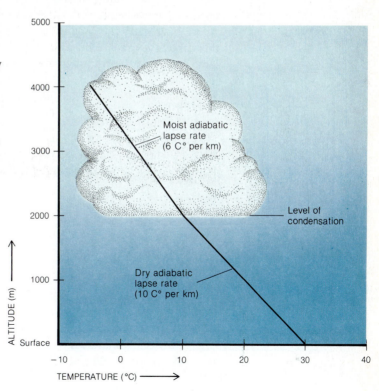

saturated air to almost 9 C° per 1000 m (5 F° per 1000 ft) for very cold saturated air. For convenience, we use an average value of 6 C° per 1000 m (3 F° per 1000 ft).

As long as an unsaturated air parcel continues to ascend, its temperature will drop and its relative humidity will approach saturation. Forces arising from density differences within the atmosphere may either enhance or suppress this vertical motion of air. The net effect depends on the stability of the atmosphere.

ATMOSPHERIC STABILITY

An air parcel is subject to buoyant forces that arise from density differences between the parcel and the surrounding (ambient) air. The warmer an air parcel is, the lower is its density. Parcels that are warmer (lighter) than the ambient air tend to rise, and parcels that are cooler (denser) than the ambient air tend to sink. An air parcel continues to rise or sink until it reaches air of equivalent temperature (or density).

We determine the **atmospheric stability** by comparing the temperature change of an ascending or descending air parcel with the temperature profile, or sounding, of the ambient air layer in which the parcel ascends or descends. As we have seen, the cooling rate of a rising air parcel depends on whether the parcel is saturated (moist adiabatic lapse rate) or unsaturated (dry adiabatic lapse rate), and the warming rate of a descending air parcel is 10 C° per 1000

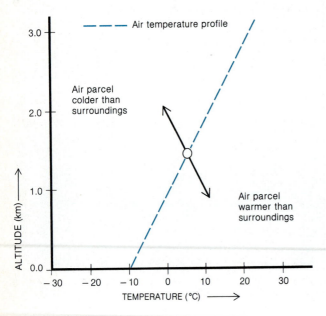

FIGURE 6.8
Upward and downward displacements of an unsaturated air parcel within stable air. The parcel is subject to a force that restores it to its original altitude. Hence, stable air inhibits vertical motion.

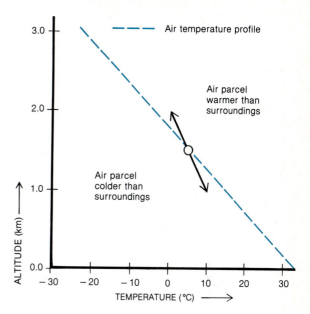

FIGURE 6.9
Upward and downward displacements of an unsaturated air parcel within unstable air. The parcel accelerates away from its original altitude. Hence, unstable air enhances vertical motion.

m. Recall from Chapter 1 that the vertical temperature profile (or sounding) of ambient air is measured by a balloon-borne radiosonde.

Within a **stable air layer,** an ascending air parcel becomes cooler than the ambient air, or a descending air parcel becomes warmer than the ambient air (Figure 6.8). For either upward or downward displacements, an air parcel in stable air is thus forced to return to its original altitude. Within an **unstable air layer,** an ascending air parcel becomes warmer than the ambient air and continues to ascend, or a descending air parcel becomes cooler than the ambient air and continues to descend (Figure 6.9).

Although lapse rates for both unsaturated and saturated air parcels are essentially fixed, the temperature profile—and hence atmospheric stability—changes significantly from season to season, from day to day, and even from hour to hour. Stability can change as a consequence of (1) local radiational heating or cooling, (2) air mass advection, or (3) large-scale ascent or descent of air.

On a clear and calm night, radiational cooling of the ground stabilizes the overlying air, whereas during the day, intense solar heating of the ground destabilizes the overlying air. This situation is illustrated in Figure 6.10. An air mass is stabilized as it travels over a colder surface (snow-covered ground, for example), and it is destabilized as it flows over a warmer surface. Whether by radiation or by advection, stability changes because the temperature profile (sounding) changes. Air is stabilized when cooling from below makes the sounding less steep, and air is destabilized when heating from below steepens the sounding.

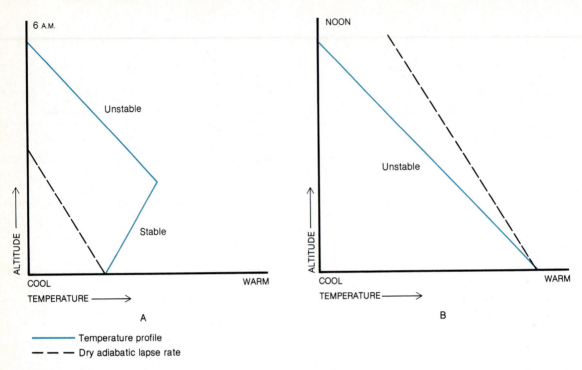

Temperature profile
Dry adiabatic lapse rate

FIGURE 6.10

(A) Nocturnal radiation cooling generates a surface temperature inversion by dawn.
(B) By noon, however, bright sunshine has warmed the ground and destabilized the
overlying air.

Generally, an air layer becomes more stable when it descends (subsides)
and less stable when it ascends. When air subsides, the upper portion undergoes more compressional warming than does the lower portion so that the
sounding becomes less steep and air layer stability increases. The lower portion
of an ascending air layer is usually more humid than the upper portion and
so achieves saturation sooner. Consequent release of latent heat in the lower
portion of the air layer steepens the sounding and reduces the stability.

A further complication is that stability can change with altitude. For
example, as shown in Figure 6.11, a stable air layer may be sandwiched
between unstable air layers. Different types of air mass advection at different
levels within the troposphere are often responsible for such variations in
stability.

Figure 6.12 summarizes a number of possible atmospheric stability conditions, along with the dry adiabatic and average moist adiabatic lapse rates.
If the sounding indicates that the temperature of the ambient air is dropping
more rapidly with altitude than the dry adiabatic lapse rate (that is, more than
10 C° per 1000 m), then the ambient air is unstable for both saturated and
unsaturated air parcels. This situation is called **absolute instability.** If the

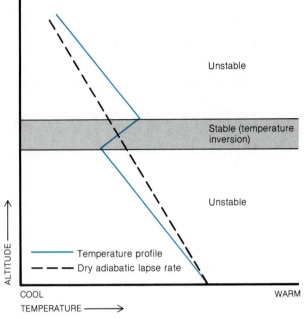

Unstable

Stable (temperature inversion)

Unstable

ALTITUDE

——— Temperature profile
– – – Dry adiabatic lapse rate

COOL   WARM
TEMPERATURE ———→

FIGURE 6.11

Air stability can change with altitude. In this example, a stable air layer (a temperature inversion) is sandwiched between unstable air layers.

FIGURE 6.12

Air stability is determined by comparing vertical temperature profiles (solid lines) with the dry adiabatic lapse rate in the case of unsaturated air parcels and with the moist adiabatic lapse rate in the case of saturated air parcels. Here a number of possible stability cases are illustrated.

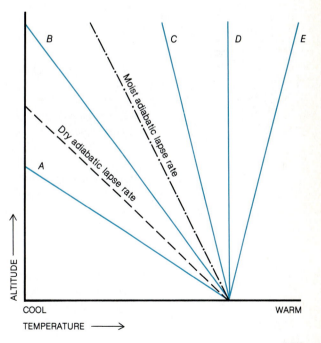

B   C   D   E

Moist adiabatic lapse rate

Dry adiabatic lapse rate

A

ALTITUDE

COOL   WARM
TEMPERATURE ———→

A   Absolute instability
B   Conditional stability
C   Absolute stability (lapse)
D   Absolute stability (isothermal)
E   Absolute stability (inversion)

145

sounding is between the dry adiabatic and moist adiabatic lapse rates, **conditional stability** prevails. That is, the air layer is stable for unsaturated air parcels and unstable for saturated air parcels. An air layer is stable for both saturated and unsaturated air parcels when the sounding indicates any of the following conditions.

1. The temperature of the ambient air drops more slowly with altitude than the moist adiabatic lapse rate.
2. The temperature is constant with altitude (called **isothermal**).
3. The temperature increases with altitude (a **temperature inversion**).

Any one of these three categories of temperature profiles is a case of **absolute stability.**

What happens when a sounding coincides with either the dry or moist adiabatic lapse rate? A sounding that equals the dry adiabatic lapse rate is neutral for unsaturated air parcels and unstable for saturated air parcels. A sounding that is the same as the moist adiabatic lapse rate is neutral for saturated air parcels and stable for unsaturated air parcels. Within a **neutral air layer,** a rising or descending air parcel always has the temperature of its surroundings. Hence, a neutral air layer neither impedes nor spurs vertical motion.

It is evident that atmospheric stability influences weather by affecting the vertical motion of air. Stable air suppresses vertical motion, and unstable air enhances vertical motion, convection, expansional cooling, and cloud development. Because stability also affects the rate at which polluted air mixes with clean air, stability must be considered in assessing air pollution potential. This point is examined in Chapter 16.

Lifting Processes

Air expands and cools as a consequence of orographic lifting, frontal uplift, and convection currents. If expansional cooling is great enough, the relative humidity approaches 100 percent and clouds develop. As described in the Special Topic, "Clouds by Mixing," saturation and cloud formation are also possible when unsaturated air masses (or air layers) mix.

Orographic lifting occurs when air is forced up the slopes of hills and mountains. A mountain range that is oriented perpendicular to prevailing winds forms a natural barrier that results in more cloudiness and heavier precipitation on one side of the range than on the other side. As air is forced up the windward slopes (the side facing the wind), expansional cooling increases the relative humidity. With sufficient lifting (that is, cooling), the relative humidity approaches 100 percent, clouds form, and rain or snow, called **orographic precipitation,** may develop. On the leeward slopes (the

Clouds by Mixing

Clouds (or fog) develop as the consequence of cooling caused by uplift of air or, as we will see in Chapter 7, by contact with a surface that has been chilled by emission of IR radiation. A third process of cloud formation involves mixing of air masses (or air layers within the same air mass) that differ in temperature and vapor pressure. Surprisingly, a cloud may form even though the two air masses (or air layers) are unsaturated to begin with.

As emphasized earlier, saturation vapor pressure increases rapidly with rising temperature. This relationship is displayed schematically in Figure 1 where tem-

perature is plotted on the horizontal axis and vapor pressure is plotted on the vertical axis. The average temperature and average vapor pressure of a specific air mass (or air layer) plot as a single point on that diagram. An air mass that plots below the saturation vapor pressure curve is unsaturated, on the curve is saturated, and above the curve is supersaturated.

Suppose that two different unsaturated air masses plot as points A and B on the diagram. It is reasonable to assume that mixing the two air masses would produce a new air mass having properties (average temperature and vapor pressure) that plot somewhere along a straight line connecting points A and B. The precise location of that point (the mixture) along the line depends on the relative volume of the two air masses. Note that in this case, the mixture of the two air masses is unsaturated so that no cloud forms.

Consider, however, two other unsaturated air masses, C and D, one cold and dry, the other warm and humid. In this case, the straight line linking C and D intersects the saturation vapor pressure curve. This means that the new air mass resulting from the mixture of air masses C and D is saturated and a cloud forms.

This description of cloud formation by mixing provides insight as to the development of aircraft contrails (Figure 16.10). Jet engine exhaust mixes with the ambient air and may form a *contrail* (condensation trail). Heat and water are combustion products of jet engines so that the exhaust is hot and humid. If the ambient (surrounding) air has the right combination of relatively low vapor pressure and low temperature, the mixture is saturated and a contrail forms. Such conditions are most likely at high altitudes.

FIGURE 1

Variation of the saturation vapor pressure with temperature. Air masses that plot as points above the curve (in the shaded area) are supersaturated, on the curve are saturated, and below the curve are unsaturated.

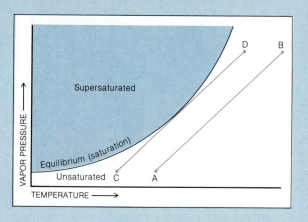

downwind side of the mountain range), air descends, compressional warming lowers the relative humidity, and clouds vaporize. The mountain range thus establishes two contrasting climatic zones: a moist climate on the windward slopes and a dry climate on the leeward slopes (Figure 6.13). Dry conditions often extend many hundreds of kilometers to the lee of prominent mountain ranges; this region is known as a **rain shadow.**

FIGURE 6.13
A mountain range that intercepts a flow of humid air may induce a cloudy, rainy climate on its windward slopes and a dry climate on its leeward slopes and beyond.

An orographically induced contrast in precipitation is especially apparent from west to east across Washington and Oregon, where the north–south Cascade Mountain Range intercepts the prevailing humid airflow from off the Pacific Ocean. The result is exceptionally rainy conditions in the western portions of these states and arid conditions in the eastern portions. Figure 6.14 shows two contrasting Pacific Northwest landscapes that underscore the

FIGURE 6.14
A rainforest in the Olympic National Park of northwestern Washington (A) is in marked contrast to the cold desert vegetation of eastern Oregon (B). Mountain ranges between the two areas account for the striking disparity in climate. [Photograph A from the National Park Service; photograph B from the Oregon Department of Transportation]

A

B

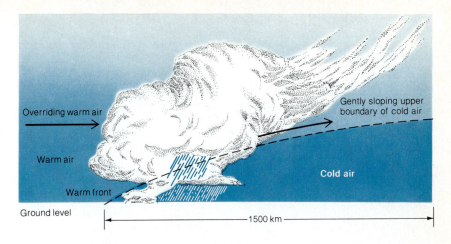

FIGURE 6.15
Warm, light air displaces cooler, denser air by overriding the cool air along a gently sloping frontal surface. The rising warm air expands and cools to the saturation point, and clouds develop. [Modified from R. A. Anthes et al., *The Atmosphere*, 3rd ed. Columbus, OH: Charles E. Merrill, 1981, p. 207.]

disparity in precipitation. Markedly different plant and animal communities are found on opposite slopes of the mountain barrier. The contrast in precipitation also has a direct impact on domestic water supply, the type of crops grown, and the type of shelter that must be built. For example, Denver, located on the leeward side of the front range of the Rocky Mountains, must import some of its water from the wet western side of the range via tunnels.

Clouds and precipitation are often triggered by **frontal uplift,** which occurs when contrasting air masses meet. Because a warm and humid air mass is less dense than a cold and dry air mass, the advancing warm air rides up and over the cold air (Figure 6.15). The leading edge of the warm air at the Earth's surface is called a **warm front.** On the other hand, denser, cold and dry air displaces warm and humid air by pushing under it (Figure 6.16).

FIGURE 6.16
Cool air displaces lighter warm air by sliding under the warm air, thereby forcing the warm air to rise along a steeply sloping frontal surface. The warm air expands and eventually cools to saturation, and clouds develop. [Modified from R. A. Anthes et al., *The Atmosphere*, 3rd ed. Columbus, OH: Charles E. Merrill, 1981, p. 210.]

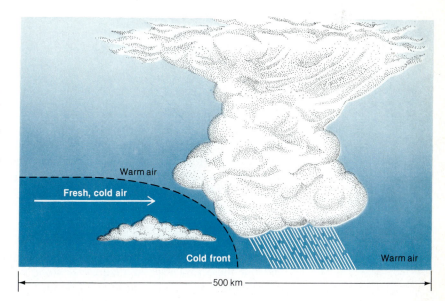

FIGURE 6.17
The sky is partly covered with cumulus clouds. Convective currents are upward where there are clouds and convective currents are downward where the sky is cloud free. [Photograph by J. M. Moran]

The leading edge of cold air at the earth's surface is called a **cold front.** The net effect of the replacement of one air mass by another air mass is the lifting of air, which, in turn, leads to expansional cooling, cloud development, and perhaps rainfall or snowfall. We have much more to say about fronts and associated weather in Chapter 11.

Convection can also lead to the formation of clouds and precipitation. It is an important means of heat transfer (Chapter 4) that develops when the sun heats the Earth's surface, and the warm surface in turn heats the air in contact with it. Neighboring cooler, denser air sweeps down and under the warmer, lighter air forcing the warm air to rise. As the warm air rises, it expands and cools. Eventually, the rising air becomes so cool and dense that it sinks back toward the surface, thus completing the convective cell circulation. Clouds may form where convection currents are surging upward, but clouds dissipate where the convection currents are downward (Figure 6.17). In general, the higher that convection currents surge into the atmosphere, the greater is the expansional cooling, and the more likely is the occurrence of clouds and precipitation.

Conclusions

The atmosphere is one of many reservoirs of water in the global hydrologic cycle. Water enters the atmosphere via vaporization processes (evaporation, transpiration, and sublimation), and the concentration of water vapor in air is described in terms of vapor pressure, mixing ratio, and relative humidity. Through expansional cooling, the relative humidity approaches saturation, and water vapor condenses (or deposits) into tiny water droplets (or ice crystals) that we see as clouds. Radiational cooling and local increases in water vapor concentration can also raise the relative humidity to saturation. We explore these processes in Chapter 7, along with the mechanics of cloud development and the classification of clouds.

Energy Conservation and the Dry Adiabatic Process

Energy cannot be created or destroyed, but it can change from one form to another. Another way to express this *law of energy conservation* is to observe that in any physical or biological system we can account for all the original energy, regardless of the energy transformations that take place. Application of this law to atmospheric processes provides us with valuable information about the workings of weather.

Heat gained by an air parcel is either added to the parcel's store of internal energy or is used to do work on the parcel. Conversely, heat released by an air parcel is either subtracted from the internal energy of the parcel or is the consequence of the parcel's work on its surroundings. In order to understand this concept, let us first examine separately (1) internal energy and (2) work done on or by an air parcel. We will then combine the two in one expression of energy conservation.

Internal energy refers to the total kinetic molecular energy of the mixture of gases composing an air parcel, that is, *heat*. A change in the internal energy, or heat energy (ΔQ), of an air parcel is therefore proportional to a change in the temperature (ΔT) of the parcel. That is,

$$\Delta Q \propto \Delta T$$

As we saw in Chapter 3, however, internal energy and temperature are related through the specific heat (C). So,

$$\Delta Q = C\Delta T$$

Actually, the specific heat of dry air can be evaluated at constant pressure ($C_p = 1005$ J per kg per K) or, as in this case, at constant volume ($C_v = 718$ J per kg per K). Hence,

$$\Delta Q = C_v\Delta T$$

Work done on or by an air parcel consists of compression or expansion of the air parcel. An illustration

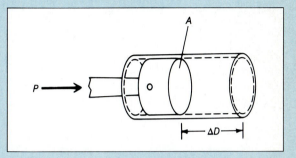

FIGURE 1

A piston is used to compress a sample of air or to allow the air sample to expand. Pressure (P) is applied over a distance (ΔD). The piston has cross-sectional area A.

will demonstrate this concept. Suppose we have a sealed cylindrical container of air, as in Figure 1, equipped at one end with a piston that can compress the air sample or allow the air sample to expand. If we compress the air, we use energy to do work on the air sample. If we then release the piston, the air sample expands and works against (pushes) the piston. When the air is compressed, work is therefore done on it, but when the air expands, the air does work on its surroundings. Energy is either supplied (during compression) or released (during expansion).

We can calculate the work (or energy) involved in the compression or expansion of air by referring to the piston example. The energy required to compress or expand the air sample is simply the product of the required force (F) times the distance over which the force is applied. Recall that pressure is a force per unit area. If we apply a pressure (P) on an air sample by moving a piston of fixed cross-sectional area (A) a variable distance (ΔD), the volume (ΔV) of the air sample changes.

$$\Delta V = A\Delta D$$

The energy required (ΔQ) for the volume change would be

$$\Delta Q = F\Delta D$$

or

$$\Delta Q = PA\Delta D$$

or

$$\Delta Q = P\Delta V$$

To this point, we have considered internal energy and the energy involved in the compression and expansion of air separately. We now combine the two components in one equation.

$$\Delta Q = C_v\Delta T + P\Delta V$$

The heat flow into and out of an air parcel is thus accounted for by changes in internal energy, by the work of expansion or compression of air, or by both.

With this statement of the law of energy conservation, we can now explore the adiabatic process. According to the *adiabatic assumption*, there is no net heat flow through the imaginary walls of an air parcel. That is, in the previous equation,

$$\Delta Q = 0$$

Hence,

$$C_v\Delta T + P\Delta V = 0$$

This means that when an air parcel undergoes an adiabatic process, a temperature change (ΔT) accompanies a volume change (ΔV). The temperature of an air parcel drops when its volume increases as when the air parcel ascends within the atmosphere. Conversely, the temperature of an air parcel rises when its volume decreases as when the air parcel descends within the atmosphere. Ascending unsaturated air expands and cools adiabatically at a constant rate of 10 C° per 1000 m of ascent. This is called the *dry adiabatic lapse rate*. Descending air is compressed and warms at a rate of 10 C° per 1000 m of descent.

Adiabatic processes can be displayed graphically on a *Stüve thermodynamic diagram* like the one in Figure 2. The Stüve diagram presents the relationships among several atmospheric variables. Horizontal lines are *isobars*, lines of constant air pressure, and vertical lines are *isotherms*, lines of constant temperature. Straight black sloping lines are *dry adiabats*, the temperature change of an unsaturated air parcel that is subjected to adiabatic expansion or compression. Dry adiabats are labeled in kelvins (K). Curved dashed lines are *moist adiabats*, the temperature change of a saturated air parcel that is subjected to expansional cooling. Blue sloping lines are lines of equal saturation mixing ratio in grams of water vapor per kilogram of dry air.

As an illustration of the usefulness of a Stüve diagram, consider an unsaturated air parcel that undergoes dry adiabatic expansion as a consequence of uplift (along a front, for example). The air parcel has a temperature of 20 °C (68 °F), a pressure of 1000 mb, and a relative humidity of 50 percent. These initial conditions are plotted as point *A* in Figure 2. On the diagram, we see that the saturation mixing ratio is about 16 g per kg, so that a relative humidity of 50 percent means that the mixing ratio of the parcel is 8 g per kg. The parcel is then lifted so that it cools dry adiabatically until it becomes saturated. The air parcel temperature thus follows the 293 K dry adiabat to point *B*, where the actual mixing ratio equals the saturation mixing ratio (8 g per kg) and the relative humidity is 100 percent. Hence, saturation is achieved when the air

Summary Statements

The hydrologic cycle is the ceaseless circulation of water among the Earth's oceanic, atmospheric, and terrestrial reservoirs.

Vapor pressure and mixing ratio are two direct measures of the water vapor concentration in air. Relative humidity indicates how close an air sample is to saturation with respect to water vapor. Relative humidity is temperature dependent.

Saturation vapor pressure and saturation mixing ratio are uniquely related to air temperature. Both are directly proportional to air temperature because higher temperatures increase the rate of vaporization of water.

The relative humidity of air increases when unsaturated air is cooled or when water vapor is added to the air.

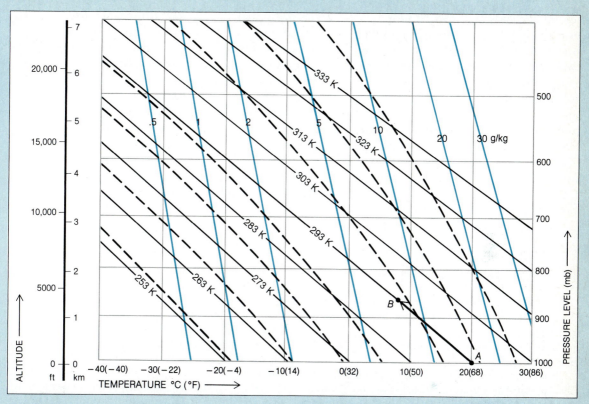

FIGURE 2

On a Stüve thermodynamic diagram, an air parcel at point *A* is subjected to a dry adiabatic expansion to point *B*.

parcel reaches an altitude of about 1500 m (4500 ft), where the parcel temperature is 8 °C (46 °F). At this altitude, water vapor in the air parcel begins to con-dense and a cloud forms. With continued ascent above this altitude, the saturated air parcel follows a moist adiabat.

In an adiabatic process, no net heat exchange occurs between an air parcel and the surrounding air. Rising air parcels are cooled by expansion, and de-scending air parcels are warmed by compression.

Atmospheric stability is determined by compar-ing the temperature of an air parcel moving verti-cally (up or down) through a layer of air with the temperature profile of the air layer. Stable air inhib-its vertical motion; unstable air enhances vertical motion.

Expansional cooling occurs through orographic lifting, frontal uplift, and convection. By these pro-cesses, the relative humidity increases, perhaps to the point of cloud development.

Key Words

hydrologic cycle	mixing ratio	compressional	conditional stability
evaporation	specific humidity	warming	isothermal
condensation	absolute humidity	adiabatic process	temperature inversion
transpiration	saturation vapor	air parcels	absolute stability
evapotranspiration	pressure	dry adiabatic lapse	neutral air layer
sublimation	saturation mixing	rate	orographic lifting
deposition	ratio	moist adiabatic lapse	orographic
precipitation	relative humidity	rate	precipitation
water budget	psychrometer	atmospheric stability	rain shadow
drainage basin	wet-bulb depression	stable air layer	frontal uplift
watershed	hair hygrometer	unstable air layer	warm front
Dalton's law	hygrograph	absolute instability	cold front
vapor pressure	expansional cooling		

Review Questions

1. Distinguish among evaporation, transpiration, and sublimation.

2. Provide some examples of sublimation.

3. What is the difference between condensation and deposition?

4. Explain how heat energy is involved in the phase changes of water.

5. How does the hydrologic cycle purify water?

6. It follows from the global water budget that there must be a net flow of water from the continents to the oceans. Why?

7. Distinguish between air pressure and vapor pressure. How do they compare in magnitude?

8. How do temperature changes influence (a) the saturation vapor pressure, (b) the saturation mixing ratio, and (c) the relative humidity?

9. Under what condition is the mixing ratio equal to the saturation mixing ratio?

10. Why does the relative humidity usually fall between sunrise and early afternoon on a clear and calm day?

11. In localities where winters are cold, some central home heating systems are equipped with humidifiers. Why?

12. Describe how humidity is measured using a psychrometer.

13. List several ways whereby the relative humidity of air is increased.

14. Describe the change in temperature as a sample of air alternately ascends and descends within the atmosphere.

15. What is an adiabatic process and how does it occur in the atmosphere?

16. Rising parcels of saturated air do not cool as rapidly as rising parcels of unsaturated air. Please explain.

17. How does the stability of the ambient air influence upward and downward movements of unsaturated air parcels?

18. How does the topography of the land influence cloud and precipitation development?

19. Why are clouds and precipitation often associated with fronts?

20. Both clear air and cloudy air are associated with convection within the atmosphere. Please explain.

Points to Ponder

1. Does a relative humidity of 25 percent measured on a cold day in January mean the same as a relative humidity of 25 percent measured on a hot day in July? Elaborate on your answer.

2. With intense radiational cooling during the early morning hours, the air temperature is observed to drop rapidly until frost forms. Thereafter, until sunrise, the air temperature either remains steady or falls very slowly. Why the change in air temperature behavior?

3. Determine the saturation mixing ratio of an air sample having a relative humidity of 25 percent and a mixing ratio of 3 g water vapor per kg dry air.

4. Determine the relative humidity of an air sample having a temperature of 20 °C (68 °F) and a vapor pressure of 6 mb.

5. Speculate on the significance of conditional stability for convection currents.

6. How might air mass advection influence air stability?

7. Speculate on how a freshly fallen layer of snow might influence the stability of the overlying air.

Projects

1. Identify some everyday examples of expansional cooling and compressional warming.

2. Check with a home appliance store to determine how a humidifier and a dehumidifier work.

3. For one or two full days keep track of the hourly temperature and relative humidity readings. Speculate on whether changes in relative humidity are due to local radiative effects (heating or cooling) or air mass advection.

Selected Readings

Bohren, C. F., and G. M. Brown. "Genies in Jars, Clouds in Bottles, and a Bucket with a Hole in It." *Weatherwise* 35 (1982):86–89. Includes a lucid explanation of the concept of saturation vapor pressure.

Bryan, K. "The Ocean Heat Balance." *Oceanus* 21 (1978):18–26. Presents a detailed description of heat exchanges at the ocean–atmosphere interface.

Forrester, F. H. "An Inventory of the World's Water." *Weatherwise* 38 (1985):84–105. Describes the various reservoirs in the global hydrologic cycle.

Kalkstein, L. S., and K. M. Valimont. "An Evaluation of Summer Discomfort in the United States Using a Relative Climatological Index." *Bulletin of the American Meteorological Society* 67 (1986):842–848. Uses a regional approach to the climatology of stress caused by a combination of high temperature and high humidity.

Leopold, L. B. *Water: A Primer.* San Francisco: W. H. Freeman, 1974. 166 pp. Presents a concise and well-illustrated treatment of the global water cycle.

Pepi, J. W. "The Summer Simmer Index." *Weatherwise* 40 (1987):143–145. Gives a critique of existing discomfort indexes and a proposal for a new one.

*I am the daughter of earth
    and water,
And the nursling of the sky:
I pass through the pores of the
    ocean and shores;
I change, but I cannot die.*

Percy Bysshe Shelly
THE CLOUD

7

Dew, Frost, Fog, and Clouds

Clouds appear in a myriad of forms. They are composed of tiny water droplets or ice crystals or a combination of the two. [NOAA photograph]

157

Clouds whisk across the sky in ever-changing patterns of white and gray; fog lends an eerie silence to a dreary day; dew and frost glisten in the morning sun. Clouds, fog, dew, and frost are all products of condensation or deposition of atmospheric water vapor. Most clouds are the consequence of saturation brought about by uplift and adiabatic cooling of air, but dew, frost, and most fogs develop when the lowest layers of air are chilled to saturation via other processes. In this chapter, we describe the formation of dew, frost, and fog and the development and classification of clouds.

Low-Level Saturation Processes

Air that is in contact with the Earth's surface can become saturated if its temperature is lowered sufficiently. Such cooling decreases the saturation vapor pressure (or saturation mixing ratio) and thus increases the relative humidity. As the relative humidity approaches 100 percent, dew, frost, or fog may form.

DEW AND FROST

Dew and frost are primarily the consequence of nocturnal **radiational cooling.** At night, an object on the Earth's surface (a plant leaf or automobile windshield, for example) emits terrestrial radiation (infrared) to the atmosphere and eventually off to space, and thereby the object cools. At the same time, the atmosphere emits terrestrial radiation back to the Earth's surface where some is absorbed by the object, and thereby the object warms. On a clear night (minimal greenhouse effect), the object emits more radiation than it receives from the atmosphere. Consequently, the surface of the object becomes cooler than the air adjacent to it and heat is conducted to the object. With sufficient cooling, the air in immediate contact with the object becomes saturated. If the air temperature remains above the freezing point, water vapor may condense on the object as **dew** (Figure 7.1); if the air temperature drops below freezing, water vapor may deposit as **frost** (Figure 7.2).

Note that dew and frost are not forms of precipitation, because they do not "fall" from clouds but, rather, develop in place on exposed surfaces. A similar phenomenon occurs when water droplets appear on the outside of a cold glass of soda on a hot summer day. The "sweat" of the glass is actually dew.

The temperature to which air must be cooled, at constant pressure, to reach saturation (relative to liquid water) is called the **dew point.** In order

FIGURE 7.1
If winds are calm, then nocturnal radiational cooling may bring a very shallow layer of air to saturation. At temperatures above freezing, water vapor condenses as dew on exposed surfaces. [Photograph by Mike Brisson]

FIGURE 7.2
If winds are calm, then nocturnal radiational cooling may bring a very shallow layer of air to saturation. At temperatures below freezing, water vapor deposits as frost on exposed surfaces. [Visuals Unlimited photograph by M. Gabridge]

for water vapor to condense as dew on the surface of an object, the temperature of that surface must drop *below* the dew point. The dew point is also a measure of the air's water vapor content. The higher the dew point, the greater is the water vapor concentration. It also follows that when the difference between the actual air temperature and the dew point is small, the relative humidity is high. Furthermore, the dew point is an index of human comfort. Although the relative humidity is an important consideration, as a rule, many people experience discomfort when the dew point rises above about 20 °C (68 °F).

Note that the dew point should not be confused with the wet-bulb temperature (described in Chapter 6); they are not the same. Recall that the wet-bulb temperature is determined by inducing evaporative cooling. Adding water vapor to the air raises the temperature at which dew will form. Hence, except at saturation, the wet-bulb temperature is higher than the dew point. At saturation, the dew-point, wet-bulb, and ambient air temperatures are the same.

The dew point can be obtained from measurements of the dry-bulb (ambient air) temperature and the wet-bulb depression (see the inside back cover). For example, if the dry-bulb temperature is 20 °C (68 °F) and the wet-bulb depression is 5 C° (9 F°), then the dew point is 11.6 °C (53 °F). Note that in this case, the wet-bulb temperature is 15 °C (59 °F), some 3.4 C° (6 F°) higher than the dew point.

When cooling at constant pressure produces saturation at temperatures at or below 0 °C (32 °F), the temperature is called the **frost point.** Water vapor deposits as frost on the surface of an object if the temperature of that surface falls *below* the frost point. Frost occurs in a variety of forms: delicate feathery patterns of crystals may develop on a windowpane during a cold winter night, or fernlike crystals of **hoarfrost** may grow to a length of several centimeters on the twigs and branches of trees (Figure 7.3).

When neither cold air advection nor warm air advection is expected, the dew point (or frost point) sometimes can be used to predict the next morning's minimum air temperature. Suppose, for example, it is autumn in New England and weather forecasters are calling for clear nighttime skies and calm winds, conditions ideal for rapid nocturnal radiational cooling. Gardeners know that all the ingredients are present for chilly temperatures by early morning, and they want to know whether they should go to the trouble of protecting their freeze-sensitive plants. As a general rule, the late afternoon's dew point is a reasonable estimate of the next morning's low temperature: if the dew point is near or below 0 °C (32 °F), freeze protection is advisable.

What is the physical basis for this rule? In response to nocturnal radiational cooling, the air temperature drops continually until the relative humidity nears 100 percent and condensation or deposition occurs (producing dew, frost, or fog). The latent heat released during condensation or deposition offsets

FIGURE 7.3
As a consequence of deposition of water vapor from a layer of humid, cold air, crystals of hoarfrost may grow to several centimeters in length on the twigs and branches of trees. [Photograph by Mike Brisson]

the radiational cooling to some extent so that the air temperature tends to stabilize near the dew point or frost point. Many other factors, however, may complicate this simple rule. For one, the length of the night is an important control of the amount of radiational cooling. Summer nights may be too short for radiational cooling sufficient to lower the air temperature to the dew point, particularly if the air is dry (low dew point).

It is a popular misconception that Florida citrus growers fear a winter frost. What they actually fear is a solid freeze, that is, a cold snap that freezes the water in plant tissues and causes potentially lethal damage to the plant. For more information on freeze protection for plants, see the Special Topic, "Freeze Control." It is also a popular misconception that frost causes leaves on deciduous trees to change color in autumn. For more on this subject, refer to the Special Topic, "Jack Frost and Autumn Color."

Freeze Control

One of the most effective strategies to protect crops against late spring or early fall freezes is to avoid cultivating sites that are on the receiving end of *cold air drainage*. In such freeze-prone areas as valley bottoms and marshes, the growing season may be several weeks shorter than that on surrounding highlands. For this reason, orchards and vineyards are often situated on hill slopes rather than on valley floors. Even site selection on hill slopes must be made carefully. Obstructions such as hedgerows, roads, or railroad embankments can impede the downslope drainage of cold air, and thus crops immediately upslope from the barrier may be damaged. To avoid this problem, growers construct channels through the barrier so that cold air can continue to drain downslope.

Based on an understanding of the conditions that favor extreme nocturnal radiational cooling, several other strategies have been devised to reduce the incidence of radiation freezes. One factor that contributes to radiation freezes is the absence of clouds. To protect crops on clear nights, growers can create their own clouds to intercept the outgoing infrared radiation. Clouds absorb infrared radiation emitted by the ground and vegetation, and subsequently reemit a portion of that radiation back to the crop. It is widely believed that a smoke cloud emanating from smudge pots or oil burners inhibits freezing temperatures by producing a greenhouse effect. However, smoke particles are actually nearly transparent to infrared radiation so that the greenhouse effect is negligible. On the other hand, a mist cloud formed by a fine water spray does produce a greenhouse effect and can provide crops with some protection. (Recall from Chapter 2 that water strongly absorbs infrared radiation.)

For small plants, other types of radiation screens are effective. For example, plastic "hot caps" can be placed over plants to create a protective microscale climate around them (Figure 1). During the day, the sun warms the soil and plants. If hot caps are placed over plants in the late afternoon, the heat gained during

the day is better conserved at night, reducing the chances of frost formation on the plants. For somewhat larger plants, other radiation screens, such as wooden slats or cheesecloth, can restrict exposure to clear night skies without significantly blocking the daytime solar radiation needed for crop growth.

A temperature inversion often develops within the lowest air layer as a result of nocturnal radiational cooling. Whereas air temperatures at ground level may fall below freezing, air temperatures near the top of the inversion (15 m or so above the ground) may be several degrees above freezing. The stability of the inversion layer prevents the warmer air aloft from mixing with the cooler air at ground level. A logical approach, then, is to use large motor-driven fans or propellers mounted on towers to circulate the warmer air aloft down to ground level. This is standard practice in citrus groves (Figure 2) where the economic value of the crop justifies the high capital investment in freeze protection equipment.

Crops are sometimes protected from freezing by spraying them with a fine water mist when the temperature of plant tissues drops to 0 °C (32 °F). Our initial reaction may be to question how a coating of ice can help the plants survive. Although water on the plant surface is at 0 °C, the latent heat of fusion released during the phase change from liquid water to ice helps stabilize plant temperatures and inhibits tissue damage. Furthermore, most plants are not damaged until their tissue temperature drops to −1 to −2 °C (28 to 30 °F). Nevertheless, this procedure does require careful monitoring. As long as the sprinkling and freezing continue, the temperature of the ice remains at about 0 °C. If, however, the sprinkling is discontinued before ambient air temperatures rise high enough to melt the ice, then heat is conducted from the leaves to the ice, and the leaf temperature drops to potentially lethal levels. In addition, care must be taken that the ice burden does not become so great that the plants are damaged by excess weight. For this reason, the sprinkling method is most suitable for low-growing vegetable crops such as cucumbers and strawberries.

Spraying reduces the threat of freeze damage by adding latent heat. Sensible heat can also be added directly through fuel combustion in heaters. In addition, heaters emit infrared radiation. Placing heaters at numerous sites throughout an orchard can raise both air and plant temperatures by several degrees. Heaters are most beneficial for those plants directly exposed to the warm plume of air, so using many small heaters is considerably more effective than using a few large ones.

These freeze-prevention strategies are not always sufficient. For three days around Christmas 1983, an arctic air mass surged into Florida dropping temperatures to well below freezing through much of the state. Damage to the citrus crop exceeded $1 billion. Again, in mid-January 1985, another cold snap of comparable severity caused further damage to surviving citrus trees. Losses from this double blow reduced citrus-producing acreage by almost 90 percent in Lake County, Florida, formerly the state's second largest citrus-producing county. Subsequently, a number of growers replanted, and the region was recovering when another deep freeze over Christmas weekend 1989 again ruined much of central Florida's citrus crop. This most recent blow may well mean an end to the citrus industry in the northern reaches of the Florida citrus belt.

FIGURE 2
Ten-meter-high wind machines pull warmer air down among the orange trees during cold nights in the San Joaquin Valley, California. [U.S. Department of Agriculture photograph]

Jack Frost and Autumn Color

As autumn approaches, leaves on many plants change from their familiar green of spring and summer to brilliant hues of purple, yellow, orange, and red (Plate 15). Autumn colors lure millions of people to the countryside to enjoy one of nature's most spectacular shows. How does nature put on this show of autumn color? Contrary to popular opinion, frosts and freezes have very little to do with the color of autumn leaves. Weather does play a role, but in order to understand that role, we must first consider the chemical changes that take place within leaves during the growing season.

In spring and summer, leaves are green because large quantities of green pigments called *chlorophylls* are present in leaf cells. Although chlorophyll molecules break down during the normal metabolic functions of a leaf and thus are continually used up, they also are continually manufactured. Consequently, the chlorophyll content of leaves remains high and leaves stay green. With the coming of autumn, however, chlorophyll production slows and the leaf's chlorophyll content declines, thereby gradually unmasking the yellow, brown, and orange pigments that are also in the leaves throughout the growing season. These pigments, which are called *carotenoids*, are common in many living things such as carrots, bananas, and canaries. In autumn, carotenoids tint the leaves of such trees as maples, ashes, birches, and cottonwoods.

The leaves of some plants turn red or purple, hues caused by another group of pigments called *anthocyanins*. Unlike the carotenoids, these pigments are not normally manufactured by leaves until late in the growing season. Anthocyanins are responsible for the color of such fruits as cranberries, purple grapes, and

FOG

Fog is a visibility-restricting suspension of tiny water droplets or ice crystals in an air layer next to the Earth's surface. Simply put, fog is a cloud in contact with the ground. By international convention, fog is defined as restricting visibility to 1000 m (0.62 mi) or less;* otherwise the suspension is called **mist.** (The popular definition of mist is a light drizzle.) Fog may develop when air becomes saturated through radiational cooling, advective cooling, the addition of water vapor, or expansional cooling.

With a clear night sky, light winds, and an air mass that is humid near the ground and relatively dry aloft, radiational cooling may cause the air layer near the ground to approach saturation. When these conditions occur, a cloud develops. A ground-level cloud formed in this way is called **radiation fog** (Figure 7.4). High humidity at low levels within the air mass is usually due to evaporation of water from a moist surface. Hence, radiation fogs are most common in marshy areas or where the soil has been saturated by a recent period of heavy rainfall or rapid snowmelt.

---

* For aviation purposes, the criterion for fog is a visibility restriction of 10 km (6 mi) or less.

strawberries. In autumn, anthocyanins are also produced in the leaves of such trees as oaks, maples, dogwoods, and persimmon. Often the colors of carotenoids and anthocyanins combine to deepen oranges and produce the fiery reds and bronzes that are typical of the autumn leaves of many trees.

What then is the role of weather in autumn colors? The leaves of deciduous plants cannot survive the dryness of cold winter air because they would lose much more water to the dry air than their roots could absorb from the frozen soil. Hence, the shorter daylengths that accompany the coming of autumn signal plants to drop their leaves as they become dormant for winter. As part of this preparatory process, the rate of chlorophyll production declines, the formerly hidden carotenoids appear, and anthocyanins are manufactured. Thus, leaves turn color.

Although these processes occur every fall, personal experience tells us that colors are more brilliant in some years than in others. Why the difference? The most brilliant leaf color usually develops when days are sunny and cool and nights are chilly but not subfreezing. For many years, it was assumed that bright sunlight triggered the synthesis of sugars in leaves and that frosty nights slowed the transport of these sugars out of the leaf. The leaf uses sugars to manufacture anthocyanins. One problem with this explanation is that as the aging leaf loses its chlorophyll, it also loses its ability to produce sugars. Scientists now know that sunlight promotes the synthesis of anthocyanins. As for the frosty nights, they probably have little to do with the brilliance of leaf color. Rather, sunny days are often followed by clear nights which enhance radiational cooling and lead to chilly air temperatures.

FIGURE 7.4
Radiation fog develops as a consequence of extreme radiational cooling of a layer of humid air on a night when there is some air movement. [Photograph by Mike Brisson]

Note that light winds rather than calm conditions favor development of radiation fog. Light winds produce a slight mixing that transfers heat to the cold ground from throughout the layer of humid air overlying the ground. Consequently, the entire air layer is chilled to below the dew point. If winds are calm, however, there is no mixing and heat transfer is by conduction alone. Because air is a poor conductor of heat, only a very thin layer of air immediately next to the ground is cooled to saturation. Hence, calm winds favor dew or frost rather than radiation fog. On the other hand, if winds became too strong, the humid air at low levels mixes with the drier air aloft, the relative humidity drops, and radiation fog does not develop.

In regions of topographic relief, air chilled by radiational cooling drains downslope and settles in low-lying areas such as river valleys. Hilltops may thus be clear of fog, while in deep valleys, the fog is thick and persistent.

Once a fog bank forms, radiational cooling becomes a little more complicated. The top of the fog bank very efficiently emits IR radiation skyward. Meanwhile, some of the IR emitted by the ground is absorbed by the fog droplets and reradiated back to the ground. Hence, the ground cools more slowly and the maximum cooling shifts to the top of the fog bank, the fog thickens, and visibility continues to drop. Typically, radiation fogs persist for a few hours after sunrise. Then, the fog gradually thins and disperses as saturated air at low levels mixes with drier air above the fog bank (Figure 7.5); the relative humidity drops and fog droplets vaporize. Mixing may be caused by convection triggered by solar heating of the ground or by a strengthening of regional winds. In winter, however, when the weak rays of the sun are readily reflected by the top of the fog layer, radiation fog may be quite persistent. For example, in the valleys of the Great Basin and in California's

FIGURE 7.5
Radiation fog dispersing at Boston's Logan International Airport as viewed from an approaching aircraft. [Photograph by J. M. Moran]

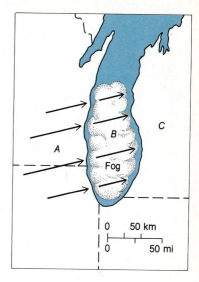

FIGURE 7.6

In summer, as warm humid air (A) flows over the relatively cool surface of Lake Michigan, the air is chilled to saturation and dense fog (B) forms. As the air flows on over the warmer land surface of lower Michigan (C), air temperatures rise, the relative humidity drops, and the fog dissipates. [From *Weather and Climate of the Great Lakes Region,* Figure pg. 109, by Val L. Eighenlaub, © 1979 by the University of Notre Dame Press.]

San Joaquin Valley, winter radiation fogs may persist for many days to weeks at a time.

Air mass advection is sometimes accompanied by fog. The temperature and the water vapor content of an air mass depend on the nature of the surface over which the air mass forms and travels. As an air mass moves from one place to another, termed **air mass advection,** these characteristics change, partly as a result of the modifying influence of the surfaces over which the air mass travels. When the advecting air passes over a relatively cold surface, the air mass may be chilled to saturation in its lower layers. This type of cooling is known as advective cooling and occurs, for example, in spring when mild and humid air flows over relatively cold, snow-covered ground. Snow on the ground may chill the air to the point that fog develops. Fog formed by advective cooling is known as **advection fog.**

Advection fog also develops when warm and humid air passes over cold ocean or lake water. Persistent and dense fogs develop in this way over Lake Michigan in summer as shown in Figure 7.6. Thick fogs over the frigid waters of the Grand Banks of Newfoundland are a consequence of mild maritime air flowing northward from over the warm Gulf Stream waters.

Nocturnal radiational cooling sometimes combines with air mass advection to produce fog. An initially dry air mass becomes more humid in its lower levels after a long trajectory over the open waters of a lake or the ocean. Over land areas, downwind from the water body, if the night sky is clear, the modified air mass is subjected to extreme radiational cooling and fog develops. San Francisco's famous fog forms in this way as onshore winds transport cool humid air into the city and nocturnal radiational cooling chills the air to saturation.

Lowering the air temperature increases the relative humidity by lowering the saturation vapor pressure. The relative humidity of air can also increase without a drop in air temperature. This usually occurs when a relatively cold air mass moves across a warm and extensive water surface such as a large lake or sea. Water evaporates into the air, raising the vapor pressure (or mixing ratio), and with sufficient evaporation, the relative humidity may approach 100 percent.

Arctic sea smoke is fog that develops in this way in winter when extremely cold, dry air flows over a large unfrozen body of water. The lower portion of the air mass reaches saturation primarily because of an increase in its water vapor concentration. Saturation occurs even though the air is heated somewhat when it comes in contact with the warmer water. The increase in relative humidity caused by rapid evaporation of the water offsets any lowering of relative humidity caused by the water's warming of the air. Because the air is warmed from below, the air is destabilized and fog appears as rising filaments or streamers that resemble smoke or steam (Figure 7.7). For this reason, a more general name for fog produced when cold air comes in contact with warm water is **steam fog.** Steam fog also develops on a cold day over a heated outdoor swimming pool or hot tub and sometimes over a wet highway when the sun comes out after a rain.

Another way of explaining the physics of steam fog is to consider the distribution of vapor pressure over the water surface. Within about 1 to 2 m of the water surface, the vapor pressure is relatively high because the air has been warmed and is saturated. Above this shallow layer the air is colder and has a much lower vapor pressure. Hence, a vertical vapor pressure gradient develops over the water body and, in response, water molecules stream up-

FIGURE 7.7
Steam fog forms when relatively cold air moves over relatively warm water. [Photograph by Mike Brisson]

FIGURE 7.8
Upslope fog forms when humid air ascends a mountainside. Expansional cooling brings the air to saturation. [Photograph by Arjen and Jerrine Verkaik/SKYART]

ward. (As we will see in Chapter 9, gases move from high toward low pressure in response to a pressure gradient.) As the humid air streams upward, it cools and the water vapor condenses into tiny droplets that we describe as steam. Eventually, usually about 5 to 10 m above the water body, the rising streamers encounter drier air and evaporate.

Fog also forms on hillsides or mountain slopes as a consequence of the upslope movement of humid air (Figure 7.8). Rising humid air undergoes expansional cooling and eventually reaches saturation. Any further ascent of the saturated air produces fog. Fog formed in this way is called **upslope fog.**

Cloud Development

Water vapor is an invisible gas, but the condensation and deposition products of water vapor are visible. **Clouds** are the visible manifestations of the condensation and deposition of water vapor within the atmosphere. They are composed of tiny water droplets or ice crystals or a mixture of both. In this section, we consider the mechanics of cloud development.

Laboratory studies have demonstrated that in clean air, air that is free of dust and other aerosols, condensation (or deposition) of water vapor requires supersaturated conditions (that is, a relative humidity greater than 100 percent). In clean air, the degree of supersaturation needed for cloud development increases rapidly as the radius of the droplets decreases. For example, formation of relatively small droplets with radii of 0.001 micrometer requires a supersaturation of nearly 340 percent. In contrast, relatively large droplets, having radii greater than 1.0 micrometer, need only slight supersaturations to form.

Why does the degree of supersaturation hinge on droplet size? The values of saturation vapor pressure listed in Table 6.3 apply only to the situation where air overlies a *flat* surface of pure water.* At equivalent temperatures, the saturation vapor pressure is higher in the air surrounding a spherical water droplet than in the air over a flat surface of pure water. As a water surface exhibits increasing curvature, it becomes easier for water molecules to escape the liquid and become vapor. Water molecules that form a curved surface have fewer neighboring molecules and hence, are more weakly held (by hydrogen bonding) than water molecules that form a flat surface. In the case of spherical water droplets, the curvature increases with decreasing radius. Thus, at the same temperature, water molecules more readily escape small droplets than large droplets. This implies that the saturation vapor pressure (and degree of required supersaturation) must increase as droplet size decreases.

We might reasonably assume that, within the atmosphere, cloud droplets grow from much smaller water droplets. However, the great supersaturations required for condensation of very small droplets do not develop in the real atmosphere. How, then, do cloud droplets form? The simple answer is that they have a head start.

In the atmosphere, at most, only slightly supersaturated conditions are necessary for cloud development. This is because the atmosphere contains an abundance of **nuclei,** tiny solid and liquid particles that provide relatively large surface areas on which condensation or deposition can take place. (In other words, air is *not* clean.) Nuclei are the products of both natural and human activity. Forest fires, volcanic eruptions, wind erosion of the soil, saltwater spray, and the discharge from domestic and industrial chimneys are all continual sources.

Some nuclei have radii greater than 1.0 micrometers. This means that the nuclei are large enough to facilitate droplet condensation at relative humidities that rarely exceed 100 percent by more than a fraction of 1 percent. More important than the presence of relatively large nuclei, however, is the fact that air contains a plentiful supply of relatively small nuclei that are hygroscopic. **Hygroscopic nuclei** have a special chemical affinity (attraction) for water molecules. Condensation begins on these nuclei at relative humidities under 100 percent. In fact, magnesium chloride, a salt in sea spray, is a hygroscopic substance that can initiate condensation at relative humidities as low as 70 percent. Because some nuclei are relatively large and many are hygroscopic, we can expect cloud development when the relative humidity nears 100 percent.

---

* Note that the saturation vapor pressure (or saturation mixing ratio) for air over a flat surface of pure water is the basis for computing relative humidity and the degree of supersaturation.

Depending on their specific function, nuclei are classified as one of two types: **cloud condensation nuclei** and **ice-forming nuclei.** Cloud condensation nuclei are active (that is, they promote condensation) at temperatures both above and below freezing because water droplets condense and remain liquid even when the cloud temperature is well below 0°C (32 °F). These are **supercooled water droplets.** Ice-forming nuclei are much less abundant than cloud condensation nuclei and become active only at temperatures well below freezing. There are two types of ice-forming nuclei: (1) **freezing nuclei,** which cause liquid droplets to freeze, and (2) **deposition nuclei** (also called **sublimation nuclei**), on which water vapor deposits directly as ice. Most freezing nuclei activate only at temperatures below −9 °C (16 °F). Deposition nuclei do not become fully active until temperatures fall below −20 °C (−4 °F), as shown in Figure 7.9.

FIGURE 7.9
Some deposition (sublimation) nuclei and their activation temperatures. Note that most nuclei are soil particles. [From V. J. Schaefer, "The Formation of Ice Crystals by Sublimation," *Weatherwise 32* (1979):256]

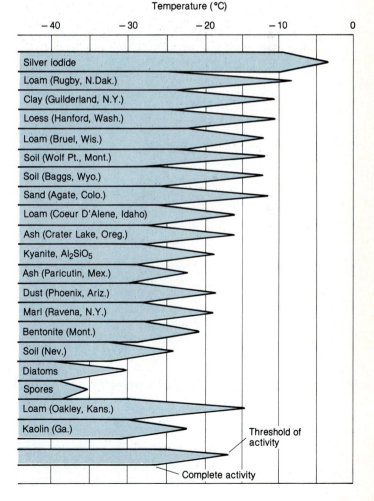

Cloud Classification

Even a casual observer of the sky notices that clouds occur in a wide variety of forms. These forms are not arbitrary. A cloud appears as a result of many processes operating in the atmosphere. In fact, keeping track of changes in cloud cover often provides clues about future weather.

The British naturalist Luke Howard was among the first to devise a classification of cloud types. Formulated in 1802-1803, the essentials of How-

Table 7.1
Cloud Classification

Genus	Altitude of cloud base above ground (km)	Shape and appearance
High clouds		
Cirrus (Ci)	6–18	Delicate streaks or patches
Cirrostratus (Cs)	6–18	Transparent thin white sheet or veil
Cirrocumulus (Cc)	6–18	Layer of small white puffs or ripples
Middle clouds		
Altostratus (As)	2–6	Uniform white or gray sheet or layer
Altocumulus (Ac)	2–6	White or gray puffs or waves in patches or layers
Low clouds		
Stratocumulus (Sc)	0–2	Patches or layer of large rolls or merged puffs
Stratus (St)	0–2	Uniform gray layer
Nimbostratus (Ns)	0–4	Uniform gray layer from which precipitation is falling
Clouds with vertical development		
Cumulus (Cu)	0–3	Detached heaps or puffs with sharp outlines and flat bases, and slight or moderate vertical extent
Cumulonimbus (Cb)	0–3	Large puffy clouds of great vertical extent with smooth or flattened tops, frequently anvil shaped, from which showers fall, with thunder

Source: M. Neiburger, J. G. Edinger, and W. D. Bonner. *Understanding Our Atmospheric Environment.* New York: W. H. Freeman, 1973, p. 11. Copyright © 1973. All rights reserved.

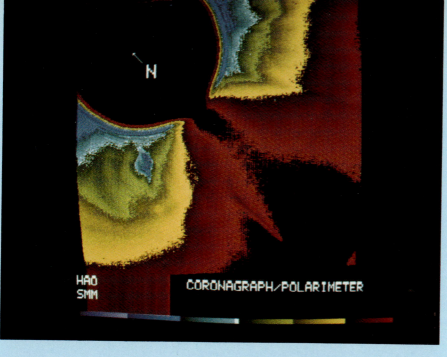

PLATE 1

The sun's corona, color-coded to levels of brightness. Photograph was taken by sensors on board the Solar Maximum Mission satellite. [National Center for Atmospheric Research/National Science Foundation]

PLATE 2

A glass prism disperses visible light into its component colors. [Photograph © David Parker, Science Source/Photo Researchers]

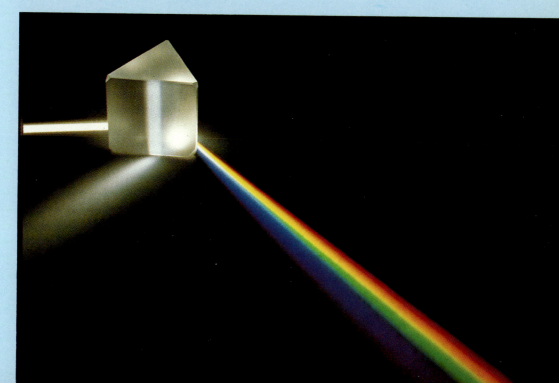

PLATE 3
The red of the setting sun. [Photograph
by Mike Brisson]

PLATE 4
Cirrus are high, thin wispy clouds
composed almost exclusively of ice
crystals. [Photograph by J. M. Moran]

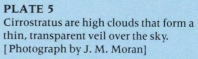

PLATE 5
Cirrostratus are high clouds that form a thin, transparent veil over the sky. [Photograph by J. M. Moran]

PLATE 6
Cirrocumulus are high clouds that exhibit a wavelike or mackerel pattern of small, white puffs. [Photograph by Arjen and Jerrine Verkaik/SKYART]

PLATE 7

Altostratus are middle-level clouds composed of ice crystals or water droplets or both. They form a uniform gray or white layer through which the sun is dimly visible. [Photograph by Arjen and Jerrine Verkaik / SKYART]

PLATE 8

Altocumulus are middle clouds consisting of patches arranged in a wavelike pattern. [Photograph by Arjen and Jerrine Verkaik / SKYART]

PLATE 9
Stratocumulus are low-level clouds consisting of relatively large, irregular rolls that form a layer. [Photograph by J. M. Moran]

PLATE 10
Stratus is a low, gray, continuous cloud layer from which drizzle may fall. [Photograph by J. M. Moran]

PLATE 15
Deciduous trees in autumn colors. [Photograph by Mike Brisson]

PLATE 16
A rainbow is caused by refraction and reflection of sunlight by raindrops. Refraction disperses sunlight into its component colors. [Photograph by Arjen and Jerrine Verkaik/SKYART]

PLATE 13
A cumulonimbus cloud billows upward and spreads laterally forming an anvil that is topped by the tropopause. [Photograph by Arjen and Jerrine Verkaik/SKYART]

PLATE 14
Altocumulus lenticularis clouds are nearly stationary mountain wave clouds that are generated by airflow that is disturbed during passage over a mountain range. [Photograph by J. M. Moran]

PLATE 15
Deciduous trees in autumn colors. [Photograph by Mike Brisson]

PLATE 16
A rainbow is caused by refraction and reflection of sunlight by raindrops. Refraction disperses sunlight into its component colors. [Photograph by Arjen and Jerrine Verkaik/SKYART]

PLATE 9
Stratocumulus are low-level clouds consisting of relatively large, irregular rolls that form a layer. [Photograph by J. M. Moran]

PLATE 10
Stratus is a low, gray, continuous cloud layer from which drizzle may fall. [Photograph by J. M. Moran]

PLATE 11
Cumulus clouds resemble puffs of cotton floating in the sky. They form as a consequence of convective currents. Airflow is upward where clouds occur and downward where air is clear. [Photograph by J. M. Moran]

PLATE 12
Cumulus clouds that show significant vertical development are called cumulus congestus; they resemble a cauliflower. [Photograph by J. M. Moran]

ard's scheme are still in use today. Cloud forms are given special Latin names and are organized by appearance and by altitude of occurrence.

On the basis of *appearance,* the simplest distinction is among cirrus, stratus, and cumulus clouds. Cirrus clouds are fibrous, stratus clouds are layered, and cumulus clouds occur as heaps or puffs. On the basis of *altitude,* the most common clouds in the troposphere are grouped into four families (Table 7.1): high clouds, middle clouds, low clouds, and clouds exhibiting vertical development. Members of the first three families are produced by gentle uplift over broad areas. These clouds spread laterally to form layers and are called **stratiform clouds.** Clouds with vertical development generally cover smaller areas and are associated with much more vigorous uplift. Consequently, these clouds are heaped or puffy in appearance and are called **cumuliform clouds.** Later chapters describe in detail the various weather systems that trigger development of stratiform and cumuliform clouds. Here, we are concerned primarily with brief descriptions of the most common clouds of each family.

HIGH CLOUDS

The base* of high clouds are at altitudes above 6 km (20,000 ft). Temperatures are so low in this region of the atmosphere (below -25 °C, or -13 °F) that clouds are composed almost exclusively of ice crystals,† a composition that gives them a fibrous or filamentous appearance. Their names include the prefix *cirro* from the Latin meaning "a curl of hair."

Cirrus clouds are nearly transparent and occur as delicate silky strands, sometimes called "mares' tails" (Plate 4). The strands are actually streaks of falling ice crystals blown laterally by strong winds. Like cirrus clouds, **cirrostratus** clouds are also nearly transparent, so the sun or moon readily shines through them. They form a thin, white veil or sheet that partially or totally covers the sky (Plate 5). **Cirrocumulus** clouds consist of small white patches arranged in a wavelike or mackerel pattern (Plate 6). None of these high clouds is thick enough to prevent objects on the ground from casting shadows during daylight hours.

MIDDLE CLOUDS

The bases of middle clouds are at altitudes between 2 and 6 km (6500 and 20,000 ft). Their names include the prefix *alto* from the Latin meaning "high." These clouds, which feature temperatures generally between 0 and -25 °C

---

* The actual altitudes of the various clouds vary seasonally and with latitude. For example, in polar regions the base of high clouds may be as low as 3 km (10,000 ft). Cloud base values quoted here are intended as general guidelines.

† In 1985, research meteorologists at Boulder, Colorado, detected supercooled water droplets in the base of a cirrus cloud layer at 8.3 km (27,000 ft) and a temperature of -36 °C (-32 °F). It is likely, however, that this was an unusual and transient event.

(32 and -13 °F), are composed of ice crystals or supercooled water droplets or a mixture of the two. **Altostratus** clouds occur as uniformly gray or white layers that totally or partly cover the sky (Plate 7). They are usually so thick that the sun is only dimly visible, as if it were viewed through frosted glass. **Altocumulus** clouds consist of roll-like patches or puffs that form in lines or waves (Plate 8). They are distinguished from cirrocumulus by the larger size of the cloud patches and by sharper edges. Sharp cloud boundaries indicate the presence of water droplets rather than ice crystals. Altocumulus may occur in several distinct layers simultaneously and rarely produce precipitation that reaches the ground.

LOW CLOUDS

The bases of low clouds range in altitude from the Earth's surface (fog) up to 2 km (6500 ft). Low clouds, which form at temperatures above -5 °C (23 °F), are composed mostly of water droplets. **Stratocumulus** clouds consist of large, irregular puffs or rolls arranged in a layer (Plate 9), and often represent the breakup of **stratus** clouds which occur as a uniform gray layer that stretches from horizon to horizon (Plate 10). Where stratus meets the ground, we experience fog. **Nimbostratus** clouds resemble stratus clouds, except that nimbostratus clouds are thicker, darker, and yield more rain and snow. Typically, only drizzle falls from stratus clouds.

VERTICAL CLOUDS*

Air surging upward as convection currents can give rise to cumulus, cumulus congestus, and cumulonimbus clouds. As unsaturated air rises, its temperature falls at the dry adiabatic lapse rate (10 C° per 1000 m), and hence, its relative humidity increases. With continued ascent and expansional cooling, the relative humidity may eventually approach 100 percent, and a cumulus cloud forms. The altitude at which condensation begins to occur through lifting is known as the **lifting condensation level (LCL)** and coincides with the altitude of the cloud base, usually between 1 and 2 km (3000 and 6500 ft).

   Cumulus clouds resemble puffs of cotton dotting the sky on a fair-weather day (Plate 11). In Figure 7.10, cumulus clouds are viewed from above, from an aircraft window. Because convection is driven by solar heating, not surprisingly cumulus cloud development often follows the daily variation of insolation. On a fair day, cumulus clouds begin forming by middle to late

---

* Typically, this category is reserved for cumuliform clouds, that is, convective clouds exhibiting vertical development. However, we may also include altostratus and nimbostratus as vertical clouds. Although usually classified as middle and low clouds, respectively (because of cloud base altitude), altostratus and nimbostratus often are sufficiently thick that they extend through more than one level.

FIGURE 7.10
Cumulus clouds viewed from above, from an aircraft window. [Photograph by J. M. Moran]

morning, after the sun has warmed the ground and initiated convection. Cumulus sky cover is most extensive by midafternoon—usually the warmest time of day. If cumulus clouds show some vertical growth, these normally "fair-weather" clouds may produce a brief, light shower of rain or snow. As sunset approaches, convection weakens, and the cumulus clouds begin dissipating (that is, they vaporize).

Where convection is suppressed, so, too, is development of cumulus clouds. Relatively cold surfaces chill and stabilize the overlying air and inhibit convection, and hence, cumulus clouds do not readily form over snow-covered surfaces. Cold water surfaces also may suppress convection. This effect is sometimes observed along the shores of the Great Lakes during late spring and early summer afternoons when, on average, lake surface temperatures are lower than the adjacent land surface. Fair-weather cumulus clouds develop over the land but not over the lake. In fact, rows of cumulus clouds may form a very distinct boundary along the shoreline (Figure 7.11).

FIGURE 7.11
Cumulus clouds form over the warm land, but not over the cold lake surface. Cold surfaces stabilize the air and inhibit convection. [Photograph by J. M. Moran]

Once cumulus clouds form, the stability profile of the troposphere determines the extent of vertical cloud development and whether cumulus clouds build into more ominous thunderstorm clouds. If the ambient air aloft is stable, vertical motion is inhibited and cumulus clouds exhibit little vertical growth. Under these conditions, the weather is likely to remain fair. On the other hand, if the ambient air aloft is unstable for saturated air, then vertical motion is enhanced, and the tops of cumulus clouds surge upward. If the ambient air is unstable to great altitudes, the entire cloud mass takes on a cauliflower appearance as it builds into a **cumulus congestus** (Plate 12) and then a **cumulonimbus** (thunderstorm) cloud (Plate 13).

SKY WATCHING

It is not at all unusual to observe both cumuliform and stratiform clouds in the sky at the same time. For example, as shown in Figure 7.12, a layer of high cirrus overlies low-level cumulus clouds. The cirrus is occurring in advance of an approaching surface warm front and is too thin to significantly weaken the incoming solar radiation. The sun heats the ground triggering convection and cumulus clouds develop.

When more than one cloud layer appears in the sky, it is instructive to visually track the direction of movement of each cloud layer. Often clouds at different altitudes move in different directions. Because clouds are displaced with the wind, this observation indicates that, at least within the troposphere, the horizontal wind changes direction with altitude. Any change in wind direction (or wind speed) with distance is known as **wind shear.** In later chapters, we have more to say about wind shear.

FIGURE 7.12
High, thin cirrus clouds overlying cumulus clouds. [Photograph by J. M. Moran]

Another feature to watch for are well-defined holes or canals in an otherwise uniform cloud layer. Holes in clouds are not unusual; numerous published reports and photographs of them have appeared over the past 60 years. Professor P. V. Hobbs of the University of Washington argues that at least some holes are caused by aircraft flying through thin clouds (probably altocumulus) composed of supercooled water droplets. Turbulence produced in the wake of an aircraft induces expansional cooling of air; water droplets freeze to ice crystals that fall out of the cloud, thereby creating a hole in the cloud. Aircraft ascending or descending through a cloud produce nearly circular holes, whereas aircraft flying horizontally through a cloud layer create canals.

Unusual Clouds

Chances are that all of us have seen most of the clouds described above, since the circumstances leading to their development are quite common. Other, more unusual clouds are formed only by rarer atmospheric conditions.

The clouds in Plate 14 are most striking in appearance. Not surprisingly, many people have mistaken clouds like them for flying saucers. Observation of this type of cloud over a period of time reveals another strange characteristic: these clouds tend to remain nearly stationary. They are **altocumulus lenticularis** clouds, that is, lens-shaped altocumulus clouds. These and other so-called **mountain-wave clouds** are generated by airflow that is disturbed by a mountain range.

As strong horizontal winds encounter a mountain range, the wind is deflected up the windward slopes and then down the leeward slopes. This is a common occurrence as the prevailing westerlies, flowing from west to east, cross the Front Range of the Rocky Mountains. If the air ascending the windward slopes is stable, then the winds describe a wavelike pattern that extends many tens of kilometers downwind of the mountain crest (Figure 7.13). Within the wave, where air flows upward, expansional cooling leads to cloud formation. Where air flows downward within the wave, compressional warming causes clouds to vaporize. Clouds thus occur at the crests of the waves and are absent in the troughs.

Mountain-wave clouds do not move, even though the winds are strong, because the wave itself is stationary. A stationary wave within the atmosphere is known as a **standing wave,** and a standing wave produced by mountains is usually called a **mountain wave.** A waterfall on a river is a good analogue. In the pool at the base of the waterfall, the water is very turbulent, but as we follow the river downstream, we quickly leave the turbulent region behind. The river water flows ceaselessly, but the disturbed segment remains stationary

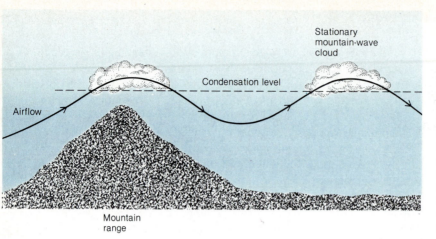

FIGURE 7.13
Mountain-wave clouds form when a mountain range deflects the horizontal wind into a wavelike pattern. Clouds develop on the wave crests where the airflow is upward and expansional cooling takes place. Clouds are absent in the wave troughs where airflow is downward and compressional warming occurs.

at the foot of the waterfall, just as the mountain wave remains stationary to the lee of the mountain range.

Interestingly, patterns exhibited by stratocumulus and cirrocumulus (as well as by varieties of altocumulus other than lenticularis) are also caused by waves propagating within the atmosphere. These waves, however, are not linked to mountain ranges and propagate horizontally at different altitudes within the troposphere. Note that atmospheric waves produce bands of clouds that are aligned either parallel or perpendicular to the wind direction.

Because almost all water vapor is confined to the troposphere, so, too, are most clouds. One exception is the occasional intense cumulonimbus cloud, the top of which may penetrate the tropopause and enter the lowest portion of the stratosphere. Another exception is the colorful **nacreous clouds** which occur in the upper stratosphere. Temperatures at these altitudes (25 to 30 km) favor water in either the solid or supercooled state. Because of their soft, pearly luster, these rarely seen clouds are also called **mother-of-pearl clouds.** They are best viewed at high latitudes when illuminated by the setting sun.

Somewhat mysterious are the wavy, cirruslike **noctilucent clouds** that occur in the upper mesosphere. They are seen rarely, and then only at high latitudes just after sunset or before sunrise. The virtual absence of water vapor in the upper mesosphere has prompted some scientists to suggest that noctilucent clouds are composed of meteoric dust.

Conclusions

Many processes operating within the atmosphere bring air to saturation. Through expansional or radiational cooling or by the addition of water vapor, the relative humidity of air approaches 100 percent. As air approaches saturation, the abundance of nuclei (and especially hygroscopic nuclei) in the atmosphere favors deposition or condensation, that is, the development of clouds. Clouds may or may not yield precipitation—most do not. The special conditions required for rain and snowfall are among the topics discussed in the next chapter.

Summary Statements

Nocturnal radiational cooling chills the Earth's surface which, in turn, cools the air in immediate contact with the surface. If the temperature of that surface drops below the dew point (or frost point), water vapor condenses as dew (or deposits as frost) on the surface.

Fog is a suspension of tiny water droplets or ice crystals in an air layer that is in contact with the Earth's surface. On the basis of mode of origin, fog is classified as radiation, advection, steam, or upslope.

Clouds, the visible manifestations of condensation and deposition within the atmosphere, are composed of minute water droplets or ice crystals or both.

Condensation and deposition occur on nuclei at relative humidities near 100 percent. Cloud condensation nuclei are much more abundant than ice-forming nuclei. Most ice-forming nuclei activate at temperatures well below the freezing point.

Many condensation nuclei are hygroscopic. That is, they have a special chemical affinity for water molecules and promote condensation at relative humidities under 100 percent.

Clouds are assigned special Latin names and are classified by appearance and by altitude of occurrence.

High, middle, and low clouds are caused by relatively gentle uplift over broad areas. Hence, these clouds have a layered appearance. Clouds with vertical development are the consequence of more vigorous uplift and are heaped or puffy in form.

Mountain-wave clouds develop downwind of prominent mountain ranges and are nearly stationary. Noctilucent clouds occur in the upper mesosphere, are rarely seen, and are probably composed of meteoric dust.

Key Words

radiational cooling	upslope fog	cumuliform clouds	cumulonimbus
dew	clouds	cirrus	wind shear
frost	nuclei	cirrostratus	altocumulus
dew point	hygroscopic nuclei	cirrocumulus	lenticularis
frost point	cloud condensation	altostratus	mountain-wave
hoarfrost	nuclei	altocumulus	clouds
fog	ice-forming nuclei	stratocumulus	standing wave
mist	supercooled water	status	mountain wave
radiation fog	droplets	nimbostratus	nacreous clouds
air mass advection	freezing nuclei	lifting condensation	mother-of-pearl
advection fog	deposition nuclei	level (LCL)	clouds
Arctic sea smoke	sublimation nuclei	cumulus	noctilucent clouds
steam fog	stratiform clouds	cumulus congestus	

Review Questions

1. What conditions favor extreme nocturnal radiational cooling?

2. Dew and frost are not forms of precipitation. Explain this statement.

3. How is the dew point a measure of the water vapor content of air?

4. Describe the circumstances that favor the development of radiation fog. Why is this type of fog usually short lived? Under what conditions might radiation fog persist for many days?

5. Identify two different situations in which fog forms as a consequence of warm air advection.

6. How does Arctic sea smoke develop?

7. What is a cloud?

8. Explain why clouds typically form without supersaturated conditions.

9. List some of the sources of cloud condensation nuclei.

10. What is the significance of *hygroscopic* nuclei?

11. Cloud condensation nuclei are much more abundant than ice-forming nuclei. What is the implication of this for the composition of clouds?

12. Distinguish between freezing nuclei and deposition nuclei.

13. How are clouds classified?

14. How and why does the composition of clouds vary with altitude?

15. What is the significance of the lifting condensation level?

16. How does ambient air stability influence the vertical growth of cumulus clouds?

17. Speculate on what might cause a cumulus or cumulonimbus cloud to appear to tilt or lean with altitude.

18. What causes the ripple or banded pattern of cirricumulus, altocumulus, and stratocumulus clouds?

19. If the temperature at the lifting condensation level is 0 °C and the temperature at the ground is 20 °C, determine the approximate altitude of the cumulus cloud base above the ground.

20. Explain why cumulus clouds tend to vaporize toward sunset. What is the effect of a snow cover on cumulus cloud development?

Points to Ponder

1. Distinguish between the dew-point temperature and the wet-bulb temperature. Under what conditions are they equivalent?

2. Fog can have a catastrophic effect on a snow cover. Explain why.

3. In late spring and early summer, fog frequently develops over Lake Michigan as southwest winds push warm, humid air over the lake. Explain this phenomenon.

4. Dew formation is an *isobaric* (constant pressure) process, but expansional cooling is not an isobaric process. Explain the difference.

5. Speculate on why frost formation on fruit may actually protect the fruit.

6. On a clear and calm night, would you expect a coating of frost to be thicker on the topside (facing the sky) or bottomside (facing the ground) of a plant leaf? Explain your choice.

Projects

1. Using the cloud photographs in this book as a guide, identify today's clouds. Can you distinguish among low, middle, and high clouds?

2. Record the changes in cloud type as a warm front or cold front approaches your area. Describe changes in cloud thickness and altitude.

3. Through the course of the day, keep track of changes in the lateral extent and vertical development of cumulus clouds. You may wish to take a series of photographs. Determine whether the changes you observe are caused by diurnal variations in insolation or an approaching weather system.

Selected Readings

Bohren, C. "An Essay on Dew." *Weatherwise* 41 (1988):226–231. Gives a fascinating explanation of the mechanics of dew (and frost) formation.

Gedzelman, S. D. "In Praise of Altocumulus." *Weatherwise* 41 (1988):143–149. Provides a well-illustrated discussion of the origins of the various forms of altocumulus clouds and associated optical effects.

Hobbs, P. V. "Holes in Clouds?" *Weatherwise* 38 (1985):254–258. Explores proposed explanations for holes and canals in clouds.

Meyer, M. B., et al. "FOG-82: A Cooperative Field Study of Radiation Fog." *Bulletin of the American Meteorological Society* 67 (1986):825–832. Describes an intensive field study of radiation fog at the Albany County Airport, New York, in the autumn of 1982.

Richards, C. J. "Focus on the Sky—The Black and White Way." *Weather* 41 (1986):214–224. Provides some very detailed advice on the advantages and techniques of photographing clouds with black and white film.

Schaefer, V. J., and J. A. Day. *A Field Guide to the Atmosphere.* Boston: Houghton Mifflin, 1981. 359 pp. Presents an exceptionally well-illustrated survey of cloud and precipitation processes.

Scorer, R., and A. Verkaik. *Spacious Skies.* London: David and Charles, 1989. 192 pp. Includes outstanding cloud photography.

Stewart, T. R., R. W. Katz, and A. H. Murphy. "Value of Weather Information: A Descriptive Study of the Fruit-Frost Problem." *Bulletin of the American Meteorological Society* 65 (1984):126–137. Reports on the use of weather information by fruit growers in the Yakima Valley, Washington, in efforts to avoid freeze damage.

8

Precipitation, Weather Modification, and Atmospheric Optics

Snow and all other forms of precipitation originate in clouds when a special set of conditions triggers rapid growth of cloud droplets or ice crystals. [Photograph by Mike Brisson]

T HIS CHAPTER continues our study of water in the atmosphere by considering several closely related topics. The first portion focuses on the development of precipitation in clouds and the forms and measurement of precipitation. This serves as the basis for our subsequent discussion of techniques to induce precipitation (cloud seeding) and to dissipate fog. The chapter closes with a description of various optical phenomena that are produced when sunlight or moonlight interacts with clouds or with precipitation.

Precipitation Processes

The development of clouds is no guarantee that it will rain or snow. Nimbostratus and cumulonimbus clouds produce the bulk of precipitation, but most clouds—even most of those associated with a large storm system—do not yield any rain or snow. This is because a special combination of circumstances, as yet not completely understood, is required for clouds to precipitate.

TERMINAL VELOCITY

The water droplets or ice crystals that compose clouds are so minute that they remain suspended indefinitely, unless they vaporize or undergo considerable growth. Updrafts within clouds are usually strong enough to prevent cloud particles from leaving the base of a cloud and falling to the Earth's surface. Even if droplets or ice crystals descend from a cloud, their downward drift is so slow that they travel only a short distance before vaporizing in the unsaturated air beneath the cloud.

The speed of a cloud droplet or ice crystal (or any other particle) falling in calm air is regulated by two forces: (1) the force of gravity, which accelerates the particle downward, and (2) an opposing force caused by the resistance of the air through which the particle descends. As the particle accelerates downward, it meets with increasing air resistance. (Meanwhile, the force of gravity remains essentially constant.) Eventually, the resisting force equals (balances) the force of gravity, and the particle drifts downward at a constant speed known as the **terminal velocity.***

For a particle to remain suspended in air, updrafts of air must be strong enough to counter the particle's terminal velocity. Generally, the terminal velocity increases with the size of the particle. Hence, the larger the particle, the more vigorous the updraft must be in order to keep the particle in suspension. Cloud droplets and ice crystals are so small (most having diameters

---

* The concept of balanced forces is discussed in greater detail in the next chapter.

of 10 to 20 micrometers) that their very low terminal velocities (typically only 0.3 to 1.2 cm per second) are readily countered by even weak updrafts. Even if there were no updrafts, such low terminal velocities would mean that it would take 24 hours or longer for cloud particles to reach the ground. Long before that happened, all the cloud particles would vaporize in the unsaturated air below the cloud base.

Cloud particles somehow must grow large enough to counter updrafts so that they survive a descent to the Earth's surface as raindrops or snowflakes without completely vaporizing. This is no minor task! It takes about 1 million cloud droplets (having diameters of 10 to 20 micrometers) to form a single raindrop (about 2 mm in diameter). How does this growth take place?

Cloud physicists have determined that condensation alone cannot cause cloud droplets to grow into raindrops. They have identified two important processes by which cloud particles grow large enough to precipitate: the collision–coalescence process and the Bergeron process.

THE COLLISION–COALESCENCE PROCESS

Within **warm clouds,** that is, clouds at temperatures above the freezing point of water, droplets may grow by colliding and coalescing (merging) with one another. The **collision–coalescence process** requires that cloud droplets be of different sizes. Droplets of uniform size have essentially the same terminal velocity, and so collisions between droplets are rare. In contrast, cloud droplets of unequal size have different terminal velocities, and collisions are frequent. Unequal droplet size within the same cloud usually stems from the presence of "giant" sea-salt nuclei that produce relatively large droplets (greater than 40 micrometers in diameter). A larger (heavier) droplet falls more rapidly than a smaller (lighter) droplet. As it descends, a larger droplet intercepts and coalesces with smaller (slower) droplets in its path (Figure 8.1). By repeated collisions and coalescence, the droplet grows, its terminal velocity increases, and collisions become more frequent. Eventually, the droplet is so large that it falls from the cloud as a raindrop.

THE BERGERON PROCESS

Although cloud droplet growth through collision–coalescence alone is important, especially in the tropics, most precipitation that falls in middle and high latitudes originates through the Bergeron process. Named for the Scandinavian meteorologist Tor Bergeron, who first described it in 1933, the **Bergeron process** applies to **cold clouds,** which are at temperatures below 0 °C (32 °F). The process requires the coexistence of water vapor, ice crystals, and supercooled liquid water droplets. Before getting into the details of the Bergeron process, we first need to examine the composition of cold clouds.

As we noted in Chapter 7, most ice-forming nuclei are not active at temperatures higher than −9 °C (16 °F). Consequently, clouds at temperatures

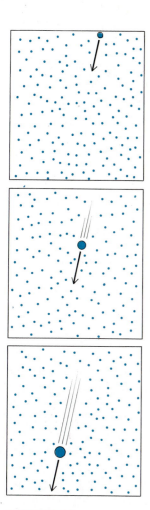

FIGURE 8.1
A relatively large water droplet falls within a cloud of much smaller droplets. The large droplet collides with the smaller droplets and grows by coalescence.

between 0 °C and − 9 °C (32 and 16 °F) are typically composed of supercooled water droplets exclusively. From − 10 °C (14 °F) down to − 20 °C (− 4 °F), clouds are mixtures of supercooled water droplets and ice crystals. Below − 20 °C (− 4 °F), the activation temperature for many deposition nuclei (refer back to Figure 7.9), clouds usually consist of ice crystals only.

The distribution of supercooled water droplets and ice crystals is somewhat more complicated in clouds that have significant vertical development. Cumulonimbus clouds, for example, have different components at different altitudes depending on the vertical temperature profile within the cloud. Typically, they are composed exclusively of ice crystals aloft where temperatures are very low and of water droplets near the milder cloud base. In between is a mixture of supercooled water droplets and ice crystals. Furthermore, vigorous convection currents within the cumulonimbus cloud transport liquid water droplets aloft where they freeze. This is a very important source of ice crystals within thunderstorm clouds.

The Bergeron process takes place in clouds that contain a mixture of ice crystals and supercooled water droplets (Figure 8.2). Initially, supercooled water droplets far outnumber ice crystals because cloud condensation nuclei are much more abundant than ice-forming nuclei. Very quickly, however, ice crystals grow at the expense of supercooled water droplets primarily because the saturation vapor pressure is greater over water than it is over ice.

At subfreezing temperatures, water molecules vaporize more readily from a liquid water surface than from solid ice because water molecules are bonded more strongly in the solid phase than in the liquid phase. Consequently, the saturation vapor pressure over water is greater than that over ice (refer back to Table 6.3). It follows that in clouds composed of a mixture of ice crystals and supercooled water droplets, a vapor pressure that is saturated for water droplets is supersaturated for ice crystals. Suppose, for example, that the vapor pressure is 2.86 mb in a cloud at a temperature of − 10 °C (14 °F). From Table 6.3, this vapor pressure translates into a relative humidity of 100 percent (saturation) for the air surrounding water droplets and a relative humidity of 110 percent (supersaturation) for the air surrounding ice crystals. In response to supersaturated conditions, water vapor is deposited on the ice crystals, and the ice crystals grow larger. When deposition removes water vapor from the cloud, the relative humidity of the air surrounding the water droplets dips below 100 percent, and the droplets vaporize. Thus, in the Bergeron process, ice crystals grow at the expense of supercooled water droplets.

As ice crystals grow larger and heavier, their terminal velocities increase, and they collide and coalesce with smaller supercooled water droplets and ice crystals in their path, thereby growing still larger. Eventually, the ice crystals become so heavy that they fall out of the cloud base. If air temperatures are below freezing at least most of the way to the ground, the crystals reach the Earth's surface in the form of snowflakes. If the air below the cloud is above freezing, snowflakes melt and fall as raindrops.

• Supercooled water droplets

* Ice crystals

FIGURE 8.2
Within a cold cloud, ice crystals grow at the expense of supercooled water droplets. As they grow larger, ice crystals fall faster and collide with droplets and other ice crystals in their paths. Eventually, they grow large enough to fall out of the cloud as snowflakes.

FIGURE 8.3
Some of the rain falling from the base of a distant thundershower never reaches the ground because the precipitation vaporizes in the relatively dry air beneath the thundercloud. [Photograph by Don Beimborn]

Once a raindrop or a snowflake leaves a cloud, it enters unsaturated air—a hostile environment in which either evaporation or sublimation takes place. In general, the longer the journey to the ground and the lower the relative humidity of the air beneath the clouds, the greater the quantity of rain or snow that returns to the atmosphere as vapor. Figure 8.3 shows rain falling from a distant thunderstorm; some of the rain evaporates as it descends and encounters drier air. This shaft of falling precipitation is called **virga.**

Forms of Precipitation

Precipitable water is the amount of water produced when all the water vapor in a column of air is condensed. The air column is assumed to extend from the Earth's surface to the tropopause, and the condensed water is described in units of depth. Condensing all the water vapor within the troposphere would produce a layer of water that would cover the entire Earth's surface to an average depth of only about 25 mm (1 in.). With the poleward decline in air temperature, precipitable water amounts also decline. (Lower temperatures reduce the rate at which water vaporizes into the atmosphere.) Average precipitable water depths decrease with latitude from more than 40 mm (1.6 in.) in the humid tropics to less than 5 mm (0.2 in.) near the poles.

As we have seen, within the troposphere some water vapor changes phase to the water droplets and ice crystals of clouds, and through the collision–coalescence and Bergeron processes, some clouds yield precipitation. **Precipitation** is water in solid or liquid form that falls to the Earth's surface. Besides the familiar rain and snow, precipitation also occurs as drizzle, freezing rain, ice pellets, and hail.

Drizzle consists of small water drops from 0.2 to 0.5 mm in diameter that drift very slowly toward the Earth's surface. Drizzle drops are relatively small because they originate in stratus clouds. Stratus clouds are so low that droplets originating within them have only a limited opportunity to grow by

187

coalescence. Drizzle is associated with fog and poor visibility, but never with convective clouds.

Rain falls mostly from nimbostratus and cumulonimbus clouds, and the bulk of rain originates as snowflakes (or hailstones), which melt on the way down as they encounter air that is above 0 °C (32 °F). Because rain originates in thicker clouds that have higher bases, raindrops travel farther than drizzle and undergo more growth by coalescence. Most commonly, raindrop diameters are in the range of 1 to 6 mm; beyond this range, drops are unstable and tend to break up into smaller drops.

Raindrops that are produced in warm clouds are usually smaller than those falling from cold clouds. Rarely do warm cloud raindrops reach 2 mm in diameter. Precipitation growth via a combination of the Bergeron and collision–coalescence processes (in cold clouds) produces larger drops than through collision–coalescence alone (in warm clouds). Recently, research meteorologists sampling warm cloud raindrops over Hilo, Hawaii, discovered an exception to this rule. They reported numerous raindrops of 4 to 5 mm diameter and some as large as 8 mm diameter (about the size of a pea). In this case, strong updrafts in convective clouds prolonged the period of coalescence, thereby enabling warm drops to grow to extraordinary sizes.

Freezing rain, or freezing drizzle, forms a coating of ice (Figure 8.4) that sometimes grows thick and heavy enough to bring down tree limbs, snap

FIGURE 8.4
Freezing rain coats subfreezing surfaces of branches, power lines, and pavement. [Photograph by Mike Brisson]

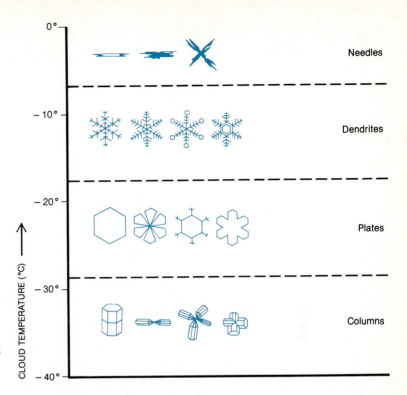

FIGURE 8.5
Snowflakes take on a variety of forms depending at least partially on cloud temperature.

power lines, and totally disrupt traffic. Freezing rain develops when rain falls from a relatively mild air layer aloft into a shallow layer of subfreezing air at ground level. The drops become supercooled and then freeze immediately on contact with cold surfaces.

Freezing rain can be localized and persistent, especially in hilly or mountainous terrain such as the Appalachians of Pennsylvania and West Virginia. Weather events may unfold as follows. During a clear winter night of extreme radiational cooling, cold dense air drains downslope and settles into deep river valleys. (Under these circumstances, a temperature inversion develops, and the lowest air temperatures occur at the lowest elevations.) Then, after sunrise, an approaching storm system spreads clouds into the region. The wind strengthens and blows perpendicular to the mountain range so that shallow pools of subfreezing air are trapped in the valley bottoms. Meanwhile, at higher elevations along the mountain slopes, air temperatures rise above freezing. Rain then develops and is chilled as it falls through subfreezing air in the valley bottoms. Raindrops then freeze on contact with cold surfaces such as roads and utility wires. Such a situation often persists until the wind shifts parallel to the valleys and flushes out the layer of subfreezing air.

Snow is an assemblage of ice crystals in the form of flakes. Although it is said that no two snowflakes are identical, all snowflakes are hexagonal (six-sided), just as the constituent ice crystals of the flakes are hexagonal. The snowflake form varies with water vapor concentration and with temperature and may consist of plates, stars, columns, or needles (Figure 8.5). Snowflake

189

size also depends in part on the availability of water vapor during the crystal growth process. At very low temperatures, water vapor concentrations are low, so snowflakes are relatively small. Snowflake size also depends on collision efficiency as the flakes drift toward the ground. At relatively high temperatures, snowflakes are wet and readily stick together after colliding, so that flake diameters may eventually exceed 5 cm (2 in.). The appearance of such large flakes is usually a signal that the snow is about to turn to rain.

Snow pellets and snow grains are closely related to snowflakes. **Snow pellets** are soft conical or spherical white particles of ice with diameters of 1 to 5 mm. They are formed when supercooled cloud droplets collide and freeze together. **Snow grains** originate in much the same way as drizzle except that they are frozen; diameters usually are less than 1 mm.

Is it ever too cold to snow? When is it too warm to snow? For answers to these questions, see the Special Topic.

Ice pellets, also called **sleet,** are actually frozen raindrops. They develop in much the same way as freezing rain with this difference: the surface layer of cold air is so deep that raindrops freeze before striking the ground. Sleet can readily be distinguished from freezing rain, because sleet bounces when it strikes the ground and freezing rain does not. Accumulations of ice pellets or freezing rain can cause very hazardous walking and driving conditions.

Hail consists of rounded or jagged lumps of ice, often characterized by concentric layering resembling the internal structure of an onion (Figure 8.6). Hail develops within intense thunderstorms as strong convection currents transport ice pellets upward, into the middle and upper reaches of a cumulonimbus cloud. Along the way, ice pellets grow larger by collecting supercooled water droplets, and eventually become too heavy to be supported by

Special Topic

When Is It Too Cold or Too Warm to Snow?

During an episode of particularly frigid winter weather, some people argue that it's too cold to snow. Indeed, the coldest weather is often accompanied by fair skies, and the climate records of the northern United States and Canada indicate that snowfall totals decline with lowering temperatures. For some midwestern United States cities (such as Bismarck and Minneapolis), March on average is both the snowiest and the mildest of the winter months (December through March). In Canada, except in the mountains, average annual snowfall declines from more than 250 cm (100 in.) over much of the relatively mild southeast to less than 60 cm (24 in.) on the frigid shores of the Arctic Ocean. Even though total snowfall does indeed decrease with

falling temperature, snow is possible even at extremely low air temperatures. Snowflakes are small, however, and accumulations are usually meager.

The relatively small amount of water vapor in very cold air means that comparatively little water is available for precipitation. Recall from Chapter 6 that the saturation vapor pressure drops rapidly as air temperature falls. For example, the water vapor concentration in saturated air at -30 °C (-22 °F) is only about 12 percent of the water vapor concentration in saturated air at -5 °C (23 °F). Hence, the water available for precipitation decreases with falling air temperature.

The heaviest snowfalls typically occur when the temperature of the lower atmosphere is within a few

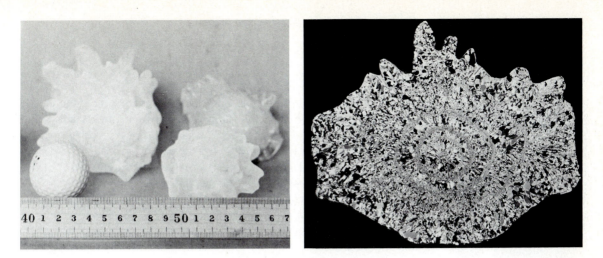

FIGURE 8.6
Hailstones sometimes grow to the size of golf balls or larger (left). A hailstone consists
of concentric layers of clear and opaque ice as shown in this cross section photographed in polarized light (right). [NOAA (left) and NCAR/NSF (right) photographs]

convective updrafts. Ice pellets then descend through the cloud, exit the cloud
base, and enter air that is typically above the freezing point. The pellets begin
to melt, but if large enough to start with, some ice will survive the journey to
the ground as hailstsones. Most hail consists of harmless granules of ice less
than 1 cm in diameter, but violent thunderstorms may spawn destructive
hailstones the size of golf balls or larger. Hail is usually a spring and summer
phenomenon that is particularly devastating to crops. We have much more to
say about hail in Chapter 13.

degrees of the freezing point because at that temperature the potential amount of water that can precipitate
in the solid form (snow) is maximum. On the other
hand, moderate or heavy snowfall is very unlikely
when the temperature of the lower atmosphere falls
below −20 °C (−4 °F). But even in the coldest regions
of the globe where precipitable water amounts are lowest, some snow falls. For example, an estimated average
annual 5 cm (2 in.) of snow falls on the high interior
plateau of Greenland where average annual temperatures are below −30 °C (−22 °F).

When is it too warm to snow? Surprisingly, snow
is possible when the air temperature near the Earth's
surface is well above the freezing point. Snow can fall
even when the near-surface air temperature is as high
as 10 °C (50 °F)! The only requirement is that the wet-
bulb temperature remain below 0 °C (32 °F), which
also means that the relative humidity is very low. For
example, if the air temperature is 5 °C (41 °F), the
relative humidity must be under 32 percent for the
wet-bulb temperature to be subfreezing (see the inside
back cover). Initially, some snowflakes vaporize and
others melt as they fall through dry, above-freezing air.
Vaporization and melting of snowflakes tap heat from
the ambient air; that is, sensible heat is converted to
latent heat, and the air temperature drops. With sufficient vaporization and melting, the air eventually cools
to the wet-bulb temperature, that is, below the freezing
point. Therefore, at the Earth's surface, what started
out as rain (or a mixture of rain and snow) turns to
snow.

Precipitation Measurement

FIGURE 8.7

A standard National Weather Service rain gauge. [Courtesy of Qualimetrics]

FIGURE 8.8

A tipping-bucket rain gauge is designed to provide a cumulative record of rainfall in increments of 0.01 in. [Courtesy of Qualimetrics]

We collect and measure precipitation today using essentially the same device that was used as long ago as the fifteenth century: a container open to the sky. The standard National Weather Service **rain gauge** is equipped with a cone-shaped funnel at the top that directs rainwater into a long, narrow cylinder, which is seated inside a larger, outer cylinder (Figure 8.7). The narrow cylinder magnifies the scale of the rainwater, so the observer can resolve rainfall in increments of 0.01 in. Total rainfall of less than 0.005 in. is recorded as a "trace." Rainwater that accumulates in the inner cylinder is measured by a stick, which, in the United States, is graduated in inches. Canada and most other nations use metric units. Rainfall is measured at some fixed time once every 24 hours, and the gauge is then emptied.

With regard to snow, we are interested in measuring (1) the depth of snow that falls during each 24-hour period between observations, (2) the meltwater equivalent of that snowfall, and (3) the depth of snow on the ground at each observation time. New snowfall is usually collected on a simple board that is left on top of the old snow cover. When new snow falls, the depth is measured to the board; the board is then swept clean and moved to a new location. The meltwater equivalent of new snowfall can be determined by melting the snow collected in a rain gauge (from which the funnel and inner cylinder have been removed). Total snow depth is usually measured with a special yardstick (graduated in tenths of an inch) or meterstick. Snow depth is determined at several representative locations and then averaged.

As a general rule, 10 cm of fresh snow melts down to 1 cm of rainwater, although this ratio varies considerably depending on the temperature at which the snow falls. "Wet" snow falling at surface air temperatures at or above freezing has a much greater water content (per centimeter) than "dry" snow falling at very low surface air temperatures. The ratio of snowfall to meltwater may vary from 3:1 for very wet snow to 30:1 for dry fluffy snow.

Monitoring the rate of rainfall is often desirable, especially in areas prone to flooding. Accordingly, some rain gauges are designed to provide a cumulative record of rainfall with time. A **weighing-bucket rain gauge** consists of a continuously recording scale that calibrates the weight of accumulating rainwater as water depth. Cumulative rainfall is recorded by either a device that punches a paper tape or a pen that marks a chart on a clock-driven drum. In subfreezing weather, antifreeze in the collection bucket melts snow as it falls into the gauge so that a cumulative meltwater record is produced.

A **tipping-bucket rain gauge** (Figure 8.8) is somewhat more accurate than a weighing-bucket rain gauge but does not perform well in subfreezing weather. This instrument features two small, free-swinging containers, each of which can collect the equivalent of 0.01 in. of rainfall. Alternating with one another, each container fills with water, tips and spills its contents, and thereby trips an electric switch that either marks a clock-driven chart or sends an electric pulse for recording by a computer or magnetic tape.

Both rainfall and snowfall are notoriously variable from one place to another, especially in convective showers. Therefore, the siting of precipitation gauges is particularly important in order to ensure accurate and representative readings. The site must not only be sheltered from strong winds but also be well away from buildings and vegetation that might shield the instrument. Generally, obstacles should be no closer than four times their height.

Weather Modification

Weather modification is any change in weather that is induced by human activity. The activity may be either intentional or inadvertent. This section describes two principal types of intentional weather modification: cloud seeding and fog dispersal. Later, we discuss hail suppression (Chapter 13), hurricane modification (Chapter 14), and the impact of air pollution on weather and climate (Chapters 16 and 18).

CLOUD SEEDING

Since World War II, considerable research has gone into methods of enhancing precipitation by cloud seeding. **Cloud seeding** is an attempt to stimulate natural precipitation processes by injecting nucleating agents into clouds. Most cloud-seeding experiments are directed at cold clouds.

The objective of seeding cold clouds is to stimulate the Bergeron process in clouds that are deficient in ice crystals. The seeding (nucleating) agent is either silver iodide (AgI), a substance with crystal properties similar to those of ice, or dry ice, solid carbon dioxide (CO_2), at a temperature of about -80 °C (-110 °F). Silver iodide crystals are freezing nuclei that are active at -4 °C (25 °F) and below. Within a cloud, dry ice pellets are cold enough to cause surrounding supercooled water droplets to freeze and the frozen droplets then function as nuclei that grow into snowflakes.*

Cloud seeding is done from aircraft by either dropping flares containing silver iodide (Figure 8.9) or dispensing dry ice pellets from a hopper. Cloud

---

* Latent heat released during freezing of water droplets increases buoyancy and thereby stimulates cloud growth.

FIGURE 8.9
This aircraft is injecting clouds with nucleating agents (silver iodide or dry ice) in an effort to stimulate precipitation formation via the Bergeron process. [NCAR/NSF photograph]

seeding from ground-based generators can be less satisfactory because the seeding agent (silver iodide) may not diffuse adequately or reach sufficient altitudes to be effective. Nonetheless, both methods continue to be used to seed cold clouds.

In warm clouds of relatively uniform droplet size, sea-salt crystals and other hygroscopic substances can be injected to trigger the development of relatively large cloud droplets. Such seeding stimulates the collision–coalescence process.

The **Sierra Cooperative Pilot Project (SCPP)** is an example of a long-term, ongoing study of cloud seeding. As of this writing, investigators have compiled more than ten years of data in efforts to seed winter orographic clouds on the windward western slopes of California's Sierra Nevada. The goal of this seeding is to enhance snowfall and thereby thicken the mountain snowpack. The consequent increase in spring runoff is intended to help California meet its increasing domestic and agricultural water demands.

The American River Basin, just west of Lake Tahoe, is the principal SCPP study site, and January through March is the primary seeding season. Clouds targeted for seeding are those rich in supercooled water droplets and deficient in ice crystals. In an attempt to identify the type(s) of cloud most suitable for precipitation enhancement through seeding, SCPP scientists employ an aircraft specially outfitted with sophisticated instruments that measure the size and concentration of cloud and precipitation particles. Another aircraft seeds clouds with silver iodide crystals or dry ice pellets, and, on the ground, an array of precipitation gauges, radar, and other weather instruments monitor the effectiveness of seeding.

Does cloud seeding work? Over a period of many years, in the 1970s and early 1980s, NOAA scientists conducted a statistically rigorous experiment designed to test the effectiveness of weather modification. The experiment was carried out over southern Florida and involved the seeding of cumulus clouds.* Test days were about evenly divided between days when clouds were seeded with silver iodide crystals and days when clouds were seeded with inert (chemically inactive) sand grains. Only after the experiment was completed and the results analyzed were participating scientists made aware of the specific days when silver iodide was used. This procedure was designed to ensure both an unbiased selection of clouds to be seeded and an unbiased interpretation of precipitation results.

The experiment was divided into two phases. Results from an initial "exploratory" effort (phase one) were to be either verified or rejected by a later "confirmatory" procedure (phase two) when the seeding was repeated. Phase one results were very encouraging, showing a 25 percent increase in rainfall on days when silver iodide was the seeding agent, as compared with days when sand was the bogus seeding agent. This finding was statistically

---

* Florida Area Cumulus Experiment (FACE).

significant at the 90 percent level, meaning that there is only a 10 percent probability that the rainfall increase was simply the result of chance. The success of phase one was not repeated, however; during phase two of the experiment, seeding showed no statistically significant increase in rainfall.

The same type of experimental design, involving separate exploratory and confirmatory phases, was carried out in Israel from 1961 to 1967 and from 1969 to 1975. In these studies, however, statistically significant rainfall enhancement during the first phase was verified in the second phase of seeding. The contrasting findings of the Florida and Israeli experiments underscore the uncertainties of cloud seeding and the need for a better understanding of precipitation processes.

Although cloud seeding is probably successful in some instances, the actual volume of additional precipitation produced by cloud seeding and the advisability of large-scale seeding efforts are matters of considerable controversy. Some cloud seeders claim to increase precipitation by 15 to 20 percent or more, but the question remains: Would the rain or snow that follows cloud seeding have fallen anyway? Even if apparently successful, cloud seeding on a large geographical scale may merely redistribute a fixed supply of precipitation, so that an increase in precipitation in one area might mean a compensating reduction in another. For example, rainmaking might benefit agriculture on the high plains of eastern Colorado but might also deprive wheat farmers of rain in the adjacent downwind states of Kansas and Nebraska. Such conflicts can cause legal wrangles between adjoining counties, states, and provinces.

In some experiments, seeding may actually have reduced precipitation. An example is a well-documented cloud-seeding effort carried out over south-central Missouri during five consecutive summers in the 1950s (called **Project Whitetop**). A possible explanation for the reduced precipitation is that the clouds were overseeded. Prior to seeding, cumulus clouds apparently contained just enough ice crystals for precipitation. The addition of more nuclei by seeding probably produced too many ice crystals competing for too few supercooled water droplets. Consequently, Project Whitetop seeding generated a large number of ice pellets that were so small that they tended to remain suspended in the clouds, or they vaporized before reaching the ground.

FOG DISPERSAL

Fog can pose a serious hazard to both surface and air travel. Many auto accidents and some ship collisions and aircraft crashes have been attributed, at least in part, to visibility restrictions caused by dense fog. Fog frequently forces flight delays, reroutings, and cancellations that cost airlines millions of dollars each year and inconvenience thousands of passengers. Although the need for an effective method of **fog dispersal** is great, especially at airports, little progress has been made in this area for both technical and economic reasons.

FIGURE 8.10
Turboclair fog dissipation system has been operating at Orly Airport, Paris, France, since 1970. Jet engines lining the runway blow a stream of warm air over the runway; the relative humidity drops so that fog droplets evaporate. [Photograph by Jean-J. Moreau, Cliche, Aeroport de Paris]

During World War II, the British had considerable success in clearing warm radiation fogs* from runways at 15 military airfields. Heat from fuel burners deployed alongside runways raised the air temperature, thereby lowering the relative humidity to below saturation so that fog droplets vaporized. First operational in late 1943, this fog dispersal system assisted in the safe landing of an estimated 2500 Royal Air Force planes, which returned a total of 15,000 airmen from combat missions. Although this thermal approach to fog dispersal was revived from time to time after the war, the fuel costs incurred were excessive.

Today, the only warm fog dispersal systems operated routinely are at Orly and De Gaulle airports outside Paris (Figure 8.10). In this so-called Turboclair system, underground jet engines direct streams of warm exhaust over runways. Warming lowers the relative humidity causing fog droplets to vaporize.

With shallow radiation fogs at airports, some clearing may be achieved by using helicopters to induce vertical circulation of air. The rotor blades force the warmer unsaturated air that is just above the fog layer to mix with the cooler saturated air below. Consequently, the relative humidity of the mixture falls below 100 percent, and fog droplets vaporize.

Another approach to fog dispersal applies techniques of cloud seeding. Warm fogs are seeded with hygroscopic substances that absorb water vapor, thereby reducing the relative humidity by reducing the vapor pressure. Fog droplets vaporize, and hygroscopic droplets grow into raindrops that fall to the surface. Cold fogs (0 to -20 °C, or 32 to -4 °F) are seeded with dry ice pellets that stimulate the Bergeron process. For the most part, however, fog dispersal by seeding is an experimental technology that requires further research and development, very little of which is now taking place.

---

* Warm fog is composed of water droplets at temperatures above 0 °C (32 °F) and is the most common type of fog. The British fog dispersal system was known by the acronym FIDO, for Fog Investigation Dispersal Operation.

Atmospheric Optics

As the sun's rays travel through the atmosphere, they may be reflected or refracted by suspended water droplets or by ice crystals or by raindrops. The consequence is a variety of optical phenomena, including halos, rainbows, coronae, and glories. The characteristics and origins of these optical effects are the subjects of this section. As a Special Topic, we also examine mirages, another optical effect, which, however, is not due to moisture in the atmosphere.

HALOS

A **halo** is a whitish ring of light surrounding the sun or, sometimes, the moon. It is formed when the sun's rays are refracted by the tiny ice crystals that compose high, thin clouds such as cirrostratus. **Refraction** is the bending of light as it passes from one transparent medium (such as air) into another transparent medium (such as ice or water). The light rays bend because the speed of light is greater in air than in ice or water. Refraction occurs whenever light rays strike the interface between two media at any angle other than 90 degrees (Figure 8.11). (Beams of light that are perpendicular to the water or ice surface are not refracted.)

FIGURE 8.11
Light rays may be refracted (bent) as they travel from one transparent medium into another. (A) A light ray is refracted as it travels from air into water. The speed of light is less in water than in air, so the light ray is bent toward a line drawn perpendicular to the water surface, and angle *r* is less than angle *i*. (B) Light rays that are perpendicular to the water surface are not refracted. (C) Refracted light can be deceptive. The shark appears to be farther away from the raft than it really is. This is because of refraction and the fact that human perception assumes that light travels in straight lines.

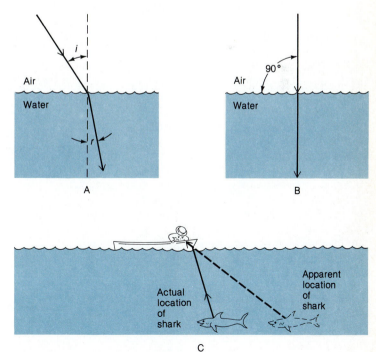

Mirages

Appearances can be deceiving—especially in the case of mirages. Distant buildings or hills may appear higher or lower than they really are. A nonexistent pool of water may suddenly appear on the highway ahead, or a sailboat viewed from shore may appear upside down. These mirages are optical illusions that are caused by the refraction (bending) of light rays within the lower atmosphere.

Light travels at different speeds through different transparent substances. Light changes speed, for example, as it travels from air into an ice crystal. For reasons discussed elsewhere in this chapter, a light ray is refracted at the interface between two different transparent media. The speed of light also varies within a single medium if the density of that medium is not uniform. Hence, light rays bend as they pass through a substance of varying density, such as the atmosphere.

If the density of the atmosphere were the same throughout, then light rays would always travel along straight paths at constant speed, and no refraction would occur. As described in Chapter 5, however, air density varies with changes in temperature and pressure. Because our concern here is with optical phenomena that occur within the lowermost troposphere and involve relatively short viewing distances, we need not be concerned with the influence of horizontal pressure gradients on air density. For the same reasons, we can also ignore the effect of horizontal temperature gradients on air density, but we cannot ignore the effect of vertical temperature profiles on the change of air density with altitude.

As a rule, light rays are refracted in the atmosphere such that the denser air is on the inside (concave side) of the bend, and the less dense air is on the outside (convex side) of the bend. Normally, air density decreases with altitude so that light reflected from a distant object follows a curved path to the observer (Figure 1). Because human perception is based on the assumption that light travels in straight paths, to the observer the object *appears* to be higher than it really is. Hence, for example, the setting or rising sun normally appears slightly higher in the sky than it is.

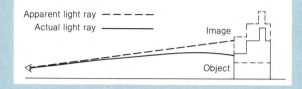

FIGURE 1
Because air density declines with altitude, objects normally appear to be slightly higher than they actually are.

Now, if the air temperature decreases with altitude at less than the usual rate or if there is a temperature inversion (an increase of temperature with altitude), air density decreases with altitude faster than normal. Hence, light rays reflected from a distant object bend more sharply than usual before reaching the viewer, and objects appear higher than normal (Figure 2). This is known as a *superior mirage.*

If, on the other hand, the lowest air layer features a steeper than usual temperature lapse rate, light rays are refracted less than normal and objects appear lower than we usually see them. This is known as an *inferior mirage.*

What happens when the vertical temperature gra-

FIGURE 2
If the temperature profile features a temperature inversion, air density declines with altitude more rapidly than normal, and objects appear significantly higher than usual. This is known as a superior mirage.

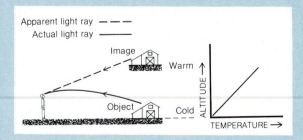

dient changes with altitude? As warm air is advected over a cold, snow-covered surface, the warm air mass is chilled from below, and a temperature inversion develops. The vertical temperature gradient is greatest in the lowest layer of air (Figure 3A). Consequently, a distant object appears to be displaced upward (superior mirage), but because the bottom of the object is within a greater air temperature gradient than is the top of the object, the bottom appears to be lifted more than the top. The object, a building, for example, not only appears to be uplifted, but it also appears shorter than it really is.

A different optical effect is observed when both the temperature and the vertical temperature gradient decrease with altitude (Figure 3B). By midday a desert surface is heated intensely by the sun, and the steepest lapse rate develops just above the ground. With such a temperature profile, a distant object appears to be displaced downward (inferior mirage), but because the bottom of the object is within a steeper air temperature gradient than is the top of the object, the bottom is displaced downward more than the top. The object not only appears to sink, but it also appears to be stretched vertically.

These are only a few examples of the many possible types of mirage. As the vertical air temperature profile becomes more complex, so, too, do the types of mirage that can appear. For example, the familiar oasis mirage in a desert is an inverted image of the sky seen below the horizon; it is caused by a complex temperature profile. All mirages, then, are displacements or distortions of something real.

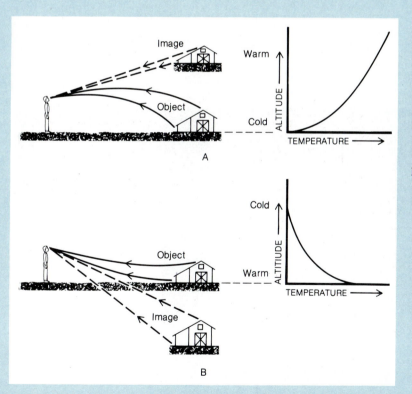

FIGURE 3

(A) A superior mirage, in which an object appears shorter than it really is, occurs when the vertical temperature gradient decreases with altitude while the temperature increases with altitude (a temperature inversion). (B) An inferior mirage, in which an object appears taller than it really is, occurs when the temperature lapse rate is steepest near the surface.

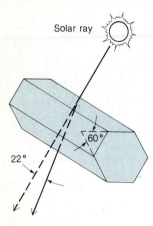

Solar ray

60°

22°

FIGURE 8.12

A ray of sunlight is refracted as it passes through an ice crystal from one side (the top side in the drawing) to another side (the bottom side in the drawing). The angle between the two sides of a hexagonal ice crystal is 60 degrees. This type of refraction produces a 22-degree halo about the sun or moon.

An analogy may help to explain refraction. Suppose you are driving your auto down a highway and suddenly encounter a patch of ice along the right side of the road. The right wheels travel over the ice while the left wheels remain on dry pavement. You slam on the brakes. The left side of the car slows while the right side slips on the ice. Your auto consequently swerves to the left. The swerving of the auto is analogous to a light ray bending toward the medium in which light travels more slowly.

Light is refracted twice as it passes through an ice crystal, that is, upon entering and upon exiting. In clouds, ice crystals occur as hexagonal (six-sided) plates or columns. If crystals are very small and randomly oriented, the light ray is refracted from side to side as it passes through an ice crystal (Figure 8.12). This side-to-side refraction focuses the light in a circle having a radius* of about 22 degrees. (Figure 8.13). For reference, the halo's radius appears to be the same as the width of this page when held at arm's length.

A much rarer halo has a radius of about 46 degrees about the sun (or moon). In this case, refraction is caused by columnar ice crystals with diameters in the 15- to 25-micrometer range. Light rays travel through an ice crystal from side to top or from side to base, rather than from side to side (Figure 8.14).

---

* The radii of halos are usually measured as angles. Suppose, for example, that you are viewing a halo about the moon. Now visualize two lines: one line joins you with the moon, and the other line joins you with any point on the halo. Depending on the type of halo, the angle formed by the two imaginary lines will be either 22 or 46 degrees.

FIGURE 8.13

A 22-degree halo about the sun is caused by refraction of sunlight by ice crystals. This photograph was taken with a wide-angle lens. [Photograph by Arjen and Jerrine Verkaik/SKYART]

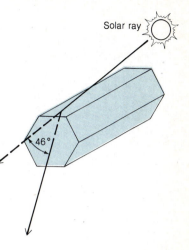

FIGURE 8.14

In a rarer situation than shown in FIGURE 8.12, a solar ray is refracted from side to top or from side to bottom as it passes through a columnar ice crystal. The angle between side and top (or bottom) is 90 degrees. This type of refraction produces a 46-degree halo about the sun or moon.

In some instances, light is concentrated in two brilliant spots about 22 degrees on either side of the sun (Figure 8.15). Called **sundogs** (because they appear to follow the sun around the sky), **mock suns,** or **parhelia,** these refractive phenomena involve relatively large ice crystals that are not randomly oriented.

FIGURE 8.15

Bright spots appearing on either side of the sun, sometimes called sundogs, are caused by refraction of sunlight by ice crystals in the atmosphere. These unusually spectacular sundogs appeared over Foam Lake, Saskatchewan. [Phototheque photograph by Daniel Comeau]

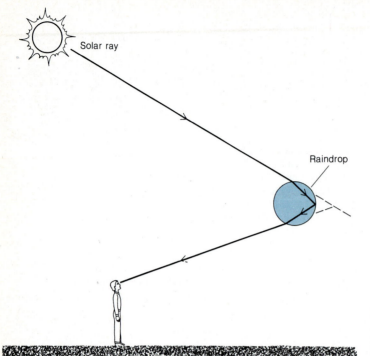

FIGURE 8.16

A solar ray is refracted and internally reflected by a raindrop. This is the basic optical process that causes a rainbow.

RAINBOWS

A **rainbow** is caused by a combination of refraction and reflection of sunlight (or, rarely, of moonlight) by raindrops. Sunlight striking a shaft of falling raindrops is refracted and internally reflected by each drop of rain. As shown in Figure 8.16, a solar ray is refracted as it enters a raindrop; then the ray is reflected by the inside back of the drop before being refracted again as it exits the drop.

A rainbow appears to an observer who has his or her back to the sun and is facing a distant rain shower (Figure 8.17).* A rainbow never forms when the sky is completely cloud covered—the sun must be shining. Because weather usually progresses from west to east, the appearance of a rainbow in the evening signals improving weather. Rain showers to the east are moving away, and clearing skies are approaching from the west, where the sun is setting.

---

* You can create your own rainbow with the spray from a garden hose. Simply direct the spray so that you observe the spray with the sun at your back.

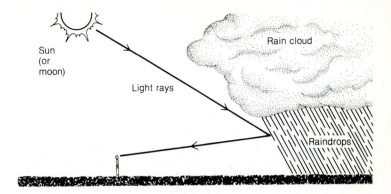

FIGURE 8.17

A rainbow appears to an observer who has his/her back to the sun and faces a distant rain shower.

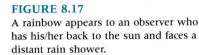

Raindrop refraction produces the concentric arcs of color in a rainbow. Indeed, raindrop refraction is very similar to the refraction of light by a glass prism: both the prism and the raindrop disperse sunlight into its component colors (Plate 2). Violet light, at the short wavelength end of the visible spectrum, is slowed the most as it passes through a prism (or raindrop) and is refracted the most. Red light, at the long wavelength end of the visible spectrum, is slowed the least and is refracted the least.

Raindrops disperse light into an arc composed of six bands of color: red, orange, yellow, green, blue, and violet, listed in order from the outer to the inner band (Plate 16). In most cases, a much dimmer, secondary rainbow appears about 8 degrees above the primary rainbow. This secondary rainbow is produced by double reflection within the raindrops (Figure 8.18). The order

FIGURE 8.18

Refraction of sunlight by raindrops plus double reflection within raindrops produces a dimmer secondary rainbow just above the primary rainbow.

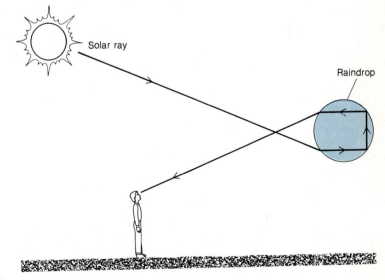

of the colors in the secondary rainbow is the reverse of that in the primary rainbow.

Since ice crystals refract light just as raindrops do, you may wonder why halos are not colored. In fact, ice crystals do disperse light into its component colors, but because the size and shape of ice crystals vary more than the size and shape of raindrops, the colors produced by an assemblage of ice crystals tend to overlap one another rather than form discrete bands. In halos, colors therefore wash out, although occasionally a reddish tinge is visible on the inside of a halo.

CORONAE

A **corona** consists of a series of alternating light and dark rings that surround the moon or, less often, the sun. Typically, a corona is only a few degrees in radius and hence, is far smaller than a halo. It is caused by diffraction of light around water droplets that compose a thin, translucent veil of altocumulus, altocumulus lenticularis, or cirrocumulus.

Diffraction is the slight bending of a light wave as it moves along the boundary of an object such as a water droplet. (Recall our discussion of the wave characteristics of light in Chapter 2.) As light waves bend, they interfere with one another. When the crests of one wave coincide with the crests of another wave, interference is constructive and a larger wave results. On the other hand, when the crests of one wave coincide with the troughs of another wave, the interference is destructive and the waves cancel each other. Where light waves interfere constructively, we see a ring of bright light, and where they interfere destructively, we perceive darkness.

If cloud droplets are of uniform size, then a corona is colored with blue-violet on the inside and red on the outside of each ring. Dependency of diffraction on wavelength causes this separation of color: the longer-wavelength red light is bent more than the shorter-wavelength blue-violet light.

GLORIES

Before the age of aircraft travel, about the only way to view a glory was from the vantage point of a lofty mountain peak. To see a glory an observer must be in bright sunshine above a cloud or fog layer, and the sun must be situated so as to cast the observer's shadow on the clouds below. The observer then sees a **glory** as concentric rings of color around the shadow of his or her head. Although less distinct, the colors of a glory are the same as those of the primary rainbow, with violet being the innermost band and red the outermost band. Today, aircraft pilots and observant passengers frequently see a glory around the shadow of their aircraft on a cloud deck below (Figure 8.19).

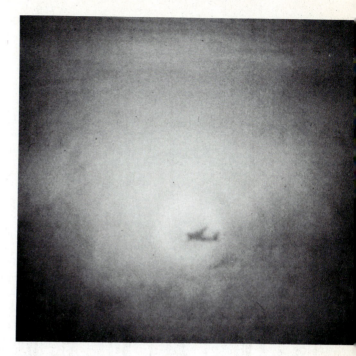

FIGURE 8.19
The glory, photographed from an aircraft, is a consequence of refraction, internal reflection of sunlight,
and diffraction by cloud droplets of uniform size.
[NCAR/NSF photograph]

The glory results from much the same optics as the primary rainbow. One
difference, however, is in the size of the reflecting and refracting particles.
Whereas rainbows occur when sunlight strikes a mass of raindrops, glories
are the consequence of sunlight interacting with a mass of smaller water
droplets of uniform size that form a cloud. In both cases, the sun's rays undergo
refraction upon entering the droplet, followed by a single internal reflection,
and another refraction upon exiting.

The optics of rainbows and glories have one other interesting difference
besides particle size. In the glory, sunlight is refracted and reflected directly
back toward the sun but, as Figure 8.17 shows, this is not the case with
rainbows. With the glory, a spherical cloud droplet diffracts light rays ever so
slightly toward the droplet so that light rays incident on a cloud droplet parallel
those returning from the cloud droplet (Figure 8.20).

FIGURE 8.20
In the special optics that produce a glory, both the incident and returning solar rays are diffracted slightly toward the cloud droplet. Consequently, the incident and
returning rays follow parallel paths.

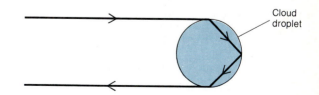

Cloud
droplet

The special optics of glories explains why it appears about the shadow of the observer who is situated in the direct path of both the incident solar rays and the returning (refracted and reflected) solar rays. This also explains an observation that must have fascinated ancient mountain mystics. Suppose that you and a friend are side by side high on a mountainslope at a site favorable for viewing a glory. On a cloud deck below, you see your own shadow beside that of your friend, but a glory appears about your head and not about your friend's head. Lest you presume that you have been singled out, note that your friend has the opposite observation, for your friend sees a glory only about his or her head. The fact is that each observer is in a position to view only one glory.

Conclusions

In this and previous chapters, we have seen how water changes phase to produce clouds and precipitation as it is transported within the global hydrologic cycle. It is evident that a special set of circumstances is necessary for clouds to yield precipitation in its various forms. An understanding of these circumstances has enabled scientists to develop techniques to stimulate natural precipitation processes through cloud seeding.

Atmospheric circulation plays a key role in bringing air to saturation and triggering cloud development. The next several chapters describe the many atmospheric circulation systems and their associated weather conditions. We begin in Chapter 9 with a discussion of the forces that drive and shape atmospheric circulation.

Summary Statements

The Bergeron and collision–coalescence processes are mechanisms whereby cloud particles grow large enough to counter updrafts and fall to the Earth's surface as precipitation. The bulk of the Earth's precipitation originates in cold clouds composed of a mixture of ice crystals and supercooled water droplets.

The collision–coalescence process occurs in warm clouds and depends on the presence of relatively large cloud droplets that grow through collisions and coalescence with smaller cloud droplets.

The Bergeron process requires the coexistence of ice, liquid water, and water vapor; it depends on the difference in saturation vapor pressure between water vapor over ice and water vapor over water. As a consequence of this difference, ice crystals grow at the expense of supercooled water droplets.

The principal forms of precipitation are rain, drizzle, freezing rain, snow, ice pellets (sleet), and hail. The form taken by precipitation depends on the source cloud and the vertical temperature profile below the cloud.

Cloud seeding is a type of weather modification intended to enhance rainfall or snowfall by stimulating natural precipitation processes. In most cloud-seeding experiments, nucleating agents are injected into cold clouds in an attempt to spur the Bergeron process. Cloud-seeding efforts are evaluated by rigorous statistical testing and are not always successful.

Fogs can be dispersed either by raising the air temperature (thereby lowering the relative humidity) or by seeding.

A halo is produced when the ice crystals composing high, thin clouds refract sunlight or moonlight. Rainbows develop when falling raindrops refract and internally reflect sunlight. Coronae are produced when water droplets composing a thin cloud layer diffract either moonlight or sunlight. Glories are the consequence of the same optical effect as a primary rainbow, except that the refracting and reflecting agents are cloud droplets instead of raindrops.

Key Words

terminal velocity	freezing rain	tipping-bucket rain	refraction
warm clouds	snow	gauge	sundogs
collision–coalescence	snow pellets	weather modification	mock suns
process	snow grains	cloud seeding	parhelia
Bergeron process	ice pellets	Sierra Cooperative	rainbow
cold clouds	sleet	Pilot Project	corona
virga	hail	(SCPP)	diffraction
precipitable water	rain gauge	Project Whitetop	glory
precipitation	weighing-bucket rain	fog dispersal	
drizzle	gauge	halo	
rain			

Review Questions

1. Distinguish between warm clouds and cold clouds.

2. Describe the collision–coalescence process of precipitation formation.

3. Describe the Bergeron process of precipitation formation.

4. Once a raindrop or a snowflake leaves a cloud it enters a *hostile* environment. What is meant by this statement?

5. Why are drizzle drops smaller than raindrops? Why do warm clouds generally produce smaller raindrops than cold clouds?

6. Distinguish between ice pellets (sleet) and freezing rain. Distinguish also between ice pellets and hail.

7. Under what conditions does freezing rain develop? Explain why freezing rain can be destructive.

8. Explain how hail may reach the ground even when surface air temperatures are well above the freezing point.

9. Why must care be exercised in siting precipitation gauges?

10. Define weather modification.

11. What is the basic objective of cloud seeding?

12. Comment on the effectiveness of cloud seeding. Does it work?

13. List some of the benefits and costs of successful cloud-seeding efforts.

14. What is the principle underlying the thermal method of fog dispersal?

15. How might cloud seeding actually reduce the amount of precipitation that falls?

16. Compare and contrast the optics of rainbows with those of the glory. Explain why you can view only your own glory.

17. What does the appearance of a morning rainbow suggest about the weather later in the day?

18. Why do raindrops and ice crystals refract light rays?

19. Distinguish between a primary rainbow and a secondary rainbow.

20. What type of cloud produces a halo about the sun or moon?

Points to Ponder

1. Explain why cold clouds at temperatures between 0 and -20 °C consist of more supercooled water droplets than ice crystals. Explain also why most clouds colder than -20 °C are made up almost entirely of ice crystals. What type of clouds are these?

2. Why is the saturation vapor pressure greater over supercooled water droplets than over ice crystals?

3. Give two reasons why rainfall is likely to be heavier on top of a mountain than in a neighboring valley.

4. Is rain possible where you are, even though you measure a relative humidity well below 100 percent? Explain your answer.

5. Identify and describe the factors that control the size of snowflakes.

6. As a general rule, 10 in. of snow melts down to 1 in. of rainwater. From this, determine the average density of snow.

7. Speculate on why in some localities the exterior of a rain gauge is painted white. Why is it desirable to shelter a rain or snow gauge from strong winds?

Projects

1. Design an experiment in which you create your own rainbow.

2. Assemble photographs of a variety of atmospheric optical effects. You may wish to take your own photographs. Provide an explanation for each optical phenomenon.

Selected Readings

Blanchard, D. C. "Science, Success and Serendipity." *Weatherwise* 32 (1979):236–241. Recounts the pioneering efforts of V. Schaefer, I. Langmuir, and T. Bergeron in cloud physics research.

Colbeck, S. C. "What Becomes of a Winter Snowflake?" *Weatherwise* 38 (1985):312–315. Presents a well-illustrated description of processes taking place within a snowbank.

Fraser, A. B. "Chasing Rainbows." *Weatherwise* 36 (1983):280–287. Describes a series of bright arcs observed within a primary rainbow.

Greenler, R. *Rainbows, Halos, and Glories.* New York: Cambridge University Press, 1980. 195 pp. Discusses the causes of a wide variety of optical phenomena in the atmosphere; includes some spectacular photography.

Ogden, R. J. "Fog Dispersal at Airfields—Part 1." *Weather* 43 (1988):20–25; "Fog Dispersal at Airfields—Part 2." *Weather* 43 (1988):34–38. Discusses the highly successful British effort to disperse radiation fogs at military airfields during World War II.

Reynolds, D. W., and A. S. Dennis. "A Review of the Sierra Cooperative Pilot Project." *Bulletin of the American Meteorological Society* 67, No. 5 (1986):513–523. Gives a detailed account of a long-term study of seeding of cold orographic clouds.

Snow, J. T., and S. B. Harley. "Basic Meteorological Observations for Schools: Rainfall." *Bulletin of the American Meteorological Society* 69, No. 5 (1988):497–507. Discusses rainfall measurement and evaluates inexpensive rain gauges suitable for classroom use.

Walker, J. *Light from the Sky.* New York: W. H. Freeman, 1980. 78 pp. Reprints *Scientific American* articles on various atmospheric optical phenomena, including mirages, halos, rainbows, and the glory.

9

The Wind

The Forces
        Air Pressure Gradients
        Centripetal Force
        Coriolis Effect
        Friction
        Gravity
        Summary
Joining Forces
        Hydrostatic Equilibrium
        Geostrophic Wind
        Gradient Wind
        Surface Winds
Continuity of Wind
Scales of Weather Systems
Wind Pressure
Wind Measurement
Conclusions

Special Topic: Wind Power
Special Topic: Wind Gusts, Wind Shear, and Atmospheric
                        Stability

Mathematical Note: Geostrophic and Gradient Winds

Many forces interact to initiate and shape air movement. Ultimately, it is the sun that supplies the energy that drives the wind. [Photograph by Mike Brisson]

SOME WEATHER systems favor clear skies, light winds, and frosty mornings, whereas others bring ominous clouds, precipitation, and biting winds. Some weather systems trigger brief showers, and others are accompanied by persistent fog and drizzle. Certain weather systems are highly localized and shortlived; others dominate the weather over thousands of square kilometers for prolonged periods. Different weather systems bring different types of weather depending on the air circulation pattern that characterizes each system.

In this chapter, we describe the various forces that control air circulation. We begin by examining each force separately as if each force acted independently of all the other forces. We then show how these forces combine to initiate and modify atmospheric circulation. In the Special Topic "Wind Power," we also consider how modern technology is tapping the energy of winds.

The Forces

Atmospheric circulation is nothing more than the wind. It is useful to distinguish between the horizontal component of the wind (east–west and north–south) and the vertical component of the wind (up–down). Although the horizontal wind is usually considerably stronger than vertical air motion, the vertical component plays the key role in cloud formation. As we will see later in this chapter, however, vertical air motion is linked to horizontal air motion.

For convenience of study, imagine the wind as a continuous stream of air composed of discrete **air parcels.** Assume that any force acting on an air parcel represents the influence of that same force on a stream of air parcels, in other words, on the wind. Now assume also that each air parcel consists of a unit mass of air—a single gram, for example. Hence, in examining each force that influences air motion, we examine the force per unit mass of air.

A force per unit mass is numerically equivalent to an acceleration. This follows from **Newton's second law of motion,*** where

$$\text{force} = \text{mass} \times \text{acceleration}$$

For this reason, we sometimes use the terms "force" and "acceleration" interchangeably when we consider the motion of air parcels. Although the terms are numerically equivalent, a change in velocity is actually a response

---

* This is the second of three fundamental laws of motion stated in 1687 by Sir Isaac Newton, the British mathematician and physicist.

Wind Power

Harnessing the energy of the wind is a technology that was well established as early as the twelfth century in portions of the Middle East where water power was not available. In North America, the energy crisis of the 1970s spurred renewed interest in this ancient technology. Today scientists are employing modern aerodynamic principles and space-age materials in designing and constructing modern wind-driven turbines that convert some of the wind's kinetic energy into electricity.

Large, modern windmills, like the one shown in Figure 1, are capable of generating 200 kW of electricity. If such machines were to sustain this level of power production continuously, each would provide enough power for nearly 100 average American homes. By the year 2000, an estimated 30,000 large wind-driven turbines and thousands of small wind-mills would be required to meet one-tenth of our total electrical needs.

In Chapter 4, we saw how the sun drives the atmosphere. Actually, only about 2 percent of the solar radiational energy that reaches the Earth is ultimately converted into the kinetic energy of wind. This is still a tremendous quantity of energy. Theoretically, windmill blades can convert 60 percent of the wind's energy into mechanical energy. Typically, however, wind-generating systems extract only about 25 percent of the wind's energy, and average wind speeds must be at least 20 km (12 mi) per hour before most wind-powered electricity-generating systems can operate economically.

The power that a windmill can extract from the wind is directly proportional to (1) air density, (2) the area swept out by the windmill blade, and (3) the cube of the instantaneous wind speed (V^3). Wind speed is by far the most important consideration in evaluating wind energy potential at any locality, for even small changes in wind speed translate into large changes in energy generation. For example, doubling the wind speed (a common occurrence) multiplies the available wind power by a factor of eight ($2 \times 2 \times 2$). Unfortunately, both wind speed and direction vary continuously with time (see Figure 9.26), and wind speed

FIGURE 1
This 200-kW wind turbine generator supplies part of the electric power used by residents of Clayton, New Mexico. [U.S. Department of Energy photograph]

also varies with exposure of the site, roughness of the terrain, elevation above the surface, and the season. As a general rule, a minimum of several years of detailed wind monitoring is needed for a preliminary evaluation of wind power potential at any location. The long-term climatic record should also be consulted to check for the frequency of potentially destructive winds. Wind data from nearby weather-observing stations can be

FIGURE 2
This windmill farm in southern California feeds electricity into existing power grids. [Photograph courtesy of Southern California Edison Company]

tem must include a means of storing the energy generated during gusty periods for use when the wind is light or calm. Several thousand dollars' worth of storage batteries are required to provide a reasonable storage capacity in a wind power system for a typical home. A 3- to 5-kW wind turbine is needed to meet the total electrical requirements, including heating, of a typical household. Today, such systems are commercially available, but the cost of materials and construction—including a tower, storage batteries, and generator—ranges from about $5000 to $20,000.

Economy of scale suggests that centralized arrays of wind turbines, called *windmill farms,* are preferable to individual household wind turbines. Windmill farms consist of 50 or more super wind power generators, each capable of producing one or more megawatts of electricity. Windmill arrays now operating in California, for example, supplement conventional energy sources by feeding electricity to existing power grids (Figure 2). Since 1982, more than 13,000 turbines have been installed in California, and by the summer of 1985, they had produced the energy equivalent of 1 million barrels of oil. By 1990, windmill farms in California were supplying 1 percent of the state's total electrical demand.

Given current technical and economic limitations, wind power has its greatest immediate potential in those regions where average winds are relatively strong and consistent in direction. In North America, such regions include the western High Plains, the Pacific Northwest coast, portions of coastal California, the eastern Great Lakes, the south coast of Texas, and exposed summits and passes in the Rockies and Appalachians. Low-power wind systems have potential in small, isolated communities and on individual farms and ranches. The environmental impact of wind systems is usually minimal: they may be somewhat noisy, detract some from the beauty of the landscape, and kill some birds.

With today's renewed interest in wind power, it is ironic to note that in the 1930s and 1940s, the Rural Electrification Administration replaced about 50,000 small wind-powered pumping and electricity-generating systems on midwestern farms.

very useful, as long as care is taken in extrapolating wind data between localities and from one elevation to another.

At ordinary speeds, wind is a relatively diffuse energy source, comparable in magnitude to insolation. Hence, a wind turbine's power generation potential also depends on the area swept out by the windmill blades. Larger windmill blades harvest more energy. However, as windmill blades get larger, design problems develop, and costs soar and soon become prohibitive. So far, the largest experimental wind turbine systems have blade diameters approaching 100 m (300 ft).

The most formidable obstacles to the development of wind power potential stem from the inherent variability of the wind. The electrical output of wind turbines varies as a consequence, and a wind power sys-

to a force. A force acts on an air parcel to bring about an acceleration or deceleration of that parcel.

Forces acting on air parcels, which either initiate or modify motion, are the consequence of (1) air pressure gradients, (2) the centripetal force, (3) the Coriolis effect, (4) friction, and (5) gravity. Actually, the centripetal force is not an independent force but a consequence of other forces. Nonetheless, it is instructive to examine the centripetal force along with the other forces.

AIR PRESSURE GRADIENTS

A **gradient** is simply a change in some property with distance. An **air pressure gradient** exists whenever air pressure changes from one place to another. As noted in Chapter 5, spatial variations in air pressure can arise from contrasts in air temperature (principally), from differing water vapor concentrations, or from both. An air pressure gradient thus develops between a mass of cold, dry air and a mass of warm, humid air. In addition, diverging and converging winds can bring about air pressure changes and thereby induce an air pressure gradient. This important process is covered in Chapter 11, when we examine the origin of highs and lows.

Air pressure gradients develop both horizontally and vertically within the atmosphere. A horizontal pressure gradient refers to air pressure changes along a surface of constant altitude (mean sea level, for example). A vertical pressure gradient is a permanent feature of the atmosphere, since air pressure is greatest at the Earth's surface and decreases rapidly with altitude.

In order to represent horizontal air pressure gradients on a weather map, the air pressure measured at each weather station is first reduced to sea level, as discussed in Chapter 5. Lines are then drawn joining localities that have the same air pressure reading; some interpolation between weather stations is always necessary. These lines are called **isobars.** The interval between successive isobars is usually 4 mb.

An isobaric analysis is used to locate centers of high and low pressure, and to determine the magnitude of the horizontal air pressure gradients between weather systems. Closely spaced isobars (Figure 9.1A) mean that air pressure changes rapidly with distance, and the pressure gradient is described as steep or strong. More widely spaced isobars (Figure 9.1B) indicate that air pressure changes less with distance, and the pressure gradient is weaker. Note that air pressure gradients are always measured in a direction perpendicular to the isobars.

How do air pressure gradients influence the movement of air? Let us examine an analogy. Suppose a bathtub is partially filled with water, as shown in Figure 9.2. As we slosh the water back and forth from one end of the tub to the other, a water pressure gradient develops along the bottom of the tub. At any instant, the water pressure is high where the water level is high, and low where the water level is low. If we stop agitating the water, the water level gradually returns to a horizontal surface, thus creating a uniform water

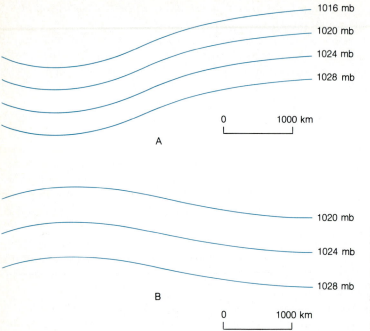

1016 mb
1020 mb
1024 mb
1028 mb

0   1000 km

A

1020 mb

1024 mb

1028 mb

B

0   1000 km

FIGURE 9.1
The horizontal air pressure gradient is relatively steep where isobars are close together (A) and weaker where isobars are farther apart (B). Isobars are lines of equal air pressure. Here the contour interval (the difference between successive isobars) is 4 mb.

pressure along the tub bottom. Hence, in response to a water pressure gradient, water flows from one end of the tub (where the water pressure is greater) to the other end (where the water pressure is less) to eliminate the pressure gradient.

Similarly, when an air pressure gradient develops, air flows in such a way as to eliminate the pressure gradient. Thus, the wind blows away from regions where air pressure is relatively high and toward locales where air pressure is relatively low. The wind is strong where the pressure gradient is steep (closely spaced isobars), and light or calm where the pressure gradient is weak (widely spaced isobars). The force that causes air parcels to move as the consequence of an air pressure gradient is known as the **pressure gradient force.**

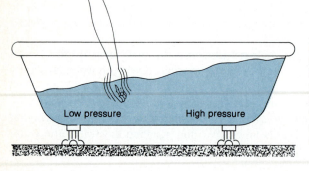

Low pressure   High pressure

FIGURE 9.2
Sloshing water back and forth in a bathtub creates a horizontal pressure gradient on the bottom of the tub. The water pressure gradient in the tub is analogous to horizontal air pressure gradients in the atmosphere. That is, in response to a pressure gradient, the water (or air) flows from an area of higher pressure toward an area of lower pressure.

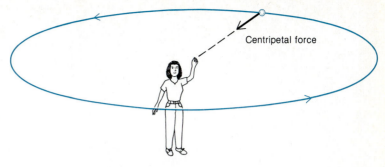

FIGURE 9.3
A rock attached to a string describes a circular path. Centripetal (for "center seeking") is the name we give to the force that confines an object to a curved path. If the string is cut, the centripetal force is eliminated and the rock flies off in a straight line (tangent to the circular path).

CENTRIPETAL FORCE

When we tie a rock to a string and swing it around, the rock describes a circular path of constant radius (Figure 9.3). If we then cut the string, the rock flies off in a straight path. This is an illustration of **Newton's first law of motion:** an object at rest remains at rest, and an object in straight-line, unaccelerated motion remains that way unless acted on by a net (unbalanced) force. Before the string is cut, the string exerts a force on the rock by confining it to a circular path. This force is directed inward, perpendicular to the direction of motion and toward the center of the circular path. For this reason it is known as the **centripetal force.***

A force brings about an acceleration. We usually think of an acceleration as merely a change in the speed of an object, as when an auto speeds up or slows down. However, an acceleration may also consist of a change in the direction of a moving object without any change in the object's speed. This is the case in our example of the rock on a string. The centripetal force is a net force that is responsible for a continuous change in direction (a curved rather than a straight path), not a change in speed. Cutting the string eliminates the centripetal force, and the rock then follows a straight path. From the rock and string illustration, it follows that whenever the wind describes a curved path, the centripetal force operates. As we will see later in this chapter, however, the centripetal force results from an imbalance of other forces and is not itself an independent force.

CORIOLIS EFFECT

Imagine that you are located far away at some fixed point in space, looking back at planet Earth. Over many hours, you follow the track of a storm, which is clearly identifiable by its slowly swirling mass of clouds (Figure 9.4). From your perspective, the storm center appears to move in a straight line at constant

---

* *Centripetal* means "center seeking."

FIGURE 9.4
In this visible satellite image, an intense midlatitude cyclone is readily identified by its swirling mass of clouds. The storm center is located just east of New Jersey. [NOAA National Environmental Satellite, Data, and Information Service]

speed. At the same time, an observer on Earth is tracking the storm. From that observer's perspective, the storm center appears to follow a curved path. Surprisingly, both descriptions of the storm's track are correct!

The two descriptions are correct because the two observers used different frames of reference in following the storm's movements. The Earthbound observer's frame of reference is the familiar north–south, east–west, and up–down coordinate system that rotates with the Earth. To the Earthbound observer, it is not obvious that this coordinate system is rotating because it and the observer rotate together. Viewed from space, however, as in Figure 9.5, the Earthbound coordinate system actually shifts as the Earth rotates. It is as if the Earth and the coordinate system rotate under the storm (or any other object moving over the Earth's surface). Meanwhile, from your distant vantage point in space, you followed the storm's movement with respect to a nonro-

FIGURE 9.5

When viewed from space, our north–south, east–west frame of reference changes as the Earth rotates on its axis. Here a south wind in the Northern Hemisphere is deflected (relative to the frame of reference) to the right and becomes a southwest wind. [Modified from F. K. Lutgens and E. J. Tarbuck, *The Atmosphere: An Introduction to Meteorology,* 4th ed., © 1989, p. 169. Adapted by permission of Prentice-Hall, Inc., Englewood Cliffs, NJ]

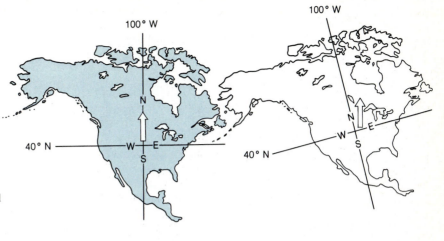

tating coordinate system, fixed in space. In summary, then, the difference in storm track (straight versus curved) is due to the difference in coordinate system (nonrotating versus rotating).

Recall our earlier discussion of Newton's first law of motion and the centripetal force. We saw that curved motion implies that a net force is operating, whereas unaccelerated, straight-line motion implies a balance of forces. If we apply this law to our storm track example, we conclude that a net force operates when we use the Earthbound rotating coordinate system, whereas forces are balanced when we use the nonrotating coordinate system. Hence, changing our frame of reference (coordinate system) from nonrotating to rotating gives rise to a net force responsible for curved motion. This deflective force is known as the **Coriolis effect,** named for Gustav Gaspard de Coriolis, who first described the phenomenon mathematically in 1835.

Wind direction and wind speed are measured with respect to the north–south, east–west, and up–down frame of reference that rotates with the Earth. Therefore, we must take the Coriolis effect into account in any explanation of air circulation. Because of the Coriolis effect, in the Northern Hemisphere, the wind is deflected to the right of its initial direction, and in the Southern Hemisphere, the wind is deflected to the left of its initial direction (Figure 9.6).

Although the Coriolis effect influences the wind regardless of its direction, the magnitude of the effect varies significantly with latitude. This is because the Coriolis effect stems from the rotation of the Earth on its axis, which imparts a rotation to our Earthbound frame of reference. The rotation of our frame of reference is maximum at the poles and declines with latitude to zero at the equator. This variation can be understood by visualizing the daily rotation of towers situated at different latitudes. In a 24-hour day, the Earth makes one complete rotation, as would a tower situated at the North or South Pole. In the same period, a tower at the equator would not rotate at all, but

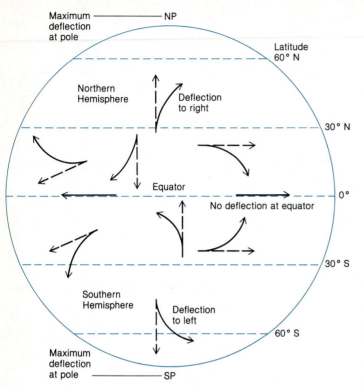

FIGURE 9.6

Large-scale winds are deflected to the right in the Northern Hemisphere and to the left in the Southern Hemisphere. This deflection is due to the Coriolis effect.

instead would describe an end-over-end motion. For a tower located at any latitude in between, some rotation of the tower occurs as the Earth rotates but not as much as at the poles. The Coriolis effect is thus latitude dependent: the Coriolis effect is zero at the equator and increases with latitude to a maximum at the poles.

The Coriolis effect also varies with wind speed; that is, the deflection increases as the wind strengthens. This is because, in the same period of time, fast air parcels cover greater distances than slow air parcels. The longer the trajectory, the greater is the shift of the coordinate system with respect to the air parcel. For practical purposes, the Coriolis effect has an important influence only on the air circulation within large-scale weather systems, that is, systems larger than thunderstorms.

A rotational motion usually accompanies the draining of water from a sink or a bathtub. A popular misconception holds that the direction of this rotation (clockwise or counterclockwise) is consistently in one direction in the Northern Hemisphere and in the opposite direction in the Southern Hemisphere, presumably because of the Coriolis effect. At the very small scale represented by the sink or bathtub, however, the magnitude of the Coriolis effect is simply too small to have a significant effect on the direction of rotation. The drainage direction is more likely a consequence of some residual motion of the water when the sink or bathtub was first filled with water and can be either clockwise or counterclockwise.

Although this book focuses mostly on weather systems in the Northern Hemisphere, it is useful to know why the Coriolis effect reverses direction from the Northern to the Southern Hemisphere, causing large-scale winds in the Southern Hemisphere to swerve to the left rather than to the right. This reversal is related to the difference in an observer's sense of the Earth's rotation in the two hemispheres. To an observer at the North Pole, the Earth rotates counterclockwise, whereas to an observer at the South Pole, the Earth rotates clockwise. This rotation reversal translates into a reversal in deflective direction between the two hemispheres.

FRICTION

We normally think of **friction** as the resistance that an object encounters as it moves in contact with other objects. We usually associate friction with solids as when we attempt to slide a heavy appliance across the floor. But friction also affects fluids, both liquids and gases. The friction of fluid flow is known as **viscosity.** One source of fluid friction is the random motion of molecules composing a liquid or gas; this type of fluid friction is called **molecular viscosity.** Considerably more important, however, is fluid friction that arises from much larger scale irregular motions, called eddies, which develop within fluids; this type of fluid friction is known as **eddy viscosity.**

The river in Figure 9.7 illustrates the effects of eddy viscosity. Rocks in the river bed break the stream flow into eddies that form to the lee of the

FIGURE 9.7
Rocks in a river bed break the current into turbulent eddies that appear as white water to the lee of rocks. [Photograph by J. M. Moran]

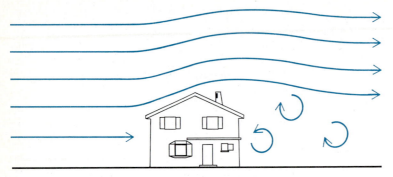

FIGURE 9.8

Turbulent eddies develop in the wind on the leeward side of a house.

rocks. Eddies, visible as white water behind the rocks, tap some of the stream's kinetic energy so that the stream slows. In an analogous manner, obstacles on the Earth's surface such as trees, houses, and telephone poles break the wind into eddies of various sizes to the lee of each obstacle (Figure 9.8). Consequently, the wind is slowed.

A snow fence (Figure 9.9) provides a practical illustration of the frictional slowing of the wind. Snow fences are designed to trap wind-blown snow, in some instances to keep the snow from being blown onto a nearby highway

FIGURE 9.9

A snow fence slows the wind, thereby decreasing the wind's ability to transport snow. Hence, snow accumulates to the lee of the snow fence. [Photograph by Mike Brisson]

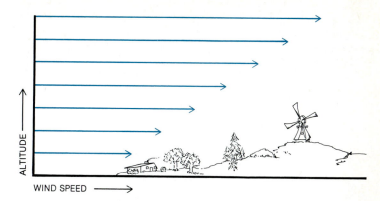

ALTITUDE ⟶

WIND SPEED ⟶

FIGURE 9.10
The horizontal wind strengthens with altitude—away from the frictional resistance offered by objects on the Earth's surface.

and in others to keep the soil snow covered.* A snow fence breaks the wind into small eddies, thereby tapping some of the wind's kinetic energy. The wind diminishes, losing some of its snow-transporting ability, and snow accumulates on the lee side of the fence.

The rougher the surface of the Earth, the greater is the eddy viscosity of the wind. A forest thus offers more frictional resistance to the wind than does the relatively smooth surface of a freshly mowed lawn. The eddy viscosity diminishes rapidly with altitude above the Earth's surface, away from the obstacles mainly responsible for frictional resistance. Hence, wind speed increases with altitude. This explains the advantage of siting a windmill at as high an elevation as possible (Figure 9.10). Above an average altitude of about 1 km (0.62 mi), friction is a minor force that merely acts to smooth the flow of air. The atmospheric zone in which frictional resistance (eddy viscosity) is essentially confined is called the **friction layer.**

Fluid flow that is characterized by eddy motion is known as **turbulence.** So far we have been considering obstacles on the Earth's surface as sources of eddies that exert a drag on the wind. Irregular fluid flow that originates in this way is known as **mechanical turbulence.** In addition, eddies develop in air as a consequence of solar heating of the ground. Irregular fluid flow that originates in this way is known as **thermal turbulence.** Convection currents are examples of thermal turbulence within the atmosphere. In actual practice, it is virtually impossible to distinguish the two sources of turbulent eddies.

Regardless of source, we experience turbulent eddies as gusts of wind. You may have noticed that the gustiness of the wind often varies with the time of day; that is, gusts tend to be strongest during the warmest hours of the day. The reasons for this are discussed in the Special Topic, "Wind Gusts, Wind Shear, and Atmospheric Stability."

---

* The snow trapped by a snow fence protects the underlying soil from deep penetration of freezing temperatures. The snow accumulation also ensures a supply of soil moisture.

Wind Gusts, Wind Shear, and Atmospheric Stability

You may have noticed that the day's strongest and gustiest winds often occur during the afternoon hours. This phenomenon is perhaps most apparent following a night of extreme radiational cooling. By dawn, a temperature inversion has developed in the lowest air layer, the surface air temperature has reached its diurnal minimum, and surface winds are very light or calm. In the hours following sunrise, the surface temperature rises steadily in response to increasing insolation, and eventually the lowest air layer is destabilized. By late morning or early afternoon, surface winds strengthen and become gusty, only to die down by sunset. This observation suggests a relationship between wind gustiness and atmospheric stability.

Wind gusts are caused by irregular whirls of air called *eddies*. We are all familiar with the analogous whirls or eddies that form in rapidly flowing streams of water (see Figure 9.7). Eddies characterize turbulent motion of fluids and, as noted elsewhere in this chapter, are either mechanical or thermal in origin.

As a rule, as the wind strengthens, it becomes more turbulent and the eddies become more energetic. Within the friction layer, horizontal wind speeds increase with altitude. Hence, strong eddies develop aloft, where winds are strong, and weaker eddies form near the Earth's surface, where winds are lighter. Eddies are transported upward and downward within the friction layer depending on atmospheric stability.

As shown in Chapter 6, atmospheric stability influences the vertical motion of air; that is, stable air

GRAVITY

The atmosphere is subject to the same force that holds all objects on the Earth's surface, **gravity.** The force of gravity is actually the net effect of two other forces working together: (1) the force of attraction between the Earth and all other objects, called **gravitation,** and (2) a much weaker centripetal force imparted to all objects because of their rotation with the Earth on its axis. The two forces combine to produce the force of gravity, which accelerates a unit mass of any object downward at the rate of 9.8 m per second each second.

The force of gravity always acts downward and perpendicular to the Earth's horizontal surface. For this reason, gravity, unlike the Coriolis effect and frictional forces, does not modify the horizontal wind. Gravity does, however, affect air that is ascending or descending, such as in convection currents, and gravity is responsible for the downhill flow of cold, dense air.

suppresses vertical motion, and unstable air enhances vertical motion. When the friction layer is unstable, strong eddies are transported from higher altitudes downward toward the surface. Consequently, winds aloft weaken because they lose some kinetic energy, and winds at the surface strengthen and become gusty because they gain some kinetic energy from eddies descending from aloft. On the other hand, when the friction layer is stable, strong eddies generated aloft remain there, and weak eddies that form near the surface remain near the surface.

It follows that vertical wind shear and stability are related. A *wind shear* occurs whenever the wind changes speed (or direction) with distance or, as in this case, with altitude. Hence, a vertical wind shear characterizes the friction layer because horizontal winds strengthen with altitude. The magnitude of this wind shear, however, depends on air stability. As noted, when the air is unstable, eddy transport reduces the vertical wind shear (the difference in horizontal wind speed between the surface and aloft). When the air is stable, the lack of eddy transport increases the vertical wind shear. In fact, although winds are typically light or calm within a low-level temperature inversion, winds are often quite strong in the air layer just above the inversion.

In summary, the greater the stability of the atmosphere's friction layer, the stronger is the vertical wind shear and the less gusty are the surface winds. An unstable friction layer features a weaker vertical wind shear and relatively energetic and gusty surface winds. Thus, in our example, as the lower troposphere is destabilized in the hours following sunrise, surface winds strengthen and become gusty.

SUMMARY

We have now examined the various forces that affect horizontal and vertical air motion, and we can draw the following conclusions.

1. Pressure gradient force accelerates air away from regions of high air pressure and toward areas of low air pressure. The acceleration is proportional to the pressure gradient; that is, it is greater with closer isobar spacing.
2. A centripetal force (an imbalance of "actual" forces) operates whenever the wind describes a curved path.
3. The Coriolis effect deflects large-scale winds to the right of the initial direction in the Northern Hemisphere and to the left of the initial direction in the Southern Hemisphere. It is zero at the equator and maximum at the poles, and it is proportional to wind speed.
4. Frictional resistance slows winds within about 1 km (0.62 mi) of the Earth's surface.
5. Gravity accelerates air downward but does not modify horizontal winds.

Joining Forces

To this point, we have examined the forces operating in the atmosphere as if each force acted independently of all the others. In reality, these forces interact with one another in governing both the direction and speed of the wind. In some cases, two or more forces achieve a balance, or equilibrium. From Newton's first law of motion, when the forces acting on an air parcel are in balance, there is no net force, and the parcel either remains stationary or continues to move along a straight path at constant speed. Therefore, when forces balance, the net acceleration is zero.

Let us now examine how forces interact in the atmosphere to control the vertical and horizontal flow of air, that is, the wind. These interactions result in (1) hydrostatic equilibrium, (2) the geostrophic wind, (3) the gradient wind, and (4) surface winds, the winds within the friction layer (Table 9.1).

HYDROSTATIC EQUILIBRIUM

We noted earlier that the atmosphere features a vertical pressure gradient. As shown schematically in Figure 9.11, the force due to this pressure gradient is directed upward from high pressure at the Earth's surface toward the lower air pressure aloft. If this force acted alone, the vertical pressure gradient force would accelerate air away from the Earth, and we would be left gasping for breath. However, except in some small-scale violent weather systems (tornadoes, for example), the atmosphere's vertical pressure gradient force is almost balanced by the equal and oppositely directed force of gravity. An actual balance of these two forces is known as **hydrostatic equilibrium.**

Whenever forces are in balance, no net acceleration occurs; that is, there is no change in velocity. Hydrostatic equilibrium, then, does not preclude vertical (up or down) motion of air in the atmosphere. Because of the balance of forces, upward-moving air parcels continue upward at *constant* velocity, and downward-moving air parcels continue downward at *constant* velocity. Slight deviations from hydrostatic equilibrium, however, cause air to change speed vertically.

Table 9.1
**Forces Involved in Large-Scale Atmospheric Circulation—
A Summary**

Forces	Hydrostatic equilibrium	Geostrophic wind	Gradient wind	Surface winds
Pressure gradient	×	×	×	×
Centripetal			×	×
Coriolis		×	×	×
Friction				×
Gravity	×			

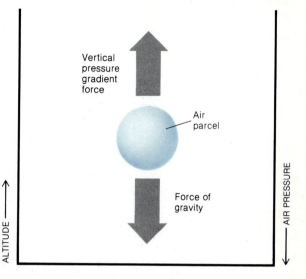

FIGURE 9.11
With hydrostatic equilibrium, the upward-directed vertical pressure-gradient force acting on an air parcel is balanced by the downward-directed force of gravity.

GEOSTROPHIC WIND

The **geostrophic wind** is an unaccelerated horizontal wind that flows along a straight path at altitudes above the friction layer. It results from a balance that develops between the horizontal pressure gradient force and the force due to the Coriolis effect.

When a horizontal pressure gradient (P_H) develops, air parcels at first accelerate directly across isobars, away from high pressure and toward low pressure (Figure 9.12). As air parcels speed up, however, the Coriolis effect (C) strengthens, causing air parcels to swerve gradually to the right of their initial flow direction (in the Northern Hemisphere). The two forces eventually

FIGURE 9.12
The horizontal air pressure gradient causes air parcels to accelerate across isobars from areas of high pressure toward areas of low pressure. The Coriolis effect then deflects air parcels to the right in the Northern Hemisphere. The Coriolis effect increases in magnitude until it balances the pressure-gradient force. The result is an unaccelerated horizontal wind blowing parallel to isobars, that is, the geostrophic wind.

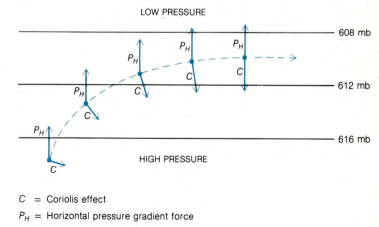

C = Coriolis effect
P_H = Horizontal pressure gradient force
- - -> Geostrophic wind

attain a balance, so that the wind blows at a constant speed in a straight path parallel to the isobars with the lowest air pressure to the left of air motion (in the Northern Hemisphere). Because the Coriolis effect is a large-scale phenomenon, the geostrophic wind develops only in large-scale weather systems.

GRADIENT WIND

The **gradient wind** has many characteristics in common with the geostrophic wind. It also is large scale, horizontal, frictionless, and parallel to isobars. The important difference between the two winds is that the geostrophic wind blows in a straight path, whereas the path of the gradient wind is curved. Because a net centripetal force constrains air parcels to a curved trajectory, the gradient wind is not the consequence of balanced forces. Recall that the centripetal force changes only the air parcel's direction of motion, not its speed. The horizontal pressure gradient force, the Coriolis effect, and the centripetal force thus interact in the gradient wind.

A gradient wind develops at altitudes above the friction layer around a dome of high air pressure, called an **anticyclone** (or high), or around a center of low air pressure, called a **cyclone** (or low). In an ideal anticyclone, isobars form a series of concentric circles about the location of highest air pressure, as shown in Figure 9.13. The horizontal pressure gradient force (P_H) is directed

FIGURE 9.13
In a Northern Hemisphere anticyclone above the friction layer, the gradient wind blows clockwise and parallel to isobars. In this idealized situation, the isobars form a series of concentric circles.

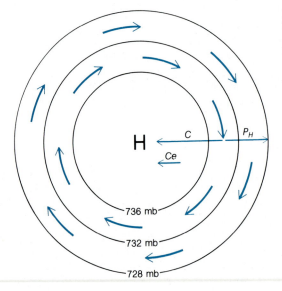

P_H = Horizontal pressure gradient force
C = Coriolis effect
Ce = Centripetal force
Gradient wind

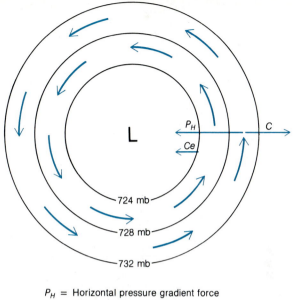

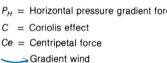

L

724 mb

728 mb

732 mb

FIGURE 9.14

In a Northern Hemisphere cyclone above the friction layer, the gradient wind blows counterclockwise and parallel to isobars. In this idealized situation, the isobars form a series of concentric circles.

P_H = Horizontal pressure gradient force

C = Coriolis effect

Ce = Centripetal force

→ Gradient wind

radially outward, away from the center of the high. The Coriolis effect (C) is directed inward. The Coriolis effect is slightly greater than the pressure gradient force, with the difference giving rise to the inward-directed centripetal force (Ce). (This is what was meant earlier when we indicated that the centripetal force results from an imbalance of other forces.) In a Northern Hemisphere anticyclone above the friction layer, the gradient wind consequently blows clockwise and parallel to isobars.

In an ideal cyclone, isobars form a series of concentric circles about the location of lowest air pressure. As indicated in Figure 9.14, the horizontal pressure gradient force (P_H) is directed inward, toward the cyclone center, and the Coriolis effect (C) is directed radially outward from the center of the low. The pressure gradient force is slightly greater than the Coriolis effect, with the difference equal to the net inward-directed centripetal force (Ce). In a Northern Hemisphere cyclone above the friction layer, the gradient wind consequently blows counterclockwise and parallel to isobars.

The geostrophic and gradient winds are conceptual models that only approximate the actual behavior of horizontal winds above the friction layer. The approximation is nonetheless quite useful, and meteorologists rely on such approximations in their analysis of isobaric patterns on weather maps. A mathematical description of geostrophic and gradient winds is presented in a Mathematical Note at the end of this chapter.

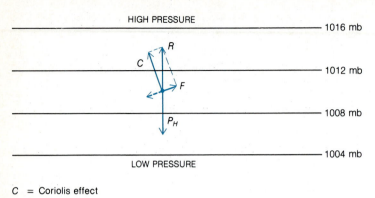

C = Coriolis effect

F = Force of friction

R = Coriolis effect + force of friction

P_H = Horizontal pressure gradient force

←–– Direction of surface wind

FIGURE 9.15

At the Earth's surface and elsewhere within the friction layer, the Coriolis effect combines with friction to balance the horizontal pressure-gradient force. As a consequence, the horizontal wind blows across isobars and toward lower air pressure. Recall that the Coriolis effect acts at right angles to the initial wind direction and friction acts in a direction opposite that of the wind.

SURFACE WINDS

Geostrophic winds and gradient winds are frictionless; that is, they occur at altitudes above the friction layer. What is the effect of friction on the horizontal winds within the friction layer, the surface winds? Intuitively, we know that friction should slow the wind, but in addition, friction interacts with the other forces to change the wind direction.

For large-scale air motion in a straight path, the frictional force (F) combines with the Coriolis effect (C) to balance the horizontal pressure gradient force (P_H), as shown in Figure 9.15. Note that the frictional force and Coriolis effect are at right angles to one another. Friction acts opposite to the direction of motion, and the Coriolis effect is at right angles to the direction of motion. Consequently, the horizontal wind slows down and shifts direction across isobars and toward low pressure. The deflection angle of surface winds crossing isobars varies from about 10 degrees over relatively smooth surfaces like the ocean, where friction is low, to almost 45 degrees over rough terrain, where friction is greater.

What is the effect of friction on the horizontal surface winds blowing in an anticyclone and cyclone? As with straight-line surface winds, friction slows cyclonic and anticyclonic winds and combines with the Coriolis effect to shift winds so that they blow across isobars and toward low pressure. At the Earth's surface, therefore, anticyclonic winds blow clockwise and outward, as shown in Figure 9.16, and surface cyclonic winds blow counterclockwise and inward, as shown in Figure 9.17.

The characteristic circulation of a cyclone enables us to formulate a simple rule of thumb for locating cyclone centers. If you stand with your back to the wind and then turn approximately 45 degrees to your right, the cyclone center

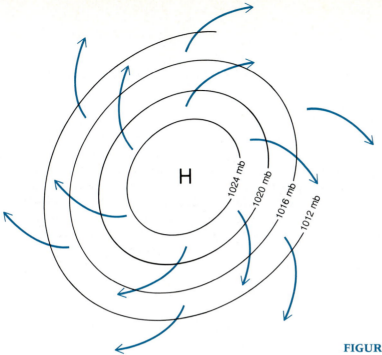

1024 mb
1020 mb
1016 mb
1012 mb

H

← Surface winds within friction layer

FIGURE 9.16
Surface winds blow clockwise and outward in a Northern Hemisphere anticyclone.

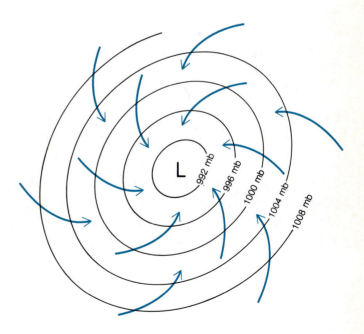

L

992 mb
996 mb
1000 mb
1004 mb
1008 mb

FIGURE 9.17
Surface winds blow counterclockwise and inward in a Northern Hemisphere cyclone.

→ Surface winds within friction layer

will be located to your left. This rule is a modification of an observation first stated in 1857 by the Dutch meteorologist Christopher H. D. Buys-Ballot. It must be applied with caution, however, because large-scale surface winds may be modified by a sea breeze or other local air circulation.

In the Southern Hemisphere, cyclonic and anticyclonic circulations are mirror images of their Northern Hemisphere counterparts. This occurs because of the change in direction of the Coriolis deflection between the two hemispheres. In the Southern Hemisphere, surface winds in a cyclone blow in a clockwise and inward direction, and surface winds in an anticyclone blow in a counterclockwise and outward direction. Above the friction layer, Southern Hemisphere cyclonic winds are clockwise and parallel to isobars, and Southern Hemisphere anticyclonic winds are counterclockwise and parallel to isobars.

FIGURE 9.18
Note how isobars on a typical surface weather map describe trough and ridge patterns. Shaded areas indicate precipitation. [NOAA weather map]

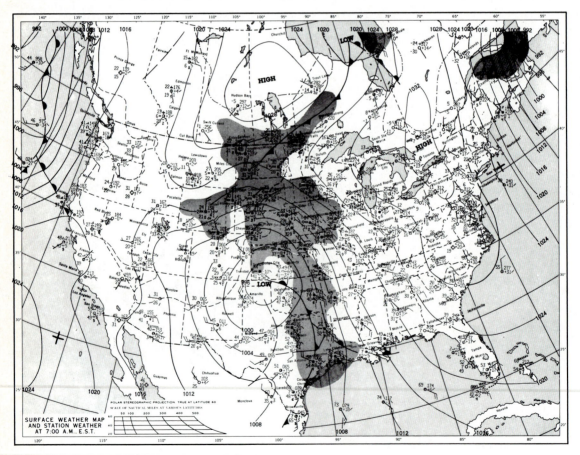

A glance at almost any national weather map (Figure 9.18) reveals that isobars seldom describe lengthy straight segments or circular patterns. Instead, isobars often form patterns of ridges and troughs. Nonetheless, in ridges and troughs, winds tend to parallel isobars above the friction layer and cross isobars toward low pressure within the friction layer.

An additional consideration in analyzing isobaric patterns for wind is the isobar spacing. As noted earlier, the steeper the air pressure gradient, the faster is the wind. Where isobars are closely spaced, the geostrophic and gradient winds are strong. Where isobars are widely spaced, these winds are weak. The same rule applies to surface winds.

Continuity of Wind

Air is a continuous fluid, and this continuity implies a link between the horizontal and vertical components of the wind. For example, surface winds are forced to follow the undulating topography of the Earth's surface, ascending hills and descending into valleys. In addition, uplift occurs along frontal surfaces as one air mass moves horizontally and either overrides or pushes under another air mass (Chapter 6). Having examined the horizontal circulation of anticyclones and cyclones, we can identify other important connections between the horizontal and vertical components of the wind.

As noted above, surface winds in a Northern Hemisphere anticyclone spiral clockwise and outward. Consequently, the horizontal surface winds diverge away from the center of the high. A vacuum does not develop at the center, however, because air slowly descends toward the Earth's surface and replaces the air that is diverging. Air is driven downward by air that converges toward the high center aloft (Figure 9.19). Recall that adiabatic compression raises the temperature and lowers the relative humidity of descending air. Skies therefore tend to be fair within anticyclones, and anticyclones are appropriately described as "fair weather" systems.

FIGURE 9.19
In this idealized vertical cross section of an anticyclone, air converges aloft, sinks, and diverges at the Earth's surface.

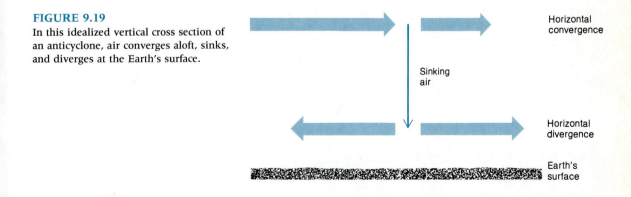

Horizontal convergence

Sinking air

Horizontal divergence

Earth's surface

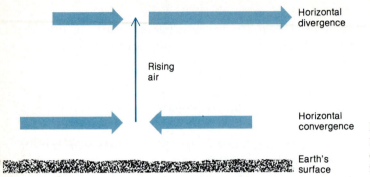

FIGURE 9.20

In this idealized vertical cross section of a cyclone, air converges at the Earth's surface, rises, and diverges aloft.

Furthermore, within an anticyclone, the horizontal air pressure gradient is typically very weak over a broad area at the center of the system. The resulting light or calm winds coupled with clear night skies favor intense radiational cooling. Hence, the ground and the air adjacent to the ground may be chilled to the point that dew, frost, or radiation fog develops.

Surface winds in a Northern Hemisphere cyclone spiral counterclockwise and inward. Surface winds therefore converge toward the center of a low. Air does not simply pile up at the center, however, because the air ascends in response to a divergent flow of air aloft (Figure 9.20). Recall that adiabatic expansion lowers the temperature and increases the relative humidity of ascending air. Clouds and precipitation may eventually develop, so that cyclones are typically stormy weather systems.

Continuity of the wind also means that vertical motion can be induced by downwind changes in frictional resistance. The rougher the Earth's surface, the more resistance it offers to horizontal winds. When the horizontal wind

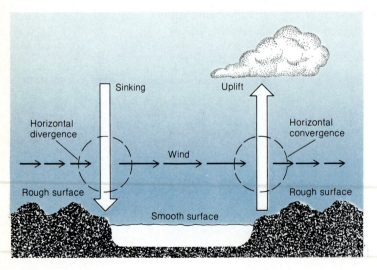

FIGURE 9.21

Surface winds undergo horizontal divergence when blowing from a rough to a smooth surface, and horizontal convergence when blowing from a smooth surface to a rough surface. Horizontal divergence of air causes air to sink, and horizontal convergence of air causes air to rise.

blows from a rough surface to a relatively smooth surface—as when it blows from land to sea—the wind accelerates. As Figure 9.21 shows, this acceleration causes the wind to diverge (stretch), thereby inducing downward motion of air. In contrast, when the horizontal wind blows from a smooth to a rough surface, the wind slows and converges (piles up), thereby inducing upward air motion. This is one reason why, along a coastline, convective clouds (cumulus) tend to develop with an onshore wind (from sea to land) and tend to dissipate with an offshore wind (from land to sea).

Scales of Weather Systems

Although the atmosphere is a continuous fluid, for convenience of study we subdivide atmospheric circulation into discrete weather systems operating at various spatial scales. The large-scale wind systems of the planet (polar easterlies, westerlies, and trade winds) are features of the **global-scale circulation. Synoptic-scale weather** is continental or oceanic. Migrating storms, air masses, and fronts, usually highlighted on television and newspaper weather maps, are important components of this scale. **Mesoscale systems** include thunderstorms and sea and lake breezes—phenomena that may influence the weather in only a portion of a city while adjacent areas are unaffected. A circulation system covering a small area—a tornado, for example—represents the smallest spatial subdivision of atmospheric circulation, **microscale weather.** In Chapters 10 through 14, we examine weather systems of all scales starting with the largest.

Each smaller-scale weather system is part of and dependent on larger-scale atmospheric circulation. That is, the various scales of atmospheric motion form a kind of hierarchy. For example, extreme nocturnal radiational cooling requires a synoptic weather pattern that favors clear skies and light or calm winds. At the microscale, such weather conditions may be accompanied by frost formation or radiation fog.

Wind Pressure

As wind speed increases, the potential for wind damage to buildings and other structures also increases. In this regard, the concept of wind pressure is useful. **Wind pressure** is defined as the force per unit area produced by the wind on an object in its path. Wind pressure is directly proportional to the square of the wind speed. Hence, each doubling of wind speed increases the wind pressure by a factor of four ($2^2 = 4$).

In designing buildings to withstand high winds, engineers and architects consult the climatic record to determine the probability of potentially destructive winds. Because wind speed increases with altitude, wind pressure is evaluated as a function of building elevation. Therefore, during a windstorm,

FIGURE 9.22
In the absence of direct measurements of wind, it is possible to reconstruct the wind speed from storm damage based on wind pressure estimates. This tree was twisted by winds estimated at 80 km per hour. [Photograph by J. M. Moran]

a building's upper stories are subject to greater wind pressure than its lower stories. A building's exposure is also an important consideration in planning for wind pressure. For example, a building situated on a coastal plain and facing the open ocean is exposed to stronger winds than a building situated among many other buildings within a congested city center.

In the reverse application of this approach, wind pressure, and hence wind speed, can be reconstructed from storm damage (Figure 9.22). In this way, for example, meteorologists have been able to estimate wind speeds in tornadoes based on the damage they produce. This is necessary because tornadic winds are so strong that they destroy conventional weather instruments.

Wind Measurement

Meteorologists are interested in monitoring both the speed and the direction of wind. Most wind-monitoring instruments are designed to measure only the horizontal component of the wind since it is usually considerably stronger than the vertical component. For some specialized research purposes, very sensitive instruments are available that measure vertical wind speeds or a combination of vertical and horizontal wind components.

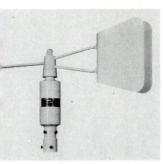

FIGURE 9.23

Horizontal wind direction is monitored by a wind vane. The instrument's arm points in the direction from which the wind blows. [Courtesy of Belfort Instrument Company]

A traditional **wind vane,** like the one shown in Figure 9.23, consists of a free-swinging horizontal shaft with a flat vertical plate at one end and a counterweight at the other end. The counterweight always points directly into the wind. Simple wind vanes may be read visually, whereas more sophisticated models employ electronic devices that permit remote reading and recording. Another design is the airport **wind sock,** which consists of a cone-shaped cloth bag that is open at both ends. The larger end of the sock is held open by a metal ring that is attached to a pole and is free to rotate. Air enters the larger opening and stretches the sock downwind (Figure 9.24).

Wind direction is always designated as the direction *from which* the wind blows. For example, a wind blowing from the east toward the west is described as an "east wind," and a wind blowing from the northwest toward the southeast is a "northwest wind." A wind vane may be linked electronically or mechanically to a dial that is calibrated to read in points of the compass or in degrees. Measured clockwise from true north, an east wind is specified as 90 degrees, a south wind as 180 degrees, a west wind as 270 degrees, and a north wind as 360 degrees. The wind is recorded as 0 degrees only under calm conditions.

Wind speed can be estimated by observing the wind's effect on lake or ocean surfaces or on flexible objects such as trees. Such observations are the basis of the **Beaufort scale,** which is a graduated sequence of wind strength ranging from 0 for calm conditions to 12 for hurricane-strength winds (Table 9.2). The scale bears the name of Sir Francis Beaufort, who developed it in the early 1800s while a ship's commander in the British Navy. Beaufort's goal was to standardize the terms used by sailors in describing the state of the sea under various wind conditions. In 1838, after some revision, the British Navy adopted the Beaufort scale, and in 1853 it was sanctioned for international use by seafarers. Later, when the scale was extended from sea to land, it was

FIGURE 9.24

An airport wind sock gives wind direction and a general indication of wind speed. The sock points downwind. [Photograph by J. M. Moran]

Table 9.2
Beaufort Scale of Wind Force

Beaufort number	General description	Land and sea observations for estimating wind speeds	Wind speed 10 m above ground (km/hr)
0	Calm	Smoke rises vertically. Sea like mirror.	Less than 1
1	Light air	Smoke, but not wind vane, shows direction of wind. Slight ripples at sea.	1–5
2	Light breeze	Wind felt on face, leaves rustle, wind vanes move. Small, short wavelets.	6–11
3	Gentle breeze	Leaves and small twigs moving constantly, small flags extended. Large wavelets, scattered whitecaps.	12–19
4	Moderate breeze	Dust and loose paper raised, small branches moved. Small waves, frequent whitecaps.	20–28
5	Fresh breeze	Small leafy trees swayed. Moderate waves.	29–38
6	Strong breeze	Large branches in motion, whistling heard in utility wires. Large waves, some spray.	39–49
7	Near gale	Whole trees in motion. White foam from breaking waves.	50–61
8	Gale	Twigs break off trees. Moderately high waves of great length.	62–74
9	Strong gale	Slight structural damage occurs. Crests of waves begin to roll over. Spray may impede visibility.	75–88
10	Storm	Trees uprooted, considerable structural damage. Sea white with foam, heavy tumbling of sea.	89–102
11	Violent storm	Very rare; widespread damage. Unusually high waves.	103–118
12	Hurricane	Very rare; much foam and spray greatly reduce visibility.	119 and over

necessary to develop wind speed equivalents for each Beaufort number; this was done in 1926. The scale is still used today. In fact, some modern mariners prefer Beaufort numbers to onboard instrument measurements of wind speed.

As a rule, a **cup anemometer,** such as the one shown in Figure 9.25, provides an accurate measurement of horizontal wind speed. This device works on the same principle as a bicycle or an automobile speedometer. The wind spins the cups (usually 3 or 4), thus generating a weak electric current, which is calibrated on a dial in meters per second, kilometers per hour, miles per hour, or knots.* Several other types of anemometer are available, including the very sensitive **hotwire anemometer.** In this instrument, the wind blows past a heated wire, or wires, and the heat lost to the air is then calibrated in terms of wind speed.

Recording a continuous trace of wind speed and direction is sometimes informative. Wind vanes and anemometers can be linked to pens that record

---

* One knot = 1 nautical mile (1.85 km) per hour; 1 knot = 0.51 meters per second; 1 mile per hour = 0.44 meters per second.

FIGURE 9.25
Wind speed is measured by a cup anemometer. The faster the wind speed, the faster the cups spin. [Courtesy of Belfort Instrument Company]

on a clock-driven drum. As shown in Figure 9.26, the trace indicates a considerable variation in both wind direction and wind speed with time. More commonly today, the output from wind vanes and anemometers is recorded by computer or on magnetic tape.

Ideally, a wind vane/anemometer system should be mounted on a tower so that the instruments monitor horizontal winds 10 m (33 ft) above the ground. It is best to avoid rooftop locations because winds tend to accelerate over buildings. In addition, the system should be sited well away from (1) structures that might shelter the instruments, and (2) any obstacles that might channel (and thus accelerate) the wind.

FIGURE 9.26
Continuous trace of time variations in wind speed and wind direction over a six-hour period.

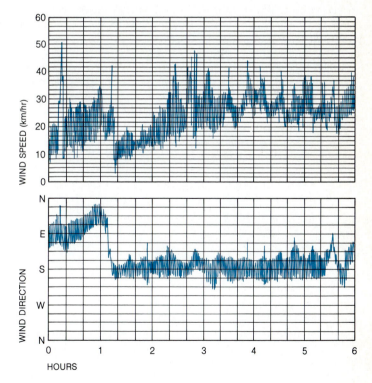

Conclusions

Much of the discussion in this chapter may seem quite abstract. From everyday experience, we are aware of the end product of the forces described, namely, the wind, but we are not readily aware of the individual forces. In learning about atmospheric forces and how they interact, we see that each force is

---

Mathematical Note

Geostrophic and Gradient Winds

The geostrophic wind is the result of a balance between the horizontal pressure gradient force and the Coriolis effect. Here we examine the geostrophic motion of a unit mass of air (1 gram, for example). The acceleration imparted to this air parcel by a horizontal pressure gradient is given by

$$- \frac{1}{\rho} \left(\frac{\Delta P}{\Delta N} \right)$$

where ρ is the density of the air and ΔP is the change in air pressure over a horizontal distance, ΔN, measured perpendicular to isobars. (Pressure gradients are always measured perpendicular to isobars.)

The acceleration imparted to the air parcel by the Coriolis effect is given by

$$(2\Omega \sin \phi)V$$

where V is the speed of the air parcel, Ω is the angular velocity of the Earth as it rotates on its axis ($\Omega = 7.29 \times 10^{-5}$ radian per second), and ϕ is the latitude.

For the geostrophic wind,

$$-\frac{1}{\rho} \left(\frac{\Delta P}{\Delta N} \right) = (2\Omega \sin \phi)V$$

Solving for V, the speed of the geostrophic wind, we have

$$V = -\frac{1}{(2\Omega \sin \phi)\rho} \left(\frac{\Delta P}{\Delta N} \right)$$

The geostrophic wind thus strengthens with an increasing pressure gradient (that is, closer spacing of isobars) and decreasing latitude. Recall that the geostrophic wind blows parallel to isobars and at right angles to the pressure gradient force and Coriolis effect (which are directed one opposite the other).

The gradient wind results from a slight imbalance between the horizontal pressure gradient force and the Coriolis effect. This imbalance is realized as a centripetal force. Again, assume that we are examining the motion of a unit mass of air. The acceleration imparted to the air parcel by the centripetal force is

$$V^2/r$$

where V is the speed of the air parcel, and r is the radius of curvature of the path described by the air parcel. The centripetal force is the force that constrains the air parcel to a curved trajectory. This force is strong where the curvature is sharp (small r) and weak where the curvature is gradual (large r).

The three forces (pressure gradient, Coriolis, and centripetal) interact such that

$$\frac{V^2}{r} + (2\Omega \sin \phi)V - \frac{1}{\rho}\left(\frac{\Delta P}{\Delta N} \right) = 0$$

The net force, which is the centripetal force, operates on the gradient wind only to change the direction of the wind as it follows a curved path, and not to change the wind's speed. We could solve the above equation for V to determine the speed of the gradient wind for some radius of curvature, latitude, and horizontal pressure gradient (measured perpendicular to isobars). Note that for wind blowing in a straight line,

$$V^2/r = 0$$

and the equation reduces to that presented earlier for geostrophic flow.

bound by certain constraints. For example, friction is important only in the lower troposphere, and the Coriolis effect always shifts the wind to the right in the Northern Hemisphere. In the following chapters, our awareness of these and other constraints will aid in understanding the characteristics of the various weather systems—for example, why hurricanes do not form at the equator, and why winds in a tornado may blow in either a clockwise or counterclockwise direction. We are now ready to begin our discussion of the characteristics of the various circulation systems—beginning with those operating at the global scale.

Summary Statements

Wind is the consequence of an interaction of forces generated by air pressure gradients, the Coriolis effect, friction, and gravity. The centripetal force is not an independent force and is the resultant of other forces.

The pressure gradient force initiates air motion and arises in part from spatial variations in air temperature and, to a lesser extent, water vapor concentration.

The centripetal force describes the imbalance in actual forces that operates whenever the wind follows a curved path, as it does in anticyclones, cyclones, ridges, and troughs.

The Coriolis effect, which is due to the Earth's rotation on its axis, deflects the wind to the right of its initial direction in the Northern Hemisphere and to the left of its initial direction in the Southern Hemisphere. The deflective force is zero at the equator and increases by latitude to a maximum at the poles.

Friction has an important influence on winds blowing within about 1 km of the Earth's surface. Obstacles on the Earth surface slow the wind by generating turbulent eddies.

The force of gravity always acts downward and perpendicular to the Earth's surface.

Hydrostatic equilibrium is the balance between the upward-directed pressure gradient force and the downward-directed force of gravity. Slight deviations from hydrostatic equilibrium cause air to accelerate upward or downward.

The geostrophic wind is an unaccelerated horizontal wind that blows along straight paths at altitudes above the friction layer and results from a balance between the horizontal pressure gradient force and the Coriolis effect. The geostrophic wind always blows parallel to *straight* isobars.

The gradient wind is a horizontal, frictionless air flow that parallels *curved* isobars. At altitudes above the friction layer in the Northern Hemisphere, the gradient wind blows clockwise about anticyclones and counterclockwise about cyclones.

In large-scale surface winds (global and synoptic scales), friction slows the wind and combines with the Coriolis effect to shift the wind direction across isobars toward low-pressure areas.

Within the friction layer, horizontal winds blow clockwise and outward in Northern Hemisphere anticyclones, counterclockwise and inward in Northern Hemisphere cyclones.

The concept of wind continuity implies a linkage between the horizontal and vertical components of the wind. In an anticyclone, horizontal winds converge aloft and diverge at the surface. Continuity of air means that air descends near the high center, the relative humidity drops, and the weather is fair.

In a cyclone, horizontal winds diverge aloft and converge at the surface. Continuity of air means that air ascends near the center of the low, the relative humidity increases, and the weather is stormy.

Key Words

air parcels	Coriolis effect	gravity	synoptic-scale
Newton's second law	friction	gravitation	weather
of motion	viscosity	hydrostatic	mesoscale systems
gradient	molecular viscosity	equilibrium	microscale weather
air pressure gradient	eddy viscosity	geostrophic wind	wind pressure
isobars	friction layer	gradient wind	wind vane
pressure gradient	turbulence	anticyclone	wind sock
force	mechanical	cyclone	Beaufort scale
Newton's first law of	turbulence	global-scale	cup anemometer
motion	thermal turbulence	circulation	hot-wire anemometer
centripetal force			

Review Questions

1. Why does the vertical component of the wind play the key role in cloud and precipitation formation?

2. What causes horizontal gradients in air pressure?

3. A weak horizontal air pressure gradient is indicated by what type of pattern of isobars?

4. How does an air pressure gradient acting alone influence the motion of air?

5. Why is it appropriate to refer to the Coriolis effect and friction as "responsive" forces rather than "active" forces?

6. Why does the Coriolis deflection reverse direction between the Northern and Southern Hemispheres?

7. Explain how the Coriolis deflection arises from a change in our frame of reference.

8. The Coriolis deflection is latitude dependent. Explain why.

9. How does wind speed change with altitude within the friction layer?

10. State Newton's first and second laws of motion.

11. What is the significance of the centripetal force? What causes the centripetal force in cyclones and anticyclones?

12. What is hydrostatic equilibrium?

13. Distinguish between the geostrophic wind and the gradient wind.

14. Describe the horizontal air circulation about a low-pressure system (a) within the friction layer and (b) above the friction layer.

15. Describe the horizontal air circulation about a high-pressure system (a) within the friction layer and (b) above the friction layer.

16. Describe how forces interact to shape the horizontal air circulation within (a) cyclones and (b) anticyclones.

17. Explain why the horizontal wind parallels isobars above the friction layer and crosses isobars within the friction layer.

18. How does the isobar spacing within cyclones and anticyclones affect horizontal wind speeds?

19. Provide several examples of how horizontal winds are linked to the vertical motion of air.

20. Explain why cyclones produce stormy weather and anticyclones are fair weather systems.

Points to Ponder

1. Speculate on how the magnitude of a typical horizontal air pressure gradient compares with that of the vertical air pressure gradient.

2. How might air mass advection influence vertical air pressure gradients?

3. Speculate on why vertical wind speeds typically are considerably weaker than horizontal wind speeds.

4. In view of Newton's first law of motion, is gradient wind the consequence of "balanced" forces? Explain your answer.

5. What type of isobaric pattern is associated with a high-pressure system? What is the implication of this pattern for weather?

6. What is meant by the statement that the Coriolis effect is more "apparent" than "real"?

7. Why is the Coriolis effect most important in global- and synoptic-scale weather systems?

8. Explain why dew or frost formation might be associated with an anticyclone.

Projects

1. Analyze several televised or newspaper national weather maps and predict the surface wind direction at various cities based on isobar patterns.

2. Predict how wind direction and speed change as a cyclone approaches your location. Do the same for an approaching anticyclone. Compare your predictions with what actually happened.

Selected Readings

Blackadar, A. "Simple Motions on the Rotating Earth." *Weatherwise* 39 (1986):99–103. Provides an excellent description of the Coriolis effect.

Forrester, F. H. "How Strong Is the Wind?" *Weatherwise* 39 (1986):147–151. Describes the origin of the Beaufort scale.

Karapiperis, P. P. "The Tower of the Winds." *Weatherwise* 39 (1986):152–154. Concerns a building in Athens, probably dating to the first century B.C., that demonstrates that the people of the day understood the linkage between wind direction and weather.

Moretti, P. M., and L. V. Divone, "Modern Windmills." *Scientific American* 254, No. 6 (1986):110–118. Gives an account of modern advances in an ancient technology.

Snow, J. T., et al. "Basic Meteorological Observations for Schools: Surface Winds." *Bulletin of the American Meteorological Society* 70 (1989):493–508. Presents a comprehensive examination of wind-measuring instruments with suggestions for construction of a wind vane and cup anemometer for school use.

The fair breezes blew, the
   white foam flew,
The furrow follow'd free;
We were the first that ever
   burst
Into that silent sea.

Down dropt the breeze, the
   sails dropt down,
'Twas sad as sad could be
And we did speak only to
   break
The silence of the sea!

All in a hot and copper sky,
The bloody Sun, at noon,
Right up above the mast did
   stand,
No bigger than the Moon.

Day after day, day after day,
We stuck, nor breath nor
   motion;
As idle as a painted ship
Upon a painted ocean.

Samuel Taylor Coleridge
THE RIME OF THE ANCIENT
MARINER

10

Global-Scale Circulation

Global-scale circulation encompasses wind belts that encircle the globe and pressure systems that cover huge areas of the Earth's oceans. [NOAA satellite photograph]

G

LOBAL-SCALE AIR circulation ultimately is responsible for the development and displacement of most smaller-scale weather systems. It is therefore, appropriate for us to begin our description of weather systems with those that are planetary in scale. In this chapter, we describe the wind belts and pressure systems that operate at the global scale, with special emphasis on circulation patterns of the midlatitude westerlies.

Idealized Circulation Pattern

Global-scale air circulation is a complex pattern of winds and pressure systems. To better understand how this circulation is shaped, we start with an idealized model of planet Earth. Picture the Earth as a nonrotating sphere with a uniform solid surface. As on the real Earth, assume that the sun heats the equatorial regions more intensely than the polar regions. On the sunlit side of the globe, a temperature gradient develops between the equator and poles. In response, two huge convective cells form—one in each hemisphere (Figure 10.1A). Cold, dense air sinks at the poles and flows at the surface toward the equator, where warm, light air rises and aloft flows toward the poles.

If our idealized planet begins to rotate, the Coriolis effect comes into play (Figure 10.1B). In the Northern Hemisphere, surface winds shift to the northeast (that is, northeast to southwest), and in the Southern Hemisphere, surface winds become southeasterly (that is, southeast to northwest). Hence, on our hypothetical planet, surface winds blow counter to the planet's rotation direction, which is from west to east. This is an impossible situation because the surface winds would have a braking effect on the rotating Earth.

Circulation is maintained in the atmosphere of our idealized Earth because the global winds split into three zones in each hemisphere, so that some winds blow with the planet's rotational direction, whereas other winds blow counter to the planet's rotational direction (Figure 10.1C). In the Northern Hemisphere, average surface winds are northeasterly from 0 to 30 degrees, southwesterly from 30 to 60 degrees, and northeasterly from 60 degrees to the North Pole. In the Southern Hemisphere, surface winds are southeasterly from 0 to 30 degrees, northwesterly from 30 to 60 degrees, and southeasterly from 60 degrees to the South Pole.

Surface winds converge along the 0-degree and 60-degree latitude circles. By causing air to rise, convergence leads to expansional cooling, cloud development, and precipitation. These convergence zones are therefore belts of low pressure (Figure 10.1D). Surface winds diverge at the poles and along the 30-degree latitude circles. In these regions, air descends, leading to compressional warming and fair weather. These are zones of high pressure.

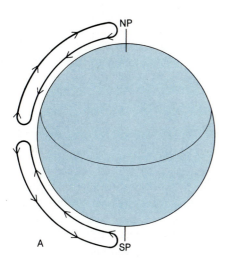

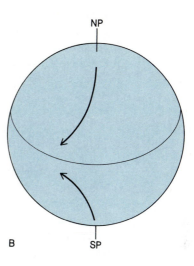

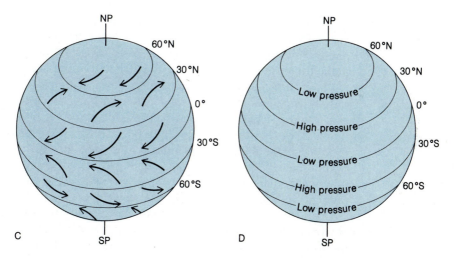

FIGURE 10.1

Global-scale air circulation on an idealized model of the Earth featuring a uniform solid surface. (A) If the sphere is nonrotating, huge convective currents develop in the atmosphere on the sunlit portion of the planet so that air circulates between the hot equator and the cold poles. (B) With a rotating Earth, surface winds become northeasterly in the Northern Hemisphere and southeasterly in the Southern Hemisphere owing to the Coriolis effect. (C) In reality, surface winds divide into three zones in each hemisphere. (D) Zones of converging and diverging surface winds give rise to east–west belts of low and high pressure.

If we add the continents and oceans to our idealized Earth, the temperature characteristics of the Earth's surface become more complex and more realistic, and so do the air pressure pattern and the global-scale circulation. Some of the pressure belts break into separate cells, and important contrasts in air pressure develop over land versus sea. In the next section, we describe the principal features of global-scale circulation on the real Earth as these features are presently understood.

Pressure Systems and Wind Belts

Maps of the average air pressure at sea level for January and July (Figure 10.2) reveal several areas of relatively high and low atmospheric pressure. These are the **semipermanent pressure systems.** Although these systems are persistent features of global circulation, they do exhibit important seasonal changes in both location and strength—hence the modifier *semipermanent*. Pressure systems include the subtropical anticyclones, the equatorial trough, and the subpolar lows.

Subtropical anticyclones are imposing features of the global-scale circulation that are centered over subtropical latitudes (on average, near 30 degrees N and S) of the North and South Atlantic, the North and South Pacific, and the Indian Ocean. These highs extend vertically from the ocean surface to the tropopause and exert a strong influence on weather and climate.

Stretching outward from the center of each subtropical high, out over its eastern flank, are extensive areas of subsiding stable air. Subsiding air undergoes compressional warming, which produces low relative humidities and sunny skies. The world's major deserts, including the Sahara of North Africa and the Sonora of Mexico and southwest United States, are located under the eastern flanks of subtropical anticyclones. On the far-western portions of the subtropical highs, however, subsidence is less and the air is not as stable. Consequently, episodes of cloudy, stormy weather are more frequent in these regions. This contrast in climate from one flank of a subtropical high to the other is particularly apparent across southern North America. The weather of the American Southwest (on the eastern side of the North Pacific high, also called the Hawaiian high) is considerably drier than the weather of the American Southeast (on the western side of the North Atlantic high, also called the Bermuda–Azores high).

As is typical of anticyclones, a subtropical high features a weak horizontal air pressure gradient over a wide area at the system's center. Hence, surface winds are very light or even calm over extensive areas of the subtropical oceans. This situation played havoc with ancient sailing ships, which were becalmed for days or even weeks at a time. Ships setting sail from Spain to the New World were often caught in this predicament, and crews were forced

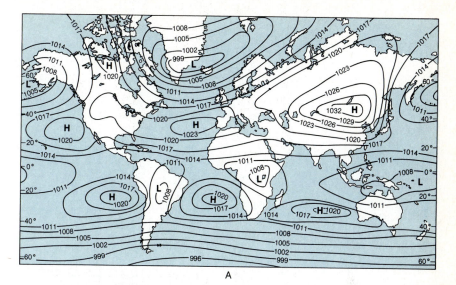

A

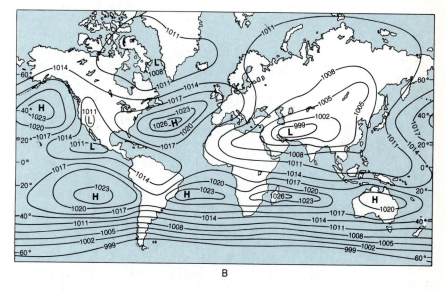

B

FIGURE 10.2
Mean sea-level air pressure for (A) January and (B) July. Contours are isobars in millibars.

to jettison their cargo of horses when supplies of food and water ran low. For this reason, early mariners referred to this area of calm winds as the **horse latitudes,** a name now applied to all latitudes under subtropical highs. The stanzas from Coleridge's *Rime of the Ancient Mariner* at the beginning of this chapter aptly describe the weather conditions of horse latitudes.

In the Northern Hemisphere, surface winds blow clockwise and outward, away from the centers of the subtropical highs forming the westerlies and

Table 10.1

Features of Global-Scale Circulation

Latitude		Global-scale systems
90° N	High	Polar anticyclones
	↙ ↙	Polar northeasterlies
60° N	Low	Subpolar cyclones
	↗ ↗	Westerlies
30° N	High	Subtropical anticyclones
	↙ ↙	Northeast trades
0°	Low	Equatorial trough (ITCZ)
	↖ ↖	Southeast trades
30° S	High	Subtropical anticyclones
	↘ ↘	Westerlies
60° S	Low	Subpolar cyclones
	↖ ↖	Polar southeasterlies
90° S	High	Polar anticyclones

trade winds (Table 10.1). Surface winds north of the horse latitudes constitute the highly variable **midlatitude westerlies** (actually southwesterlies). Surface winds blowing out of the southern flanks of the anticyclones are known as the northeast **trade winds.** The trades are the most persistent winds on the planet, in some regions blowing from the same direction more than 80 percent of the time. Analogous winds develop in the Southern Hemisphere. Recall, however, that the Coriolis deflection reverses in the Southern Hemisphere. A counterclockwise and outward surface airflow thus causes southeast trade winds on the northern flanks of the Southern Hemisphere subtropical highs and a belt of northwest winds on the southern flanks.

Interestingly, the belts of westerlies and trades over the North Atlantic were known to mariners at least as far back as the fifteenth century. On his venture to find a route to India, Christopher Columbus took advantage of what we now know is the circulation about the Bermuda–Azores anticyclone. In his westward voyage, Columbus first sailed southward and picked up the southeast trade winds that eventually took his ships to the Caribbean. On his return trip to Spain, he sailed north and picked up the westerlies.

The trade winds of the two hemispheres flow into a broad east–west belt of low pressure near 0 degrees latitude, called the **equatorial trough,** where rising air motion induces cloudiness and rainfall. The most active weather develops along the **intertropical convergence zone (ITCZ),** a discontinuous belt of thunderstorms paralleling the equator (Figure 10.3). The mean position of the ITCZ is at the latitude of the Earth's highest mean surface temperature, the so-called **heat equator.** For reasons discussed in Chapter 17, the heat equator is in the Northern Hemisphere, just north of the geographical equator.

FIGURE 10.3

This satellite photograph shows a discontinuous east–west band of cumulonimbus clouds (white blotches) near the equator. The cloud band marks the convergence of the surface trade winds of both hemispheres and is known as the intertropical convergence zone (ITCZ). [NOAA satellite photograph]

On the poleward side of the subtropical highs, surface westerlies flow into regions of low pressure. In the Northern Hemisphere, there are typically two separate **subpolar lows**—the **Aleutian low** over the North Pacific Ocean and the **Icelandic low** over the North Atlantic. Because of the counterclockwise and inward circulation of surface winds in these pressure cells, the midlatitude southwesterlies converge with the polar northeasterlies. In contrast, in the Southern Hemisphere the midlatitude northwesterlies and the polar southeasterlies converge along a nearly continuous belt of low pressure surrounding the Antarctic continent.

The surface westerlies meet and override the polar easterlies along the **polar front.** A **front** is a narrow zone of transition between air masses of contrasting density, that is, air masses of different temperatures, different water vapor contents, or both. In this case, dense, cold air masses flowing toward the equator meet milder, lighter midlatitude air masses moving poleward. The polar front is not continuous around the globe; rather, it is well defined in some areas and not in others, depending on the temperature contrast across the front. Where that temperature gradient is steep, the front is well defined and is a potential site for the development of synoptic-scale storms. On the other hand, where the air temperature contrast is minimal the polar front is poorly developed and inactive.

This brief description is a generalized view of the major components of global-scale circulation at the Earth's surface. The distribution of these surface circulation features is summarized schematically in Figure 10.4.

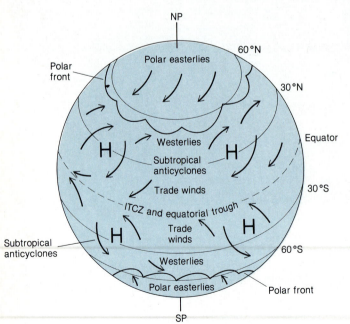

FIGURE 10.4
A schematic representation of the global-scale surface circulation of the atmosphere.

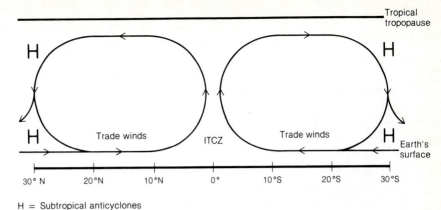

FIGURE 10.5

An idealized representation of the Hadley cell circulation in tropical latitudes of the Southern and Northern hemispheres. Air rises over the intertropical convergence zone (ITCZ) and sinks in the subtropical anticyclones.

H = Subtropical anticyclones

WINDS ALOFT

What is the pattern of the global winds aloft, in the middle and upper troposphere? As noted earlier, air subsides in subtropical anticyclones, sweeps toward the equator as the surface trade winds, and then ascends in the ITCZ. Aloft, in the middle troposphere, air flows poleward, away from the equatorial trough and into the subtropical highs. The Coriolis effect shifts these upper-level winds toward the right in the Northern Hemisphere (southwest winds), and toward the left in the Southern Hemisphere (northwest winds). In the tropics, therefore, the winds aloft actually blow in a direction opposite that of the surface trade winds. The vertical profile of this circulation, shown schematically in Figure 10.5, resembles a huge convective cell and is known as the **Hadley cell,** named for the scientist who first proposed its existence in 1735. Hadley cells are thus situated on either side of the ITCZ and extend poleward to the subtropical highs.

It was once proposed that separate cells, similar to the Hadley cells, occurred in both midlatitudes and polar latitudes. Detailed upper-air monitoring, however, has failed to provide definitive evidence of such cells. Aloft, midlatitude winds blow from west to east in a wavelike pattern of ridges and troughs, as shown in Figure 10.6. These winds are responsible for the devel-

FIGURE 10.6

Aloft, in the middle and upper troposphere, the westerlies flow in a wavelength pattern of ridges and troughs.

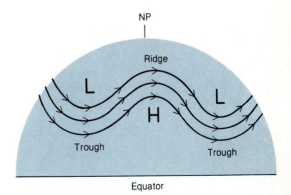

253

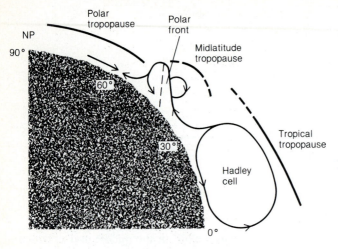

FIGURE 10.7

Vertical cross section showing the north–south (meridional) winds in the Northern Hemisphere troposphere. Note that the tropopause occurs in three segments and that the vertical scale is considerably exaggerated.

opment and displacement of synoptic-scale weather systems (highs, lows, and air masses) discussed in Chapter 11, and for the poleward heat transport described in Chapter 4. Because these winds are so important to our mid-latitude weather, we examine the upper-air westerlies in more detail in a separate section of this chapter.

In polar regions, air subsides and flows away at the surface from shallow, cold anticyclones. In the Northern Hemisphere, these highs are well developed only in winter over the continental interiors. In the Southern Hemisphere, cold highs persist over the glacier-bound Antarctic continent year-round. Aloft, polar winds are westerly.

Figure 10.7 shows the vertical profile of Northern Hemisphere winds in the troposphere from the equator to the poles. In this perspective, we are viewing only the north–south and up–down components of the wind and are neglecting the west–east component. Note that the tropopause is not continuous from pole to equator but occurs in discrete segments. Thus, the polar tropopause is at a lower altitude than the midlatitude tropopause, which, in turn, is a lower altitude than the tropical tropopause.

SEASONAL SHIFTS

Between winter and summer important changes take place in planetary circulation. Pressure systems, the polar front, the global wind belts, and the ITCZ follow the sun, shifting toward the poles in spring and toward the equator in autumn. Because the seasons are reversed in the two hemispheres, the global-scale systems of both hemispheres move north and south in tandem. In addition, the strength of pressure cells varies between seasons. Subtropical anticyclones exert lower surface pressures in winter than in summer. The icelandic low deepens in winter and greatly weakens in summer, and, though well developed in winter, the Aleutian low disappears in summer.

Seasonal reversals in surface air pressure also occur over the continents. These pressure shifts stem from the contrast in solar heating between land and sea. For reasons detailed in Chapter 3, the sea surface exhibits smaller temperature variations over the course of a year than does the Earth's land surface. The maps of global surface temperature in Plate 17 vividly display this land–sea contrast in thermal stability. Temperatures were derived from infrared radiation sensors onboard weather satellites.* Parts A and B are color-enhanced images of January and July temperatures, respectively. In the Northern Hemisphere, these are usually the coldest and warmest months of the year. The final image (part C) is the temperature difference between the two months. Note that the seasonal temperature contrast is considerably greater over the continents (up to 30 C°, or 54 F°) than over the oceans (typically 8 to 10 C°, or 14 to 18 F°).

As a consequence of the temperature contrast, the continents are sites of relatively high pressure in winter and of relatively low pressure in summer. In winter, cold anticyclones appear over northwestern North America and over the interior of Asia, the most prominent of which is the massive Siberian high. In summer, belts of low pressure form across North Africa and from Arabia eastward into Southeast Asia. Warm low-pressure cells also develop in summer over the desert country of Mexico and southwestern United States.

Seasonal shifts in the location of the global-scale wind belts, pressure systems, and the ITCZ leave their mark on the world's climates. For example, northward migration of the ITCZ triggers summer monsoon rains in Central America, North Africa, India, and Southeast Asia (see Chapter 12). As shown in Figure 10.8, north–south movements of the ITCZ are greater over continents

---

* Recall from Chapter 2 that radiation emitted by any object is temperature dependent, so temperature can be measured remotely by infrared sensors onboard Earth-orbiting satellites.

FIGURE 10.8

Seasonal shifts of the intertropical convergence zone (ITCZ). The ITCZ follows the sun, reaching its most northerly location in July and its most southerly location in January. As we will see in Chapter 12, these shifts contribute to the monsoon circulations of tropical latitudes.

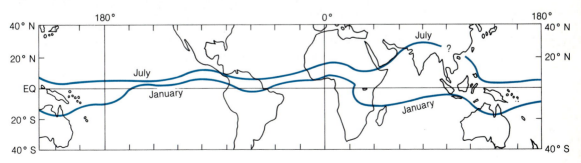

than over the oceans. Over the oceans the mean latitude of the ITCZ varies by only 4 degrees through the year (from 4 degrees N in April to 8 degrees N in September). The anchoring of the ITCZ over the ocean is due to the ocean's relatively great thermal stability.

As a subtropical anticyclone shifts north and south with the sun, its dry eastern flank influences some localities in winter and other localities in summer, thereby imposing a pronounced seasonality on rainfall. We have much more to say about the influence of the planetary circulation on world patterns of climate in Chapter 17.

SINGULARITIES

A **singularity** is a weather event that occurs on or near a certain date with unusual regularity. The frequency of these events in the climatic record is greater than would be expected on the basis of chance alone. Most singularities are linked to regular changes in features of the global-scale atmospheric circulation. For example, in regions of seasonal precipitation, the onset and ending of the rainy season may constitute singularities. Consider some other examples.

The **January thaw** is the most widely recognized and perhaps the only real singularity in a statistically rigorous sense. It is a period of relatively mild weather around January 20 to 23 and occurs primarily in the New England states. The thaw is caused by a flow of warm air on the back side (western flank) of the Bermuda–Azores anticyclone, which, for some unexplained reason, temporarily shifts north of its usual midwinter location.

Weather episodes that do not fit precisely the definition of a singularity, but nonetheless are fairly regular, are the July rainfall maximum in Arizona and the so-called Indian summer weather in the northeastern United States. In late June, the North Pacific subtropical anticyclone shifts abruptly northward, allowing the Bermuda–Azores high to extend its influence westward, across subtropical North America. The anticyclone's clockwise circulation pumps warm, humid air into the southwestsern United States and brings a rainy end to Arizona's dry spring. For example, in Phoenix, Arizona, the mean monthly rainfalls for April, May, and June are only 8 mm (0.30 in.), 3 mm (0.12 in.), and 2 mm (0.08 in.), respectively. For July, however, the mean monthly rainfall jumps abruptly to 20 mm (0.80 in.).

There is no precise date for **Indian summer,** but it usually develops in October or November after the autumn's first freeze. Large, warm anticyclones stagnate over the eastern United States, displacing the principal storm track northward along the St. Lawrence River Valley. Typical Indian summer weather consists of an episode of mild, sunny days with hazy skies, cool nights, and frosty mornings.

Upper-Air Westerlies

The midlatitude westerlies of the Northern Hemisphere merit special attention here because they govern the weather in the United States and Canada. As we noted earlier, in the middle and upper troposphere, the westerlies flow about the hemisphere in wavelike patterns of ridges and troughs (refer back to Figure 10.6). Winds exhibit a clockwise (anticyclonic) curvature in the ridges, and counterclockwise (cyclonic) curvature in the troughs. Between two and five waves typically encircle the hemisphere at any one time. These long waves are called **Rossby waves,** after Carl G. Rossby, the Swedish-American meteorologist who discovered them in the late 1930s.

Rossby waves characterize the tropospheric westerlies above the 500-mb level, that is, above the altitude where the pressure drops to 500 mb. Below this, the waves are distorted somewhat by friction and topographic irregularities of the Earth's surface.

The wavelike configuration of the circulation allows us to describe the westerlies by wavelength (distance between successive troughs or, equivalently, successive ridges), amplitude (north–south extent), and the number of waves encircling the hemisphere. The westerlies exhibit changes in all three measures, and, as a direct consequence, the weather changes.

The westerlies are more vigorous in winter than in summer. In winter, they strengthen and exhibit fewer waves of longer length and greater amplitude. This seasonal difference stems from the north–south pressure gradient, which is steeper in winter because of the greater temperature contrast between north and south at that time of year. In summer, north–south temperature differences are less, pressure gradients are weaker, and, as a consequence, so are the westerlies.

LONG-WAVE PATTERNS

The "weaving westerlies" consist of two components of motion: a north–south wind superimposed on a west–east wind. We refer to the north–south airflow as the westerlies' **meridional component,** and the west–east airflow as the **zonal component.** The meridional component of Rossby waves brings about a north–south exchange of air masses and poleward transport of heat. In the Northern Hemisphere, winds from the south carry warm air masses northward, and winds from the north transport cold air masses southward. Cold air is thus exchanged for warm air, and heat is transported poleward. As Rossby waves change in length, amplitude, and number, however, concurrent changes take place in the advection of air masses. Consider some examples.

Occasionally, the westerlies flow almost directly from west to east, nearly parallel to latitude circles, with a weak meridional component (Figure 10.9). This is a **zonal flow pattern** in which the north–south exchange of air masses is minimal. Cold air masses stay to the north, and warm air masses

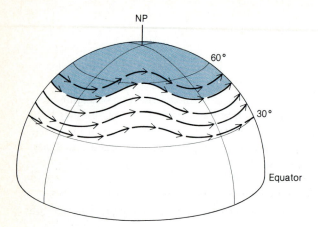

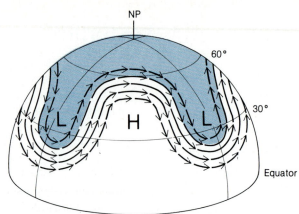

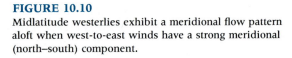

FIGURE 10.9
Midlatitude westerlies exhibit a zonal flow pattern aloft when winds blow almost directly west to east, with only a small meridional (north–south) component.

FIGURE 10.10
Midlatitude westerlies exhibit a meridional flow pattern aloft when west-to-east winds have a strong meridional (north–south) component.

remain in the south. At the same time, the United States and southern Canada are flooded by air that originated over the Pacific Ocean. Pacific air dries out to some extent as it passes through the western mountains, and then the air is compressed and warmed adiabatically as it descends onto the Great Plains—spreading uniformly mild and generally fair weather east of the Rocky Mountains.

At other times, the westerlies exhibit considerable amplitude, flowing in a pattern of deep troughs and sharp ridges (Figure 10.10). In this **meridional flow pattern,** masses of cold air surge southward, and warm air streams northward. Greater temperature contrasts develop over the United States and southern Canada. Where contrasting air masses collide, warm air overrides cold air and the stage is set for the development of storms that are then swept along by the westerlies.

These two illustrations of Rossby wave configurations are the opposite extremes of a wide range of possible westerly wind patterns, each featuring different components of meridional and zonal flow. The situation may be further complicated by a **split flow pattern,** in which westerlies to the north (over Canada, for example) have a wave configuration that differs from that of westerlies to the south (over the lower United States).

The westerly wind pattern typically shifts back and forth between dominantly zonal and dominantly meridional flow. For example, zonal flow might persist for a week and then give way to a more meridional flow that lasts for a few weeks, and then it's back to zonal flow again. The transition from one wave pattern to another is usually abrupt, sometimes taking place within a day. This abruptness poses a challenge to weather forecasters because a sudden shift in the upper-air winds may divert a storm toward or away from a locality, or cause an unanticipated influx of colder air that could, for example, change rain to snow.

Unfortunately for the long-range weather forecaster, the shifts between westerly wave patterns have no regularity. That is, no predictable zonal/meridional cycle has been discerned. The only observation useful to forecasters is that meridional patterns tend to persist for longer periods than zonal patterns. During the winter of 1976–77, for example, the strong meridional wave pattern shown in Figure 10.11 persisted from late October through mid-February. Northwesterly flow aloft brought surge after surge of bitterly cold arctic air into the midsection of the United States, resulting in one of the coldest winters of this century for that area. Meanwhile, southwesterly winds aloft brought unseasonably mild air to the far-western United States, including Alaska.

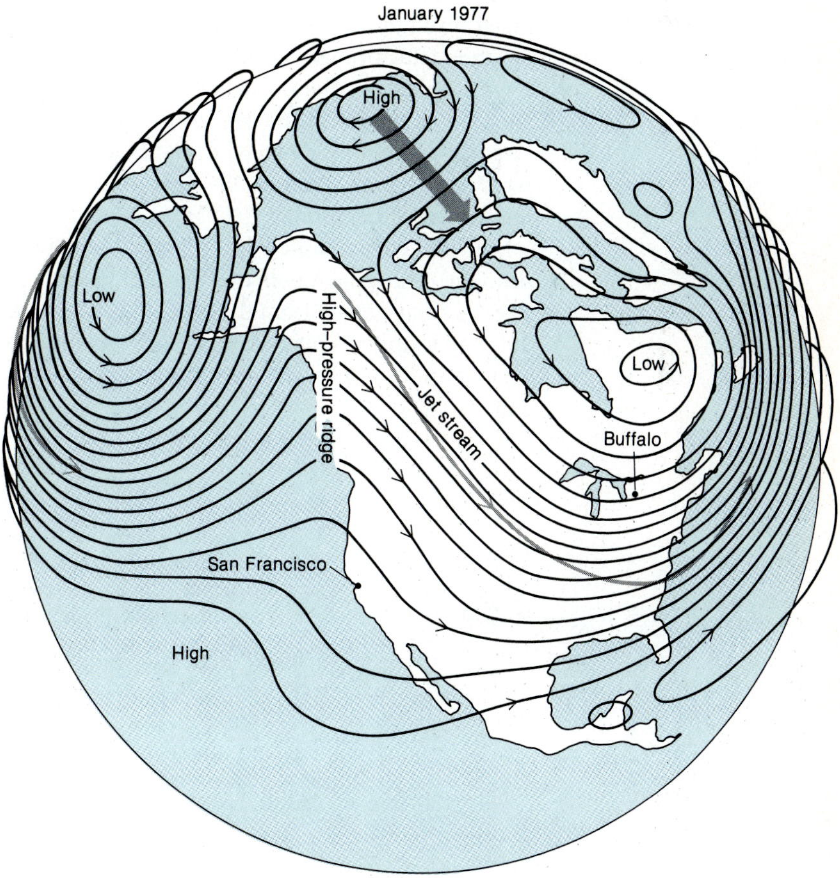

FIGURE 10.11

The strong meridional westerly wind pattern that persisted through much of the winter of 1976–77 brought record cold to the Midwest and East and record drought and warmth to the West. [Modified after T. Y. Canby, "The Year the Weather Went Wild," *National Geographic* 152 (1977):801. Redrawn from artwork by William H. Bond, © National Geographic Society.]

BLOCKING SYSTEMS

For the continent as a whole, North American weather is more dramatic when the westerly wave pattern is strongly meridional. Sometimes undulations of the westerlies become so great that huge, whirling masses of air actually separate from the main westerly airflow. This situation, shown schematically in Figure 10.12, is analogous to the whirlpools that form in rapidly flowing rivers. In the atmosphere, cutoff masses of air whirl in either a cyclonic or an anticyclonic direction. A **cutoff low** or a **cutoff high** that blocks the usual west-to-east progression of weather systems is referred to as a **blocking system.** Because blocking systems tend to persist for extended periods of time (usually several weeks or more), extremes of weather such as drought or anomalous heat or cold may result.

In July and August 1976, a large blocking high, stationed to the west of the British Isles, deflected the flow of cool, humid maritime air far to the north of its usual path. As a consequence, residents of Great Britain and of most of Europe experienced one of the hottest and driest summers on record. Fires raged through the parched forests of Germany and Italy, and drought severely cut the yields of many food crops.

Blocking circulation patterns were also responsible for the serious 1972 shortfall in the wheat harvest of the Soviet Union. In winter, stationary anti-cyclones over the Soviet wheatlands favored prolonged periods of dry, cold weather. Because of unusually light snowfall and thin snow cover, subfreezing temperatures penetrated deeply into the soil and killed much of the winter wheat. Wheat replanted in the spring did not fare much better because of drought. Through much of the summer of 1972, a blocking high diverted storm systems away from the wheat-growing region, so rainfall totals for May, July, and August were less than half of normal. This drought, following a relatively dry winter, meant that much of the wheat withered in desiccated soil.

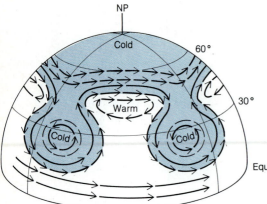

FIGURE 10.12
Aloft, the midlatitude westerlies sometimes exhibit an extreme meridional flow pattern in which huge pools of rotating air are cut off from the main west-to-east circulation. The pool of cold air rotating counterclockwise is a blocking cyclone, and the pool of warm air rotating clockwise is a blocking anticyclone. The latter is sometimes referred to as an "omega block" because of its resemblance to the Greek letter.

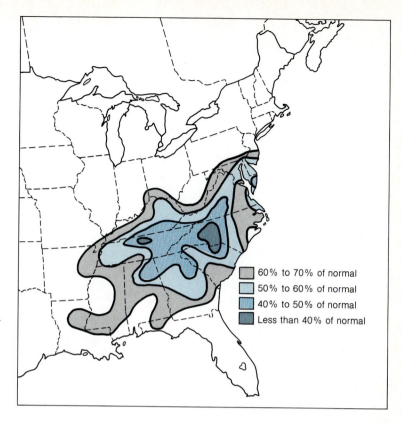

FIGURE 10.13
Percentage of the long-term (1951–1980) average December–July precipitation over portions of the southeastern United States for the period December 1985 through July 1986. [From K. H. Bergman et al., "The Record Southeast Drought of 1986," *Weatherwise* 39 (1986):262]

60% to 70% of normal
50% to 60% of normal
40% to 50% of normal
Less than 40% of normal

The 1972 failure of the Soviet wheat crop had worldwide economic ramifications because it forced the Soviets to enter the international grain market. As a result, the price of wheat rose sharply, and the reserves of the grain-exporting nations were depleted. In the United States, this increased demand caused the price of wheat to more than triple between early 1972 and early 1974.

Another example of the association of weather extremes with blocking circulation patterns comes from the southeastern United States. In the spring and early summer of 1986, an unusually severe and prolonged drought affected a ten-state region. In the hardest hit areas precipitation totals for the eight months preceding August 1986 were less than 40 percent of the long-term average (Figure 10.13). The primary reason for the May-to-early-July episode of moisture deficit was a persistent westerly long-wave pattern that featured a trough anchored off the East Coast near longitude 65 degrees W and, upstream (to the west), a ridge over the Southeast. In effect, the Bermuda–Azores anticyclone extended its influence well west of its usual location. The persistence of this circulation pattern meant day after day of subsiding stable air over the Southeast and suppression of the convective activity that normally brings the region abundant spring and summer rains.

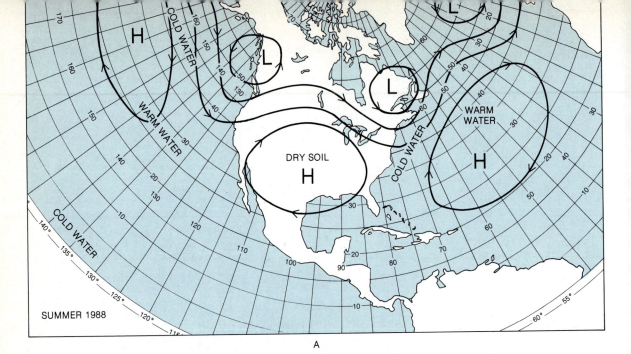

A

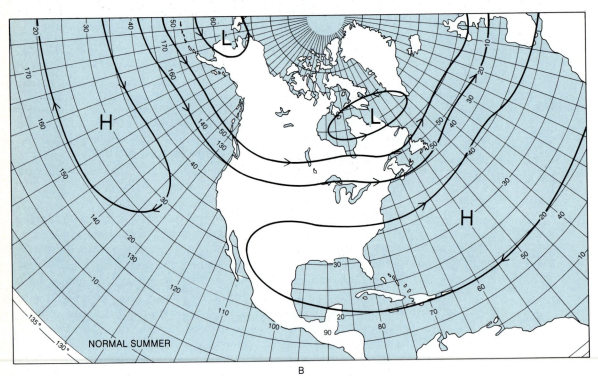

B

FIGURE 10.14

Prevailing upper-air winds (A) during the summer of 1988 as contrasted with (B) the long-term average pattern. A blocking warm anticyclone over the nation's midsection contributed to a severe drought during much of the summer of 1988. [From J. Namias, "Written in the Winds, The Great Drought of '88," *Weatherwise* 42 (1989):86–87]

The drought that affected the Midwest and Great Plains during the spring and summer of 1988 is a more recent example of the linkage between a weather extreme and a blocking weather pattern. In Figure 10.14, the major upper-air circulation features that dominated the summer of 1988 are contrasted with long-term average conditions. Between early May and mid-August, prevailing westerlies were more meridional than usual and featured a huge stationary warm high-pressure system over the nation's midsection and troughs over both the west and east coasts. The belt of strongest westerlies was displaced north of its usual position so that moisture-bearing weather systems were diverted into central Canada and well north of their usual paths. In the Corn Belt, the May through June period was the driest since 1895. By late July, drought was categorized as either severe or extreme over 43 percent of the land area of the contiguous United States. By the end of the growing season, the impact on the nation's grain harvest was severe: Corn production was down by 33 percent, soybeans by 20 percent, and spring wheat by more than 50 percent.

In addition to drought, stagnant anticyclones in summer bring extended periods of searing heat that can take many human lives. In the summer of 1980, more than 1200 people in the southern Mississippi River Valley died from heat-related stress. A stalled, warm anticyclone caused maximum daily air temperatures to hover near 38 °C (100 °F) for over a month. A similar episode in the Midwest two years later contributed to an estimated 200 deaths.

JET STREAM

Embedded in the upper-level westerlies is a relatively narrow corridor of very strong winds called the **jet stream.** The jet stream is not always a single stream of air but may splinter into separate ribbons. In midlatitudes, the jet stream is situated in the upper troposphere between the midlatitude tropopause and the polar tropopause, directly over the polar front (Figure 10.15). For this reason, it is known as the **polar front jet stream.** This jet stream, which follows the meandering path of the planetary westerly waves, attains wind speeds that frequently exceed 160 km (100 mi) per hour. Westbound aircraft understandably avoid the jet stream because it is a head wind, whereas eastbound flights seek it because it is a tail wind.

Why is a jet stream associated with the polar front? Recall that air temperature influences air density and hence air pressure. Where temperature contrasts are great, pressure gradients are steep and winds are strong. The polar front, situated between the cold polar easterlies and the mild westerlies, is a zone of significant temperature contrast, which causes a relatively steep horizontal air pressure gradient aloft. The strongest winds (that is, the jet stream) would be anticipated over the zone of greatest surface temperature contrast, the polar front. The relationship between the jet stream and the polar front is examined in more detail in the Mathematical Note at the end of this chapter.

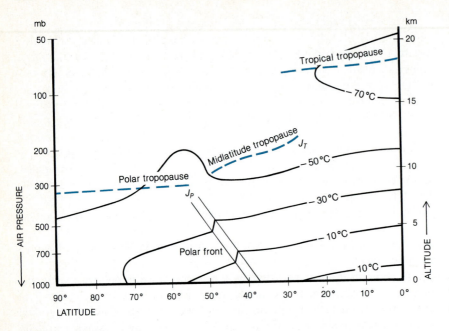

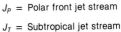

FIGURE 10.15

A vertical cross section (meridional profile) through the Northern Hemisphere troposphere showing the polar front jet stream (J_P) above the polar front and near the break in the tropopause between midlatitudes and polar latitudes. Also, the subtropical jet (J_T) is located near 25 degrees N and near the break in the tropopause between midlatitudes and tropical latitudes. In this perspective, both jet streams blow perpendicular to and into the page. Solid lines are isotherms in Celsius degrees. Note the considerable vertical exaggeration.

Like the polar front, the jet stream is not uniformly well defined around the globe. Where the polar front is well defined, that is, where surface temperature gradients are particularly steep, jet stream winds accelerate. Such a segment, in which the wind may accelerate by as much as 100 km (62 mi) per hour, is known as a **jet maximum.** The strongest jet maxima develop in winter along the east coasts of North America and Asia where the land-to-sea temperature contrast is particularly great. In those areas, wind speeds in the jet maximum on rare occasions have topped 350 km (217 mi) per hour. A typical jet maximum is 160 km (100 mi) wide, 2 to 3 km (1 to 2 mi) thick, and several hundred kilometers in length.

What, then, is the role of the polar front jet stream in the generation and maintenance of synoptic-scale storms? Air flowing through a jet maximum changes in both speed and direction, and these changes induce a complex pattern of horizontal divergence and horizontal convergence aloft. In Figure 10.16, a jet maximum (viewed from above) is divided into four quadrants: left rear, right rear, left front, and right front. Horizontal divergence occurs in the left front and right rear quadrants, and horizontal convergence occurs in

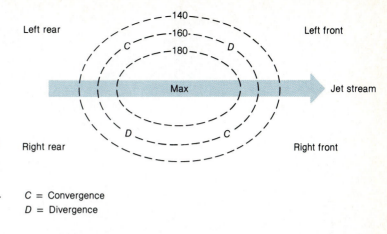

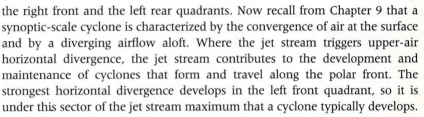

FIGURE 10.16

A jet stream maximum is divided into quadrants. The jet maximum is outlined by isotachs (dashed lines) of equal wind speed (in kilometers per hour). Air movement into and out of the jet maximum induces areas of horizontal divergence (*D*) and areas of horizontal convergence (*C*).

C = Convergence
D = Divergence

the right front and the left rear quadrants. Now recall from Chapter 9 that a synoptic-scale cyclone is characterized by the convergence of air at the surface and by a diverging airflow aloft. Where the jet stream triggers upper-air horizontal divergence, the jet stream contributes to the development and maintenance of cyclones that form and travel along the polar front. The strongest horizontal divergence develops in the left front quadrant, so it is under this sector of the jet stream maximum that a cyclone typically develops.

Although the polar front jet stream weaves with the westerlies, a jet maximum typically progresses from west to east at a faster pace than does the west to east displacement of the troughs and ridges in the westerlies (Figure 10.17). The strongest divergence aloft occurs when a jet maximum is on the east side of a trough so that cyclone development is much more likely when the jet maximum is in that position.

FIGURE 10.17

An upper-level trough is shown progressing from west to east across the United States. Meanwhile, a jet maximum (dashed outline) travels eastward at a more rapid pace than does the trough. Dashed lines are isotachs, lines of equal wind speed. [From F. S. Sechrist and E. J. Hopkins, *Meteorology: Weather and Climate.* Madison, WI: University of Wisconsin Extension, 1974, p. 146.]

Monday

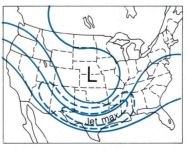

Tuesday

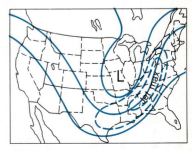

Wednesday

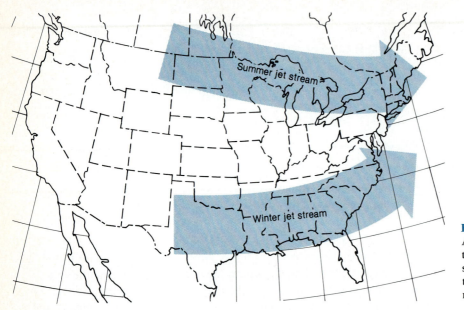

FIGURE 10.18

Approximate average locations of the polar front jet stream in winter (December through March) and in summer (June through October).

Like the polar front, the jet stream undergoes important seasonal shifts. The jet stream strengthens in winter, when north–south temperature contrasts are great, and weakens in summer, when temperature contrasts are less. As shown in Figure 10.18, the average summer location of the jet stream is across southern Canada, and the average winter position is across the southern United States. These locations represent long-term averages; the jet stream actually weaves over a considerable range of latitude from week to week, and even from day to day. As a general rule, when the polar front jet stream is south of us, the weather tends to be relatively cold, and when the polar front jet stream is north of us, the weather tends to be relatively warm.

The polar front jet stream is not the only jet stream. The **subtropical jet stream** occurs near the break in the tropopause between tropical and middle latitudes and on the poleward side of the Hadley cell (Figure 10.15). It is stronger and less variable in latitude than its northerly counterpart. Jet streams also occur in the Southern Hemisphere, but so far have not received as much study as those to the north.

SHORT WAVES

Short waves are another feature of the upper-level westerlies. These waves, which are ripples superimposed on Rossby long waves, are important to surface weather systems. Although Rossby waves drift very slowly eastward, short waves travel rapidly through the Rossby waves. As five or fewer long waves encircle the hemisphere, there may be a dozen or more short waves.

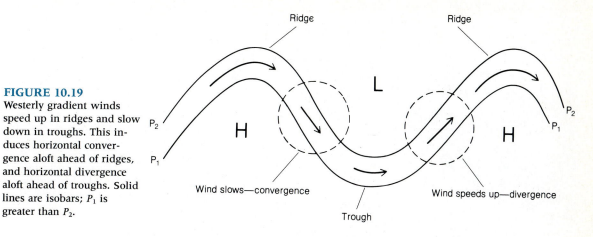

FIGURE 10.19
Westerly gradient winds speed up in ridges and slow down in troughs. This induces horizontal convergence aloft ahead of ridges, and horizontal divergence aloft ahead of troughs. Solid lines are isobars; P_1 is greater than P_2.

Both short waves and long waves in the westerlies can contribute to cyclone development. For the same isobar spacing (air pressure gradient), gradient and geostrophic wind speeds are not equal. Anticyclonic gradient winds are stronger than geostrophic winds, and cyclonic gradient winds are weaker than geostrophic winds. Hence, as shown in Figure 10.19, westerly winds tend to strengthen in a ridge and weaken in a trough. The result is horizontal divergence of air to the east of the trough (and west of the ridge). Short and long waves thus favor storm development in the same way as the polar front jet stream—by inducing diverging airflow aloft. Consequently, conditions aloft are most favorable for storm development when the jet stream maximum appears on the east side of a trough (and west of a ridge).

El Niño/Southern Oscillation (ENSO)

Earlier we saw how blocking patterns in the westerlies can trigger extremes in weather. Some midlatitude weather extremes also may be linked to circulation anomalies that develop in tropical latitudes. Perhaps one of the most extensively studied of these circulation anomalies involves both the tropical atmosphere and ocean and is known as **ENSO,** the acronym for El Niño/Southern Oscillation.

El Niño is the term that fishermen applied to a period of reduced fish catch that accompanies a rise in sea-surface temperature off the coast of Ecuador and northern Peru. El Niño is an annual event occurring around the Christmas season, and hence the name.* Normally, off the Pacific coast of

---

* El Niño is the Spanish equivalent of Christ child.

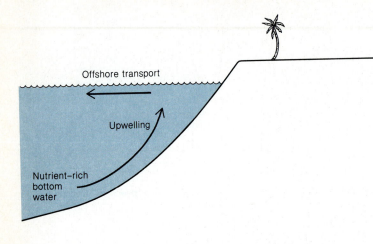

Offshore transport

Upwelling

Nutrient–rich
bottom
water

FIGURE 10.20

Schematic cross section of upwelling in a coastal area whereby cold, nutrient-rich bottom waters circulate up toward the surface and replace surface waters that are transported offshore by winds.

South America, cold nutrient-rich bottom water is transported to the surface in a process known as **upwelling** (Figure 10.20). Nutrients spur plankton growth, which, in turn, supports the Peruvian anchovy fishery, the world's largest fishery. In good years, the anchovy catch can exceed 11 million metric tons. During El Niño, however, a warm surface current spreads from the equator southward along the coast and suppresses upwelling. Deprived of an adequate nutrient supply, anchovy production declines and fishing is temporarily disrupted.

Fortunately, a typical El Niño is shortlived, lasting only several months, and affects primarily the waters off northernmost Peru. Occasionally, however, an intense El Niño* develops that persists for a year or longer and spreads anomalously warm surface waters along the entire coast of Peru and over large areas of the east and central equatorial Pacific. During an intense El Niño, the anchovy fishery collapses. An intense El Niño is usually accompanied by weather extremes in widely separated regions of the world.

An important step in understanding the link between an intense El Niño and weather extremes came in 1966 when Jacob Bjerknes, a researcher at the University of California at Los Angeles, demonstrated a relationship between El Niño and the southern oscillation. The **southern oscillation** is a seesaw variation in air pressure between the eastern and western tropical Pacific.[†] Specifically, air pressure over Indonesia and Australia varies inversely with air pressure over the tropical east Pacific. The horizontal pressure gradient thus changes as air pressure to the west rises and air pressure to the east falls (and

---

* In recent years, some members of the scientific community have used "El Niño" only in reference to an intense event.

[†] Sir Gilbert Walker discovered the southern oscillation in 1924. Walker noted that when air pressure is low at Darwin (on the north coast of Australia), it is high at Tahiti (a south Pacific island at about 18 degrees S and 149 degrees W), and vice versa.

vice versa). Bjerknes found that El Niño begins when the air pressure gradient between the eastern and western Pacific begins to weaken. This finding spurred scientists to attempt to model El Niño.

Under normal conditions, southeast trade winds trigger upwelling off the west coast of South America. The trades drive warm surface waters westward away from the coast, and, in response, cold nutrient-rich bottom water wells up to the surface. Strong trade winds, and hence vigorous upwelling, are

---

Special Topic

The ENSO Event of 1982–83

In 1982, an ENSO event began in May, and by December, the warming of surface ocean waters was unprecedented in magnitude and extent (see Plate 18). The unusual warming spread westward along the equatorial Pacific to near the international dateline (180 degrees W), and, in places, sea-surface temperature anomalies (departures from normal) reached +6 C° (+10.8 F°). This anomalously warm water over vast areas triggered major changes in atmospheric circulation patterns over tropical and middle latitudes. Circulation changes, in turn, led to weather extremes in many areas.

The winter storm track was displaced hundreds of kilometers southeast of its usual position, resulting in destructive high winds and heavy rains in California. Excessive rains between mid-November 1982 and late January 1983 caused the worst flooding of the century in Ecuador. French Polynesia was struck by no less than six tropical storms in a span of only three months. At the other extreme, severe drought parched Australia, Indonesia, and southern Africa. Australia suffered a $2 billion loss in crops, and millions of sheep and cattle died. Meanwhile, drought in the African Sahel grew even worse.

The 1982–83 ENSO also had a devastating impact on marine life in the eastern tropical Pacific, particularly off the coast of South America. With suppression of upwelling, the growth of microscopic plankton that normally flourish in upwelling areas diminished sharply. The 1982–83 ENSO was so intense that it triggered a 20-fold reduction in the quantity of plankton off the South American west coast. This decline reduced the numbers of anchovy, which feed heavily on plankton, to a record low. Other fish dependent on the plankton, such as jack mackerel, also decreased in number. With the decline of fish populations, marine birds (such as frigate birds and terns) and marine animals (such as fur seals and sea lions) also suffered major population declines because of breeding failures and loss of their food sources.

Even wildlife on remote Christmas Island (2 degrees N, 157 degrees W) in the Central Pacific suffered from the effects of the 1982–83 ENSO. Hundreds of thousands of sea birds, which feed on fish and squid, normally nest on this island. However, scientists arriving on the island in November 1982 discovered that virtually all the adult birds had deserted the site, leaving their young behind to starve. In addition to reduced food sources, heavy rains and flooding probably contributed to the reproductive failures of the birds. By July 1983, with the return of more normal atmospheric and oceanic conditions near Christmas Island, the few surviving birds began to nest again.

The 1982–83 ENSO event is a remarkable example of the vast complex of interconnections between atmosphere and ocean, and the many forms of life, not only in the equatorial Pacific, but around the entire globe as well. Scientists will be studying these interconnections for many years in an attempt to better understand and predict the origin and impact of future ENSO events.

maintained by a steep air pressure gradient between the eastern and western Pacific. However, when the air pressure gradient declines as part of the southern oscillation, the southeast trades, and hence upwelling, also weaken. Slackening trades allow the warm surface water that had been driven westward to drift slowly eastward along the equator and then southward off the Peruvian coast. Upwelling is suppressed, and El Niño is in progress.

During an intense ENSO (El Niño/Southern Oscillation), the southeast trades eventually shift direction and become equatorial westerlies. The westerlies drive warm surface waters eastward over a much larger area of the Pacific than during a typical El Niño. In addition, reversal of the trades has ramifications for the weather of Indonesia and the west coast of South America. Normally, prevailing winds over Indonesia are onshore (southeast trades) so that warm, humid air surging inland brings abundant rainfall. During an intense ENSO event, however, winds become offshore (westerly) and the weather is generally dry. Meanwhile, on the other side of the tropical Pacific, anomalously high sea-surface temperatures off the west coast of South America spur convection and heavy rainfall in normally arid coastal localities. As a consequence, the desert blooms.

For reasons not fully understood, an intense ENSO may be accompanied by weather extremes in midlatitudes as well as in tropical latitudes. Seven moderate-to-intense ENSO events have occurred since the early 1950s, the most recent of which was a moderate episode that began in August 1986 and ended by February 1988. The impact of the 1982–83 ENSO, probably the most intense of the century, is summarized in this chapter's Special Topic.

Recently, scientists have coined the term **La Niña** (the girl) for oceanic/atmospheric conditions that are opposite to those of El Niño. La Niña features strong trade winds and unusually low sea-surface temperatures in the central and eastern tropical Pacific. An intense La Niña may also be linked to weather extremes in midlatitudes. For example, the persistent atmospheric circulation pattern that led to the U.S. drought of 1988 (discussed earlier in this chapter) coincided with an intense La Niña. We have more to say about possible connections between sea-surface temperature anomalies and prevailing atmospheric circulation patterns in Chapter 15.

Conclusions

Features of the global-scale circulation are important controls of weather and climate. This chapter focused on the linkages between the westerly winds aloft and the weather in midlatitudes. Specifically, the westerlies steer storms, control air mass exchange and poleward heat transport, and are the source of divergence aloft for the development of cyclones. Our examination of atmospheric circulation continues in the next chapter with a focus on synoptic-scale weather systems.

Mathematical Note

The Polar Front and the Midlatitude Jet Stream

Why is a jet stream found over the polar front? To understand the reasons for this important association, we must return to the concept of *hydrostatic equilibrium*. We learned in Chapter 9 that hydrostatic equilibrium is the balance between the upward-directed pressure gradient force and the downward-directed force of gravity. We now express this relationship mathematically.

The air parcel in Figure 1 has density ρ, a cross-sectional area A, and a volume V. The air pressure acting on the top of the parcel, P_2 (at altitude Z_2), is due to the weight of the total atmospheric column

above Z_2. The air pressure acting on the bottom of the parcel, P_1 (at altitude Z_1), is due to the weight of the total atmospheric column above Z_1. The pressure difference between the parcel top and the parcel bottom is the vertical pressure gradient and depends on the weight of the parcel itself. The parcel's weight, W, is equal to its mass, M, times the acceleration of gravity, g, and is distributed over the area A. Hence,

$$P_2 - P_1 = W/A$$

but since

$$W = Mg = \rho V g$$

and

$$V = A(Z_2 - Z_1)$$

then

$$P_2 - P_1 = \rho g(Z_2 - Z_1)$$

That is,

$$\Delta P = -\rho g \Delta Z$$

This is known as the *hydrostatic equation*. The minus sign on the right side of the equation accounts for the inverse relationship between air pressure and altitude. That is, pressure drops as altitude increases.

According to the hydrostatic equation, the pressure change (ΔP) that accompanies an altitude change (ΔZ) depends on the density of air and on the acceleration of gravity. For practical purposes, we may assume that gravity is constant with altitude, at least for several kilometers. Air density is thus the ultimate determinant of the rate at which air pressure drops with increasing altitude. Because cold air is denser than warm air, air pressure drops more rapidly with altitude in cold air than in warm air.

FIGURE 1

In a column of air, a change in pressure ($P_1 - P_2 = \Delta P$) is associated with a change in altitude ($Z_2 - Z_1 = \Delta Z$).

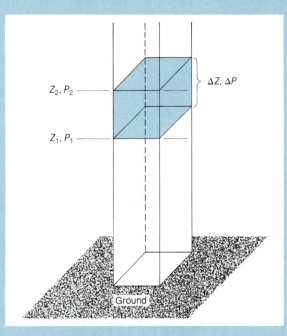

The hydrostatic equation allows us to compute the thickness (ΔZ) of an air layer bounded above and below by constant pressure (isobaric) surfaces; say, for example, the thickness of the 700- to 500-mb layer. Air layer thickness is inversely proportional to the average air density within the layer. Air layer density, in turn, depends on temperature and humidity (Chapter 5). Vertical variations in humidity are not nearly as important as temperature in determining air density so that we can formulate this rule of thumb: the thickness of an air layer between successive isobaric surfaces is directly proportional to the average temperature within the air layer. Hence, for example, the 700- to 500-mb thickness is greater in warm air than in cold air.

Now suppose that a sharp horizontal air temperature gradient develops at the Earth's surface, as across a well-defined segment of the polar front. The hydrostatic equation tells us that on the warm (less dense) side of the front the air pressure drops less rapidly with altitude than it does on the cold (more dense) side of the front. Equivalently, an air layer bounded by isobaric

surfaces is thicker on the warm side than on the cold side of the front. Consequently, isobaric surfaces slope as shown in Figure 2, and the slope steepens with altitude. The steepening slope means that the horizontal air pressure gradient strengthens with altitude, so that the westerly wind speed must increase with altitude. The wind reaches its maximum speed just below the tropopause and over the area of maximum horizontal temperature gradient (the polar front); this is the jet stream.

Westerly wind speeds peak near the tropopause because, within the stratosphere, the latitudinal (north–south) temperature gradient reverses. (That is, within the stratosphere, air temperatures rise from equator to pole.) This weakens the latitudinal pressure gradient so that the westerlies slow with altitude within the stratosphere.

In summary, a jet stream is associated with steep horizontal air temperature gradients at the Earth's surface. Where the temperature contrast across a front is particularly great, we would expect to encounter a jet maximum aloft.

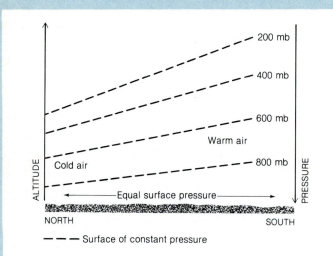

FIGURE 2
Air pressure declines more rapidly with altitude in a column of cold air than in a column of warm air. Hence, the horizontal air-pressure gradient steepens with altitude when cold air is situated next to warm air. A strengthening of the horizontal pressure gradient, in turn, means that the horizontal wind speed increases with altitude.

Summary Statements

The principal features of the global-scale circulation are semipermanent pressure cells, polar easterlies, the westerlies, trade winds, the intertropical convergence zone, and the polar front. These features shift poleward during spring and toward the equator during autumn.

Trade winds blow out of the equatorward flank of the subtropical anticyclones, and the westerlies blow out of the poleward flank. The east side of a subtropical anticyclone features subsiding air and dry conditions, whereas the west side is more humid.

Aloft, global-scale winds form Hadley cells in the tropics, westerly waves in the midlatitudes, and westerly winds over polar regions.

Contrasts in surface temperatures favor relatively high pressure over continents and low pressure over seas in winter, and relatively low pressure over continents and high pressure over seas in summer.

Upper-air westerlies describe a wavelike pattern that, with time, undergoes changes in length, amplitude, and wave number. These changes, in turn, influence the degree of meridional and zonal flow and the north–south exchange of air masses.

Cutoff anticyclones and cutoff cyclones within the westerlies block the usual west to east progression of weather systems and may lead to extreme weather conditions such as drought, excessive rainfall, or periods of unusually high or low temperatures.

The midlatitude jet stream is a corridor of very strong winds within the westerlies and is situated at the top of the troposphere over the polar front. A jet maximum is a region of particularly strong jet stream winds over a well-defined segment of the polar front.

Cyclone development is most likely under the left front quadrant of a jet stream maximum on the east side of a trough. It is here that horizontal divergence aloft is strongest.

ENSO refers to a lengthy period of anomalously high sea-surface temperatures over a vast area of the equatorial and eastern tropical Pacific. An ENSO is accompanied by weather extremes in tropical latitudes and in other areas of the globe.

Key Words

semipermanent pressure systems	subpolar lows	meridional component	jet maximum
subtropical anticyclones	Aleutian low	zonal component	subtropical jet stream
horse latitudes	Icelandic low	zonal flow pattern	short waves
midlatitude westerlies	polar front	meridional flow pattern	ENSO
trade winds	front	split flow pattern	El Niño
equatorial trough	Hadley cell	cutoff low	upwelling
intertropical convergence zone (ITCZ)	singularity	cutoff high	southern oscillation
heat equator	January thaw	blocking system	La Niña
	Indian summer	jet stream	
	Rossby waves	polar front jet stream	

Review Questions

1. Why are global-scale pressure systems described as "semipermanent"?

2. Why do some of the world's major deserts occur on the eastern side of the subtropical anticyclones?

3. Explain the origin of the name "horse latitudes."

4. How are the trade winds linked to the subtropical anticyclones?

5. How are the westerlies linked to the subtropical anticyclones?

6. Describe the weather along the intertropical convergence zone (ITCZ).

7. What is a front?

8. How is the Hadley cell like a giant convective current?

9. Describe how the tropopause slopes with latitude.

10. The major features of the global-scale circulation "follow the sun" through the course of a year. Explain this statement.

11. Give an example of a singularity.

12. Is Indian summer a true singularity?

13. Distinguish between the meridional component and the zonal component of midlatitude westerly waves. Which component is responsible for poleward heat transport?

14. Distinguish between a zonal flow pattern and a meridional flow pattern.

15. Why is the weather more extreme when the westerly wave pattern is strongly meridional?

16. What is the relationship between the polar front jet stream and the midlatitude westerlies?

17. Define jet stream maximum.

18. Describe the seasonal changes in the polar front jet stream.

19. What is the relationship between short waves and Rossby long waves?

20. In westerly waves, why does horizontal divergence of air occur east of a trough and horizontal convergence of air occur east of a ridge?

Points to Ponder

1. Speculate on why the tropical tropopause is at a greater altitude than the polar tropopause.

2. Explain how the jet stream and short waves contribute to cyclone development.

3. Suppose that the prevailing westerly circulation pattern shifts from dominantly zonal to dominantly meridional. How will this change influence north–south air mass exchange? How will the polar front be affected?

4. Describe the relationship of blocking systems to extremes in weather.

5. Speculate on how a split flow pattern might influence the weather over North America.

6. Why is a jet stream associated with the polar front?

7. The Southern Hemisphere westerlies are stronger than their Northern Hemisphere counterpart. Explain why.

Projects

1. From your analysis of televised temperature summaries for the nation on selected days, speculate on whether the westerly wave pattern is dominantly zonal or dominantly meridional. Support your choice.

2. From your analysis of surface temperatures across the nation, locate the polar front and the polar front jet stream.

Selected Readings

Bergman, K. H., et al. "The Record Southeast Drought of 1986." *Weatherwise* 39 (1986):262–266. Contains some data on a severe drought that was triggered by a blocking circulation pattern.

Chang, J. *Atmospheric Circulation Systems and Climates.* Honolulu: Oriental Publishing Company, 1972. 328 pp. Provides an excellent detailed discussion of atmospheric circulation systems.

Kerr, R. A. "Another El Niño Surprise in the Pacific, But Was It Predicted?" *Science* 235 (1987):744–745. Reviews the apparent success of an atmosphere–ocean interactive model in predicting the 1986–87 ENSO event.

Lorenz, E. N. "A History of Prevailing Ideas About the General Circulation of the Atmosphere." *Bulletin of the American Meteorological Society* 64 (1983):730–733. Briefly surveys the principal people and ideas involved in developing our understanding of global-scale circulation.

Ramage, C. S. "El Niño." *Scientific American* 254, No. 6 (1986):77–83. Presents a detailed review of the ENSO mechanism and some of the problems involved in modeling the phenomenon. Also see followup letters to the editor in the November 1986 issue of *Scientific American*.

11

Synoptic-Scale Weather

Chapter Written by

Professor Patricia M. Pauley

DEPARTMENT OF METEOROLOGY
UNIVERSITY OF WISCONSIN, MADISON

Synoptic-scale circulations are a subdivision of the global-scale circulation and include air masses, fronts, cyclones, and anticyclones. This water-vapor satellite image portrays a midlatitude cyclone in the North Atlantic with its characteristic comma cloud, as well as a ridge over the southeastern United States. [NOAA satellite photograph]

T

HE CEASELESS succession of synoptic-scale weather systems is directly responsible for day-to-day variations in weather. This chapter examines the weather conditions that accompany the major weather systems of midlatitudes: air masses, fronts, cyclones, and anticyclones. We learn how these systems are interrelated and how their development and movements are governed by the global-scale circulation.

Air Masses

An **air mass,** a huge volume of air covering thousands of square kilometers, is relatively uniform horizontally in temperature and water vapor concentration, although both quantities typically decrease with altitude. The properties of an air mass are determined by the type of surface over which it develops, that is, its source region. A source region features nearly homogeneous surface characteristics over a wide area, such as a great expanse of snow-covered ground or a vast stretch of ocean. In order for an air mass to become uniform in temperature and water vapor concentration, it must reside in its source region for several days to weeks.

Simply put, air masses are either cold (polar, abbreviated as P) or warm (tropical or T), and either dry (continental or c) or humid (maritime or m). Air masses that form over the cold, often snow-covered surfaces of high latitudes are relatively cold, and those that develop over the warm surfaces of low latitudes are relatively warm. Air masses that form over land tend to be relatively dry, and those that develop over the oceans are relatively humid. Thus, we have four basic types of air masses: cold and dry continental polar (cP), cold and humid maritime polar (mP), warm and dry continental tropical (cT), and warm and humid maritime tropical (mT). A fifth type, arctic (A) air, is cold and dry like continental polar air but is distinguished by its bitter cold.

NORTH AMERICAN TYPES

All of these air mass types occur over North America and are formed over the continent and surrounding waters in certain characteristic regions. Figure 11.1 shows the locations of the principal source regions for each type.

Continental tropical air develops over the subtropical deserts of Mexico and southwestern United States primarily in summer and is hot and dry. **Maritime tropical air** is very warm and humid, and its source regions are over tropical and subtropical oceans. It retains these properties year-round and is responsible for oppressive summer heat and humidity east of the Rocky

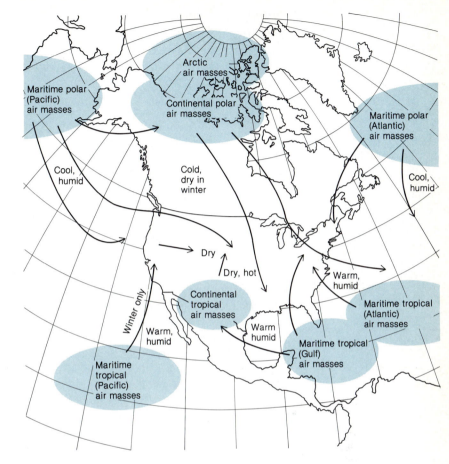

FIGURE 11.1
Air mass source regions for North America.

Mountains. The source regions for **maritime polar air** are over the cold ocean waters of the North Pacific and North Atlantic. Along the West Coast, mP air brings heavy winter rains (snows in the mountains) and persistent coastal fogs in summer. Dry **continental polar air** develops over the northern interior of North America. In winter, cP air is typically very cold because the ground in its source region is often snow covered, the days are short, insolation is weak, and radiational cooling is extreme. In summer, when the snow-free source region warms in response to long days of bright sunshine, cP air is quite mild and pleasant. **Arctic air** forms over the snow- or ice-covered regions of Siberia, the Arctic Basin, Greenland, and North America north of about 60 degrees N in much the same way as cP air but in a region that receives almost no insolation in winter, although it still radiates strongly in the infrared. These air masses are exceptionally cold and dry and are responsible for the bone-numbing winter cold waves that sweep across the Great Plains and at times penetrate as far south as the Gulf of Mexico and Florida.

Table 11.1

Stability, Temperature, and Humidity Characteristics of North American Air Masses

Air mass type	Source region stability		Characteristics	
	Winter	*Summer*	*Winter*	*Summer*
A	Stable		Bitter cold, dry	
cP	Stable	Stable	Very cold, dry	Cool, dry
cT	Unstable	Unstable	Hot, dry	Hot, dry
mP (Pacific)	Unstable	Unstable	Mild, humid	Mild, humid
mP (Atlantic)	Unstable	Stable	Cold, humid	Cool, humid
mT (Pacific)	Stable	Stable	Warm, humid	Warm, humid
mT (Atlantic)	Unstable	Unstable	Warm, humid	Warm, humid

For example, the so-called Siberian Express is an extremely cold Arctic air mass that forms over Siberia, crosses over the North Pole into Canada, and plunges into the Great Plains and on southward.

Air masses may differ not only in temperature and humidity but also in stability. As we noted in Chapter 6, stability is an important property of air because it influences vertical motion and the consequent development of clouds and precipitation. Table 11.1 lists the usual stability, temperature, and humidity characteristics of North American air masses within their source regions.

MODIFICATION

Air masses do not remain in their source regions indefinitely but move from place to place. As they travel, their properties modify, those of some air masses changing more than those of others. Changes may occur in temperature, humidity, and/or stability. Air masses modify primarily in three ways: (1) by exchanging heat or moisture, or both, with the surface over which the air mass travels; (2) by radiative cooling (infrared emission); and (3) by processes associated with large-scale vertical motion. Let us now examine examples of these mechanisms.

In winter, as a mass of cP air travels southeastward from Canada into the United States, its temperature usually modifies quite rapidly. Although temperatures in the Northern Plains might dip well below -18 °C (0 °F), by the time the polar air reaches the southern United States, temperatures may not drop much below the freezing point. This rapid air mass modification occurs because, outside its source region, polar air is usually colder than the ground over which it travels. The sun heats snow-free ground, and the warmer ground heats the bottom of the air mass, destabilizing it and triggering convective currents that distribute heat vertically throughout the air mass. A similar process of heating from below and destabilization also occurs when a cP air mass crosses the Atlantic Coast and travels over the warm waters of the western Atlantic. In addition, evaporation from the sea surface increases the moisture content of the air mass. Saturation is easily achieved in the cold, relatively unstable air, leading to extensive low-level cloudiness and fog. When cP air travels over snow-covered ground, however, modification is much less rapid because most of the incoming solar radiation is reflected away, rather than being absorbed and heating the ground. The resulting cold surface increases the stability of the air and thereby reduces convection. As previously mentioned, radiative cooling can play a key role in this situation, leading to a decrease in temperature when days are short and solar heating and convection are minimal. In addition, since air masses are identified with low-level high-pressure systems, they can be warmed adiabatically by the large-scale subsidence that characterizes these high-pressure systems.

A tropical air mass does not modify as readily as a polar air mass because, outside its source region, tropical air is often warmer than the ground over which it travels. The bottom of the air mass cools by contact with the ground; this cooling stabilizes the air and suppresses convective currents. As a result, cooling is restricted to the lowest portion of the air mass. In contrast, if the ground is warmer than the tropical air mass moving over it, the air mass can become even warmer in the same way described in the previous paragraph. Thus, a cold wave loses much of its punch as it pushes southward, but a summer heat wave can retain its warmth as it journeys from the Gulf of Mexico well into Canada.

As a consequence of orographic lifting, air masses undergo significant changes in temperature and humidity. When cool and humid mP air sweeps inland off the Pacific Ocean, the air is forced up the windward slopes of the coastal mountain ranges and so cools adiabatically. Cooling leads to condensation (or deposition) and the development of clouds and precipitation. Condensation is also accompanied by an increase in temperature resulting from the release of latent heat, which offsets the adiabatic cooling to some extent. The air then warms adiabatically and clouds dissipate, as the air descends the leeward slopes into the Great Basin. Some decrease in temperature through evaporative cooling is associated with the dissipation of clouds, but a net heating (and drying) occurs since the water vapor that was condensed and precipitated out is no longer available for evaporation. The same processes are repeated as the air mass is forced to flow up and over the Rockies. Eventually, the air mass emerges on the Great Plains considerably milder and drier than the original mP air. East of the Rockies, such an air mass is described simply as **Pacific air.**

When the westerly wave pattern aloft is predominantly zonal (Chapter 10), Pacific air floods the eastern two-thirds of the United States and southern Canada. Polar air masses stay far to the north, and tropical air masses keep to the south. Consequently, much of the region experiences an episode of mild and generally dry weather.

Frontal Weather

A **front** is a narrow zone of transition between air masses that differ in density. Density differences are usually due to temperature contrasts; for this reason we use the nomenclature "cold" and "warm" fronts. However, differences may also arise from contrasts in humidity or some combination of temperature and humidity. Since the transition zone associated with an actual front often occurs over a distance of a few hundred kilometers, traditionally we draw the front on a map along the warm (less dense) side of the transition zone. The temperatures are therefore nearly constant on the warm side of the front and decrease behind the front to a region of nearly constant temperatures in the cold air mass.

A front is also associated with a trough in the sea-level pressure pattern, a corresponding wind shift, and convergence. As we saw in Chapter 6, the warmer air is uplifted where contrasting air masses meet, and uplift may cool the air sufficiently for clouds and precipitation to develop along the front. Depending on the slope of the front in the vertical plane and the air motions relative to the front, frontal weather may be confined to a very narrow band, or it may extend over a broad region. In this section, we discuss the four basic fronts: stationary, warm, cold, and occluded.

STATIONARY FRONT

A **stationary front** is just that—a front that exhibits essentially no movement. For example, a stationary front frequently develops along the Front Range of the Rocky Mountains when a shallow pool of polar air surges south and southwestward out of Canada. The cold dry air mass abuts the mountain range and can push no further westward; therefore, its leading edge is marked by a stationary front paralleling the mountain range. Similarly, a stationary front can form when any type of preexisting front loses its motion as it becomes parallel to the upper-level flow pattern. Under the proper conditions, a stationary front can also form along a boundary in surface thermal characteristics, such as a coastline or snow-cover boundary.

Many features of a stationary front are common to all fronts. As seen in a vertical cross section such as that shown in Figure 11.2, a front slopes back from the Earth's surface toward colder air, or more precisely toward denser air. A front lies in a trough in the pressure pattern on any horizontal surface intersecting the front; this is especially evident in the sea-level isobars. Recall that the wind is approximately parallel to the isobars, with friction turning the wind somewhat toward lower pressure. The wind will typically change direction rather abruptly across the front; this condition is called a wind shift. The change in wind direction and speed across a front is usually associated with convergence, which is also related to upward motion, clouds, and precipitation, as explained in Chapter 9.

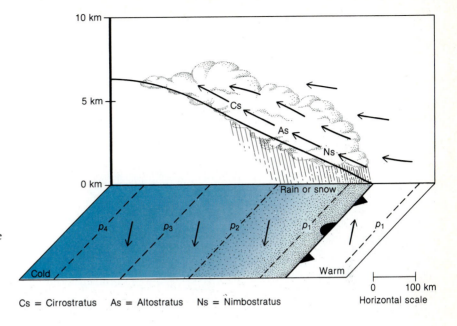

FIGURE 11.2

A stationary front has surface winds essentially parallel to the front and often a wide region of clouds and rain or snow to the cold side of the front. The clouds and precipitation result from overrunning. Note the considerable vertical exaggeration in the cross section.

Cs = Cirrostratus   As = Altostratus   Ns = Nimbostratus

A stationary front does not always have a broad region of associated clouds and precipitation as depicted in Figure 11.2. The details can vary considerably from case to case, depending on the supply of moisture and the specifics of the motion relative to the front. However, in the cases that do produce precipitation, the rain or snow is predominantly on the cold side of the stationary front. Warm moist air flows upward over the cooler air mass, more or less along the frontal surface. The air cools through adiabatic expansion, producing condensation and precipitation. This situation is often referred to as **overrunning** and can result in an extended period of relatively widespread cloudiness and drizzle, light rain, or light snow.

WARM FRONT

If a stationary front begins to move in such a way that the warm (less dense) air advances while the cold (more dense) air retreats, the front changes in character and becomes a **warm front.** The overall characteristics of a warm front are very similar to those of the stationary front, as can be seen in Figure 11.3.

The typical differences between warm fronts and stationary fronts can be examined by comparing Figures 11.3 and 11.2. The slope (the ratio of vertical rise to horizontal distance) of the warm front is shallower near the Earth's surface because friction has retarded the front. The airflow on the warm side of the front is quite similar in both instances, but that on the cold side of the front is retreating in the instance of the warm front. Thus, the warm air can

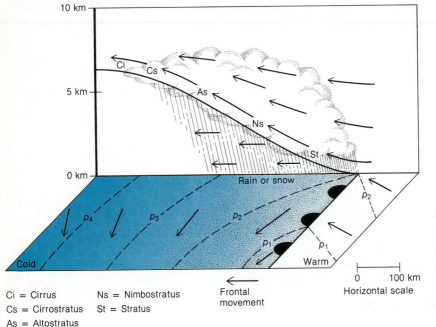

FIGURE 11.3

As with the stationary front, overrunning along a warm front triggers cloud development. However, the warm front has a shallower slope at low levels, and the surface winds on the cold side of the front are retreating.

Ci = Cirrus
Cs = Cirrostratus
As = Altostratus
Ns = Nimbostratus
St = Stratus

advance relative to the Earth's surface, rather than just gliding up over the cold air.

As a warm front advances toward a locality, clouds develop and gradually lower and thicken in the following sequence: cirrus, cirrostratus, altostratus, nimbostratus, and stratus. The initial wispy cirrus clouds (Figure 11.4) may appear more than 1000 km (620 mi) in advance of the surface warm front. Slowly, clouds spread laterally, forming thin sheets of cirrostratus that turn the sky a bright milky white. The tiny ice crystals comprising these high clouds (bases above 6 km) may reflect and refract sunlight to produce various images

FIGURE 11.4

Thin, wispy cirrus clouds may appear more than 1000 km (620 mi) in advance of a surface warm front. [NOAA photograph]

such as halos or sundogs (Chapter 8). The appearance of these optical phenomena may herald the approach of stormy weather several days in advance.

In time, cirrostratus clouds give way to altostratus, thin clouds with bases at altitudes of 2 to 6 km. Light rain or snow begins soon after altostratus clouds thicken enough to block out the sun. Steady precipitation falls from low, gray nimbostratus clouds and persists until the warm front finally passes, a period that may exceed 24 hours. Thus, copious amounts of rain may fall ahead of the surface warm front, and since the precipitation intensity is usually only light to moderate, much of the water infiltrates the soil (if the ground is not frozen). This is the type of soaking rain that farmers appreciate. If it is cold enough for the precipitation to fall in the form of snow, then accumulations may be substantial.

Just ahead of the surface warm front, steady precipitation usually gives way to drizzle falling from low stratus clouds (bases below 2 km) and sometimes to fog. So-called **frontal fog** develops when rain falling into the shallow layer of cool air at the ground evaporates and increases the moisture content to the saturation point. After the warm front finally passes, frontal fog dissipates and skies at least partially clear, since the zone of overrunning has also passed. The weather typically turns warmer and more humid.

The cloud and precipitation sequence just described for a warm front applies when the advancing warm air is stable. The sequence changes somewhat when the warm air is unstable. In that situation, uplift is more vigorous and gives rise to cumulonimbus clouds (thunderstorms) embedded within the zone of overrunning ahead of the surface warm front. Hence, brief periods of heavy rainfall, or perhaps snowfall, may punctuate the otherwise steady fall of light-to-moderate precipitation.

COLD FRONT

A stationary front becomes a **cold front** if it begins to move in such a way that colder (more dense) air displaces warmer (less dense) air. Over North America, the temperature contrast across a cold front is typically greater than that across stationary or warm fronts. However, in summer the temperatures on either side of the cold front can sometimes be nearly identical. When this occurs, the density contrast results from differences in water vapor content rather than differences in temperature. Following the passage of a cold front, we usually notice a drop in temperature in winter and a drop in moisture content in summer.

Although the influence of friction causes the slope of a warm front to become more shallow, it steepens a cold front close to the Earth's surface into the characteristic "nose" shape, shown in Figure 11.5. The slope of a cold front is steeper (1:50 to 1:100) than the slope of a warm front (1:150 on average). Because of the steep frontal slope and the typical flow aloft across the front from the cold to the warm side, uplift is restricted to a narrow zone

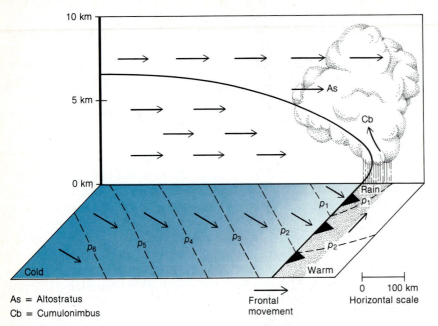

As = Altostratus
Cb = Cumulonimbus

Frontal
movement

0   100 km
Horizontal scale

FIGURE 11.5

A cold front has a characteristic "nose" shape in cross
section. The surface winds
on the cold side of the front
blow toward the front, and
clouds and precipitation are
limited to a narrow band at
and just ahead of the front.

at or near the front's leading edge. The low-level air motions are also quite
different from those for warm and stationary fronts; the low-level motion in
the cold air is at least in part toward the front, forcing the warm air aloft.

If the cold front advances slowly but steadily, say 30 km per hour (20
mph), the type of frontal weather depends on the stability of the warmer air.
Any precipitation is likely to be showery and brief, appearing in a narrow
band at and just ahead of the front. If the warm air is relatively stable, then
nimbostratus and altostratus clouds may form. If the warm air is unstable, the
uplift is more vigorous, giving rise to cumulonimbus clouds towering above
nimbostratus. These thunderstorms can be accompanied by strong and gusty
surface winds, hail, or other violent weather.

If the cold front moves along at a rapid pace, say 45 km per hour (30
mph), then a **squall line,** a band of intense thunderstorms, may develop
either right at the front, or up to 300 km (180 mi) ahead of the front. Squall
lines are discussed in more detail in Chapter 13, where we consider severe
weather systems.

A typical cold front trails southward from an extratropical cyclone, as will
be discussed later in this chapter, and sweeps along from west to east. Many
cold fronts also drop southward out of Canada into the Great Plains, even
traveling as far as the Gulf of Mexico. The occasional cold front that moves
from east to west is called a "back-door" cold front. This type is most often
seen in New England in the winter, ushering in mP air from over the Atlantic
Ocean.

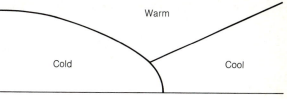

FIGURE 11.6
Schematic vertical cross section of a cold-type occlusion. This is the most common type of occlusion over North America.

OCCLUDED FRONTS

A cold front typically travels about twice as fast as a warm front and eventually may catch up with the warm front to form an **occluded front,** sometimes called simply an **occlusion.** If the air behind the advancing cold front is colder than the cool air ahead of the warm front, the cold air slides under and lifts the warm air, the cool air, and the warm front (Figure 11.6). The resulting **cold-type occlusion** has the characteristics of a cold front at the surface, but the temperature contrast between the cold and cool air is typically less than that for a cold front. The weather ahead of the occlusion is similar to that which occurs in advance of a warm front; the actual frontal passage may be marked by more showery conditions, such as those associated with a cold front.

A second and less common type of occlusion occurs when the air behind the advancing cold front is not as cold as the air ahead of the warm front (Figure 11.7). Such a **warm-type occlusion** often occurs in the northerly portions of western coasts, such as in Europe or in the Pacific Northwest. In this case, the air behind the cold front is relatively mild (mP), having traversed ocean waters, whereas the air ahead of the warm front is relatively cold (cP), having traveled over land. With this type of occlusion, the cool air behind the cold front slides under the warm air but rides over the cold air. The weather ahead of a warm-type occlusion is similar to that ahead of a warm front, with the surface front behaving as a warm front. Both types of occlusions can be difficult to locate from surface data, because the temperature contrast across the front is often small, the precipitation occurs over a broad region masking the front, and the trough in the pressure pattern associated with the front is at times not as pronounced as that for cold and warm fronts.

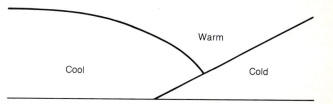

FIGURE 11.7
Schematic vertical cross section of a warm-type occlusion.

CHANGES IN FRONTS

The characteristics that define a front—differences in temperature and moisture content, wind shift, convergence, and the like—change with time in much the same way that air masses modify with time. We define a front primarily by the density contrast across the front. If processes in the atmosphere act to increase the density contrast, for example, through convergence, then a front becomes definable or grows stronger. This process is called **frontogenesis.** On the other hand, if the density contrast weakens, the front also weakens and may become indistinguishable, a process called **frontolysis.** The precipitation associated with the front also tends to increase or diminish in intensity as the front strengthens or weakens.

SUMMARY

In midlatitudes, the atmosphere tends to be comprised of air masses that differ in temperature and moisture content. These air masses are separated by transition zones called fronts, which are classified primarily by the temperature behind the front as it passes by relative to the temperature ahead of the front. The motion of a front is related to the low-level motion on the cold side of the front. The cold air retreats ahead of a warm front, advances behind a cold front, and moves parallel to a stationary front. Clouds and precipitation develop along fronts only when a significant density contrast exists between air masses given that an adequate water vapor content is present. If there is little difference in temperature and humidity across the front, then the front may pass virtually unnoticed except for a shift in wind direction. The strength of the density contrast marking a front varies with time, leading to the formation and dissipation of fronts.

Frontal weather occurs in combination with cyclones. We will explore this relationship next.

Midlatitude Cyclones

The extratropical **cyclone,** or low-pressure system, is the principal weather maker of midlatitudes. The counterclockwise (in the Northern Hemisphere) and inward low-level flow of air is associated with convergence and therefore rising motion, cloudiness, and precipitation. We will describe the life cycle and characteristics of these large-scale storms in this section.

LIFE CYCLE

If conditions are favorable in the middle and upper troposphere, that is, if there is upper-air support, an extratropical cyclone can form and grow. **Cyclogenesis,** the birth of a midlatitude cyclone, usually takes place along the polar front directly under an area of strong horizontal divergence aloft. As noted in Chapter 10, strong divergence aloft occurs to the east of an upper-level trough and under the left-front quadrant of an upper-level jet stream maximum, or jet streak. If divergence aloft removes more mass from a column of air than is brought in by convergence near the surface, the surface air pressure at the bottom of the column drops (a process also referred to as **deepening**), a horizontal pressure gradient develops, and a cyclonic circulation begins. In other words, a storm is born. The westerly flow aloft then steers and supports the cyclone as it goes through its life cycle (Figure 11.8).

FIGURE 11.8
A midlatitude cyclone passes through its life cycle: (A) incipient cyclone, (B) wave cyclone, (C) beginning of occlusion, and (D) bent-back occlusion. As a wave cyclone, the storm is east of the upper-level trough; at occlusion, the storm is under the upper-level trough. [Modified from G. T. Trewartha and L. H. Horn, *An Introduction to Climate*, 5th ed. New York: McGraw–Hill, 1980, p. 165.]

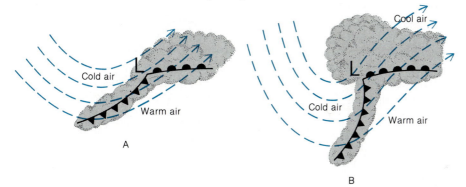

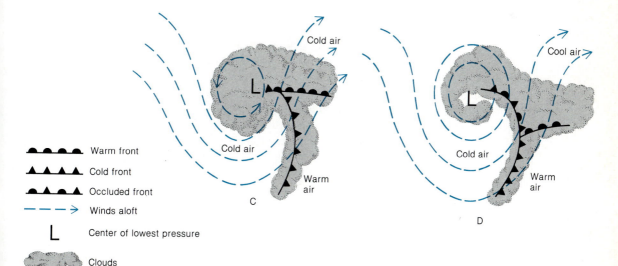

Just prior to the formation of our prototypical cyclone, the polar front is stationary, and surface winds are directed parallel to the front. As the surface air pressure drops, surface winds begin to converge and the front begins to move (Figure 11.8A). West of the low center,* the front advances toward the southeast as a cold front. East of the low center, the front moves northward as a warm front. The minimum pressure for the incipient low might be 1000 mb at this time, with a single closed isobar on a standard surface map, which uses a 4-mb increment between isobars. Satellite imagery shows that the narrow cloud band associated with the stationary front develops a bulge at the low center and extends along the warm front. The upper-level pattern depicted in Figure 11.8A has a trough to the west of the surface low, a position that favors further development of the system. An example of a cyclone at this stage can be seen in Nevada in Figure 15.4A. The behavior of the system is somewhat complicated by the Rocky Mountains, and so neither a warm front nor a simple cloud pattern is evident in the satellite image at this time. Figure 15.6A shows that the upper-air pattern at this time has a large trough along the Pacific Coast in a position to yield divergence aloft over the incipient low center and so future deepening of the surface cyclone.

If the upper-air support is not favorable for further development of the cyclone, the low center is typically seen to ripple along the stationary front without deepening further, producing light precipitation as it goes. Such cyclones affect only a small region and typically move along the front at 50 to 70 km per hour (30 to 45 mph).

If conditions are favorable for the incipient cyclone to continue to mature, its central pressure drops further and the associated horizontal pressure gradient and counterclockwise winds strengthen. The upper-level trough also frequently deepens while remaining to the west of the low center. Because the cold front generally moves faster than the warm front, the angle between the two fronts gradually closes. That is, the so-called warm sector of the storm, situated between the warm and cold fronts (and occupied by warm air at the surface), becomes better defined. At this stage in the storm's life cycle, the front forms a pronounced wave pattern (Figure 11.8B)—hence the descriptive name **wave cyclone.** The surface cyclone may now have a central pressure of 992 mb, with perhaps three closed isobars, and is moving eastward or northeastward at 40 to 55 km per hour (25 to 35 mph). A **comma cloud** is apparent in satellite images at this stage, reflecting the increase in the storm's circulation. The head of the comma extends from the low center to the northwest, with its tail trailing along the cold front. Clouds associated with overrunning are also present north of the warm front. Figure 15.4B shows the storm from Figure 15.4A one day later; a well-defined warm sector is evident in the southern Plains. The secondary front coming in behind the cold

---

* The low center is at the point of lowest sea-level pressure. The minimum sea-level pressure value is known as the *central pressure* of the cyclone.

front marks the boundary between mP air to the south and an arctic air mass to the north. The upper-air trough (Figure 15.6B) is very well defined and is positioned to yield further development of the system.

As the cold front continues to advance toward the warm front, an occluded front begins to form near the low center, forcing the warm air aloft as the warm sector occupies a smaller area. Figure 11.8C represents the beginning of the **occlusion stage** of the cyclone. Note that the upper-level pattern has developed a closed circulation almost above the surface cyclone. When the upper-level low center is directly over the surface low center, the system is said to be vertically stacked. Dry air descending behind the cold front is drawn nearly into the center of the cyclone as a "dry slot" that separates the cloud band along the cold front from the comma head, now more west than northwest of the low center. The central pressure of the storm has decreased significantly to perhaps 985 mb in the time that has elapsed between Figures 11.8B and 11.8C.

Some cyclones continue to deepen after an occluded front is formed at the surface. Typically, the upper-air low center is not directly over the surface low center while deepening is occurring, but it may be fairly close. These storms develop a closed, vertically stacked (or nearly so) circulation that is troposphere-deep and that typically moves much more slowly than previously, say 30 km per hour (20 mph), and may even stall completely. The circulation tends to draw the occluded front around the low center into a configuration that is sometimes called a **bent-back occlusion** (Figure 11.8D). The central pressure of the low may now be 980 mb, with a half dozen or more closed isobars quite close together, associated with winds that can exceed 75 km per hour (45 mph). At the Earth's surface, the warm sector is still present but is detached from the cyclone itself. The cold, warm, and occluded fronts all come together at the **point of occlusion,** or **triple point,** a location associated with conditions favorable to the formation of a new cyclone, sometimes called a secondary cyclone. (Note the similarity between the point of occlusion with its fronts in Figure 11.8D and the incipient low with its fronts in Figure 11.8A.) At this stage of development, the cloud pattern typically becomes a spiraling swirl with enhanced bands associated with the fronts. A very intense system with a central pressure of 960 mb or less can cause the spiraling cloud band to circle the low center several times. Figure 15.4C shows our sample cyclone one day after Figure 15.4B. Note the large number of closed isobars, the tight pressure gradient indicating strong winds, and the large swirling cloud mass in the satellite image. A low center is also evident aloft (Figure 15.6C), but not quite directly over the surface low.

Eventually, the cyclone weakens, its winds lessen, and its central pressure increases; this process is called **cyclolysis** or **filling.** The low can weaken during any of the stages described above, if its upper-air support diminishes. As the central pressure increases, the storm may lose its identity in the sea-level pressure field and be indicated only by cloudy skies and drizzle. Such weakening is inevitable once the system is vertically stacked, but the filling process may not begin for many hours and may proceed slowly, allowing an intense circulation to persist for days.

These basic stages in the life cycle of cyclones were first formulated during World War I by Norwegian researchers in Bergen. This conceptual model (Chapter 1) is therefore referred to as the **Norwegian cyclone model.** Major advances in weather monitoring, especially those involving remote sensing by satellite, have verified the Norwegian model. Amazingly, it remains a remarkably close approximation of our current understanding of midlatitude cyclones, even though individual cyclones may not follow the model exactly.

The life cycle described above may occur over several days as in the example from Chapter 15, or it may develop over a much shorter period. If the upper-air support is less favorable, the storm may spend a longer time in any one of the early stages, even weaken temporarily, and still become fully occluded. Sometimes storms develop with meager upper-air support (weak divergence aloft) and are weak and poorly defined. At other times, widespread cloudiness and precipitation are linked to an upper-air or surface trough, and not to a closed cyclonic circulation at the surface. When the upper-level conditions are ideal, the entire life cycle from incipient cyclone to bent-back occlusion can occur in less than 24 hours.

A rapidly intensifying extratropical cyclone, or ''bomb,'' is defined as a cyclone whose central pressure decreases by at least 24 millibars in 24 hours

FIGURE 11.9
Visible satellite image of the cloud pattern associated with an intense occluded midlatitude cyclone centered over the North Atlantic. [NOAA satellite photograph]

adjusted to 60 degrees N, the latitude of Bergen, Norway, where the criterion was developed. Few cyclones meet this criterion, and most of those that do satisfy it occur over warm ocean currents such as the Kuroshio Current off the coast of Japan and the Gulf Stream off the East Coast of the United States. Since these oceanic storms are not well observed on a day-to-day basis, recent field experiments have attempted to use research aircraft and other special instrumentation to investigate them. The **Genesis of Atlantic Lows Experiment (GALE),** sponsored by a number of agencies including the National Science Foundation, the Office of Naval Research, and the National Oceanic and Atmospheric Administration, took place from January to March 1986; its primary goal was to measure the initial formation of these oceanic cyclones in a region off the coast of North Carolina. The **Experiment on Rapidly Intensifying Cyclones over the Atlantic (ERICA)** was sponsored primarily by the Office of Naval Research and was conducted from December 1988 to February 1989 over the North Atlantic. The goal of ERICA was to observe rapidly intensifying cyclones throughout their life cycle.

An extreme example of a developing cyclone was observed in ERICA on 4–5 January 1989. An incipient cyclone with a central pressure of 996 mb was first identified off Cape Hatteras, North Carolina, at 7:00 P.M. EST on 4 January. Twenty-four hours later, the storm was located 700 km (450 mi) south of Newfoundland with a central pressure of 936 mb. The storm deepened by 60 mb in 24 hours, which represents 2.5 times the criterion for a ''bomb.'' A satellite image for this storm as it neared peak intensity is shown at the beginning of the chapter and in Figure 11.9.

CYCLONE WEATHER

As an illustration of typical cyclone weather, consider a wave cyclone that has developed a well-defined warm sector as it moves into the upper Midwest. The storm is likely still intensifying but already affects a wide region with its circulation, clouds, and precipitation. Figure 11.10 is a schematic representation of the winds, temperature distribution, cloud shield, and precipitation patterns associated with our typical midlatitude low in winter.

Ideally, based on surface weather, we can divide a wave cyclone into four sectors about the storm center. The lowest air temperatures are to the northwest of the storm center where strong, gusty northwest winds advect cP air or A air southeastward. Strong winds make the air feel colder than the temperature might suggest (windchill effect). West of the center, precipitation tapers off to showers, and there is a tendency toward clearing skies as the storm center moves toward the northeast. The leading edge of the cold air mass is south of the storm and is accompanied by a narrow band of showers and embedded thunderstorms. The southwest sector has generally clear skies and sinking motion. The mildest air is in the southeast (warm) sector of the storm where south and southeast winds advect mT air from the Gulf of Mexico toward the north. Skies are generally partly cloudy, dew points are high, and

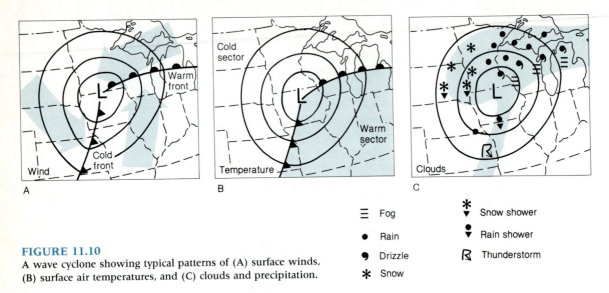

A

B

C

☰ Fog	✱▼ Snow shower
● Rain	▼ Rain shower
◗ Drizzle	℞ Thunderstorm
✳ Snow	

FIGURE 11.10
A wave cyclone showing typical patterns of (A) surface winds,
(B) surface air temperatures, and (C) clouds and precipitation.

scattered convective showers are possible, especially during the early after-noon. Cloudiness increases, and showers become more prevalent, as the cold front progresses eastward. To the north and northeast of the storm center is a zone of extensive overrunning as mT air surges over a wedge of cool air maintained by east and northeast winds at the surface. Skies in the northeast sector are cloudy, and precipitation is steady and substantial.

Note the weather conditions in the various sectors of a wave cyclone are consistent with our earlier description of cold and warm frontal weather. While the flow in and around cyclones is predominantly horizontal, the uplift of air along frontal surfaces leads to the characteristic cloud and precipitation pat-terns shown in Figure 11.8. A conceptual model that combines the horizontal and vertical air motions into a three-dimensional depiction describes fronts and cyclones in terms of three interacting airstreams, often referred to as conveyor belts. Just as mechanical conveyor belts transport goods (or even people) from one location to another, these atmospheric conveyor belts trans-port air with certain characteristics from one location to another and are named for the type of air they transport.

According to the conveyor-belt model, the warm conveyor belt at low levels roughly corresponds to the arrow over Illinois in Figure 11.10A. In wave cyclones such as this, the warm conveyor belt is located just to the warm side of the cold front near the Earth's surface, south and east of the storm center. Typically, this warm moist airstream may ascend slightly as it progresses northward in the cyclone's warm sector at low levels, then ascends rapidly as it glides northward over the sloping warm frontal surface (see Figure 11.3) north of the surface warm front. As the air ascends, it cools adiabatically, its water vapor condenses, forming clouds, and rain or snow falls. The latent heating associated with this process causes the air in the warm conveyor belt to ascend even more rapidly. However, recall that the flow aloft at this stage of the cyclone's development is more or less from the southwest, with the surface cyclone located between the upper-level trough and ridge. The warm conveyor belt thus turns from a southerly to a southwesterly or even westerly direction as it ascends over the warm front, as can be inferred from Figure 11.8B. This airstream therefore helps explain the broad region of cloudiness and precipitation north of the warm front.

While the warm conveyor belt is gliding up over the warm frontal surface, the flow underneath it is colder and directed from the east or northeast. This flow forms the cold conveyor belt and at low levels corresponds to the arrow over northern Wisconsin in Figure 11.10A. Like the warm conveyor belt, this airstream is also ascending as it progresses toward the west. The air is moistened by the evaporation of part of the warm frontal precipitation falling through it and so becomes saturated. The cold conveyor belt begins to turn south as it passes around the north side of the low-level storm center. However, recall that the flow aloft is generally from the southwest over the surface cyclone. As the cold conveyor belt continues to ascend, it changes direction again to join this southwesterly flow aloft. The clockwise turning of the saturated air in the cold conveyor belt, first toward the south and then swinging around until it is heading east, leads to the distinctive "comma cloud" to the northwest of the cyclone center as depicted in Figure 11.8B.

The third airstream or conveyor belt hypothesized to be important in the development of a cyclone is the so-called dry conveyor belt, approximately represented near the Earth's surface by the arrow over Nebraska in Figure 11.10A. While the cyclone center and fronts are typically regions of upward motion, the region west of the cold front experiences downward motion. This descending flow aloft originates in the upper troposphere and lower stratosphere upstream of the upper-level trough and so is quite dry, especially in comparison to the humid air in the warm conveyor belt. As the dry conveyor belt descends, it also begins to move in toward the low center. This dry airstream is responsible for the typically clear skies behind the surface cold front as well as the dry slot that is seen to delineate the comma cloud, especially in the later phases of the cyclone's development (Figure 11.8C and D).

CYCLONE TRACKS

The typical weather pattern associated with a midlatitude wave cyclone indicates that storms have a warm side and a cold side. Note that, in Figure 11.10, the coldest air is located to the northwest of the low center (where winds are northwesterly) and the warmest air is to the southeast of the low center (where winds are southerly). Hence, as a cyclone traverses the continent, the weather to the left (cold side) of the storm track is quite different from the weather to the right (warm side) of the track.

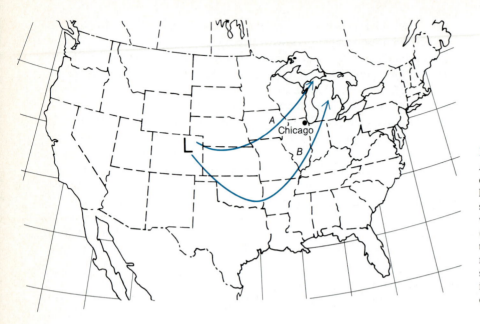

FIGURE 11.11
A cyclone that developed over eastern Colorado follows a path that takes the storm center either to the west (track *A*) or to the east (track *B*) of Chicago. For track *A*, Chicago is on the relatively warm side of the storm; for track *B*, Chicago is on the relatively cold side of the storm.

Table 11.2
Sequence of Surface Weather Conditions in Chicago as Mature Winter Storm Tracks West (Track *A*) and East (Track *B*) of City

			Track A			
Wind direction	E	SE	S	SW	W	NW
Frontal passage	—	WF	—	CF	—	—
Advection	—	Warm	Warm	Cold	Cold	Cold
Air pressure tendency	F	F	F	R	R	R
			Time →			

			Track B			
Wind direction	E	NE	N	NW		
Frontal passage	—	—	—	—		
Advection	—	—	Cold	Cold		
Air pressure tendency	F	F	R	R		
			Time →			

WF = Warm front
CF = Cold front
F = Pressure falling
R = Pressure rising

Consider, for example, Figure 11.11. A winter cyclone develops in eastern Colorado. As the storm matures, it moves northeastward toward the Great Lakes region and may take either track *A* or track *B*. In either case, the storm center passes within 150 km (90 mi) of Chicago, but track *A* takes the storm west of Chicago and track *B* takes it east of Chicago. The storm's effect on Chicago's weather depends on its track, as summarized briefly in Table 11.2.

If the storm takes track *A*, Chicago residents experience the warm side of the storm. Steady rain (or perhaps snow, briefly, at the onset) gives way to drizzle and fog after 12 to 24 hours. As the surface warm front passes over the city, skies partially clear and winds shift abruptly from east to south, advecting warm and humid (mT) air at the surface. Clearing is short-lived, however, as scattered showers and thunderstorms herald the arrival of colder air. As the surface cold front passes through the city, winds **veer** (turn clockwise), blowing first from the southwest, then west, and finally northwest. Skies clear again, and the air temperature and dew point fall.

In contrast, if the storm takes track *B*, Chicago residents experience the cold side of the storm and no frontal passages. Gusty east and northeast winds drive steady snow or rain (depending on how cold the air is) for 12 hours or longer. Then winds gradually **back** (turn counterclockwise) to a northerly direction, precipitation tapers off to snow flurries or showers, and air temperatures begin to drop. Finally, winds shift slowly to northwesterly, skies begin to clear, and temperatures continue to drop.

In summary, then, if you are on the warm side of a cyclone, the wind direction veers with time and a warm front is followed by a cold front. If you are on the cold side of a synoptic-scale cyclone, the wind direction backs with time without the passage of fronts.

The specific track taken by any synoptic-scale cyclone depends on the pattern of the upper-level westerlies in which the storm is embedded. The storm center tends to move in the direction of the wind blowing directly above the storm at the 500-mb level. As a rule, the forward motion of the storm is about one-half of the 500-mb wind speed. However, keep in mind that the upper-level flow pattern also changes with time, as can be seen in Figure 11.8.

Figure 11.12 shows the principal storm tracks across the lower 48 states of the United States. Cyclones following some of these tracks have nicknames. For example, the Panhandle Hook starts in the lee of the Rockies and "hooks" almost due north, and the Alberta Clipper starts in the lee of the Canadian Rockies and moves very rapidly across the northern tier of states. All storms tend to converge toward the northeast; their ultimate destination is usually the Icelandic low* of the North Atlantic or Western Europe. Although many storm tracks appear to begin just east of the Rocky Mountains, in reality they have their origins over the Pacific Ocean. As cyclones travel through the

---

* Actually, the Icelandic low is a statistical feature of the large-scale circulation. On any given day, it may or may not be evident on the maps, and it may or may not be over Iceland.

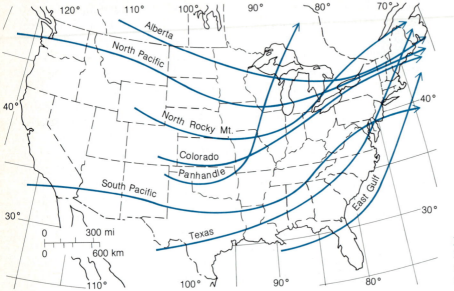

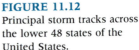

FIGURE 11.12
Principal storm tracks across the lower 48 states of the United States.

mountains, they often lose their identity temporarily but redevelop on the Great Plains just east of the Front Range of the Rockies. What actually happens to these cyclones is explained in more detail in the Special Topic, "The Case of the Missing Storm."

The notion that storms in middle latitudes generally move from west to east was first suggested in 1703 by Daniel Defoe, the English journalist and novelist. Defoe drew this conclusion from his study of a great storm that lashed the British Isles on 7–8 December of that year. He received reports that, several days earlier, a similar storm had ravaged the East Coast of North America. In the United States, Benjamin Franklin is generally credited with discovering that storms usually move in an easterly or northeasterly direction. In 1743 Franklin, a resident of Philadephia, learned through correspondence with his brother, who lived in Boston, that cloudiness from an intense coastal storm reached Boston many hours after clouds had reached Philadelphia, located about 500 km (300 mi) to the southwest.

It may be confusing to think of a storm as moving *toward* the northeast while the winds in the northeast sector of the storm blow *from* the northeast. In fact, New Englanders long assumed that the powerful "nor'easters" that pounded their shores moved down the coast from the northeast because the storm winds were northeasterly. Actually, "nor'easters" usually are storms similar to the ERICA storm described previously; they intensify off the North Carolina coast and then track in a northeasterly direction up the coast. In effect, there are two motions: (1) the movement of the storm center up the coast and (2) the counterclockwise circulation of winds about the storm center. The circulation of a storm is for the most part independent of the storm's path, much as the spin of a frisbee is independent of its trajectory.

Generally, storms that form in the south yield more precipitation than those that develop in the north because southern storms are closer to the primary source of moisture, maritime tropical air. For example, Alberta cyclones typically yield only light amounts of rain or snow, whereas Colorado- and Gulf-track storms (Figure 11.12) often produce heavy accumulations of rain or snow. The speed of the storm also affects the total precipitation it

---

Special Topic

The Case of the Missing Storm

After cyclones sweep ashore along the West Coast of North America, they disappear in the sea-level pressure pattern as they move inland over the mountainous West. Later, cyclogenesis is observed east of the Front Range, typically on the plains of Alberta or eastern Colorado. What actually happens to these storms as they cross the mountains?

Visualize a storm moving ashore as a huge cylinder of air spinning about a vertical axis in a cyclonic (counterclockwise) direction (Figure 1). As the cylinder moves up the windward slopes of a mountain range, the column is forced to shrink in height. As it shrinks it also widens, and the spin of the cylinder slows so that, in effect, the storm's circulation weakens. As the cylinder of air then descends the leeward slopes, the air column stretches, contracts horizontally, and the cyclonic spin strengthens. The weakening of cyclonic circulation upslope and the strengthening of the circulation downslope account for the seeming disappearance and reap-

pearance of a storm as it crosses mountainous terrain.

The situation is analogous to that of an ice skater performing a spin. The skater changes the rate of spin by extending or drawing in his or her arms. When the arms are extended (analogous to horizontal divergence and the widening of our cylinder), the spin rate slows. When the arms are brought as close as possible to the skater's body (analogous to horizontal convergence and the contracting of our cylinder), the spin rate increases. In effect, both the skater with arms extended and the storm passing through mountainous terrain conserve angular momentum. Simply put, the *conservation of angular momentum* means that a change in the radius of a rotating mass is balanced by a change in its rotational speed. An increase in radius (horizontal divergence) is accompanied by a reduction in rotational speed, and a decrease in radius (horizontal convergence) is accompanied by an increase in rotational speed.

FIGURE 1

The cyclonic spin (circulation) of an air cylinder weakens with horizontal divergence on the windward slopes of the mountain range and strengthens with horizontal convergence on the leeward slopes of the mountain range.

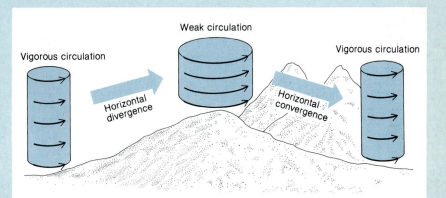

Vigorous circulation

Weak circulation

Vigorous circulation

Horizontal divergence

Horizontal convergence

produces at a particular location. A storm that is fast moving may yield rain or snow for only a few hours; a slower moving storm may precipitate for 12 hours or more.

From our earlier discussion of the linkage between winds aloft and migratory synoptic-scale cyclones, we can deduce that storms should exhibit seasonal variability. In summer, when the mean position of the polar front and jet stream is across southern Canada, very few well-organized cyclones occur in the United States, and the Alberta storm track shifts northward across central Canada. In winter, however, when the mean positions of the polar front and jet stream shift southward, cyclogenesis is more frequent in the United States. Alberta-track storms are the most common, since they occur year-round, whereas storms with more southerly tracks develop primarily in winter. In fact, one sure sign of the beginning of winter circulation patterns is the appearance of Colorado lows.

COLD- AND WARM-CORE SYSTEMS

Occluded cyclones such as those shown in Figure 11.8D are cold-core systems; that is, the lowest temperatures on a horizontal surface are found at the center of the low. In vertical cross section, constant pressure (isobaric) surfaces within **cold-core lows** are concave upward (Figure 11.13), and the depth of the low increases with altitude, implying that the circulation is cyclonic throughout the atmosphere and most intense at higher levels. Recall from the mathematical note in Chapter 10 that the thickness (Δz) between pressure surfaces is directly proportional to the layer–mean temperature. The requirement that the thickness (mean temperature) be lowest at the center of the low produces the characteristic pattern.

FIGURE 11.13
Vertical cross section shows sloping isobaric (constant-pressure) surfaces within a cold-core low-pressure system.

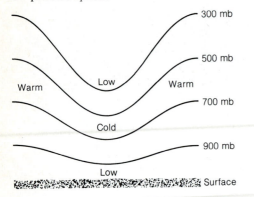

FIGURE 11.14
Vertical cross section through a wave cyclone.

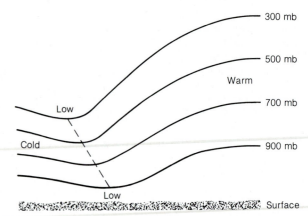

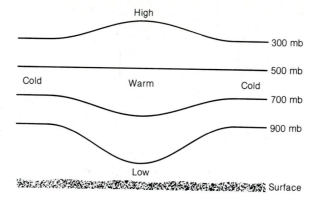

FIGURE 11.15
Vertical cross section shows sloping isobaric (constant pressure) surfaces within a warm-core low-pressure system. The shallow near-surface cyclonic circulation weakens and then reverses with altitude.

A wave cyclone, on the other hand, has the lowest temperatures to the northwest of the storm and the highest temperatures to the southeast. Figure 11.14 portrays a cross section through the low from northwest to southeast. Thickness (mean temperature) arguments lead to the conclusion that the low center aloft is not located above the low center near the surface, but rather is displaced to the cold side of the storm, implying a tilt with height for the system. This condition is consistent with Figure 11.8B, which shows the upper-level trough lagging behind the surface cyclone.

A different type of low that sometimes appears on midlatitude surface weather maps has characteristics markedly different from those of cold-core or tilted lows. Cyclones of this type are stationary, have no fronts, and are associated with fair weather. They develop over arid or semiarid deserts, including the interior of Mexico and the southwestern United States, when the hot summer sun heats the ground, which in turn heats the overlying air. Intense heating lowers the air density over an area wide enough for a synoptic-scale low to appear. This **warm-core low,** or **thermal low,** is very shallow, and its circulation weakens rapidly with altitude, that is, away from the heat source (the ground). The surface counterclockwise circulation frequently reverses at some altitude, and the thermal low is overlain by an anticyclone, as indicated by the upper-level pressure surface bulging upward in Figure 11.15.

In this section, we have focused on the general weather conditions associated with cyclones. In the daily march of weather in midlatitudes, cyclones are followed by anticyclones. Let us now examine these fair-weather systems.

Anticyclones

An anticyclone is, in many ways, the opposite of a cyclone. Cyclonic circulation favors the convergence of contrasting air masses and the development and maintenance of fronts. In anticyclones, subsiding air and diverging surface winds favor the formation of a uniform air mass and fair skies. Like cyclones, however, anticyclones, can have either cold or warm cores.

COLD- AND WARM-CORE SYSTEMS

Cold-core anticyclones are actually domes of continental polar (cP) or arctic (A) air and, depending on the specific air mass, are labeled either **polar highs** or **arctic highs.** They are the product of extreme radiational cooling over the often snow-covered continental interior of North America well north of the polar front. Cold anticyclones are shallow systems in which the clockwise circulation weakens with altitude and often reverses. A cold trough is therefore typically situated over a cold anticyclone.

Cold anticyclones are most intense (that is, they exhibit the highest surface pressures) in winter when the associated air mass is coldest. The air is extremely stable in these systems, and soundings indicate a temperature inversion in the lowest kilometer or two, associated with strong subsidence and adiabatic heating above the inversion. Massive arctic or polar anticyclones with very high central pressures tend to remain stationary over their source regions. However, lobes of cold air (smaller cold highs) often break out of the source region and slide southeasterly across Canada and into the United States. These cold air surges interact with the circulation of migrating cyclones, helping to maintain and strengthen the temperature contrast across the cyclone's cold front. This explains why winter storms are usually followed by clearing skies and sharply colder temperatures.

On rare occasions in the winter, a particularly strong arctic high brings a surge of bitterly cold air that sweeps as far south as southern Florida and can even traverse the Gulf of Mexico into Central America. The resulting subfreezing temperatures can spell disaster for citrus growers. As an illustration, the synoptic weather pattern that caused the costly Florida freeze of December 1983 is shown in Figure 11.16. Note that the cold front has already pushed all the way south to Cuba. The cloudiness over the Gulf of Mexico results from the cold air being warmed and moistened by the warm Gulf waters, generating fog and low clouds. Much of what appears to be white over the continent east of the Rockies is not cloudiness but, rather, snow cover.

Warm-core anticyclones form south of the polar front and consist of extensive areas of subsiding warm, dry air. Thickness arguments show that,

FIGURE 11.16 *(opposite)*
Satellite view of much of North America on 24 December 1983 and the corresponding pattern of surface air pressure. Solid lines are isobars labeled in millibars. An unusually extensive arctic high-pressure system is centered over eastern Montana where surface pressure readings exceed 1060 mb. The frigid air mass associated with the high covers almost the entire United States except the Southwest. Early that morning, temperatures plunged to record low levels over the Plains, Midwest, and Deep South. Subfreezing temperatures in the citrus- and vegetable-growing areas of South Texas and South Florida caused more than $1 billion in crop damage. In the satellite image, much of the white in the northern United States is snow cover rather than clouds. The bright white over the Gulf is cloudiness caused by cold air streaming over the relatively warm ocean water. (Dashed line off the East Coast is a trough.) [Photograph from NOAA/Satellite Data Services; map from NOAA]

like cold-core cyclones, warm-core anticyclones stengthen with altitude. They are massive systems with a circulation extending from the Earth's surface up to the tropopause. The semipermanent subtropical anticyclones, such as the Bermuda High, are examples of warm-core highs, but other warm-core anticyclones may develop over the interior of North America, especially in summer.

A cold anticyclone produces high surface air pressures because cold air is relatively dense, but how does a warm anticyclone produce high surface pressures? After all, in equal volumes, warm air is lighter than cold air. The answer is that the volumes are not equal. The high surface pressure of warm-core anticyclones results from essentially a larger volume of air over the high center, related to a higher tropopause. A layer 200 mb deep contains the same amount of mass regardless of its thickness (Δz). In a warm-core anticyclone these individual layers would have the greatest thickness in the center of the high, and the intensity of the high would increase with altitude.

Cold-core anticyclones modify as they travel and may eventually become warm-core systems. A cold-core anticyclone is actually a dome of relatively cold air, and as that air mass traverses land that is bare of snow, it is heated from below and moderates considerably, its pressure decreasing significantly. As a cold-core high drifts southeastward over the United States, air mass modification may be sufficient to allow the pressure system to merge with the warm-core subtropical high.

ANTICYCLONE WEATHER

As noted in Chapter 9, an anticyclone is a fair-weather system. This is because surface winds blowing in a clockwise and outward pattern (in the Northern Hemisphere) induce subsidence of air over a broad area. Because subsiding air is compressionally warmed, relative humidities drop and clouds usually dissipate or fail to develop. In addition, as noted earlier, the horizontal pressure gradient is weak over a wide region near the center of an anticyclone; prevailing winds are therefore very light or calm. At night, clear skies and light winds favor rapid radiational cooling and, in some instances, the development of dew, frost, or fog (Chapter 7).

The horizontal pressure gradient strengthens away from the central region of an anticyclone, and so do the winds. With strengthening winds, significant advection occurs. Typically, well to the east of the high center, northerly winds advect cold air southward, whereas to the west of the high center, southerly winds advect warm air northward. These advection patterns help to increase the temperature contrast across the trough that separates highs and lows and thereby influence frontogenesis.

An understanding of the basic circulation characteristics of an anticyclone helps us to anticipate the sequence of weather events as an anticyclone travels into and out of a midlatitude location. Consider what happens in winter as a cold anticyclone slides southeastward out of southern Canada and into the northeastern United States. The following sequence may take several days to a week, depending on the anticyclone's forward speed.

Ahead of the anticyclone, strong northwest winds bring a surge of cold continental polar or arctic air. Strong winds and falling temperatures produce low windchill temperatures and make more work for home furnaces. To the lee of the Great Lakes, heavy lake-effect snow showers break out (discussed in Chapter 12). Even hundreds of kilometers downwind of the lakes, showers* can bring scattered light snowfalls. However, as the center of the anticyclone drifts closer, winds slacken, skies clear, and nocturnal radiational cooling produces very low surface temperatures. Under these conditions, air temperatures dip to their lowest levels, especially if the ground is snow covered. Then, as the anticyclone center moves away toward the southeast, winds again strengthen, but this time from the south, and warm air advection begins. The first sign of warm air advection is the appearance of high, thin cirrus clouds in the western sky.

In summer, a Canadian high-pressure system causes the same advection patterns as in winter except that the temperature contrast between air masses is considerably less. Air advected ahead of the high on northwesterly winds is seldom much cooler than the air advected behind the high on southwesterly winds. Often, the most noticeable difference between the northwesterlies and southwesterlies is a contrast in water vapor content. Air advected ahead of the high is often less humid, and therefore more comfortable, than the air advected behind the high.

The pattern of temperature advection associated with an anticyclone also applies to a ridge. Cold air advection usually occurs ahead (to the east) of a ridge, and warm air advection occurs behind (to the west of) a ridge.

The circulation pattern of an anticyclone (or ridge) does not occur in isolation from that of a cyclone (or trough). The atmosphere is, after all, a continuous fluid, with anticyclones following cyclones and cyclones following anticyclones. In our illustration, therefore, northwest winds develop ahead of the high and on the back (west) side of a retreating low. The southerly airflow behind the high develops to the east of a low, which is approaching from the west. In both cases, winds are caused by horizontal pressure gradients that develop between migrating anticyclones and cyclones. We consider the circulation patterns of cyclones and anticyclones in more detail in the Special Topic, "Vorticity."

---

* As the cold air travels over the warmer lakes, it is warmed and moistened by the surface and its stability decreases. The rain or snow showers that develop can be advected for a fairly long distance before their moisture supply is depleted and they dissipate.

Vorticity

We have been examining the circulation of air in global-scale and synoptic-scale weather systems. A powerful but difficult concept used in discussing such circulation patterns is vorticity. *Vorticity* is a measure of the rotation of an air parcel about an axis through the air parcel and perpendicular to the Earth's surface. By convention, cyclonic vorticity is a counterclockwise rotation (in the Northern Hemisphere) and has positive values. Anticyclonic vorticity is a clockwise rotation (in the Northern Hemisphere) and has negative values.

If a group of neighboring air parcels all have positive vorticity, that does not mean that each is spinning about its own axis like a group of spinning figure skaters. Rather, the group of air parcels is spinning around a common axis, such as a low-pressure center. To extend the analogy, this would be like a group of figure skaters holding hands and skating in a large circle. If our circle of skaters were skating at constant speed, each would have the same vorticity. If the skaters then decreased the radius of their circle by half but kept skating at the same speed, they would be completing one trip around the circle in half the time and so would double their vorticity.

In the atmosphere, vorticity arises in two ways. Similar to the analogy above, if the wind is following a curved path it will have either positive or negative vorticity. Flow around anticyclones or ridges has anticyclonic or negative vorticity; flow around cyclones or troughs has cyclonic or positive vorticity. All other things being equal, the vorticity values will be largest in the centers of cyclones or anticyclones, at the crest of a ridge, and in the base of a trough. Figure 1A depicts vorticity due to curvature. If pinwheels were placed horizontally in the flow as shown, the one in the ridge would begin to rotate anticyclonically (indicating negative vorticity), whereas the one in the trough would begin to rotate cyclonically (indicating positive vorticity).

The second way to produce vorticity is to have a change in the horizontal wind in the direction perpendicular to the direction of the wind, or *wind shear.* The wind shear is said to be anticyclonic if it has anticy-

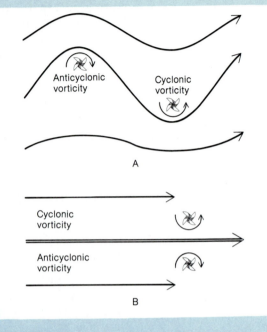

FIGURE 1
Vorticity is produced either by curvature of the wind (A) or by wind shear (B).

clonic vorticity, and to be cyclonic if it has cyclonic vorticity. Figure 1B portrays wind shear. If north is to the top of the page, the long arrow in the center depicts the greatest wind speed, whereas the arrows to the north or south depict slower wind speed. The pinwheel to the north would gain a counterclockwise rotation, indicating positive vorticity, whereas the pinwheel to the south would gain a clockwise rotation, indicating negative vorticity. Note that there is no curvature in the flow in this example; the vorticity results entirely from the wind shear.

The previous discussion examines vorticity due to both curvature and wind shear, but leads to the conclusion that flow without either curvature or shear would have zero vorticity. This is because we are examining the concept relative to the Earth's surface. This type of vorticity is therefore called *relative vorticity*. However, the Earth is also rotating about its own axis and so also exhibits vorticity. Thinking back to our discussion of the Coriolis effect (Chapter 9), we can understand that the Earth's velocity, or *planetary vorticity* (about a vertical axis perpendicular to the Earth's surface), must be dependent on latitude, increasing from zero at the equator to a maximum at the poles. As the world turns, a telephone pole at the equator appears to an observer in space to sweep out a circle but does not spin about its own axis (zero planetary vorticity), whereas a telephone pole at the North Pole spins with the Earth in a counterclockwise direction (maximum positive planetary vorticity). At the South Pole, a telephone pole spins with the Earth in a clockwise direction (maximum negative planetary vorticity).

FIGURE 2

Development of a lee-side trough. Air motions aloft are deflected first northward as the flow traverses the windward side of the mountain range, then southward as the flow traverses the lee side of the range. These large-scale nearly horizontal motions lead to the formation of a trough in the lee of the mountains, which in turn can lead to the formation of a surface low-pressure system.

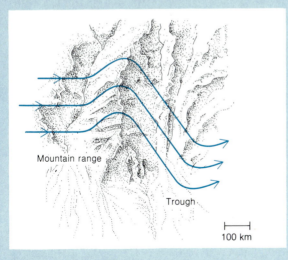

Mountain range

Trough

100 km

Absolute vorticity is defined as the sum of the relative vorticity plus the planetary vorticity. The absolute vorticity concept helps explain why a trough frequently develops in the westerlies just east of the Colorado Front Range (Figure 2). Assume that west of the mountains, winds aloft are zonal; that is, the westerlies blow directly from west to east, and the relative vorticity is zero. As the air is forced up the windward slopes, however, the tropopause acts as a lid on vertical motion, and the air is squeezed into a shallow layer, as discussed in the Special Topic, "The Case of the Missing Storm." Then, as the wind descends the leeward slopes, the air layer increases in depth. From a theoretical perspective, the ratio of the absolute vorticity to the depth of this layer of air should remain constant as the layer traverses the mountains. Hence, as the air layer travels over the mountain barrier, its depth and absolute vorticity must alternately decrease and increase together.

On the windward slopes, a decrease in absolute vorticity (and air layer depth) means a decrease in relative vorticity, so that the initially zonal winds begin to exhibit negative or anticyclonic vorticity. That is, the wind turns clockwise, a ridge forms over the mountain range, and northwest winds descend the leeward slopes. On the leeward slopes, an increase in absolute vorticity (and air layer depth) translates into an increase in relative vorticity. At the same time, the presence of northwest winds on the leeward slopes means that the airflow is into lower latitudes, and, consequently, the planetary vorticity decreases. The increase in relative vorticity prevails over the decrease in planetary vorticity, so that the northwest wind begins to exhibit positive or cyclonic vorticity. That is, the wind direction gradually turns from northwest to southwest and a trough forms east of the mountain range. This trough provides the upper-air support (horizontal divergence) for the Colorado cyclones (Figure 11.11). As the flow continues toward the south, the planetary vorticity continues to increase and the depth of the air layer changes little. The relative vorticity must therefore decrease, forming the base of the trough and changing the direction of the flow back to northwesterly.

Conclusions

We have now examined the features of the princpal weather makers of the midlatitudes: air masses, fronts, cyclones, and anticyclones. We have seen how these synoptic-scale systems interact and how they are linked to the global-scale circulation described in Chapter 10. Also recall from our discussion in Chapter 4 that synoptic-scale systems play an important role in poleward heat transport. In Chapter 12, we continue our analysis of atmospheric circulation by focusing on some special regional and local systems.

Summary Statements

Air masses are classified on the basis of their source region characteristics. Hence, there are four basic types of air masses: cold and dry, cold and humid, warm and dry, warm and humid. Air masses form over land (continental) and bodies of water (maritime), and at high altitudes (polar or arctic) and low latitudes (tropical). Arctic air is distinguished from polar air by its bitter cold.

As an air mass moves from one place to another, it modifies. The degree of modification depends on air mass stability and the nature of the surface over which the air travels, namely, whether the surface is warmer or colder than the air mass.

A front is a narrow zone of transition between air masses that contrast in temperature or water-vapor concentration, or both. There are four types of fronts: stationary, cold, warm, and occluded. Occluded fronts may be of either the cold type or the less common warm type.

Weather along or ahead of a cold front usually consists of a narrow band of clouds and brief rain or snow showers, or thunderstorms. Weather associated with warm or stationary fronts typically consists of a broad cloud and precipitation shield that can extend hundreds of kilometers ahead of the surface front.

As a midlatitude cyclone progresses through its life cycle, it is supported and steered by the upper-level flow toward the east and northeast. The storm typically begins as a wave along the polar front and deepens as surface air pressures continue to drop, winds strengthen, and frontal weather develops. The storm finally occludes as the faster moving cold front "catches up" with the slower warm front and the upper-level low center becomes vertically stacked over the surface low center.

The track followed by a cyclone is critical to the type of weather experienced at a given locality. On the cold side of a storm track, winds back with time and there are no frontal passages. On the warm side of a storm track, winds veer with time and a warm frontal passage is followed by a cold frontal passage.

Alberta cyclones occur most frequently, but Colorado and coastal storm tracks bring the heaviest precipitation.

Thermal lows are stationary, have no fronts, and are associated with very hot, dry weather.

Cold-core anticyclones are shallow systems that coincide with domes of continental polar or arctic air. Warm-core anticyclones, such as the subtropical highs, extend high into the troposphere and are accompanied by broad areas of subsiding, warm, dry air.

Key Words

air mass	warm front	comma cloud	Experiment on
continental tropical	frontal fog	occlusion stage	Rapidly Intensifying
(cT) air	cold front	bent-back occlusion	Cyclones over the
maritime tropical	squall line	point of occlusion	Atlantic (ERICA)
(mT) air	occluded front	triple point	veer
maritime polar (mP)	occlusion	cyclolysis	back
air	cold-type occlusion	filling	cold-core lows
continental polar (cP)	warm-type occlusion	Norwegian cyclone	warm-core low
air	frontogenesis	model	thermal low
arctic (A) air	frontolysis	Genesis of Atlantic	cold-core anticyclones
Pacific air	cyclone	Lows Experiment	polar highs
front	cyclogenesis	(GALE)	arctic highs
stationary front	deepening		warm-core
overrunning	wave cyclone		anticyclones

Review Questions

1. What factors determine the humidity and temperature characteristics of an air mass?

2. Distinguish among the four basic types of air masses.

3. How and why do the properties of continental polar air change between winter and summer?

4. What causes air mass modification and in what ways do air masses modify?

5. Explain why continental polar air modifies more rapidly than does maritime tropical air. What is the significance of this difference for midlatitude cold waves and heat waves?

6. What types of clouds are associated with a cold front?

7. Describe the cloud sequence as a warm front approaches your locality.

8. Explain why the surface wind flow about a center of low pressure favors the development of fronts, whereas surface anticyclonic winds do not.

9. Explain why the appearance of a halo about the sun or moon may signal the approach of stormy weather.

10. Describe the stages in the life cycle of a midlatitude cyclone.

11. A wave cyclone features a warm side and a cold side. Please explain.

12. A wave cyclone tracks northeastward along the East Coast, and the storm center passes out to sea about 80 kilometers to the east of Boston. Describe the wind shifts and changes in temperature advection at Boston. Do the same for the case where the storm center passes well to the west of Boston.

13. Distinguish between winds that back and winds that veer.

14. An Alberta cyclone usually produces much less precipitation than a storm that tracks out of eastern Colorado. Why?

15. Describe what happens when a cyclone occludes.

16. Explain why the site of principal cyclone activity shifts from the United States into Canada from winter to summer.

17. Distinguish between a warm-core high and a cold-core high.

18. Sketch a cross section through a warm-core high, similar to Figure 11.13. Sketch one for a cold-core high.

19. Distinguish between a warm-core low and a cold-core low.

20. Why is frontal weather not generated by a thermal low?

21. A warm anticyclone produces relatively high surface air pressures, and yet it is composed of a column of warm (light) air. Explain.

22. Sketch a wave cyclone with a well-defined warm sector and indicate the associated conveyor belts with arrows. Describe how the conveyor belts account for the cloud pattern present at this stage in the cyclone's development.

Points to Ponder

1. Speculate on the lowest air temperatures likely to be associated with maritime polar air in its source region.

2. Describe how Pacific air modifies as it travels from west to east across the United States.

3. Why is the slope of a cold front steeper than that of a warm front?

4. In summer we may experience passage of a cold front, and yet the air ahead of the cold front has about the same afternoon temperatures as the air behind the cold front. Explain.

5. Why is air mass stability an important factor in the type of weather that occurs along a front?

6. Under what circumstances will a frontal passage not be accompanied by cloudiness?

7. List the conditions at the surface and aloft required for cyclogenesis.

8. In your own words, explain the development of a lee mountain trough. What is the significance of this trough for Great Lakes weather?

9. Speculate on the evolution of the warm, cold, and dry conveyor belts as a cyclone occludes and develops a closed circulation aloft. Your description should be consistent with the cloud patterns in Fig. 11.8.

Projects

1. Determine the tracks of winter cyclones that influence the weather of your locality. Are you usually on the warm side or the cold side of these storms?

2. As a major winter storm approaches your area, keep an hourly log of changes in (a) cloud cover, (b) cloud type, (c) air temperature, (d) wind speed, (e) wind direction, (f) air pressure, and (g) precipitation.

Selected Readings

Carlson, T. N. "Airflow Through Midlatitude Cyclones and the Comma Cloud Pattern." *Monthly Weather Review,* 108 (1980):1498–1509. Describes the concept of conveyor belts as related to frontal cyclones.

Dorr, B. "Bombs and Ultrabombs, Exploring Explosive Ocean Storms." *Weatherwise* 43, No. 2(1990):76–83. Describes an ERICA mission into a rapidly developing extratropical cyclone.

Eagleman, J. R. "Conceptual Models of Frontal Cyclones." *Journal of Geography* 80 (1981):87–91. Distinguishes among three basic types of frontal cyclones.

Hughes, P. "The Blizzard of '88." *Weatherwise* 40, No. 6 (1987):312–320. Describes one of the most famous and disruptive East Coast snowstorms.

Kocin, P. J., and L. W. Uccellini. *Snowstorms Along the Northeastern Coast of the United States: 1955 to 1985.* Boston, MA: American Meteorological Society, 1990, 280 pp. Presents case studies of 20 of the most intense snowstorms to affect the urban Northeast.

Namias, J. "The History of Polar Front and Air Mass Concepts in the United States—An Eyewitness Account." *Bulletin of the American Meteorological Society* 64 (1983):734–755. Provides a historical perspective on a key concept of midlatitude weather systems.

Royal Meteorological Society. "The Storm of 15–16 October 1987." *Weather* 43, No. 3 (1988):65–144. Devotes an entire issue to a discussion of a severe synoptic-scale storm that ravaged portions of England.

*Let me snuff thee up, sea
    breeze!
and whinny in thy spray.*

Herman Melville
WHITE JACKET

Local and Regional Circulations

Monsoon Circulation
Land and Sea (or Lake) Breezes
Lake-Effect Snows
Heat Island Circulation
Katabatic Winds
Chinook Winds
Desert Winds
Mountain and Valley Breezes
Conclusions

Special Topic: Monsoon Failure and Drought in Sub-
              Saharan Africa

Along a coastline such as this we
can often anticipate a refreshingly
cool sea breeze on a summer afternoon. A sea breeze is just one of a
wide variety of local and regional
winds. [Photograph by J. M.
Moran]

I N THIS chapter, we describe several specialized atmospheric circulation systems. Although these systems cover different spatial scales and persist for varying periods, each is strongly influenced by land–sea temperature contrasts, topography, or other features of the Earth's surface.

Monsoon Circulation

A **monsoon circulation** characterizes regions where seasonal reversals in prevailing winds cause wet summers and dry winters. Although a weak monsoon develops over central and eastern North America, we are concerned here with the much more vigorous monsoon circulations over Africa and Asia, where 2 billion people depend on monsoon rains for their drinking water and agriculture.

What causes monsoons? A complete explanation is not yet available, but they are linked to seasonal shifts in the global-scale circulation, specifically, the north–south shifts of the intertropical convergence zone (ITCZ). As first proposed in 1686 by Edmund Halley, monsoons also depend on seasonal contrasts in the heating of the land and sea. As we have seen, the ocean has a greater thermal stability than does the land. Beginning in spring, relatively cool air over the ocean and relatively warm air over the land give rise to a horizontal air pressure gradient directed from sea to land,* initiating a flow of humid air inland. Over the land, intense solar heating triggers convection. The humid air is forced to rise, and consequent expansional cooling leads to condensation, clouds, and rain. Release of latent heat intensifies the buoyant uplift, triggering even more rainfall. Aloft, the air spreads seaward and subsides over the relatively cool ocean surface, thus completing the monsoon circulation.

By early autumn, radiational cooling chills the land more than the adjacent sea, setting up a horizontal air pressure gradient directed from land to sea. Air subsides over the land, and dry surface winds sweep seaward. Air is forced to rise over the relatively warm sea surface and, aloft, drifts landward—completing the winter monsoon circulation. Over land, therefore, the summer monsoon is wet, and the winter monsoon is dry.

Monsoon winds have sufficiently long trajectories and persist long enough to be influenced by the Coriolis effect. As shown in Figure 12.1, January and July surface monsoon winds are deflected to the right in the Northern Hemisphere and to the left in the Southern Hemisphere.

---

* Recall that air pressure gradients are always directed from areas of high pressure toward areas of low pressure.

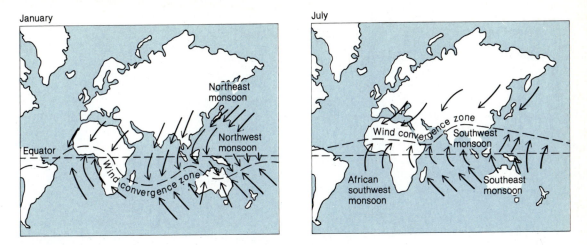

FIGURE 12.1
Surface air streams during the monsoon circulations of January and July are deflected by the Coriolis effect to the right in the Northern Hemisphere and to the left in the Southern Hemisphere. [After P. J. Webster, "Monsoons," *Scientific American* 245, No. 2 (1981):112. Copyright © 1981 by Scientific American, Inc. All rights reserved.]

Monsoon rains usually commence abruptly several weeks before the summer solstice and cease just after the autumnal equinox, but rainfall is neither uniform nor continuous. On the contrary, the rainy season typically consists of a sequence of active and dormant phases. During a **monsoon active phase,** the weather is cloudy with frequent deluges of rain, but during a **monsoon dormant phase,** the weather is sunny and hot. The monsoon shifts from active to dormant phases as bands of heavy rainfall surge inland. Heavy rains first strike coastal areas and soak the ground. As the soil becomes saturated with water, more of the available solar radiation is used for evaporation and less for sensible heating. Coastal areas cool as a consequence, the uplift weakens, and skies partially clear. Meanwhile, the area of maximum heating, vigorous uplift, and heavy rains shifts inland. Back at the coastal areas, however, the hot sun eventually dries the soil, sensible heating intensifies, uplift strengthens, and the rains resume (monsoon active phase). This sequence of active and dormant phases is repeated about every 15 to 20 days during the wet monsoon.

Insolation, land and water distribution, and topography thus impose some regularity on the monsoon circulation; that is, summers are wet and winters are dry. The global-scale circulation (especially shifts of the ITCZ) and the strength and distribution of convective activity, however, vary from one year to the next. These variations mean that the intensity and duration of monsoon rains change from year to year. As a consequence, drought is a distinct possibility in monsoon climates. The Special Topic, "Monsoon Failure and Drought in Sub-Saharan Africa," considers a particularly tragic example.

Monsoon Failure and Drought in Sub-Saharan Africa

Monsoon climates feature pronounced wet and dry seasons. The length of rainy season and the quantity of rainfall vary both temporally, from year to year, and geographically. Perhaps nowhere in the world has this inherent variability of monsoon rains caused more human misery than in sub-Saharan West Africa. A combination of low mean annual rainfall and high variability in rainfall has meant frequent prolonged drought and famine.

Sub-Saharan West Africa (usually called the Sahel) is a transition zone between the Sahara Desert to the north and the humid savanna to the south. This zone encompasses all or part of Mauritania, Senegal, Mali, Burkina Faso, Niger, Chad, and Sudan (Figure 1). Iso-

lines of mean annual rainfall run essentially east–west and increase in magnitude southward, from less than 100 mm (4.0 in.) at the Sahara's southern edge to more than 600 mm (23.6 in.) in the southern Sahel.

Rainfall in the sub-Sahara is governed by north–south shifts of the intertropical convergence zone (ITCZ), part of the seasonal adjustments of the planetary-scale circulation. As the ITCZ follows the sun, its northward surge in spring triggers convective rains, and its southward shift in fall brings the rainy season to an end. However, as the northward decline in mean annual rainfall suggests, convective activity weakens and the rainy season shortens from south to north.

FIGURE 1

Sub-Saharan Africa is the transition zone between the Sahara desert to the north and humid savanna to the south.

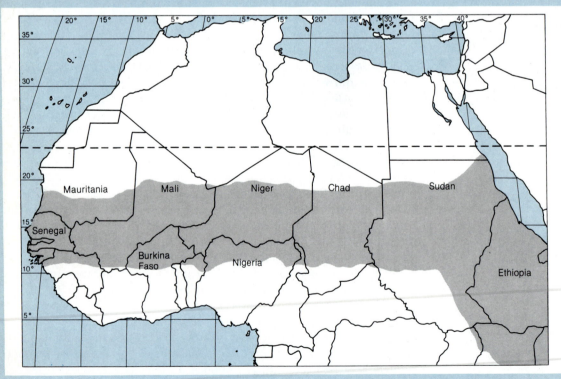

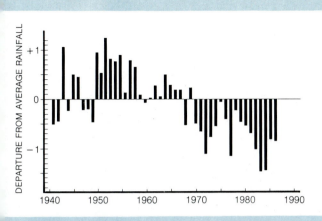

FIGURE 2

An index of the yearly average April-to-October rainfall for 20 sub-Saharan locations that lie between 11 and 18 degrees N and west of longitude 10 degrees E. Above-average rainfall plots above the horizontal line and below-average rainfall plots below the horizontal line. [From P. J. Lamb, "On the Development of Regional Climatic Scenarios for Policy-Oriented Climatic-Impact Assessment," *Bulletin of the American Meteorological Society* 68, No. 9 (1987):1122]

In summers when the ITCZ does not penetrate as far north as usual, or arrives late, or moves southward early, rainfall is below average. A succession of such summers means drought. In this century, the Sahel has endured three major droughts, the most recent of which began in 1968 and persisted into the mid-1980s, albeit with a few brief interruptions (Figure 2).

The cause of monsoon failure is much debated in the scientific community. The most recent episode of Sahelian drought is variously attributed to ENSO events (Chapter 10), the global increase in atmospheric CO_2 and other infrared-absorbing trace gases (Chapter 18), or anomalies in tropical Atlantic Ocean surface temperatures. There is even some suggestion that modification of the land by human activities (overgrazing and deforestation) may contribute to aridity by altering the regional radiation balance.

Although the cause of monsoon failure is unknown, in 1984 its devastating effects were strikingly brought home worldwide on television news (Figure 3). Pictures of starving children, emaciated livestock, and withered crops spurred a massive international relief effort. The people of the Sahel are particularly vulnerable to drought because 80 percent of them depend on agriculture for their livelihood. Drought forced them off lands that even in the best of times are marginal for crops and livestock. People migrated to urban areas in search of food and work, and many ended up in refugee camps.

In the Sahel and elsewhere, drought is a complex problem that involves more than just an anomalous atmospheric circulation pattern. Social, economic, and political factors also affect the impact of drought. As a further complication, in the sub-Sahara, the mix of these nonatmospheric factors varies from country to country.

With famine a perennial risk in the Sahel, the U.S. Agency for International Development (USAID) is attempting to shorten the response time of international relief organizations. To this end, USAID operates the *Famine Early Warning System (FEWS)*. FEWS utilizes imagery from the NOAA polar-orbiting weather satellite (Chapter 15) to track crop and vegetation patterns over the sub-Sahara. The goal is to detect early signs of drought. This information is combined with various social data acquired from the ground (such as population movements and health problems) and is issued as monthly reports to African government officials and international planning agencies.

In the late 1980s, more abundant rainfall coupled with agricultural reforms at least temporarily improved the food supply for much of the Sahel. Nonetheless, the region remains very vulnerable to renewed drought and famine.

FIGURE 3

The human tragedy of the sub-Saharan drought. [Photograph courtesy of FAO]

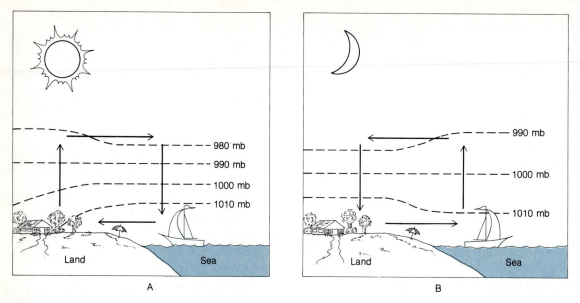

FIGURE 12.2
Vertical cross sections of (A) a sea (or lake) breeze and (B) a land breeze. Arrows indicate wind direction, and dashed lines are isobaric (constant-pressure) surfaces.

Land and Sea (or Lake) Breezes

For those of us lucky enough to live near the ocean or a large lake, sea or lake breezes bring welcome respite from the oppressive heat of a summer afternoon.* On warm days, a weak synoptic-scale air pressure gradient allows a cool wind to sweep inland from the sea or from a large lake. Depending on the source, this refreshing wind is called either a **sea breeze** or a **lake breeze.** Both breezes are caused by differential heating of land and water.

As we noted in Chapter 3, when both land and water are exposed to the same intensity of solar radiation, the land surface warms more than the water surface. The relatively warm land heats the overlying air, thereby lowering air density. Compared to the land, the water is relatively cool, as is the air overlying the water. Consequently, as shown in Figure 12.2A, a local horizontal air pressure gradient develops between land and water, with the highest pressure over the water surface. In response to this gradient, cool air sweeps inland. Aloft, continuity requires a return airflow directed from the land to the water, with air rising over the land and sinking over the water.

Sea (or lake) breezes are shallow systems, generally confined to the lowest kilometer of the troposphere. Typically, the breeze begins near the shoreline several hours after sunrise and gradually expands both inland and out over the body of water, reaching maximum strength by midafternoon. The inland

---

* In subtropical locations, such as peninsular Florida, sea and land breezes occur at any time of year.

extent of the breeze varies from only a few hundred meters to many tens of kilometers.

By late evening the sea (or lake) breeze circulation reverses direction, and surface winds blow offshore as a **land breeze.** The change in wind direction is caused by a reversal in heat differential between land and water. At night, radiational cooling chills the land surface (and the air over the land) more than the water surface (and the air over the water). The land thus becomes cool relative to the water surface. A horizontal gradient in air density gives rise to a horizontal air pressure gradient that is directed from land to sea (or lake). A cool offshore breeze develops, along with a return airflow aloft, and air subsides over the land and rises over the water (Figure 12.2B). Land breezes reach maximum strength just before sunrise but tend to be weaker than sea (or lake) breezes.

Land and sea (or lake) breezes typically develop and diminish so rapidly that they are not significantly influenced by the Earth's rotation. In some localities, however, the Coriolis effect is responsible for a gradual shift in the direction of a sea breeze through the course of a day, but the magnitude of the Coriolis is always weaker than the pressure gradient force.

The shoreline of Lake Michigan is often the site of lake breezes in spring and summer. During daylight hours between May and August, the surface waters of Lake Michigan, on average, are cooler than the surface of the adjacent land. If synoptic-scale winds are relatively weak, then a shallow mesoscale high-pressure system develops over the lake. As shown in Figure 12.3, surface winds diverge from the high center toward the shoreline. The leading edge of this surge of cool air, the lake breeze, forms a miniature cold front, which, by midafternoon, may move inland many kilometers. In so doing, the lake breeze front forces the warm air over the land to rise so that convective clouds may mark the leading edge of the lake breeze front.

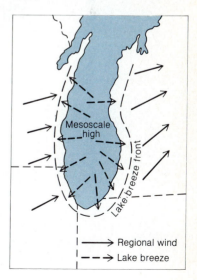

FIGURE 12.3

On days when synoptic-scale winds are weak and Lake Michigan surface waters are cooler than the adjacent land surface, a shallow mesoscale high develops over the lake. Surface winds diverge outward from the high and give rise to lake breezes. [From W. A. Lyons, "Some Effects of Lake Michigan upon Squall Lines and Summertime Convection," *Proceedings, Ninth Conference on Great Lakes Research.* Ann Arbor, MI: International Association for Great Lakes Research, 1966, p. 262.]

Lake-Effect Snows

A highly localized fall of snow immediately downwind from an open lake is known as **lake-effect snow** (Figure 12.4). Typically, such snows extend inland only a few tens of kilometers and often fall over such a small area that the event is not detected by the regular network of weather stations. Residents of the affected area, however, may be swamped by snow. For example, in February 1976, Barnes Corner, New York, at the eastern end of Lake Ontario, received 137 cm (54 in.) of road-clogging snow in less than 24 hours. Such an extreme lake-effect snowfall is sometimes called a **snowburst.**

Lake-effect snows are most frequent in autumn and early winter when lake surface temperatures are still relatively mild. As early season outbreaks of cold (arctic) air stream over the lake, water readily evaporates and raises the vapor pressure of the lowest portion of the advecting air mass. (Recall from Chapter 7 that this same process may produce steam fog.) In addition, the milder lake water heats the advecting cold air from below, reducing its stability, and enhancing convection and cloud development. Often this is enough to trigger snowfall over the lake.

As the modified (milder, more humid, and less stable) air flows toward the lake's lee shore, the contrast in surface roughness between the lake and land becomes important. The rougher land surface slows onshore winds, and the consequent horizontal convergence (Chapter 9) induces uplift of air, further development of clouds, and lake-effect snows. The topography of the

FIGURE 12.4

Landsat imagery of a sharply defined snowfall pattern on Michigan's Upper Peninsula on 20 October 1972. Cold northwest winds blowing across Lake Superior induced lake-effect snows. [From U.S. Geological Survey, EROS Data Center, Sioux Falls, SD]

shore also affects the amount of precipitation: hilly terrain forces greater uplift and heavier snowfalls than does flat terrain (Figure 12.5).

Ultimately, the frequency and intensity of lake-effect snows hinge on the degree of air mass modification, which, in turn, depends on (1) the temperature contrast between the mild lake surface and overlying cold air and (2) the over-water trajectory, called **fetch,** of the advecting cold air. As the lake-surface-to-air temperature contrast increases, so does the potential for lake-effect snow. Hence, to the lee of the Great Lakes, the bulk of lake-effect snow falls between mid-November and mid-January, the usual period of maximum temperature difference between the lake surface and overlying air.

Cold air usually sweeps into the Great Lakes region on northwest winds. Hence, considering the maximum possible fetch, the greatest potential for substantial lake-effect snows is along the downwind southern and eastern shores of the Great Lakes (Figure 12.6). In these so-called **snowbelts,** lake-effect snow accounts for a substantial percentage of the total seasonal snowfall. Occasionally, however, lake-effect snow develops on the normally upwind western shores of the lakes. For example, an early winter cyclone tracking through the lower Great Lakes region may produce strong northeast onshore winds. Hence, lake-effect snows may supplement the snow produced by frontal overrunning, and the Milwaukee–Chicago metropolitan area ends up with paralyzing accumulations of snow.

We in North America have come to associate lake-effect snows almost exclusively with the Great Lakes, but the same snow-making mechanism also

FIGURE 12.5
Some of the heaviest lake-effect snows fall in the rugged terrain of upper Michigan's Keweenaw Peninsula. This snow pole documents past seasonal snowfalls. [Photograph by S. I. Dutch]

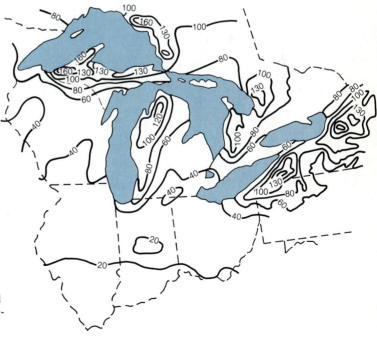

FIGURE 12.6
Average yearly snowfall in the Great Lakes region (in inches). Note that the greatest snowfall totals occur downwind of the Great Lakes and are caused by frequent lake-effect snows.

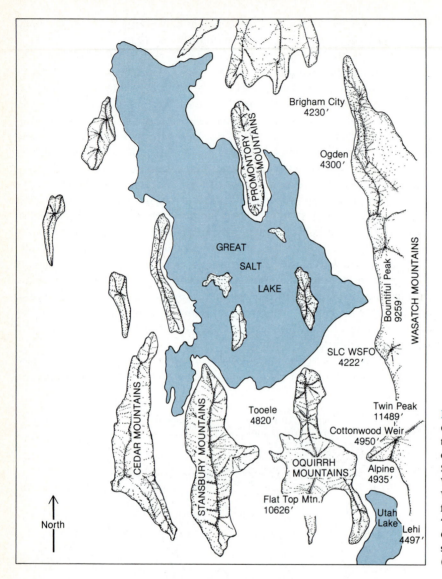

Brigham City
4230'

Ogden
4300'

PROMONTORY MOUNTAINS

GREAT

SALT

LAKE

Bountiful Peak
9259'

WASATCH MOUNTAINS

SLC WSFO
4222'

Twin Peak
11489'

Cottonwood Weir
4950'

Tooele
4820'

CEDAR MOUNTAINS

STANSBURY MOUNTAINS

OQUIRRH
MOUNTAINS

Alpine
4935'

Flat Top Mtn.
10626'

Utah
Lake

Lehi
4497'

North

FIGURE 12.7
Great topographic relief con-
tributes to the development
of lake-effect snows in the
Salt Lake Valley. (SLC
WSFO = Salt Lake City
Weather Service Forecast Of-
fice.) [From D. M. Carpenter,
"Utah's Great Salt Lake—A
Classic Lake-Effect Snow-
storm," *Weatherwise* 38, No. 6
(1985):309]

affects the valleys southeast of Utah's Great Salt Lake (Figure 12.7). Each year
perhaps a half dozen significant lake-effect snowfalls occur in the Tooele and
Salt Lake valleys, bowl-shaped depressions that slope up and away from Great
Salt Lake. Both valleys parallel the northwest-southeast-trending long axis of
the lake. Which of the two valleys receives the heavier snowfall depends on
the wind direction. Cold air streaming over the lake from the northwest is
moderated by contact with the relatively warm lake surface, and, downwind,
the mountain valley topography forces horizontal convergence and uplift.
Clouds billow upward and locally release bursts of heavy snow.

On 17–18 October 1984, for example, weather observers measured a record 46.7 cm (18.4 in.) of lake-effect snow at Salt Lake City International Airport, about 5 km (3 mi) southeast of the lake's southern shore. In the southeastern suburbs of Salt Lake City, snowfall topped 60 cm (24 in.) during the same period. Wet snow clung to trees (still in full leaf) and many tree limbs snapped under the weight of thick accumulations of snow. Falling limbs brought down power lines and caused an estimated $1 million in damage.

Great Salt Lake snowbursts are generally less intense than their Great Lakes counterparts. A major reason is the smaller area and shorter fetch of Great Salt Lake, so that there is less air mass modification over Great Salt Lake. The waters of Great Salt Lake cover an area that is only about 13 percent of that of Lake Ontario, the smallest of the five Great Lakes. The maximum fetch on Great Salt Lake is only 120 km (75 mi), whereas maximum fetches on the Great Lakes are many hundreds of kilometers.

Lake-effect snows also develop downwind of Canadian lakes (including Hudson Bay) north of the Great Lakes. Typically, snowfall peaks in autumn (October and November), prior to lake freeze-over. For example, Lake Winnipeg, in southern Manitoba, usually is ice covered by late November. Two weather stations downwind of the lake (Norway House and Berens River) usually receive more snowfall in November than in either December or January.

Heat Island Circulation

Travelers approaching large cities may be greeted by an unsightly veil of dust, smoke, and haze. This dome of polluted air is the product of a convective circulation of air that concentrates pollutants over the city. The convective circulation, in turn, is related to the temperature contrast between the city and the surrounding rural areas.

The average annual air temperature of a city typically is only slightly higher than that of the surrounding countryside, but on some days the temperature contrast may be 10 C° (18 F°) or greater. Consequently, snow melts faster and flowers bloom earlier in a city. This climatic effect is known as the **urban heat island.** Figure 12.8 illustrates the urban heat island of Washington, D.C., and Plate 19 is a satellite view of heat islands in southeastern New England.

Several factors contribute to the development of an urban heat island. One is the relatively high concentration of heat sources in cities (for example, people, cars, industry, air conditioners, and furnaces). Because all the heat from every source eventually escapes to the atmosphere, the air of a large city receives a considerable input of waste heat. On a very cold winter day in New York City, heat from urban sources may approach 100 W per m^2, about 8 percent of the solar constant.

The daytime absorption of solar radiation and the conduction of heat into city air are also facilitated by the thermal properties of urban building materials. Concrete, asphalt, and brick conduct heat more readily than do the soil and vegetative cover of rural areas. The heat loss at night by infrared radiation to the atmosphere and to space is thus partially compensated for in cities by a release of heat from the buildings and streets. In addition, during the day, a city's canyonlike terrain of tall buildings and narrow streets traps solar energy because of multiple reflections of sunlight off the sides of buildings.

The temperature contrast between city and country is further accentuated by a city's typically lower rate of evapotranspiration. Urban drainage (sewer) systems rapidly and efficiently remove most runoff from rain and snowmelt, so that less of the available solar radiation is used for evaporation (latent heating), and more solar radiation is used to heat the ground and air directly (sensible heating). On the other hand, the moister surfaces of the countryside (lakes, streams, soil, vegetation) increase the fraction of absorbed solar radiation that is used for latent heating. That is, the Bowen ratio is greater in the city than in the country.

An urban heat island is most likely to develop when synoptic-scale winds are weak. (Strong winds would mix the city air and country air and diminish the temperature contrast.) Under such conditions, in some large metropolitan

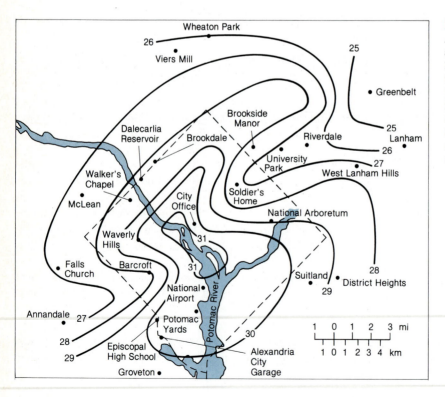

FIGURE 12.8
Average winter low temperatures (in °F) in Washington, DC, and vicinity, illustrating the urban heat island effect. [From C. A. Woolum, "Notes from a Study of the Microclimatology of the Washington, DC, Area for the Winter and Spring Seasons," *Weatherwise* 17, No. 6 (1964):6]

FIGURE 12.9
The heat island of a large metropolitan area may give rise to a convective circulation that transports air pollutants into the city. [After W. P. Lowry, "The Climate of Cities," *Scientific American* 217, No. 2 (1967):20. Copyright © 1967 by Scientific American, Inc. All rights reserved.]

areas, the relative warmth of the city compared to its surroundings can promote a convective circulation of air, as shown in Figure 12.9. Warm air at the city's center is forced to rise by cooler, denser air flowing in from the countryside. The rising columns of air gather aerosols into a **dust dome** over the city. In this way, aerosols may become as much as a thousand times more concentrated over an urban-industrial area than in the air over the rural countryside. If regional winds strengthen to more than about 15 km (10 mi) per hour, the dust dome elongates downwind in the form of a **dust plume** and spreads the city's pollutants over the countryside. The Chicago dust plume, for example, is sometimes visible 240 km (150 mi) from its source.

Katabatic Winds

Under the infuence of gravity, a shallow mass of cold, dense air can slide downhill. This **katabatic wind** usually originates in winter over extensive snow-covered plateaus or over other highlands. Although adiabatic compression warms the air to some extent, the air is so cold to start with that these winds are still quite cold when they reach the lowlands.

Among the best-known katabatic winds are the **mistral,** which descends from the snow-capped Alps down the Rhone River Valley of France and into the Gulf of Lyons along the Mediterranean coast, and the **bora,** which originates in the high plateau region of Yugoslavia and cascades onto the narrow Dalmatian coastal plain along the Adriatic Sea. Both the mistral and the bora are winter phenomena.

In some places, such as inlets of the coastal range of British Columbia, katabatic winds are channeled by mountain valleys, and this constricted flow sometimes accelerates the wind to potentially destructive speeds. Steep slopes also accelerate katabatic flow. Along the edge of the massive Greenland and Antarctic ice sheets, for example, katabatic winds frequently top 100 km (62 mi) per hour.

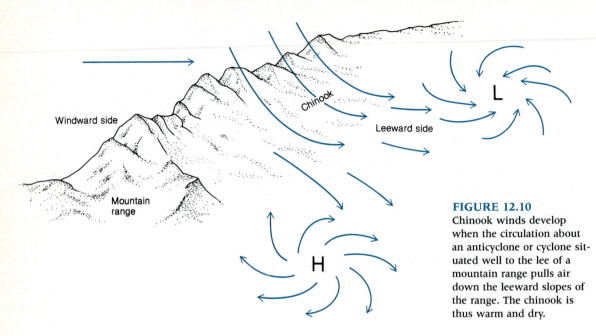

FIGURE 12.10
Chinook winds develop when the circulation about an anticyclone or cyclone situated well to the lee of a mountain range pulls air down the leeward slopes of the range. The chinook is thus warm and dry.

Chinook Winds

Like the katabatic wind, the chinook is a downslope breeze. But while the katabatic wind is cold and dry, the chinook wind is warm and dry.

Chinook winds develop when relatively mild air aloft is adiabatically compressed as it descends the leeward slopes of mountain ranges. For every 1000 m of descent, the air temperature rises about 10 C° (Chapter 6). Air that flows down the slopes of high mountain ranges such as the Rockies thus undergoes considerable warming.

Because the air is relatively warm, the chinook does not flow downslope under the influence of gravity, as does a katabatic wind; rather, the chinook wind is pulled downslope. Typically, a chinook develops when strong winds force a layer of stable air in the lower troposphere to ascend the windward slopes of a mountain range. When the wind reaches the leeward slopes, the stability of the air causes it to return (descend) to its original altitude. Further descent of the air is caused by the larger-scale circulation. For example, chinook winds descending the leeward slopes of the Rocky Mountain Front Range are drawn downslope by strong west winds associated with synoptic-scale pressure systems located well east of the mountains (Figure 12.10).

At the onset of a chinook, surface air temperatures often climb abruptly tens of degrees in response to compressional warming. For example, on 6 January 1966, at Pincher Creek, Alberta, a chinook sent the temperature soaring 21 C° (38 F°) in only four minutes. An even more dramatic temperature surge was recorded at Spearfish, South Dakota, on 22 January 1943: the air temperature rose from $-20\ °C$ ($-4\ °F$) at 7:30 A.M. to 7 °C (45 °F) at 7:32 A.M., that is, 27 C° (49 F°) in only 2 minutes! The sudden springlike warmth may just as quickly give way to severely cold conditions. For example, at the foot of the Rockies, a shift of the synoptic-scale circulation from west to north brings an abrupt end to the chinook and the return of polar or arctic air.

Chinook is a Native American word that, according to tradition, means "snow eater." The term is appropriate because of the wind's catastrophic effect on a snow cover. As the chinook is compressionally warmed, the relative humidity drops dramatically. Because the chinook is both warm and very dry, a snow cover melts and vaporizes rapidly. It is not unusual, for example, for a half meter of snow to disappear in this way in only a few hours.

Chinook winds are gusty and, locally, may reach destructive magnitudes, especially along the foothills of the Front Range of the Rocky Mountains. At Boulder, Colorado, in the foothills just northwest of Denver (Figure 12.11), violent downslope winds, sometimes gusting to 160 km (100 mi) per hour or higher, unroof buildings and topple power poles. On average, the community experiences $1 million in property damage each year because of these destructive winds. Currently, research is under way at the National Center for Atmospheric Research (in Boulder) to improve prediction of violent chinook winds.

Researchers do not agree on the precise mechanism that triggers destructive chinook winds, but according to one view, downslope winds are linked to the interaction of the Rockies with the planetary-scale westerlies. Recall from Chapter 7 that a north–south mountain range such as the Rockies deflects a strong westerly airflow into a **standing wave,** a stationary pattern of vertically oriented crests and troughs that stretches downwind of the mountain range and gives rise to mountain-wave clouds. The chinook is actually a segment of the wave that dips down the leeward slopes of the mountain

FIGURE 12.11
Boulder, Colorado, in the foothills of the Rockies, experiences some particularly violent downslope winds. [Photograph by J. M. Moran]

range. In a violent chinook, very energetic turbulent eddies that are generated aloft are transported downward into the foothills so that surface winds become very strong and gusty.

Chinook-type winds are not restricted to the leeward slopes of the Rockies. A similar wind, called the **foehn,** flows into the Alpine valleys of Austria and Germany. The same type of warm, dry wind is drawn down the eastern slopes of the Andes in Argentina, where it is known as the **zonda.** In southern California, the notorious **Santa Ana wind** is another chinook wind that typically develops in autumn and winter. A strong high-pressure system centered over the Great Basin sends northeast winds over the southwestern United States, driving air downslope from the desert plateaus of Utah and Nevada, around the Sierra Nevada, and as far west as coastal southern California (Figure 12.12). The consequent adiabatic compression produces hot and dry, gusty winds that desiccate vegetation and contribute to outbreaks of forest and brush fires. Santa Ana winds sometimes gust to 130 to 140 km (80 to 90 mi) per hour and almost always cause some property damage.

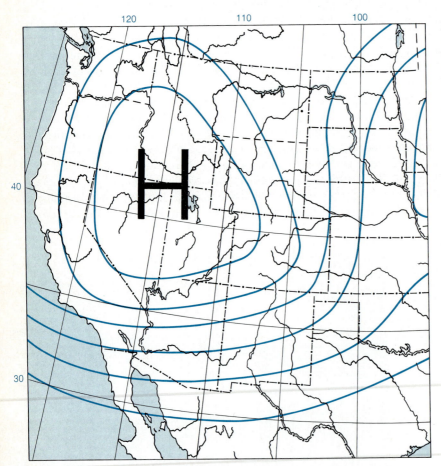

FIGURE 12.12
Schematic representation of the synoptic-scale weather pattern that favors the development of Santa Ana winds over Southern California. Solid lines are isobars.

Desert Winds

Deserts typically are very windy places primarily because of the intense solar heating of the ground. In deserts, most of the absorbed solar radiation goes into sensible heating because little is used to evaporate the scarce water. In some deserts the midday temperature of the surface exceeds 55 °C (130 °F). This hot surface produces a steep temperature lapse rate in the lowest air layer. A steep temperature lapse rate, in turn, means great instability (Chapter 6), vigorous convection, and gusty surface winds. The strength and gustiness of the wind vary with the intensity of solar radiation so that wind speeds and gustiness usually peak in the early afternoon and during the warmest months.

Local variations in the surface characteristics of a desert (albedo, moisture, topography) cause some spots to become hotter than others. When synoptic-scale winds are relatively weak, local surface winds converge toward hot spots, and, as the hot air rises, the converging surface winds begin to rotate about a vertical axis. In the process, dust is lifted from the ground, and the circulation is visible as a whirling mass of dust-laden air known as a **dust devil.** Dust devils are very small circulation systems that average about 10 m (30 ft) in diameter and less than 2000 m (6000 ft) in altitude and have a life expectancy of only several minutes. Normally, dust devil winds are not very strong and cause no property damage.

Larger-scale winds are produced in deserts by thunderstorms or migrating cyclones. Surface winds associated with these weather systems can give rise to either dust storms or sand storms, the difference between the two hinging on the size range of the loose surface sediment that is lifted by the wind. Dust consists of very small particles (less than 0.06 mm in diameter) that can be carried by the wind to great altitudes. Sand, which typically covers only a small fraction of desert terrain, consists of larger particles (0.06 to 2.0 mm in diameter) that are transported by the winds within about a meter of the ground.

One of the most spectacular dust storms, known as a **haboob,** is generated by the strong, gusty downdraft of a thunderstorm. In a desert, thunderstorm rains often evaporate completely in the dry air beneath cloud level and do not reach the ground. The thunderstorm downdraft, however, exits the thunderstorm as a surge of cool, gusty air. Dust picked up from the ground fills the air, severely restricting visibility, and the mass rolls along the ground as a huge ominous black cloud. A haboob may be more than 100 km (62 mi) wide and may reach altitudes of several kilometers. These dust storms are most common in the Sudan of North Africa but also occur in the American Southwest deserts.

Under some conditions, winds blow out of deserts and into quite different climatic regions. For example, in spring, when the subtropical anticyclones shift poleward, hot and dry winds stream northward from the Sahara Desert of North Africa, out over the Mediterranean Sea, and then into southern

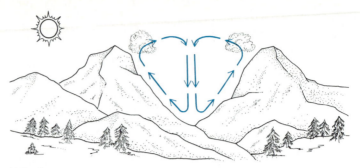

Valley breeze

Mountain breeze

FIGURE 12.13
A schematic representation of valley and mountain breeze circulation.

FIGURE 12.14
Cumulus clouds developing near the peak of a mountain as a consequence of the upslope current of air associated with a valley breeze. [Photograph by J. M. Moran]

Europe. These desert winds are little modified after crossing the Mediterranean, except in the west, where winds approaching Spain have a longer fetch over water and consequently are more humid.

Mountain and Valley Breezes

Along mountain slopes that are exposed to intense solar heating, a localized air circulation may develop that reverses direction between day and night (Figure 12.13). After the winter snows have melted, bare valley walls facing the sun absorb solar radiation, and the air in contact with the walls is heated. This warm, light air is forced by neighboring cooler, denser air to flow upslope as a **valley breeze.** As a consequence, cumulus clouds may develop near the mountaintop (Figure 12.14). At night, with clear skies and intense radiational cooling, the mountain slopes cool rapidly and the air in contact with the slopes is chilled. This cold, dense air then flows downslope as a gusty **mountain breeze** and accumulates in the valley floors where further radiational cooling may result in the formation of fog or low stratus clouds.

Conclusions

In this chapter we described several local and regional circulation systems, and in subsequent chapters we will consider others. The domination of these circulation systems by larger-scale circulation is apparent; that is, the global- and synoptic-scale patterns set boundary conditions for any smaller-scale circulation. Hence, these circulation systems are vulnerable to changes in large-scale patterns. In some cases, synoptic-scale winds reinforce mesoscale winds, as when regional winds blow in the same direction as a sea or lake breeze. In other cases, synoptic-scale winds negate or overwhelm mesoscale winds, as when northerly winds sweep along the edge of the Rocky Mountains and eliminate the possibility of chinook winds. Even the large-scale monsoon winds are influenced by shifts of the ITCZ and the subtropical anticyclones.

Summary Statements

Monsoon circulations in tropical latitudes illustrate the interplay of climate controls. Differences in solar heating between land and sea, and seasonal shifts in global-scale circulation play important roles in monsoon development.

Although summers are predictably wet and winters are predictably dry, the wet monsoon is punctuated by active and dormant phases. The intensity and duration of monsoon rains vary from one year to the next.

Lake-effect snows develop on the downwind (leeward) shore of the Great Lakes and Great Salt Lake when cold air streams across a relatively mild lake surface. Fetch and lake shore topography play important roles in governing the intensity of lake-effect snows.

Human activities influence the climate of large cities by altering the local radiation balance. This gives rise to an urban heat island and, in large metropolitan areas, to a convective circulation pattern that concentrates air pollutants in a dust dome over the city.

A variety of mesoscale circulations are favored by certain synoptic-scale conditions and are triggered by localized pressure gradients. These include land and sea, or lake, breezes and mountain and valley breezes.

A katabatic wind is a gravity-driven flow of dense, cold air. A chinook wind, on the other hand, consists of compressionally warmed air pulled down the leeward slopes of mountains by synoptic-scale circulation.

Key Words

monsoon circulation	lake-effect snow	katabatic wind	zonda
monsoon active phase	snowburst	mistral	Santa Ana wind
monsoon dormant phase	fetch	bora	dust devil
sea breeze	snowbelts	chinook winds	haboob
lake breeze	urban heat island	standing wave	valley breeze
land breeze	dust dome	foehn	mountain breeze
	dust plume		

Review Questions

1. Distinguish between the wet and dry monsoon circulations.

2. In what areas of the globe are monsoon circulations dominant features of the climate?

3. Why does the land surface warm up more than the sea surface in response to the same intensity of solar radiation?

4. Explain the shifts between active and dormant phases of a wet monsoon.

5. How is the monsoon circulation linked to seasonal shifts of the intertropical convergence zone (ITCZ) and the subtropical anticyclones?

6. The Coriolis effect influences the monsoon circulation but not most land and sea breezes. Why the difference?

7. Why does a sea breeze develop on some summer days but not on others?

8. Speculate on how topography affects the inland extent of sea breezes or lake breezes.

9. Sea breezes, lake breezes, and land breezes are driven by a horizontal air pressure gradient. What causes this pressure gradient to develop?

10. Describe the surface air circulation about a mesoscale high. Why is it a daytime phenomenon only?

11. Explain why a lake breeze develops along the Lake Ontario shoreline in summer but not during the winter.

12. Explain why sighting steam fog off the western shores of Lake Michigan suggests that it may be snowing on the lake's eastern shores.

13. What factors contribute to the development of an urban heat island?

14. Speculate on steps that city planners might take to reduce the intensity of urban heat islands, thereby decreasing the likelihood of dust dome development.

15. How is an urban dust dome linked to an urban heat island?

16. Compare and contrast the thermal properties of vegetation with the thermal properties of materials composing a city (brick, asphalt, concrete).

17. What force drives katabatic winds?

18. A katabatic wind is not a warm wind in spite of adiabatic compressional heating. Explain why.

19. How do chinook winds differ from katabatic winds?

20. What special conditions are needed for development of a valley breeze, a dust storm, and a dust devil?

Points to Ponder

1. Describe in detail the mechanics of the dry and wet monsoons.

2. In portions of India that are influenced by the monsoon circulation, the warmest month of the year is usually May. Explain why.

3. In view of your understanding of atmospheric stability, explain the following observation: In spring, in the Great Lakes region, convective cumulus clouds tend to develop over land but not over lake waters.

4. Explain why a mesoscale anticyclone sometimes develops over Lake Michigan. Is this a cold-core high or a warm-core high? Support your choice.

5. Speculate on the type of weather that might develop along a sea breeze front.

6. How does the Bowen ratio of an urban area compare with that of a rural countryside? Elaborate on your response.

Project

1. Determine whether your area is affected by any local or regional air circulating patterns. You may wish to consult the NOAA Climatic Summary for your area.

Selected Readings

Carpenter, D. M. "Utah's Great Salt Lake—A Classic Lake Effect Snowstorm." *Weatherwise* 38 (1985):309–311. Summarizes conditions required for lake-effect snows to the lee of Great Salt Lake.

Eichenlaub, V. *Weather and Climate of the Great Lakes Region.* Notre Dame, IN: University of Notre Dame Press, 1979, 335 pp. Presents a lucid discussion of local circulation phenomena in the Great Lakes region.

Ellis, W. S. "Africa's Sahel, The Stricken Land." *National Geographic* 172, No. 2 (1987):140–179. Vividly portrays the plight of people trying to survive the erratic monsoon.

Glantz, M. H. "Drought in Africa." *Scientific American* 256, No. 6 (1987):34–40. Argues that drought in sub-Saharan Africa entails social, economic, political as well as meteorological aspects.

Idso, S. B. "Arizona Weather Watchers: Past and Present." *Weatherwise* 28 (1975):56–60. Includes descriptions of desert weather systems.

Kerr, R. A. "Chinook Winds Resemble Water Flowing over a Rock." *Science* 231 (1986):1244–1245. Summarizes possible mechanisms responsible for destructive chinook winds.

Matson, M., et al. "Satellite Detection of Urban Heat Islands." *Monthly Weather Review* 106 (1978):1725–1734. Discusses the detection of heat islands using IR sensors on NOAA-5 satellite.

Monteverdi, J. P. "The Santa Ana Weather Type and Extreme Fire Hazard in the Oakland-Berkeley Hills." *Weatherwise* 26 (1973):118–121. Focuses on the synoptic weather pattern associated with Santa Ana winds.

Webster, P. J. "Monsoons." *Scientific American* 245, No. 2 (1981):109–118. Considers in some detail the energetics involved in monsoon circulations.

Zipser, E. J., and A. J. Bedard, Jr. "Front Range Windstorms Revisited: Small-Scale Differences Amid Large-Scale Similarities." *Weatherwise* 35 (1982):82–85. Details the characteristics of a locally destructive downslope wind.

*Blow, winds, and crack your
    cheeks; rage, blow.
You cataracts and hurricanes,
    spout
Till you have drench'd our
    steeples, drown'd the cocks.
You sulph'rous and thought-
    executing fires,
Vaunt-couriers of oak-cleaving
    thunderbolts,
Singe my white head. And
    thou, all-shaking thunder,
Strike flat the thick rotundity
    o' th' world
Crack nature's moulds, all
    germens spill at once,
That makes ingrateful man.*

William Shakespeare
KING LEAR

13

Thunderstorms

Lightning is one of many hazards associated with thunderstorms. Hail, torrential rains, and strong surface winds are other hazards that can accompany thunderstorms. [Photograph by Arjen and Jerrine Verkaik/SKYART]

335

OST OF US are familiar with typical thunderstorm weather—the blackening sky and abrupt freshening of wind followed by bursts of torrential rain, flashes of lightning, and rumbles of thunder. Often the cool breezes and rains bring us welcome, albeit temporary, relief on hot, muggy summer afternoons. For a farmer whose crops are wilting under the summer sun, the heavy rains may be an economic lifesaver. Some thunderstorms become violent, however, and wreak havoc. Lightning starts fires, strong winds level trees and buildings, heavy rains cause flooding, and, in some cases, thunderstorms spawn destructive hail or tornadoes.

At this very moment, an estimated 2000 thunderstorms are in progress somewhere on Earth, most of them in tropical and subtropical latitudes. In this chapter, we discuss the life cycle and characteristics of thunderstorms and hazards posed by severe thunderstorms. In the next chapter, we describe the most destructive of thunderstorm progeny, the tornado.

Thunderstorm Life Cycle

A **thunderstorm** is a small-scale system and thus affects a relatively small area and is short-lived. It is the product of very strong convection that extends deep into the troposphere, sometimes reaching to the tropopause or higher. Upward-surging air currents are made visible by billowing cauliflower-shaped cumuliform clouds, as shown in Figure 13.1.

FIGURE 13.1

As convection currents surge upward within the troposphere, (A) cumulus clouds begin to show vertical development. In some instances, (B) cumulus clouds merge and build vertically into cumulus congestus and eventually cumulonimbus (thunderstorm) clouds. Photograph B was taken about one hour after photograph A. [Photographs by Arjen and Jerrine Verkaik/SKYART]

A   B

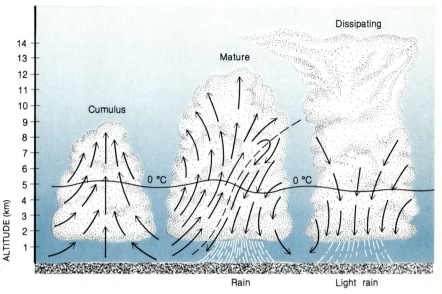

FIGURE 13.2
A thunderstorm's life cycle consists of cumulus, mature, and dissipating stages.

A thunderstorm typically consists of several convection cells, each a few kilometers in diameter. Each thunderstorm cell goes through a life cycle that is divided into three stages: cumulus, mature, and dissipating (Figure 13.2).

CUMULUS STAGE

Recall from Chapter 3 that convection currents begin when heat is conducted from the relatively warm Earth's surface to cooler air immediately above the surface. We can visualize the upward-moving branch of a convection current as a continuous stream of bubbles (or parcels) of warm, unsaturated air. The rising bubbles cool at the dry adiabatic lapse rate (10 C° per 1000 m) until they reach the **convective condensation level (CCL),** where water vapor condenses and cumulus clouds start forming. The more humid the air is to begin with, the less the expansional cooling that is needed for the bubbles to achieve saturation and the lower is the base of cumulus clouds. Hence, the base of cumulus clouds usually is lower in Florida, where relative humidities are high, than in New Mexico, where relative humidities are low.

The latent heat that is released during condensation adds to the buoyancy of the saturated bubbles, and they surge upward while cooling at the moist adiabatic lapse rate (averaging 6 C° per 1000 m). Convective bubbles continue ascending for as long as they are warmer, and thus lighter, than the surrounding air, that is, as long as the ambient air is unstable. Some of the saturated bubbles surge through the cloud top and evaporate in the relatively dry air above the cloud. As a consequence, the water vapor content of the air above the cloud increases. Because the air above the cloud is now more humid, subsequent bubbles are able to ascend higher before evaporating. As this process is repeated, the cumulus cloud billows upward. This billowing character of cumulus cloud growth is evident in Plate 12; at this stage of vertical development, the cloud is called **cumulus congestus.** If vertical growth continues, the cumulus congestus cloud builds into a **cumulonimbus** cloud, a thunderstorm cloud (Plate 13).

The **cumulus stage** of the thunderstorm life cycle begins when towering cumulus clouds merge to form a thunderstorm cell. Saturated air streams upward throughout the cell as an **updraft,** which frequently reaches altitudes of 8000 m (26,000 ft) or higher. The updraft is strong enough to keep water droplets and ice crystals suspended in the upper reaches of the cloud.

MATURE STAGE

By definition, the cumulus stage ends and the **mature stage** begins when precipitation reaches the Earth's surface. The cumulative weight of water droplets and ice crystals eventually becomes so great that they begin to fall faster than the updraft. Rain and snow descend through the cloud and drag the adjacent air downward, creating a strong **downdraft** alongside the updraft. At the same time, unsaturated air near the edge of the cloud is drawn into the cloud, a process known as **entrainment.** (Actually, entrainment continues throughout the life cycle of a thunderstorm.) The entrained air mixes with the saturated-cloud air, causing some of the water droplets and ice crystals to vaporize. The consequent evaporative cooling further weakens the buoyant uplift and strengthens the downdraft.

The downdraft exits the base of the cloud and spreads out along the ground, well in advance of the parent cell, as a mass of relatively cool and gusty air. The downdraft is relatively cool because of evaporative cooling beneath the cloud base that offsets to some extent compressional warming of the downdraft. At the surface, the arch-shaped leading edge of downdraft air resembles a miniature cold front and is called a **gust front.** Convective clouds sometimes develop as a consequence of uplift along the gust front and may initiate the formation of secondary thunderstorm cells ahead of the parent cell.

Ominous-appearing low clouds are sometimes associated with thunderstorm gust fronts. A **roll cloud** is an elongated cylindrical dark cloud that appears to rotate slowly about its horizontal axis. The roll cloud occurs behind the gust front and beneath, but detached from, a cumulonimbus cloud. Why this cloud forms is not fully understood. Although appearances might suggest otherwise, roll clouds are seldom accompanied by severe weather. This is *not* the case with shelf clouds.

A **shelf cloud** is a low, elongated cloud that is wedge-shaped with a flat base (Figure 13.3). This cloud occurs at the edge of a gust front and beneath and attached to a cumulonimbus cloud. The shelf cloud is thought to develop as a consequence of uplift of stable warm, humid air along the gust front.

FIGURE 13.3
A shelf cloud such as this one may develop along a thunderstorm gust front. Often shelf clouds are accompanied by strong and gusty surface winds and may be associated with a severe thunderstorm. [Photograph by Arjen and Jerrine Verkaik/SKYART]

Damaging surface winds may occur under the shelf cloud, and sometimes this cloud is associated with a severe thunderstorm.

The thunderstorm cell reaches its maximum intensity during the late mature stage, when cloud tops can exceed altitudes of 18,000 m (60,000 ft). Strong winds at these high altitudes distort the cloud top into the anvil shape shown in Figure 13.4. The flat top of the anvil indicates that convection

FIGURE 13.4
When the billowing cumulonimbus cloud reaches the tropopause, it spreads out and forms a flat anvil top. [NCAR/NSF photograph]

FIGURE 13.5
In this satellite image, clusters of intense thunderstorm cells appear as bright white blotches over Texas and Northeast Oklahoma. [NOAA satellite photograph]

currents have reached the extremely stable air of the tropopause. Only in severe thunderstorms will convection currents overshoot this altitude, causing clouds to billow into the lower stratosphere before collapsing. Temperatures within the upper portion of the cloud are so low that the anvil is composed exclusively of ice crystals, which give it a fibrous appearance. During this phase, the heaviest rains fall and the darkened sky is streaked with lightning. In addition, hail, strong surface winds, and even tornadoes may develop.

Viewed from space by satellite, clusters of mature thunderstorm cells appear as bright white blotches, such as those over central Texas and northeast Oklahoma in Figure 13.5. The brightness of these clusters is due to the high albedo (reflectivity) of the cloud tops (Chapter 2). Much of the solar radiation that does penetrate the cloud is absorbed, so that the sunlight emerging at the cloud base is very weak. For this reason, from our perspective on the Earth's surface, the sky darkens as thunderstorm clouds approach.

DISSIPATING STAGE

As precipitation spreads throughout the thunderstorm cell, so does the downdraft, heralding the demise of the cell. During the **dissipating stage,** upward air motion is entirely replaced by sinking air that is adiabatically warmed by compression. The relative humidity decreases, precipitation tapers off and ends, and convective clouds gradually vaporize.

Typically, a thunderstorm cell completes its life cycle in 30 minutes to an hour, but sometimes lightning, thunder, and bursts of heavy rain persist for many hours. This is because a thunderstorm usually consists of a cluster of cells. Each cell may be at a different stage in its life cycle, and new cells form and old cells dissipate continuously. A succession of many cells is thus responsible for prolonged periods of thunderstorm weather. Although a locality may be in the direct path of a distant, intense cell, the relatively brief life span of an individual cell means that the storm may dissipate long before reaching that locality.

The multicellular character of many thunderstorms complicates the motion of the weather system. A thunderstorm cluster may track at some angle to the paths of its constituent cells. For example, as illustrated schematically in Figure 13.6, a thunderstorm cell cluster tracks from west to east, whereas its five component cells head off toward the northeast. In this idealized case, new cells form in the southern sector of the storm, and old cells dissipate in the northern sector.

FIGURE 13.6
In this idealized situation, individual thunderstorm cells travel at about 20 degrees to the direction of movement of the multicellular thunderstorm. As they move, the individual cells go through their life cycle. [After K. A. Browning and F. H. Ludlam, "Radar Analysis of a Hailstorm," *Technical Note* No. 5, Meteorology Department, Imperial College, London]

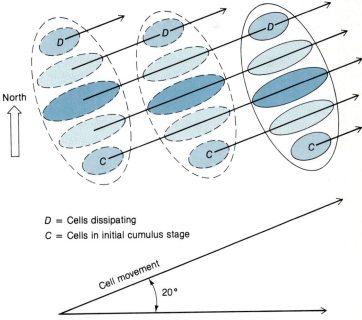

D = Cells dissipating
C = Cells in initial cumulus stage

Cell movement
20°
Storm movement

Thunderstorm Genesis

Most thunderstorms develop within masses of warm, humid maritime tropical (mT) air when the air mass is destabilized. Uplift is the key to destabilizing mT air. Because mT air is usually conditionally stable,* it becomes unstable and surges upward only when it is lifted to the condensation level. If the air is very humid to begin with, minimal lifting (that is, expansional cooling) is needed for the mT air to achieve saturation and for cumuliform clouds to develop. Uplift of air is initiated by (1) frontal activity, (2) orographic effects, (3) convergence of air at the surface, or (4) intense solar heating of the ground. Any one or a combination of these mechanisms may be sufficient to destabilize a mass of maritime tropical air. In addition, the potential instability of mT air is enhanced by cold air advection aloft or by warm air advection at the surface. Either one or both of these processes steepens the air temperature lapse rate, as shown in Figure 13.7, and reduces the ambient air stability.

Outside of the tropics, thunderstorms are classified as air mass thunderstorms, frontal thunderstorms, or mesoscale convective complexes, depending on the specific triggering mechanisms.

---

* Recall from Chapter 6 that conditionally stable air is *stable* for unsaturated air parcels and *unstable* for saturated air parcels.

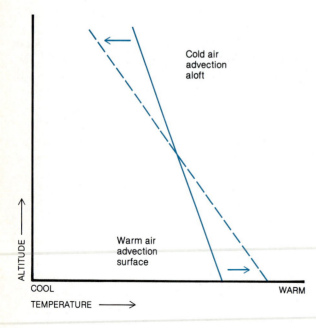

FIGURE 13.7
Cold air advection aloft, warm air advection near the surface, or a combination of the two steepens the vertical temperature profile and thereby destabilizes the air. This increases the likelihood of convection and development of thunderstorms.

AIR MASS AND FRONTAL TYPES

Both air mass and frontal thunderstorms are usually associated with the warm, humid air occupying the southeast sector of a mature midlatitude cyclone (Chapter 11). Convergence of surface winds and the consequent upward motion of air associated with the cyclone contribute to convection and thunderstorm genesis.

Air mass thunderstorms pop up almost randomly anywhere within a mass of warm, humid air. Usually, they are relatively weak systems. Because they are caused by convection currents driven by intense solar heating, most air mass thunderstorms develop in the afternoon, during the warmest hours of the day. A noteworthy exception to this general rule is the Missouri River Valley and adjacent portions of the upper Mississippi River Valley, where air mass thunderstorms are more frequent at night than during the day. Several explanations have been proposed for this nocturnal thunderstorm maximum. One idea centers on the possible role of a low-level jet stream of maritime tropical air that flows northward up the Mississippi River Valley. This jet stream strengthens at night and causes warm air advection at low levels that destabilizes the air and spurs convection.

As the name implies, **frontal thunderstorms** are associated with uplift along frontal surfaces. Most are triggered by the vigorous uplift of maritime tropical air along or ahead of a well-defined cold front, but sometimes thunderstorms break out in the overrunning zone ahead of a surface warm front if the advecting warm air is unstable.

Frontal thunderstorms are generally more energetic than air mass thunderstorms because the synoptic-scale circulation contributes more to their development. Usually, they parallel the surface front, and because frontal activity typically persists for many days, these thunderstorms may develop at any time of the day or night. In some instances, a line of thunderstorms forms perhaps 100 to 300 km (60 to 180 mi) in advance of a sharply defined cold front. These thunderstorms are aligned parallel to the cold front as a **squall line** and are often severe.

MESOSCALE CONVECTIVE COMPLEX

In recent years, study of satellite imagery has revealed a third mode of thunderstorm occurrence in addition to the air mass and frontal types. A **mesoscale convective complex (MCC)** is a nearly circular cluster of many interacting thunderstorms extending over an area that may be a thousand times larger than that of an individual air mass thunderstorm (Figure 13.8). In fact, it is not unusual for a single MCC to cover an area equal to that of the state of Iowa. These weather systems are primarily warm season (March through September) phenomena that generally develop at night and occur chiefly over the eastern two thirds of the United States, where more than 50 may be expected in a single season.

FIGURE 13.8
An enhanced infrared satellite image taken on 20 July 1977 at 3:30 A.M. E.D.T. show-
ing a mesoscale convective complex over Pennsylvania that resulted in a flash flood at
Johnstown. [Photograph courtesy of NOAA National Satellite, Data, and Information
Service]

A mesoscale convective complex is not associated with a front and usually
develops under conditions of weak synoptic-scale flow. New thunderstorms
develop continuously within an MCC so that the life expectancy of the system
is at least 6 hours and often 12 to 24 hours. The longevity of an MCC coupled
with its typically slow movement (15 to 30 km per hour) means that rainfall
is widespread and substantial. An MCC also has the potential of producing
severe weather. For example, NOAA reports that the 59 MCCs that occurred
between February and October of 1985 produced 20 tornadoes, 16 flash floods
and claimed 14 lives.

Geographical Distribution

Convective activity is dependent on intense solar heating of the Earth's surface
so that thunderstorms occur with the greatest frequency in the continental
interiors of tropical latitudes. The steamy Amazon Basin of Brazil, the Congo
Basin of equatorial Africa, and the islands of Indonesia experience the greatest
frequency of thunderstorms in the world. These localities are likely to experience thunderstorm activity on at least 100 days of the year. Because the
surfaces of large bodies of water do not warm as much as adjacent land
surfaces, thunderstorms are far less frequent over tropical oceans.

In some portions of subtropical and tropical latitudes, intense solar heating
combines with converging surface winds to trigger thunderstorm development.

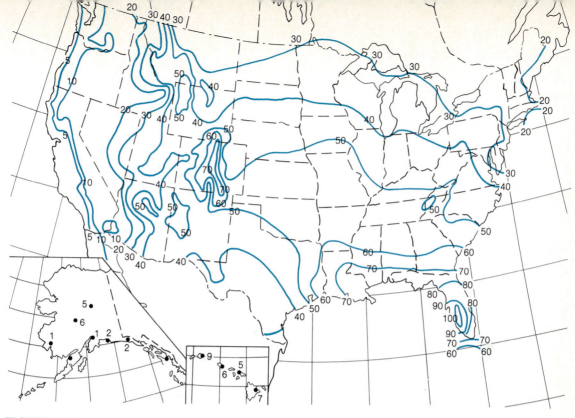

FIGURE 13.9

Thunderstorm frequency across the United States in average number of days per year. [NOAA data]

As we saw in Chapter 10, this combination characterizes the ITCZ, which is actually a discontinuous line of thunderstorms that surges north and south seasonally with the sun.

Central Florida is the most frequent site of thunderstorms in North America (Figure 13.9). In some interior localities of that state, thunderstorms can be expected on more than 90 days of the year. The Florida thunderstorm maximum is due to the sea breeze circulation that characterizes both the east and west coasts of the Florida peninsula. Convergence of the sea breezes over the interior (Figure 13.10) induces upward motion of maritime tropical air and development of cumulonimbus clouds.

Portions of the Rocky Mountain Front Range rank second only to Florida in frequency of thunderstorms in North America. On average, more than 60 thunderstorm days occur per year in a band from southeastern Wyoming southward through central Colorado and into north-central New Mexico. This high thunderstorm frequency is linked to differences in heating that result from variations in topography.

Recall from our discussion of mountain breezes in Chapter 12 that mountain slopes facing the sun absorb the direct rays of the sun and become relatively warm. The warm slopes, in turn, heat the air in immediate contact with the slopes. At the same time, air at the same altitude, but located to the

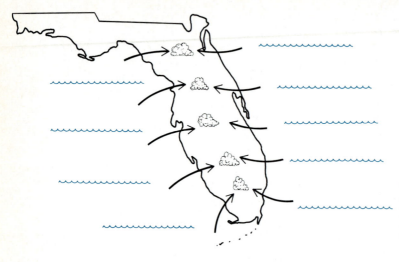

FIGURE 13.10
The relatively high frequency of thunderstorms in peninsular Florida is linked to the horizontal convergence of sea breezes blowing inland from both the Gulf and East coasts.

east of the mountains over the relatively flat terrain of the western Great Plains, is much cooler. If the cooler, denser air over the Great Plains sweeps westward, it forces the warm, light air over the mountain slopes to rise. Warm air rising over the mountain produces convective clouds that frequently billow upward to form thunderstorms. This process of thunderstorm development is enhanced whenever the synoptic-scale pressure pattern favors east winds over the western Great Plains.

The distinction is sometimes made between forced convection and free convection. In the case just described, topographic relief helps to force air upward to form convection currents. This situation is one of **forced convection.** On the other hand, convection that is triggered by intense solar heating of relatively flat terrain is known as **free convection.**

To this point we have considered those conditions conducive to deep convection and thunderstorm development. Under other circumstances, convection is inhibited and thunderstorms do not form. This situation usually occurs when air masses reside or travel over relatively cold surfaces and are thereby stabilized. Snow-covered ground cools and stabilizes the overlying air. Convection is thus suppressed, so thunderstorms are rare in middle and high latitudes during winter. Another reason for the rarity of winter thunderstorms is the relatively low water vapor concentration in cold air. As a rule, thunderstorms are unlikely when the dew point is below 13 °C (55 °F).

Thunderstorms are unusual over coastal areas that are situated downwind from relatively cold ocean waters. For example, thunderstorms are infrequent in coastal California, where prevailing winds are onshore and a shallow layer of maritime polar air often flows inland from off the relatively cold California current. Cool mP air at low levels stabilizes the air and suppresses deep convection and thunderstorm development.

Severe Thunderstorms

By definition, a **severe thunderstorm** is accompanied by locally damaging winds, frequent lightning, or large hail.* As a rule, the greater the altitude of a thunderstorm top, the more likely is the thunderstorm to produce severe weather. Why, then, do some thunderstorm cells surge to great altitudes and trigger severe weather, whereas others do not? One plausible explanation is that in severe thunderstorm cells, strong wind shear tilts the updraft, as shown in Figure 13.11. The tilt deflects much of the precipitation away from the updraft so that the precipitation falls alongside, rather than through, the updraft. Consequently, precipitation does not drag down the updraft, and the updraft continues to build the cell to greater altitudes.

In the United States and Canada, most severe thunderstorms break out over the Great Plains and are associated with mature synoptic-scale cyclones. Severe thunderstorm cells usually form a squall line within the cyclone's warm sector, ahead of and parallel to a fast-moving, well-defined cold front. The squall line appears as an ominous, rolling, and twisting mass of low, black

---

* The official National Weather Service criterion for designating a thunderstorm as "severe" includes any one or combination of the following: hailstones 1.9 cm (0.75 in.), or larger, in diameter; tornadoes or funnel clouds; surface winds stronger than 93 km (58 mi) per hour.

FIGURE 13.11
When a thunderstorm updraft is tilted, precipitation does not fall against (and thereby weaken) the updraft. Hence, the thunderstorm cell can billow upward to great altitudes and perhaps become severe. [From G. T. Trewartha and L. H. Horn, *An Introduction to Climate,* 5th ed. New York: McGraw-Hill, 1980, p. 180. Used with permission.]

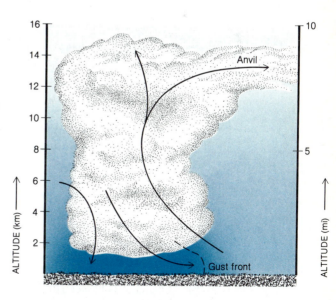

FIGURE 13.12
An ominous-appearing, rolling, twisting mass of low black clouds marks a squall line. [Photograph by S. I. Dutch]

clouds (Figure 13.12), often hundreds of kilometers long. The squall line moves very rapidly; speeds may approach 80 km (50 mi) per hour.

The midlatitude jet stream is an important ingredient in the development of severe thunderstorm cells—sometimes called **supercells.** The jet is responsible for producing the wind shear that tilts the updraft, thereby favoring great vertical development of the cell. In addition, the jet helps to produce a stratification of air that is especially favorable for formation of intense thunderstorm cells.

As we saw in Chapter 10, a midlatitude jet stream maximum induces both divergence and convergence of air aloft. Recall that diverging airflow triggers cyclone development under the left front quadrant of the jet stream maximum. Meanwhile, the air converges in the right front quadrant of the jet stream maximum, causing weak subsidence of air over the warm sector of the cyclone. The subsiding air is compressionally warmed, and its relative humidity decreases. The subsiding air is prevented from reaching the surface, however, by a shallow layer of maritime tropical air at the surface. The mT air surges rapidly northward, out of the Gulf of Mexico, as a tongue perhaps 3000 m (10,000 ft) deep, and is often described as a **low-level jet stream.** The warm, humid air is pumped northward by the circulation on the western flank of the Bermuda–Azores subtropical anticyclone.

This synoptic situation is shown schematically in Figure 13.13. Note how the midlatitude (polar front) jet and the low-level mT jet cross to the southeast of the cyclone center. Note also that the air ahead of the cold front and to the west of the warm, humid tongue is usually quite dry. For this reason, the western boundary of the mT air mass is called the **dry line.** The subtropical jet stream (Chapter 10) is sometimes also present at high levels in the troposphere, blowing from west to east over the warm sector of the cyclone. In this

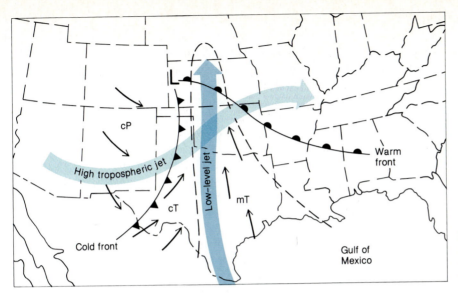

FIGURE 13.13
Synoptic situation most favorable for the development of severe thunderstorms.

case, intense squall lines are likely to form between the two upper-level jet streams.

As a consequence of compressional warming, the air subsiding from aloft becomes warmer than the underlying layer of mT air. A zone of transition develops between the two air masses as a temperature inversion (Figure 13.14). As we learned in Chapter 6, a temperature inversion is extremely

FIGURE 13.14
Vertical temperature profile that is most favorable for the sudden eruption of severe thunderstorm cells. An elevated temperature inversion separates subsiding dry air aloft from warm, humid air at the surface.

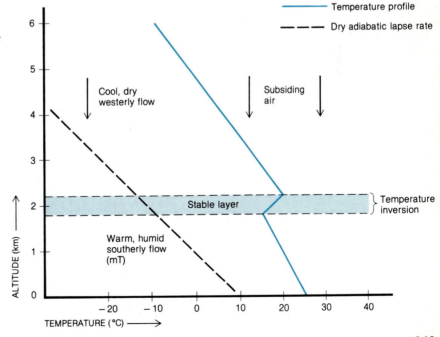

349

stable, so the two air masses do not mix and convection is confined to the surface mT air layer. As long as this stratification persists, the contrast between the two air masses mounts. The subsiding air becomes drier, and the underlying air becomes more humid. The potential for severe weather continues to grow, and all that is needed is a trigger that will force the convection currents to penetrate the elevated temperature inversion. The needed upward impetus may be supplied by the intense solar heating of midafternoon or by the lifting of air associated with the approaching cold front. By either mechanism, convection currents eventually break through the temperature inversion, and cumulus clouds billow upward at speeds that may exceed 100 km (60 mi) per hour. Such explosive updrafts can even penetrate the tropopause and surge into the lower stratosphere.

Occasionally, ominous pouchlike mammatus clouds appear on the underside of the anvil top of a cumulonimbus cloud (Plate 20). Contrary to popular belief, **mammatus clouds** do not always indicate a severe thunderstorm. In fact, in the mountainous western United States, they often are associated with relatively weak convective showers. Their unusual appearance is attributed to blobs of cold cloudy air that descend from the anvil into the unsaturated clear air beneath the anvil. Ice crystals at the margin of a blob vaporize (sublimate), causing further cooling of the blob. This cooling is sufficient to offset the compressional warming of the descending air so that the blob sinks further into the dry air as a curious-looking protuberance. Ultimately, the size of the protuberance is limited by the increasing rate of vaporization of the constituent ice crystals.

Thunderstorm Hazards

Thunderstorm hazards include lightning, downbursts, torrential rains, hail, and the spawning of tornadoes. Because tornadoes are an especially severe hazard, they are considered separately in the next chapter.

LIGHTNING

By definition, a convective rain or snow shower is a thunderstorm if accompanied by lightning. Lightning is a weather phenomenon that is directly hazardous to human life, killing perhaps a hundred people in the United States and Canada each year. This death toll may seem surprisingly high, perhaps because fatal lightning strokes are typically isolated events that do not make headlines. A single, fatal stroke of lightning is not as newsworthy as a single disastrous tornado that takes many lives, but so many lightning bolts occur daily through much of North America that injuries and fatalities quickly add up. For some tips on lightning safety, refer to the Special Topic, "Lightning Safety."

Lightning is feared not only because it can kill and injure people, but also because it ignites forest and brush fires. In the Rocky Mountain region, for example, lightning is the most common cause of forest fires, starting more than 9000 fires per year. One might think that the heavy rains associated with thunderstorms would quickly quench a lightning-induced fire. In the western basins, however, the base of a cumulonimbus cloud is usually so far above the ground, and the air below the cloud so dry, that much of the rain evaporates before reaching the fire.

Special Topic

Lightning Safety

Lightning kills and injures, and a blinding flash of lightning followed by a crash of thunder frightens many people. In the United States between 1960 and 1985, an average of 96 people died each year as a result of lightning, and many times this number were injured. Although the danger of lightning cannot be ignored, some simple safeguards will minimize the risk of injury when a thunderstorm threatens.

The odds of being struck and killed by lightning are actually quite slim, about 350,000:1, and these odds improve when precautions are taken. By comparison, the odds of being struck and killed by an auto are 50 times greater. Although no place is absolutely safe, the risk can be minimized in even the most lightning-prone area of the nation, south and central Florida, where each square kilometer of land is struck by an estimated 10 lightning bolts yearly.

When a thunderstorm approaches, the best action is to seek shelter in a house or other building, avoiding contact with conductors of electricity that provide pathways for lightning. These include pipes (don't shower), stoves (don't cook), and wires (don't use the telephone). Electrical appliances pose no hazard if properly grounded, but why tempt fate by using them?

Some confusion surrounds the safety of motor vehicles during a lightning storm. A metal car or truck is a good shelter. Cars with cloth convertible tops and the backs of pickup trucks, on the other hand, are not. In Texas, in 1979, people riding in the back of a pickup truck were struck and killed by lightning whereas passengers in the cab of the truck were unhurt.

If a building or an auto is not accessible when a thunderstorm approaches, find shelter under a cliff, in a cave, or in a low area, such as a ravine, a valley, or even a roadside ditch. Avoid (1) tall, isolated objects, such as telephone poles and flagpoles, (2) metallic objects, such as wire fencing, rails, wire clotheslines, bicycles, and golf clubs, (3) high areas, such as hilltops and rooftops, and (4) bodies of water such as swimming pools and lakes. Stay off mowers and tractors. Individual trees in open spaces are hazardous, but a thick grove of small trees may offer safe haven. A group of people in the open should spread out, keeping several meters apart.

If your hair stands on end, lightning may be about to strike. In this unlikely event, immediately drop to your knees and, placing your hands on your knees, bend forward. In this way, you make your body the smallest possible target. One should not lie flat on the ground.

Contrary to popular belief, lightning can strike the same place more than once. New York City's Empire State Building typically is struck more than 20 times a year and on one occasion was hit 15 times in only 15 minutes. Such sites should be avoided during a thunderstorm.

Fortunately, two of every three persons "struck" by lightning recover fully. Most survivors are jolted by a nearby lightning bolt and not hit directly. Immediate mouth-to-mouth resuscitation or cardiopulmonary resuscitation may revive victims who are not breathing. Those who appear merely stunned may require treatment for burns or shock. Victims of lightning bolts carry no electrical charge and can be handled safely.

Lightning is very costly to electrical utilities, each year causing tens of millions of dollars in damage to equipment (power lines, transformers) and service restoration expenses. A new tool in utilities' efforts to cope with the lightning hazard is the **lightning detection network (LDN),** a system that provides real-time information on the location and severity of lightning strokes. Each monitoring station in a network consists of direction-finding equipment that sense electromagnetic fields associated with cloud-to-ground lightning. A computer superimposes the locations of lightning strokes on a map of the region. A sample lightning stroke density map is shown in Figure 13.15.

Currently, an LDN consisting of 26 stations monitors the East Coast states as far west as a line from Erie, Pennsylvania, to Mobile, Alabama. Network stations send data to a processing center at the State University of New York at Albany, where lightning locations are computed with an accuracy of a few kilometers. An electrical utility that subscribes to the LDN service accesses lightning data through a computer link. Such information enables the utility to mobilize repair crews more efficiently and speeds up restoration of disrupted

FIGURE 13.15
A sample terminal display of lightning strikes over Wisconsin and Lake Michigan obtained via a lightning detection network. [Photograph courtesy of R*Scan Corporation, Minneapolis, MN]

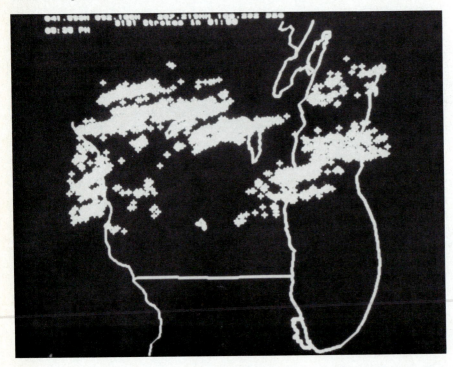

service. As of this writing, a national lightning detection network is nearing completion.

What is lightning and why is it so dangerous? **Lightning** is a brilliant flash of light produced by an electrical discharge of about 100 million volts. The potential for an electrical discharge exists whenever a charge difference develops between two objects. A normally neutral object becomes negatively charged when it gains electrons (negatively charged subatomic particles) and positively charged when it loses electrons. That is, the object is ionized. When differences in electrical charge develop within a cloud or between a cloud and the ground, the stage is set for lightning.

On a clear day, the Earth's surface is negatively charged, and the upper troposphere is positively charged. This charge distribution changes as a cumulonimbus cloud develops. Within the cloud, charges separate so that the upper portion and a much smaller region near the cloud base become positively charged. In between, a pancake-shaped zone of negative charge forms that is a few hundred meters thick and several kilometers in diameter. At the same time, the developing cumulonimbus induces a positive charge on the ground directly under the cloud.

Air is a very good electrical insulator, and so, as a thunderstorm forms and electrical charges separate and build, a tremendous potential soon develops for an electrical discharge. When the thunderstorm enters its mature stage, the electrical resistance of air breaks down and lightning flows, thereby neutralizing the electrical charges. Lightning may forge a path between oppositely charged regions of a cloud, or between clouds, or between a cloud and the ground.

The cause of charge separation within cumulonimbus clouds is not well understood, but recent field measurements and laboratory simulations offer a promising explanation focused on the role of graupel and the convective circulation within thunderstorms. **Graupel** (German for "soft hail") consists of millimeter- to centimeter-size ice pellets formed when supercooled water droplets collide and freeze on impact. Within the cloud, as falling graupel strikes smaller ice crystals in its path, opposite charges develop on the graupel and the ice crystals. For collisions that take place at temperatures below about $-15\,°C$ (5 °F), graupel becomes negatively charged and ice crystals acquire a positive charge. Vigorous updrafts separate the particles, carrying the smaller positively charged ice crystals to the upper portion of the cloud, whereas the larger negatively charged graupel concentrate mostly in the lower portion of the cloud. This mechanism explains the positive charge of the upper cloud region and the negative charge of the pancake-shaped zone.

What accounts for the positive charge near the cloud base? Typically, temperatures near the cloud base are above $-15\,°C$ (5 °F). At those temperatures collisions between graupel and ice crystals induce a positive charge on graupel and negative charge on ice crystals. The heavier graupel accumulate near the cloud base, giving that region a positive charge.

We are most concerned about lightning discharges between a cloud and the ground (Figure 13.16) because this path poses the greatest hazard. (Nonetheless, these discharges represent only about 20 percent of all lightning bolts.) Using high-speed photography, scientists have determined that a lightning flash consists of a regular sequence of events. Initially, streams of electrons surge from the cloud base and toward the ground in discrete steps, each of which is about 20 to 100 m (60 to 300 ft) long. These so-called **stepped leaders** describe a branching path and produce a narrow ionized channel. When a branch of the stepped leaders comes within about 100 m (300 ft) of the ground, it is met by a positively charged **return stroke** from the ground. The return stroke follows a path of least resistance and often emanates from tall, pointed structures such as metal flagpoles or towers. Now an ionized channel only a few centimeters in diameter links the Earth's surface with the cloud. Electrons flow, neutralization occurs, and the channel is illuminated.

Following this initial electrical discharge, subsequent surges of electrons from the cloud, called **dart leaders,** follow the same conducting path. Each dart leader is met by a return stroke (from the ground), and the conducting path is again illuminated. Typically, a single lightning discharge consists of two to four dart leaders plus return strokes. Sometimes, a dart leader is met by a return stroke that forges a new conductive path from the ground. The result is a forked lightning bolt that strikes the ground in more than one place.

The sequence just described is, by far, the most common occurrence of cloud-to-ground lightning. In less than 10 percent of cases, a positively charged leader emanates from the cloud and initiates a lightning discharge. Much more rarely, positive or negative stepped leaders propagate upward from the ground and meet a return stroke surging downward from a cloud. This ground-to-cloud lightning usually is initiated from mountaintops or tall structures such as antenna towers.

Electricity flows at the astonishing rate of nearly 50,000 km (31,000 mi) per second; the entire lightning sequence therefore occurs in less than two tenths of a second. The human eye has difficulty separating the individual flashes of light that constitute a single lightning bolt, so that we perceive a lightning flash as a flickering light. **Sheet lightning** consists of bright flashes across the sky and indicates cloud-to-cloud discharges. **Heat lightning** is simply light reflected by clouds from distant thunderstorms that occur beyond the horizon.

Where there is lightning, there is **thunder,** although sometimes we may see distant lightning and not hear the thunder. Lightning heats the air along the narrow conducting path to temperatures that may exceed 25,000 °C (45,000 °F). For this reason, people can be burned severely by a lightning bolt that strikes nearby. Such intense heating expands the air violently and initiates a sound wave that we hear as thunder.

Because light travels about a million times faster than sound, we see the lightning almost instantaneously, but we hear the thunder later. The closer we are to the thunderstorm cell, the shorter is the time interval between the lightning flash and thunder. As a rule, thunder takes about 3 seconds to travel 1 km (and 5 seconds to travel 1 mi). Thus, if you must wait 9 seconds between lightning flash and thunderclap, the thunderstorm cell is about 3 km (1.8 mi) away.

DOWNBURSTS

Severe, and sometimes not so severe, thunderstorms can produce a **down-burst,** a strong downdraft that, upon reaching the surface, diverges horizontally as an outburst of destructive winds. We can simulate this flow by aiming a garden hose nozzle downward so that the stream of water strikes the ground and bursts outward. Downbursts blow down trees, flatten crops, and wreck buildings in a starburst pattern* of destruction (Figure 13.17). On the basis of size, a downburst is classified as either a macroburst or a microburst.

A **macroburst** cuts a swath of destruction that is longer than 4.0 km (2.5 mi) and features surface winds that may top 210 km (130 mi) per hour. The leading edge of a macroburst may be marked by a gust front, and the system lasts up to 30 minutes. On 4 July 1977, an estimated 25 macrobursts struck several northern Wisconsin counties, felling trees and farm buildings over a 27 km (17 mi) by 267 km (166 mi) area. Based on damage observations, surface winds probably peaked in the range of 180 to 250 km (110 to 155 mi) per hour.

A **microburst** is a smaller and shorter-lived system than a macroburst. By convention, its path of destruction is 4 km (2.5 mi) or less, top surface

---

* A starburst pattern of destruction may be elongated in one direction, depending on the angle at which the downburst strikes the ground. Often, downburst damage is attributed to unseen tornadoes.

FIGURE 13.17
Damage to a forest caused by a small downburst called a microburst. [Photograph courtesy of T. Theodore Fujita, Professor of Meteorology, The University of Chicago]

winds may be as high as 270 km (168 mi) per hour, and its life expectancy is less than 10 minutes. Microbursts are particularly dangerous at airports where their small size defies detection, and they can play havoc with the aerodynamics of aircraft. Microbursts trigger **wind shear,** an abrupt change in wind speed or direction with distance, that can disturb the lift forces acting on an aircraft and may cause an abrupt change in altitude. For an aircraft that flies into a microburst during takeoff or landing, the result may be tragic.

The existence of downbursts was first proposed by T. Theodore Fujita of the University of Chicago. Fujita, who has conducted extensive field investigations of the phenomenon, argues that microbursts probably contributed to two commercial aircraft accidents in 1975. Apparently, in both cases, the pilots unwittingly flew their jet planes through the center of a microburst. One pilot did so on takeoff and the other while attempting to land. Both planes abruptly lost altitude and crashed. In retrospect, perhaps 27 or more civil airline accidents since 1964 have stemmed from encounters with microbursts.

Fujita and his colleagues conducted the first intensive field study of downbursts during the spring and summer of 1978. The project was dubbed **NIM-ROD,** for Northern Illinois Meteorological Research on Downbursts, and focused on the western suburbs of Chicago. During the 42-day observational period, an array of weather instruments and radar units detected 50 microbursts. Another field study of microburst activity was conducted during the summer of 1982 over a 1600-km^2 (615-mi^2) area near Denver's Stapleton International Airport. Using experimental aircraft and a dense grid of meteorological sensors, this **Joint Airport Weather Studies (JAWS)** project detected 186 microbursts in only 86 observational days. Data from both NIMROD and JAWS indicate that microbursts may occur with or without rain. Where no rain falls, the microburst usually occurs under a curtain of virga.

One of the most important findings of JAWS is the apparent inadequacy of the **Low-Level Wind Shear Alert System (LLWSAS),** an array of ground-level anemometers currently operated by the Federal Aviation Administration at more than 100 airports. Many of Denver's microbursts were either too small or too far off the ground to be detected by LLWSAS. Indeed, microbursts may be so small as to affect only a portion of one runway. The most promising microburst detection system is an airport-based Doppler radar. We have much more to say about Doppler radar in Chapter 14.

FLASH FLOODS

A stationary or slow-moving intense thunderstorm can produce torrential rains that trigger dangerous **flash flooding,** a sudden overflow of rivers or other drainageways. Typically, a thunderstorm is stationary or slow moving because (1) it is imbedded in weak steering winds aloft or (2) the system is generated by a persistent flow of humid air up a mountain slope. The second process was implicated in the disastrous flash flood that claimed 130 lives (mostly campers) in the Big Thompson Canyon of Colorado on 31 July 1976.

Prolonged heavy rains (more than 7.5 mm, or 0.3 in., per hour) can greatly exceed the infiltration capacity of the ground. Simply put, the ground cannot absorb all the rainwater. Excess water runs off to creeks, streams, rivers, or sewers, or collects in other low-lying areas. If a drainage system cannot accommodate the sudden input of huge quantities of water, flash flooding is the consequence.

Flash flooding is especially hazardous in mountainous terrain where steep slopes channel runoff into narow stream and river valleys. The water level of streams and rivers may rise so rapidly that campers and other visitors are caught by surprise and trapped. This is what happened at Big Thompson Canyon. Indeed, the post-World War II upswing in flood fatalities in the United States is largely due to more people visiting remote areas prone to flash flooding. During the 1970s, floods took an average of 200 lives annually, twice the flood fatalities of the 1960s and three times those of the 1940s. In the mid-1980s, R. E. Hallgren, director of the United States National Weather Service, estimated that, in an average year, floods take the lives of 200 Americans and cause up to $3 billion in property damage.

Because of their design and composition, urban areas are also prone to flash floods during intense downpours. Concrete and asphalt render the surfaces of a city virtually impervious to water, so elaborate sewer systems are installed to transport runoff to nearby natural drainageways. Sewer systems have a limited capacity for water, however, and may be unable to handle the excess water produced during a torrential rainfall. Water backs up and collects under viaducts and in other low-lying areas. Sometimes water levels rise so fast in these areas that motorists are trapped in their vehicles (Figure 13.18). For example, on the evening of 1 August 1985, a slow-moving severe thunderstorm drenched Cheyenne, Wyoming, with 15.4 cm (6.06 in.) of rain in

FIGURE 13.18
Motor vehicles and occupants may be trapped by the sudden rise of water accompanying flash floods such as this one in Montgomery County, Maryland, on 15 July 1975. [NOAA photograph]

Table 13.1
Some Flash Flood Safety Tips[a]

1. Avoid driving on flooded roadways or bridges. Flood waters only a half meter deep can carry away most autos. Furthermore, flooded roads may be undermined.
2. If your vehicle stalls in high water, abandon it and seek high ground.
3. On foot, never try to cross a stream if the water level is above your knees.
4. Keep children away from drainage ditches and culverts.
5. Find out the elevation of your property and the flood history of the area. If there is a risk of flooding, plan an evacuation route.
6. Be especially cautious when camping in remote areas. Avoid camping in mountain valleys or near dry stream beds. Take a battery-operated radio along to monitor changing weather conditions.

[a] Based on NOAA recommendations.

less than 4 hours. Flash flooding filled city streets with up to 2 m (6 ft) of water, 12 people lost their lives, and estimates of property damage were in excess of $65 million.

Severe thunderstorms are not the only culprits in triggering flash flooding. The breaching of a dam or levee, or the sudden release of water during breakup of a river ice jam, can also cause an abrupt rise in water level. In any event, when in flood-prone areas, we are well advised to take the precautions listed in Table 13.1.

HAIL

Hail is precipitation in the form of balls or lumps of ice, usually called **hailstones.** Hail falls from intense thunderstorm cells that are characterized by strong updrafts, great vertical development, and an abundant supply of supercooled water droplets. Hailstones range from pea size to the size of a golfball or even larger. In the United States, the largest hailstone on record was collected at Coffeyville, Kansas, on 3 September 1970. It weighed 758 g (1.67 lb) and measured 44.5 cm (17.5 in.) in circumference and 14 cm (5.6 in.) in diameter—about the size of a softball. Canada's heaviest authenticated hailstone weighed 290 g (10.23 oz) and measured 10.2 cm (4 in.) in diameter. It fell at Cedoux, Saskatchewan, on 27 August 1973.

A hailstone forms when an ice pellet is transported through portions of a cumulonimbus cloud in which there are variable concentrations of super-cooled water droplets. The ice pellet may descend slowly through the entire cloud, or it may follow a more complex pattern of ascent and descent as it is caught alternately in updrafts and downdrafts. In the process, the ice pellet grows by accretion (addition) of freezing water droplets. Eventually, when it becomes so large and heavy that updrafts can no longer support it, the ice particle falls out of the cloud base. If the ice does not melt completely during its journey through the above-freezing air beneath the cloud, it reaches the ground as hail.

When an ice pellet enters portions of the cloud containing a relatively high concentration of supercooled water droplets, water collects on the ice pellet as a liquid film, which freezes slowly to form a transparent layer, or **glaze.** When the ice pellet travels through portions of the cloud where the concentration of water droplets is relatively low, droplets freeze immediately on contact with the ice pellet. As the droplets freeze, many tiny air bubbles are trapped within the ice, producing an opaque whitish layer of granular ice, or **rime.** The result is alternating lamina of clear (glaze) and opaque (rime) ice, which, in cross section, resembles the internal structure of an onion. This concentric layering is evident when a hailstone is sliced open (refer back to Figure 8.6).

The accumulation of hailstones over the landscape often exhibits a long, narrow pattern known as a **hailstreak.** A typical hailstreak may be 2 km (1.2 mi) wide and 10 km (6.2 mi) long, and a single large thunderstorm may

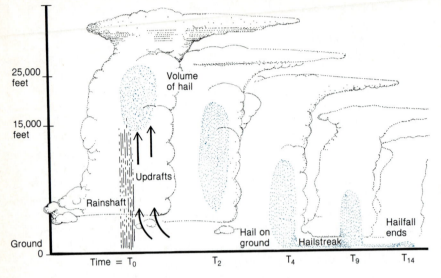

FIGURE 13.19

A model of hailstreak development. [From S. A. Changnon and J. L. Ivens, *Hail in Illinois*. Illinois State Water Survey, 1987, p. 5]

produce several hailstreaks. From a study of numerous hailstorms and hailstreaks, scientists at the Illinois State Water Survey devised a conceptual model to explain hailstreak development. As shown in Figure 13.19, a discrete volume of hail forms within the upper portion of a cumulonimbus cloud. After several minutes of hail formation, the updraft weakens, allowing the hail volume to begin descending; about 4 minutes later, the first hailstones reach the ground. In the ensuing 10 minutes, as the thunderstorm moves along, the entire hail volume is released along the ground as a hailstreak.

On rare occasions, the fall of hail is so great that snowplows must be called out to clear highways. This happened in Milwaukee, Wisconsin, on 4 September 1988, when pea-size hail formed drifts to 46 cm (18 in.) in several north-side neighborhoods. On 6 August 1980 at Orient, Iowa, drifts of hail were reported to be 1.8 m (6 ft) deep. One of the nation's most destructive hailstorms struck the western suburbs of Denver on the afternoon of 13 June 1984. Hailstones as large as golfballs fell for up to 1.5 hours, producing accumulations of 25 cm (10 in.), with drifts greater than a meter. Damage exceeded $350 million.

Hailstones may be large enough to smash windows and dent automobiles, but the most costly damage is to crops. Hail usually falls during the growing season, and, in only a matter of minutes, large hail can wipe out the fruits of a farmer's year of labor. Traditionally, farmers cope with the hail hazard by purchasing insurance. Illinois farmers, for example, lead the United States in crop-hail insurance, purchasing an average annual liability coverage exceeding $600 million. It is not surprising, then, that efforts have been made to suppress or prevent hail. A brief historical sketch of these activities is presented in the Special Topic "Hail Suppression."

Hail Suppression*

Efforts to suppress hail have deep historical roots. Indeed, in fourteenth-century Europe, church bells were rung and cannons fired in the belief that the attendant noise would somehow ward off hail. A period of particularly intense hail suppression activity took place in the grape-growing regions of Austria, France, and Italy during the late nineteenth century. M. Albert Stiger, a wine grower and burgomaster of Windisch-Feistritz, Austria, designed and built a special funnel-shaped hail suppression cannon (Figure 1). Stiger believed that smoke particles in the cannon fire would inhibit hailstone development. Amazingly, in experimental firings in 1896 and 1897 at Windisch-Feistritz, Stiger reported no hail, although severe hail damage occurred in neighboring areas.

Word of Stiger's apparent success spread throughout the vineyard regions of Europe, and hail cannons soon became commonplace. There were so many cannons that the accidental shooting of people became a problem in some localities. After Stiger's much-heralded success was not duplicated elsewhere, however, interest in hail cannons waned rapidly, and by 1905 this early attempt at weather modification ended.

FIGURE 1

A hail suppression cannon popular in the grape-growing regions of Austria, France, and Italy during the late nineteenth century. [Photograph courtesy of S. A. Changnon and J. L. Ivens, Illinois State Water Survey; © American Meteorological Society]

FIGURE 2

Large hailstones devastated this cornfield in only a matter of minutes. [NCAR/NSF photograph]

The modern era of hail suppression experimentation began after World War II. Although founded on a much better, albeit not yet complete, understanding of cloud physics, the new techniques shared some similarities with earlier efforts. For example, until the practice was outlawed in the early 1970s, Italian farmers regularly fired explosive rockets into threatening clouds in an attempt to shatter developing hailstones. In the Soviet Union today, scientists fire silver iodide crystals, a cloud-seeding agent (see Chapter 8), into thunderclouds. They theorize that silver iodide crystals stimulate the formation of large numbers of small hailstones, which will melt long before they reach the ground, instead of the normal development of small numbers of larger hailstones, which could devastate crops.

In the United States, annual agricultural losses due to hail exceed $700 million (Figure 2). It is not surprising, then, that scientists in this country set out to test the Soviet hypothesis of hail suppression and to learn more about hail-producing thunderstorms. To these ends the *National Hail Research Experiment* was launched over northeastern Colorado in 1972. Although much was learned about hailstorms, three years of seeding potential hailstorms failed to confirm the Soviet hypothesis. Some scientists questioned the experimental design and whether it was a viable test of the Soviet technique. In an interesting parallel with events in nineteenth-century Europe, declining public confidence in the effectiveness of modern hail suppression methods brought an end to federal funding of hail suppression research in 1979.

* This discussion is based on S. A. Changnon, Jr., and J. L. Ivens. "History Repeated: The Forgotten Hail Cannons of Europe." *Bulletin of the American Meteorological Society* 62 (1981):368–375.

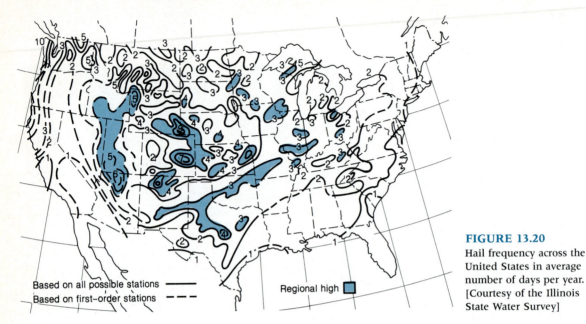

FIGURE 13.20
Hail frequency across the United States in average number of days per year. [Courtesy of the Illinois State Water Survey]

Based on all possible stations ——
Based on first-order stations – – –
Regional high ▣

Perhaps surprisingly, hail frequency is not necessarily related to thunderstorm frequency. Although Florida experiences the greatest frequency of thunderstorms in the United States, hail is very unusual in that state. In North America, hail is most likely on the High Plains just east of the Rockies, where it can be expected to fall from about 10 percent of all thunderstorms. Figure 13.20 is a map of hail frequency in the United States.

Conclusions

Thunderstorms are the products of convection currents that surge to great altitudes within the troposphere. As such, thunderstorms channel excess heat at the Earth's surface into the atmosphere. Hence, thunderstorms are most frequent in the warmest regions of the Earth.

Thunderstorms may be accompanied by damaging lightning, winds, torrential rains, and hail. These, however, are not the only hazardous progeny of thunderstorms. Although less than 1 percent of all thunderstorms produce tornadoes, in some areas of the country, the possibility of a tornado is the principal reason why people fear thunderstorms. We discuss tornadoes in the next chapter. In addition, we also examine the hurricane, the most violent of tropical storms, which often begins as a cluster of thunderstorms over tropical oceans.

Summary Statements

A thunderstorm is a mesoscale weather system produced by strong convection currents that surge high into the troposphere.

The life cycle of a thunderstorm cell consists of a three-stage sequence: cumulus, mature, and dissipating. The storm's maximum intensity is reached during the mature stage.

A thunderstorm usually consists of a cluster of cells, each of which may be at a different stage of its life cycle. A cluster of thunderstorm cells may track in a different direction than the individual cells.

Thunderstorms usually develop in maritime tropical air as a consequence of uplift caused by one or more of the following: (1) frontal activity, (2) orographic effects, (3) surface convergence, or (4) intense solar heating of the Earth's surface. They are classified as air mass thunderstorms, frontal thunderstorms, or mesoscale convective complexes. Frontal thunderstorms are usually more energetic than air mass thunderstorms.

Severe thunderstorm cells typically form a squall line ahead of a fast-moving, well-defined cold front associated with a mature midlatitude cyclone. The midlatitude jet stream causes dry air to subside over a surface layer of maritime tropical air. This produces a layering of air that can lead to explosive convection and the development of severe thunderstorms.

Lightning is light produced by the discharge of electricity within a cloud, between clouds, or between a cloud and the ground. Cloud-to-ground lightning can be resolved into a very rapid sequence of events involving stepped leaders, return strokes, and dart leaders.

Some thunderstorm cells produce downbursts—very intense downdrafts that spread out at or near the surface as potentially destructive winds. Based on size, downbursts are categorized as either macrobursts or microbursts.

Flash flooding is a special hazard in mountainous terrain, where steep slopes channel excess runoff into narrow stream and river valleys, and in urban areas, where the impervious surface causes water to collect in low-lying areas.

Hail develops in intense thunderstorm cells characterized by strong updrafts, great vertical development, and an abundant supply of supercooled water droplets.

Key Words

thunderstorm
convective
    condensation level
    (CCL)
cumulus congestus
cumulonimbus
cumulus stage
updraft
mature stage
downdraft
entrainment
gust front
roll cloud
shelf cloud

dissipating stage
air mass
    thunderstorms
frontal thunderstorms
squall line
mesoscale convective
    complex (MCC)
forced convection
free convection
severe thunderstorm
supercells
low-level jet stream
dry line
mammatus clouds

lightning detection
    network (LDN)
lightning
graupel
stepped leaders
return stroke
dart leaders
sheet lightning
heat lightning
thunder
downburst
macroburst
microburst

wind shear
NIMROD
Joint Airport Weather
    Studies (JAWS)
Low-Level Wind
    Shear Alert System
    (LLWSAS)
flash flooding
hail
hailstones
glaze
rime
hailstreak

Review Questions

1. Describe the characteristics of each stage in the life cycle of a thunderstorm cell.

2. What is a gust front? Explain how a gust front can spur development of secondary thunderstorm cells.

3. What is the significance of the relatively short life span of thunderstorms for local weather forecasting?

4. Distinguish between air mass thunderstorms and frontal thunderstorms. Which is likely to be more intense and why?

5. Why do air mass thunderstorms usually develop in the afternoon, during the warmest hours of the day?

6. What is a squall line?

7. Describe how thunderstorms develop over the interior of the Florida peninsula.

8. Why are thunderstorms most frequent in the continental interiors of tropical latitudes?

9. Under what conditions is a thunderstorm considered to be severe?

10. In what sector of a mature cyclone are thunderstorms most likely to develop? Explain why.

11. Describe the synoptic situation in middle latitudes that is most favorable for the formation of severe thunderstorms.

12. What causes lightning? Why is it dangerous?

13. What causes thunder?

14. What is a microburst, and how might it pose a hazard for aircraft?

15. Speculate on why the damage produced by a downburst is sometimes mistaken for tornado damage.

16. How might the flash flood hazard vary with the season in middle and high latitudes?

17. Why is flash flooding a particular hazard in mountainous terrain? In urban areas?

18. What type of thunderstorm may produce hail?

19. Explain the internal concentric layering of hailstones.

20. Where in the United States is hail most frequent?

21. Today, little research is directed at hail suppression. Explain why.

Points to Ponder

1. Compare and contrast a sea breeze front with a gust front.

2. What causes the anvil shape of a thunderstorm top?

3. Thunderstorms may track at an angle to the paths of their constituent cells. What is the significance of this behavior for thunderstorm forecasting?

4. Maritime tropical air is conditionally stable. How is this significant for convection and thunderstorm development?

5. Explain the diurnal and seasonal variations in thunderstorm occurrence. Also explain why thunderstorms are rare along the coast of southern California and over snow-covered terrain.

6. Identify the roles of the midlatitude jet stream and the low-level mT jet stream in the formation of severe thunderstorms.

7. Explain the formation of mammatus clouds.

Projects

1. Review the available climatic summary data for your area for the frequency of (a) thunderstorms, (b) hail, and (c) flooding.

2. If you live in a hail-prone area, you may wish to construct a hail collection pad. Merely wrap aluminum foil around a Styrofoam black (1 m² by 5 cm thick) and set outside during thunderstorm weather. Hailstones will break through the aluminum foil, enabling you to measure the range in diameters of the hailstones.

Selected Readings

Blanchard, D. O. "Mesoscale Convective Patterns of the Southern High Plains." *Bulletin of the American Meteorological Society* 71 (1990):994–1005. Describes the classification of mesoscale convective systems into three basic patterns.

Changnon, S. A., Jr., and J. L. Ivens. "History Repeated: The Forgotten Hail Cannons of Europe." *Bulletin of the American Meteorological Society* 62 (1981):368–375. Provides a fascinating discussion of the parallels between modern and early weather modification efforts.

Few, A. A. "Thunder." *Scientific American 233*, No. 1 (1975):80–90. Discusses the relationship between thunder and lightning flash.

Fujita, T. T. *The Downburst*. Chicago: Department of the Geophysical Sciences, University of Chicago, 1985. 122 pp. Presents a well-illustrated account of macrobursts and microbursts; features many case studies.

Fujita, T. T., and F. Caracena. "An Analysis of Three Weather-Related Aircraft Accidents." *Bulletin of the American Meteorological Society* 58 (1977):1164–1181. Focuses on the possible role of microbursts in aircraft accidents.

Maddox, R. A., and J. M. Fritsch. "A New Understanding of Thunderstorms: The Mesoscale Convective Complex." *Weatherwise* 37 (1984):128–135. Discusses the recent discovery of large clusters of interacting thunderstorms.

Schulz, L. W. "The Central Kansas Flash Floods of June 1981." *Bulletin of the American Meteorological Society* 65 (1984):228–234. Describes the synoptic weather conditions that led to two flash floods only eight days apart.

Williams, E. R. "The Electrification of Thunderstorms." *Scientific American* 259 (1988):88–99. Provides a well-illustrated discussion of the causes of charge separation within thunderstorms.

Wheeling, the careening
  winds arrive with lariats
  and tambourines of rain.
  Torn-to-pieces, mud-dark
  flounces of Caribbean
  cumulus keep passing,
  keep passing. By afternoon
  rinsed transparencies begin
  to open overhead.
  Mediterranean
  windowpanes of clearness.

Amy Clampitt
THE KINGFISHER, ''THE EDGE OF
  THE HURRICANE''

14

Tornadoes and Hurricanes

Visible satellite image of Hurricane Hugo as its eye slammed ashore near Charleston, South Carolina, about midnight on 21 September 1989. This violent storm took 21 lives and caused an estimated $7 billion in property damage on the United States mainland. [NOAA National Environmental Satellite, Data, and Information Service]

367

T HIS CHAPTER covers the genesis and characteristics of two special types of severe weather systems, the tornado and the hurricane. They are singled out because of their great threat to human life and the tremendous property damage they can cause. Actually, the two systems are quite different. Tornadoes are small-scale, short-lived disturbances that usually occur over certain continental localities at midlatitudes. Hurricanes are much larger and much longer-lived storms that spend much of their life cycle over tropical seas.

Tornadoes

Tornadoes are by far the most violent of weather systems. Fortunately, they are small and short-lived storms that usually affect sparsely populated regions. Occasionally, however, a major tornado outbreak causes incredible devastation, death, and injury. On 3–4 April 1974, 148 tornadoes left 315 people dead and 5484 people injured and caused extensive property damage in 13 states (Figure 14.1). Major tornado outbreaks are listed in Table 14.1.

STORM CHARACTERISTICS

A **tornado** is a small mass of air that whirls rapidly about an almost vertical axis. It is made visible by clouds and by dust and debris sucked into the system. Tornadoes are approximately funnel shaped, although a variety of forms have been observed, ranging from cylindrical masses of nearly uniform lateral dimensions (Figure 14.2A) to long, slender, ropelike pendants (Figure 14.2B). By convention, when the circulation remains aloft, it is termed a **funnel cloud,** but when it touches down on the ground, it is called a tornado.

A weak tornado's path on the ground is less than 1.5 km (1 mi) long and 100 m (324 ft) wide, and the system has a life expectancy of only 1 to 3 minutes. At the other extreme, an intense tornado can produce damage along a path more than 160 km (100 mi) long and hundreds of meters wide, and the lifetime of the system may be several hours. The most deadly tornado in North American history, the Tri-State tornado of 18 March 1925, lasted 3.5 hours and produced a 353-km (219-mi) path of devastation stretching from southeastern Missouri through the southern tip of Illinois and into southwest Indiana. Along the tornado's path, fatalities totaled 695, injuries topped 2000, and 11,000 people lost their homes.

Tornadoes are formed by and travel with severe thunderstorm cells. Tornadoes and their parent thunderstorm cells usually (about 90 percent of the time) travel from southwest to northeast. Trajectories are often erratic, however, with many tornadoes exhibiting a hopscotch pattern of destruction as

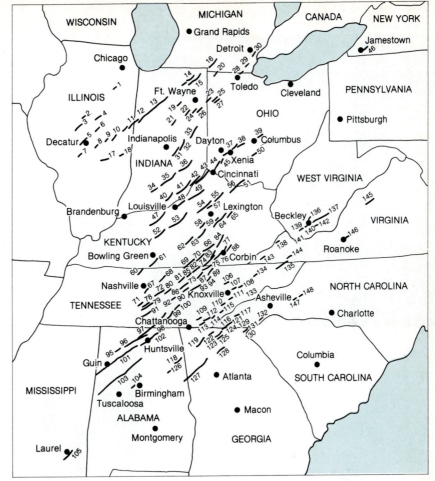

FIGURE 14.1

Tracks of the 148 tornadoes that struck a 13-state area on 3–4 April 1974. [Courtesy of T. Theodore Fujita, Professor of Meteorology, The University of Chicago]

Table 14.1

Five Greatest Tornado Outbreaks (in Number of Tornadoes per Outbreak) since 1950

	Date	Tornadoes	Deaths	Injuries	Area
1	3–4 April 1974	148	315	5484	Area between Mississipi River and Appalachian Mts.: Illinois to New York; Mississippi to Virginia
2	11 April 1965	51	256	over 1500	Southern Great Lakes; Iowa to Ohio
3	4 May 1959	46	0	2 (minor)	Great Plains: Oklahoma to Minnesota and Wisconsin
4	21 April 1967	43	58	1068	Central Mississippi Valley to southern Great Lakes: Missouri to Michigan
5	9–10 January 1975	42	11	287	Southern Plains, lower Ohio Valley and south-eastern states: Texas to Alabama; Oklahoma to southern Indiana

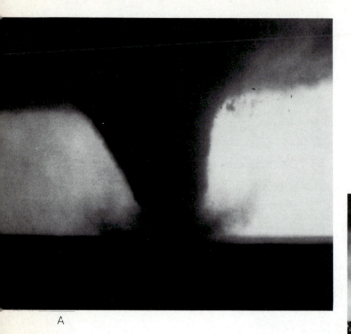

FIGURE 14.2
A tornado may appear as (A) a cylindrical mass of relatively uniform lateral dimensions or (B) a long, slender ropelike pendant. [NCAR/NSF photographs]

they alternately touch down and lift off the ground (Figure 14.3). The average forward speed is around 55 km (34 mi) per hour, although there are reports of tornadoes racing along at speeds approaching 240 km (150 mi) per hour.

An extremely steep horizontal air pressure gradient between the tornado center and edge is the force ultimately responsible for a tornado's violence. The air pressure drop over a distance of only 100 m (300 ft) or so may be equivalent to the normal air pressure drop between sea level and an altitude of 1 km, that is, about a 10 percent reduction. The Coriolis effect is also at work, but the system is so small that this effect is negligible. Thus, winds in a tornado may rotate in either a clockwise or a counterclockwise direction, although the latter dominates by far in Northern Hemisphere tornadoes. This counterclockwise bias may be inherited from the parent thunderstorm in which the Coriolis effect is stronger.

When a tornado strikes, property damage is caused by (1) very high winds, (2) a strong updraft, (3) subsidiary vortices, and (4) an abrupt pressure drop. Winds, sometimes estimated to approach 500 km (310 mi) per hour, blow down trees, power poles, buildings, and other structures. Flying debris causes much of the death and injury associated with tornadoes as broken glass, lumber, and even vehicles become lethal projectiles. In intense tornadoes, the updraft near the center of the storm's funnel may top 160 km (100 mi) per hour, strong enough to lift a house off its foundation. The most intense tornadoes usually include smaller subsidiary vortices, small whirlwinds that swirl about within the tornado. The strongest winds in the system may be associated with these whirlwinds.

It was once widely believed that a tornado causes buildings to explode. The air pressure within the building supposedly could not adjust rapidly enough to the abrupt pressure drop associated with the tornado. People were

FIGURE 14.3

The pattern of destruction caused by a tornado is often erratic, as shown by this aerial photograph of a residential area in Birmingham, Alabama, hit by a tornado on 4 April 1977. Note how some houses are completely destroyed while neighboring houses are still standing. [NOAA photograph]

therefore advised, in the event of a tornado sighting, to open windows to help equalize the internal and external air pressure. However, most buildings have sufficient air leaks so that a potentially explosive pressure differential never really develops. The destruction of buildings is due, instead, to very strong currents of air that stream over roofs and cause the structure to lift, much as air induces lifting as it flows over the curved upper surface of an airplane wing. As the roof is lifted, the walls collapse or are blown in, and the building disintegrates.

Professor T. Theodore Fujita of the University of Chicago has devised a six-point intensity scale for evaluating tornado strength and the potential damage to structures. Called the **F-scale,** and presented as Table 14.2, it is based on estimated wind speeds and categorizes tornadoes as weak (F0, F1), strong (F2, F3), or violent (F4, F5).

An F0 tornado on Fujita's scale produces minor damage, snapping twigs and small branches and breaking some windows. F1 and F2 tornadoes can cause moderate to considerable property damage and take lives. An F1 tornado can down trees and shift mobile homes off their foundations, and an F2 tornado can rip roofs off frame houses, demolish mobile homes, and uproot large trees. An F3 tornado can partially destroy even well-constructed build-ings and lift motor vehicles off the ground. At the violent end of Fujita's scale, destruction is described as devastating to incredible with the potential for many fatalities. An F4 tornado can level sturdy buildings and other structures and toss automobiles about like toys. In an F5 tornado, sturdy frame houses are lifted and transported some distance before disintegrating.

Fortunately, F5 tornadoes are rare. Of the average 778 tornadoes that strike the United States each year, perhaps only one will be rated an F5. When one does occur, however, the results can be catastrophic. In about 1 minute, an F5 tornado leveled the village of Barneveld, Wisconsin, on 8 June 1984; 100 homes were totally destroyed and nine lives were lost. The Xenia, Ohio, tornado, which was part of the massive tornado outbreak of 3–4 April 1974, rated F5 over a portion of its 51-km (31-mi) path and claimed 34 lives.

Table 14.2
The Fujita Tornado Intensity Scale

		Estimated wind speed	
F-scale	Category	km/hr	mi/hr
0	Weak	65–118	40–73
1		119–181	74–112
2	Strong	182–253	113–157
3		254–332	158–206
4	Violent	333–419	207–260
5		420–513	261–318

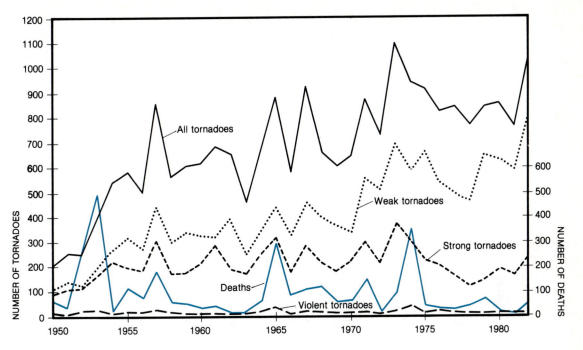

FIGURE 14.4
Annual totals of tornadoes by intensity category and tornado-related deaths in the
United States. [From J. T. Snow, "The Tornado," *Scientific American* 250, No. 4
(1984):95. George V. Kelvin Science Graphics; copyright © 1984 by Scientific American, Inc. All rights reserved.]

Fujita reports that, of the tornadoes that occurred in the United States
between 1950 and 1978, about 63 percent were rated as weak and only 2
percent were violent. The violent systems were responsible for 68 percent of
the total fatalities, however. These statistics are updated through the early
1980s in Figure 14.4. An apparent upward trend in tornado frequency can be
attributed to a better observation network rather than to any real increase in
tornado number.

In addition to reports of the terrible devastation caused by tornadoes,
strange and even bizarre events associated with tornadoes are sometimes
recounted. Reports of straws driven deeply into trees are readily explained.
As a tree is bent by the wind, its grain opens, thus allowing a straw to enter.
When the wind slackens, the grain closes, and the straw is trapped. Rains of
fish or frogs are probably the result of a tornado updraft pulling the water
and its inhabitants out of a nearby pond or lake. Among the oddest reports is
the deplumation of chickens and other fowl during a tornado. This phenomenon has variously been attributed to the pressure drop within the tornado,
strong wind (perhaps aided by natural molting), or the relaxation of the feather
follicles, a reaction brought on by nervous stress.

GENESIS AND DISTRIBUTION

The previous chapter considered the special synoptic-scale weather pattern favorable for the development of severe thunderstorms. To this pattern must be added another requirement for tornado development: tornadoes are most likely to form over relatively flat and dry terrain. They are rare in areas of great topographic relief such as the Rocky Mountain region. Flat terrain offers a minimum of frictional resistance, and dry conditions mean that most of the absorbed solar radiation is channeled into sensible heating, thereby spurring deep convection.

The central United States is one of only a few places in the world where synoptic weather conditions and terrain are ideal for tornado development; part of Australia is another. Although tornadoes have been reported in all 50 states and throughout southern Canada, most occur in **tornado alley,** a corridor stretching from eastern Texas and the Texas Panhandle northward through Oklahoma, Kansas, and portions of Nebraska. Central Oklahoma has the highest annual incidence of tornadoes, whereas local tornado frequency maxima occur in central Illinois and Indiana and in southern Louisiana (Figure 14.5).

FIGURE 14.5
Tornado frequency in number per year within areas defined by 91-km (56.5-mi) radius circles, based on 29 years of data collected since 1 January 1950. An *X* indicates a relative maximum, and an *N* denotes a local minimum. [From NSSFC, "Tornadoes," *Weatherwise* 33, No. 2 (1980):54]

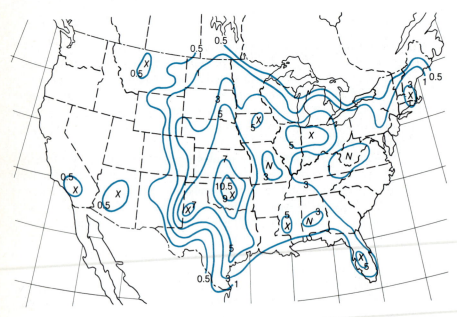

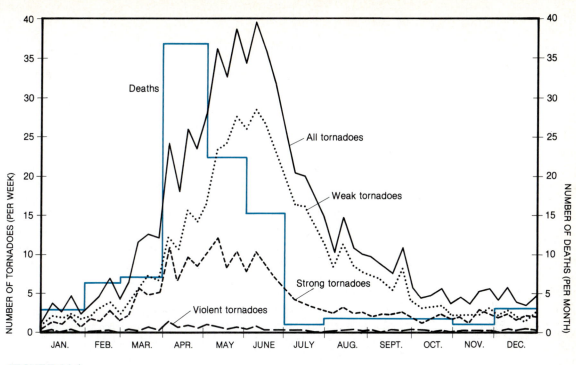

FIGURE 14.6
Average weekly tornado occurrence and monthly tornado fatalities in the United States between 1950 and 1982. [From J. T. Snow, "The Tornado," *Scientific American* 250, No. 4 (1984):95. George V. Kelvin Science Graphics; copyright © 1984 by Scientific American, Inc. All rights reserved.]

THE SPRING MAXIMUM

Almost two-thirds of all tornadoes develop during the warmest hours of the day, and almost three quarters of tornadoes in the United States occur from March to July (Figure 14.6). The peak months of tornado frequency are April (13 percent), May (22 percent), and June (20 percent).* During these times, weather conditions are optimal for spurring vigorous convection and the severe thunderstorms that spawn tornadoes.

One factor that contributes to the spring peak in tornado frequency is the relative instability of the lower atmosphere during that time of year. During the transition from winter to summer, days lengthen and insolation increases, thereby warming the ground. Heat is transported from the ground into the troposphere (Chapter 4), but it takes time for the entire troposphere to adjust to the heating from below. The upper troposphere, in fact, usually retains its winterlike coldness well into spring. The result is a relatively steep air temperature lapse rate that is favorable for severe thunderstorm development.

---

* Figures are based on the 1959–1988 period.

Another factor that contributes to the spring tornado maximum is the greater likelihood that ideal synoptic weather conditions will occur at that time of year. Recall from Chapter 13 that severe thunderstorms typically develop in the warm southeast sector of a strong midlatitude cyclone. Such cyclones achieve their greatest intensity when sharp temperature contrasts develop across the nation, that is, in spring when the polar front is well defined.

During spring there is a general northward progression of tornado occurrences. In effect, the center of maximum tornado frequency follows the sun, as do the midlatitude jet stream, the principal storm tracks, and northward incursions of maritime tropical air. In late winter, the maximum tornado frequency, on average, is along the Gulf Coast states. By May, the maximum frequency shifts to the southern Great Plains, and by June, the highest tornado incidence is usually in the northern plains and the prairie provinces east of the Rockies.

Perhaps 80 percent of all North American tornadoes are linked to midlatitude cyclones. Most of the others are the product of convective instability triggered by hurricanes. In fact, most hurricanes that strike the southeastern United States are accompanied by tornadoes. The tornadoes usually develop on the northwest flank of a hurricane, after the system has turned toward the northeast.

What are your chances of experiencing a tornado? Very slim. Less than 1 percent of all thunderstorms produce tornadoes, and even in the most tornado-prone regions of North America, a tornado is likely to strike only once every 250 years. There are, of course, exceptions to these statistics.

Table 14.3
What to Do If a Tornado Is Approaching[a]

1. Seek shelter in a tornado cellar, an underground excavation, or a steel-framed or substantial reinforced-concrete building.
2. Avoid auditoriums, gymnasiums, supermarkets, or other structures that have wide, free-span roofs.
3. In an office building or school, go to an interior hallway on the lowest floor or a designated shelter area. Lie flat on the floor with your head covered.
4. At home, go to the basement. If there is no basement, go to a small room (closet, bathroom, or interior hallway) in the center of the house on the lowest floor. Seek shelter under a mattress or a sturdy piece of furniture.
5. Stay clear of all windows and outside walls.
6. In open country, do not try to outrun a tornado in an auto. Tornadoes often move too fast and their paths are too erratic to avoid even if you drive at right angles to the apparent track of the storm. Instead, stop and seek shelter indoors, and if this is not possible, lie flat in a ravine, creek bed, or open ditch.
7. Do not seek shelter in mobile homes or motor vehicles.

[a] Modified slightly from NOAA recommendations.

Tornadoes have hit Oklahoma City no less than 26 times since 1892. To be prepared in the unlikely event that a tornado should strike your area, study the recommendations in Table 14.3. These precautions will reduce the hazard posed by tornadoes.

THE TORNADO–THUNDERSTORM CONNECTION

The most intense tornadoes usually appear on the rain-free rear portion of a severe thunderstorm. They develop in the thunderstorm's strong updraft (Figure 14.7), although the precise relationship between the two systems is not completely understood.

A tornadic circulation apparently stems from an interaction between the updraft in the thunderstorm and the larger scale horizontal wind. The horizontal wind must exhibit strong vertical shear in both speed and direction. That is, wind speed must increase with altitude, and wind direction must veer (turn clockwise) with altitude—from southeast at the surface to southwest or west aloft. The shear in wind speed causes air to rotate about a horizontal axis. When this rotation interacts with the updraft, the region of rotating air is tilted to a vertical position. The shear in the horizontal wind direction also adds to this rotation of air about a vertical axis. As a consequence, the entire updraft spins as a cylinder 10 to 20 km (6.2 to 12.4 mi) in diameter. This

FIGURE 14.7
Features of a severe thunderstorm that spawns a tornado. [From J. T. Snow, "The Tornado," *Scientific American* 250, No. 4 (1984):91. George V. Kelvin Science Graphics; copyright © 1984 by Scientific American, Inc. All rights reserved.]

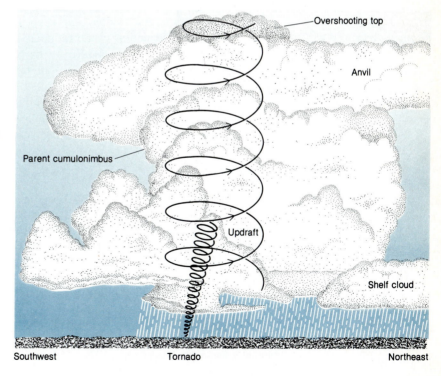

circulation system is known as a **mesocyclone,** and the odds are about 1 in 2 that a mesocyclone will evolve into a tornado.

The mesocyclone circulation actually begins in the midtroposphere and, from there, builds upward and downward. Meanwhile, the updraft strengthens as more air converges toward the base of the thunderstorm. The updraft may eventually become so strong that it overshoots the top of the thunderstorm and produces a cloud bulge on top of the thunderstorm anvil.

For reasons again not well understood, the mesocyclone in a tornadic thunderstorm narrows and spirals downward toward the ground as a funnel. As the spinning mass of air narrows, its speed increases, perhaps to violent levels. This increased speed is analogous to that of an ice skater performing a spin: the skater's rate of spin increases as the arms are brought closer to the body.

When a mesocyclone spawns a tornado, the intensity might be anywhere on the F-scale, from weak to devastating. On the other hand, tornadoes sometimes spin off the gust front associated with a severe thunderstorm; these tornadoes are almost always weak.

Our description of tornado genesis must remain tentative because it is not possible to make direct measurements of tornadoes. Traditional weather instruments are usually destroyed by tornadic winds. Much of what is known about the circulation of tornadoes and tornadic thunderstorms has been derived from analyses of motion picture footage and from laboratory and computer simulations. The future of tornado monitoring is promising, however. As we will see later in this chapter, a new type of radar, called Doppler radar, better resolves the circulation inside thunderstorms than does conventional radar. Moreover, since 1981, scientists at the National Severe Storms Laboratory in Norman, Oklahoma, have been attempting to obtain direct data on tornadoes using a special tornado-resistant instrument package, called TOTO,* that is designed to measure air temperature, air pressure, and wind speed. The plan is to place the 180-kg (400-lb) TOTO in the path of an oncoming tornado. As of this writing, however, no tornado has yet passed directly over TOTO.

TORNADO LOOK-ALIKES

A **waterspout** is a tornadolike disturbance that occurs over the ocean or over a large inland lake (Figure 14.8). It is so named because it consists of a whirling mass of water that appears to stream out of the base of the parent cumulonimbus cloud. A waterspout is usually considerably less energetic, smaller, and shorter lived than a tornado. The rare intense waterspout may well be a tornado that formed over land and then moved over a body of water. In any event, boaters should avoid waterspouts.

---

* TOTO is the acronym for Totable Tornado Observatory—as well as the name of Dorothy's dog in *The Wizard of Oz*.

A tornado has other look-alikes. Distant **virga** (rain or snow that vaporizes before reaching the ground) may be mistaken for a tornado because of its cylindrical or funnel shape (refer back to Figure 8.3). The absence of any rotary motion, however, clearly distinguishes virga from a funnel cloud. In deserts, or wherever exposed soil dries out, intense solar heating often gives rise to swirling masses of dust, called **dust devils,** which resemble tornadoes but form at ground level, are not associated with any clouds, and cause little if any damage (see Chapter 12).

Hurricanes

Hurricanes, named for Huracan, a Carib god of evil, are violent oceanic cyclones that originate in tropical latitudes, usually in the late summer and fall of the year. By definition, a hurricane has a maximum wind speed greater than 119 km (74 mi) per hour. Wind gusts often top 250 km (155 mi) per hour, although winds quickly diminish once the storm makes landfall. Most hurricane wind damage is confined to within about 200 km (125 mi) of the coast, and, as noted earlier in this chapter, it is common for hurricanes approaching coastal areas to spawn destructive tornadoes.

Hurricanes also produce torrential rains with amounts typically in the range of 13 to 25 cm (5 to 10 in.). Even if the storm tracks well inland, heavy rains often persist and may trigger flash flooding. Although much can be said about the destructive aspects of hurricanes, their rains can be beneficial. For example, in the southeastern United States, tropical storms (hurricanes and their precursors) account for an average of 10 to 15 percent of the June through October rainfall.

More than strong winds and heavy rains, the most devastating feature of hurricanes that strike coastal areas is the **storm surge,** a flood of ocean water that accompanies the storm. Low air pressure in a hurricane causes sea level to rise (about 0.5 m for every 50 mb drop in air pressure), and strong hurricane winds drive ocean waters over low-lying coastal areas, ravaging property and sometimes taking many lives.

A storm surge is superimposed on normal ocean tidal oscillations. Hence, the greatest potential for flooding and shoreline erosion occurs when the storm surge coincides with high tide in localities of great tidal range (the height difference between high and low tides). Tidal range is relatively small along the Gulf Coast (less than 1 m) and greater along the eastern seaboard (typically several meters).

A storm surge of 1 to 2 m (3 to 6 ft) can be expected with a weak hurricane, whereas the storm surge accompanying a severe hurricane may top 5 m (16 ft). Hurricane Camille, an exceptionally destructive storm, produced a maximum storm surge of about 8 m (25 ft) when it struck the Mississippi coast on 17 August 1969 (refer back to Figure 1.2). One of history's most disastrous storm surges hit the Bay of Bengal coast of Bangladesh on 13 November 1973. A storm surge of nearly 7 m (23 ft) flooded a vast coastal plain, claiming an estimated 300,000 lives by drowning.

Like tornadoes, hurricanes have an intensity scale. Known as the **Saffir–Simpson Hurricane Intensity Scale** after its designers, H. S. Saffir and R. H. Simpson, it assigns ratings to hurricanes, from 1 to 5 corresponding to increasing intensity. As shown in Table 14.4, each intensity category includes (1) a range of central air pressure, (2) a range of wind speed, and (3) the potential for storm surge and property damage. Of the 129 hurricanes that struck the United States Gulf and Atlantic coasts between 1900 and 1978, 53 (41 percent) were classified as "major"; that is, they rated 3 or higher on the Saffir–Simpson scale. The 10 deadliest and 10 costliest United States hurricanes since 1900 are identified in Tables 14.5 and 14.6, respectively.

The last category 5 hurricane to make landfall on the Atlantic Coast was Gilbert in September 1988. (Camille was the next previous one.) Gilbert originated as a tropical wave off the West African coast on 3 September. The wave followed a steady west–northwest track and, seven days later, strengthened to a hurricane while south of Puerto Rico. Gilbert then passed over the length of Jamaica on the 12th, intensified over the northwest Caribbean, and on 14 September, made landfall as a category 5 hurricane near Cozumel on Mexico's Yucatan Peninsula. At landfall sustained winds were estimated at

Table 14.4
Saffir–Simpson Hurricane Intensity Scale

Scale number (category)	Central pressure		Wind speed		Storm surge		Damage
	mb	in.	mi/hr	km/hr	ft	m	
1	≥980	≥28.94	74–95	119–154	4–5	1–2	Minimal
2	965–979	28.50–28.91	96–110	155–178	6–8	2–3	Moderate
3	945–964	27.91–28.47	111–130	179–210	9–12	3–4	Extensive
4	920–944	27.17–27.88	131–155	211–250	13–18	4–6	Extreme
5	<920	<27.17	>155	>250	>18	>6	Catastrophic

275 km (170 mi) per hour. The hurricane then continued northwestward into the Gulf of Mexico and made its final landfall (as category 3) on the coast of Mexico about 200 km (125 mi) south of the Texas border. Gilbert brought much death and destruction to both Mexico and Jamaica. The death toll was 202 in Mexico and 45 in Jamaica, and property damage in the two countries topped $4 billion.

Table 14.5

Ten Deadliest Hurricanes to Strike the United States Since 1900

Hurricane	Year	Category	Deaths
1 Texas (Galveston)	1900	4	6000
2 Florida (Lake Okeechobee)	1928	4	1836
3 Florida (Keys and S. Texas)	1919	4	600–900[a]
4 New England	1938	3	600
5 Florida (Keys)	1935	5	408
6 Audrey (Louisiana and Texas)	1957	4	390
7 Northeast U.S.	1944	3	390[b]
8 Louisiana (Grand Isle)	1909	4	350
9 Louisiana (New Orelans)	1915	4	275
10 Texas (Galveston)	1915	4	275

[a] Over 500 of these lost on ships at sea.
[b] Some 344 of these lost on ships at sea.
Source: P. J. Hebert and G. Taylor. "The Deadliest, Costliest, and Most Intense United States Hurricanes of This Century (and Other Frequently Requested Hurricane Facts)." NOAA Technical Memorandum, NWS, NHC 18, 1983.

Table 14.6

Ten Costliest Hurricanes to Strike the United States Since 1900

Hurricane	Year	Category	Damage (millions of dollars)
1 Hugo (South Carolina)	1989	4	7,000
2 Frederic (Ala. and Miss.)	1979	3	2,300
3 Agnes (Fla. to Northeast U.S.)	1972	1	2,100
4 Alicia (Texas)	1983	2	2,000
5 Camille (Miss. and La.)	1969	5	1,420
6 Betsy (Fla. to La.)	1965	3	1,420
7 Elena (Miss. and La.)	1985	2	1,250
8 Diane (Northeast U.S.)	1955	1	831
9 Eloise (Northwest Fla.)	1975	3	550[a]
10 Carol (Northeast U.S.)	1954	3	461

[a] Includes $60 million in Puerto Rico.
Source: NOAA.

STORM CHARACTERISTICS

Perhaps the most convenient way to describe a hurricane is to contrast it with midlatitude, or "extratropical," cyclones, which we examined in Chapter 11. Hurricanes develop in a uniform mass of very warm and humid air, so they have no associated fronts or frontal weather. Air pressure is distributed symmetrically about the system center, and thus, as shown in Figure 14.9, isobars form a series of closely spaced concentric circles. Typically, the central pressure is considerably lower and the horizontal air pressure gradient much steeper in a hurricane than in an extratropical cyclone. Reconnaissance aircraft extrapolated a surface air pressure of 888 mb (26.22 in.) at the center of Hurricane Gilbert while the storm was over the northwest Caribbean Sea on 13 September 1988. This is the lowest sea-level air pressure ever recorded in the Western Hemisphere. In addition, a hurricane is usually much smaller, averaging a third of the diameter of a midlatitude cyclone. Rarely do hurricane-force winds extend much beyond 120 km (75 mi) from the storm center.

Structurally, a mature hurricane is a warm-core, low-pressure system that weakens rapidly with altitude, especially above 3 km (9000 ft). In the upper troposphere, at an altitude of about 15 km (45,000 ft), it is usual to find anticyclonic airflow above the hurricane. At the hurricane center is an area of almost cloudless skies, light winds, and subsiding air, called the **eye** (Figure 14.10). The eye ranges from 20 to 200 km (12 to 125 mi) across, shrinking in diameter as the hurricane intensifies and winds strengthen. Many people

FIGURE 14.9

Surface isobar pattern on the morning of 10 August 1980 as Hurricane Allen tracked northwestward from the Gulf of Mexico and into extreme southeast Texas. Isobars (in mb) form closely spaced concentric circles about the hurricane center. See Figure 14.11 for the corresponding satellite image. [Redrawn from NOAA weather map]

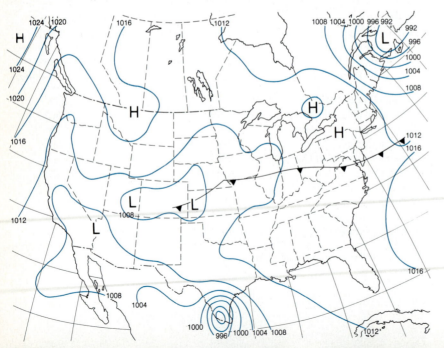

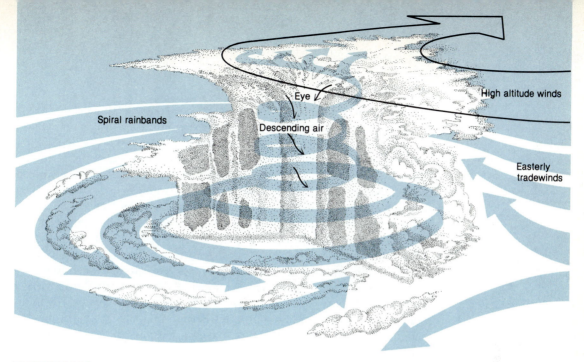

FIGURE 14.10
The internal structure of a hurricane as determined by radar and satellite monitoring. In this artist's conception, the vertical dimension is greatly exaggerated. Actual hurricanes are less than 15 km in altitude and have diameters of several hundred kilometers. [From NOAA, *Hurricane.* Washington, DC: Superintendent of Documents, 1977, p. 11.]

have been deceived into thinking the storm to be at an end when clearing skies and slackening winds follow a hurricane's initial blow. They may well be experiencing the hurricane's eye—heavy rains and ferocious winds will soon resume but from a different direction.

Bordering the eye of a mature hurricane is the **eye wall,** a ring of cumulonimbus clouds that produce heavy rains and very strong winds. The most dangerous and destructive part of the hurricane is near the eye on the side where winds blow from sea to land. Here winds are strongest and the storm surge is maximum. Cloud bands accompanied by hurricane-force winds and heavy convective showers spiral inward to the eye wall. All this is surrounded by an outer region of high clouds (cirrus or cirrostratus) and cyclonic winds. A typical cloud pattern associated with a hurricane is shown in the satellite photograph in Figure 14.11.

FIGURE 14.11
Satellite image of the cloud pattern associated with Hurricane Allen over southeast Texas on 10 August 1980. [NOAA photograph]

LIFE CYCLE

Two conditions must be met before a hurricane can develop. One of these is very warm surface ocean water. It appears that hurricane formation requires sea-surface temperatures to be at least 26.5 °C (80 °F) through a depth of 60 m (100 ft) or more. Such exceptionaly warm water sustains the hurricane circulation by the latent heat released when water evaporates from the ocean surface and subsequently condenses within the storm. As we saw in Chapter 6, temperature governs the rate of vaporization of water. Thus, when a hurricane passes over land or over cold ocean water, it loses its energy source. Consequently, the system weakens rapidly and winds diminish abruptly.

A second prerequisite for hurricane formation is a significant Coriolis effect. The influence of the Earth's rotation must be strong enough to induce and sustain a cyclonic circulation. As noted in Chapter 9, the Coriolis effect weakens toward lower latitudes and becomes zero at the equator. The minimum latitude where the Coriolis effect is strong enough for hurricane formation is about 4 degrees, and most hurricanes form in the 10- to 20-degree latitude belt.

The required combination of sufficient Coriolis effect and warm sea surface temperatures occurs only over certain portions of the world's oceans, identified in Figure 14.12. Major hurricane breeding grounds are (1) the western tropical North Pacific, where hurricanes are called **typhoons,*** (2) the Indian Ocean (including the Bay of Bengal and the Arabian Sea) and the tropical waters north of Australia, where hurricanes are termed simply cyclones, (3) the tropical North Atlantic west of the bulging west coast of Africa, and (4) the Pacific Ocean just southwest of Central America. The absence of hurricanes off either South American coast is noteworthy; it is due to relatively low sea-surface temperatures.

Only hurricanes spawned over the tropical Atlantic pose a serious threat to coastal North America. The West Coast is rarely a target of hurricanes for two reasons: (1) the prevailing winds (northeast trades) are offshore and, hence, they steer tropical storms that form west of Central America away from the coast, and (2) surface waters just off the southern California (and Baja California) coast normally are too cold to sustain hurricanes. During unusual circulation regimes, however, tropical storms have been known to strike coastal southern California and even track into the desert Southwest. For example, on 9–10 September 1976, Hurricane Kathleen crossed Baja California and tracked near Yuma, Arizona, bringing torrential rains and floods to the desert.

The requirement of relatively high sea-surface temperature makes hurricane occurrence distinctly seasonal. Because of the great thermal stability of

---

* By convention, intense tropical storms over the Pacific Ocean are called "typhoons" when they occur west of 180 degrees longitude, and "hurricanes" when they occur east of 180 degrees longitude.

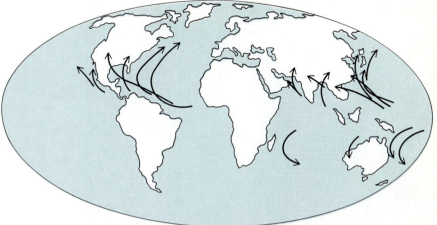

FIGURE 14.12
Hurricane breeding grounds are located only over certain portions of the world's oceans. Arrows indicate average hurricane trajectories.

ocean water, sea-surface temperatures reach a seasonal maximum long after the time of peak solar radiation. Most Atlantic hurricanes consequently develop in late summer and autumn; the official hurricane season extends from 1 June to 30 November.*

The first sign that a hurricane may be in the making is the appearance of a cluster of thunderclouds over tropical seas. This region of convective activity is labeled a **tropical disturbance** if a center of low pressure is detectable at the surface. Chances are that the tropical disturbance was triggered by the ITCZ, by a trough in the westerlies intruding into the tropics from midlatitudes, or by a wave (or ripple) in the easterly trade winds. Most North Atlantic hurricanes develop out of convective cloud clusters associated with tropical waves that continually move westward off the African coast. Storms formed in this way are sometimes referred to as Cape Verde–type hurricanes. We do not know why very few of these cloud clusters evolve into full-blown hurricanes. We do know, however, that hurricane formation is favored by low vertical wind shear (the change of horizontal wind with altitude) and high relative humidity in the middle troposphere.

If conditions favorable to hurricane genesis persist, a cyclonic circulation develops and the central air pressure begins to fall. Water vapor condenses within the storm, releasing latent heat of vaporization, and the heated air is then forced to rise by neighboring cooler air. Expansional cooling of the rising air triggers more condensation, release of even more latent heat, and a further increase in buoyancy. Rising temperatures, coupled with an anticyclonic outflow of air aloft, cause a sharp drop in air pressure, which, in turn, induces convergence of air at the surface. The consequent uplift around the developing eye leads to additional condensation and release of latent heat.

---

* In the Pacific, the hurricane season is 15 May to 30 November.

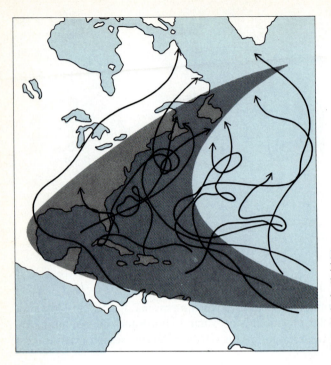

FIGURE 14.13
Hurricane trajectories are often erratic, as shown by these sample tracks. As indicated by the shaded area, however, hurricanes initially describe a generally westerly drift and eventually curve toward the northeast when they reach the western Atlantic. [From NOAA, *Hurricane*. Washington, DC: Superintendent of Documents, 1977, p. 13.]

Through these processes, the tropical disturbance intensifies and winds strengthen. When maximum wind speeds top 37 km (23 mi) per hour, the developing storm is called a **tropical depression.** When maximum wind speeds reach 63 km (39 mi) per hour, the system is classified as a **tropical storm.** If winds exceed 119 km (74 mi) per hour, the storm is officially designated a hurricane. As a hurricane decays, the storm is downgraded using this same classification scheme.

Hurricanes that threaten the eastern and southern coasts of the United States usually drift very slowly westward with the trade winds across the tropical North Atlantic and into the Caribbean. At this stage in the storm's trajectory, it is not unusual for the storm to move at a mere 10 to 20 km (6 to 12 mi) per hour. Once over the western Atlantic, however, the storm usually speeds up and begins to curve north, and eventually northeastward, as it is caught in the midlatitude westerlies. Precisely where this trajectory change takes place determines whether the hurricane enters the Gulf of Mexico and tracks up the Mississippi River Valley, moves up along the eastern seaboard, or curves back out to sea.

By the time a coastal hurricane reaches a latitude of 30 degrees N, it may begin to acquire extratropical (midlatitude) characteristics, as colder air is drawn into the circulation of the system and fronts develop. From then on, the storm follows a life cycle similar to that of any other midlatitude cyclone and often ends by occluding over the North Atlantic.

Many hurricanes depart significantly from the track just described. The hurricane tracks shown in Figure 14.13, for example, are quite erratic. Sometimes a hurricane describes a complete circle or reverses direction. In addition, some hurricanes maintain their tropical characteristics even after traveling far north along the Atlantic Coast where they are fueled by warm Gulf Stream waters. The eye of Hurricane Hazel, for example, was still discernible when the storm passed over Toronto in October 1954. New England, located more than 25 degrees of latitude north of the usual hurricane breeding ground, has been the target of many full-blown hurricanes. Perhaps the most noteworthy of these was the hurricane of 21 September 1938 (category 3). A storm surge ravaged the New England coast. Winds gusting over 200 km (120 mi) per hour severely damaged interior forests, and torrential rains caused flash flooding by rivers and streams. Fatalities were estimated at 600.

During an average hurricane season, six hurricanes and four tropical storms occur over the tropical Atlantic, Caribbean Sea, or Gulf of Mexico. And, on average, two hurricanes strike the United States coast each year. The Florida Keys is the most hurricane-prone region of the United States, having recorded 43 hurricanes in the past 100 years. For your use, we have included a hurricane tracking chart in Appendix III.

Hurricane Agnes* illustrates the life cycle of a particularly disastrous storm. It claimed 122 lives and caused more than $2.1 billion in property damage in the northeastern states, and yet much of the destruction was caused by water rather than wind. For more about Hurricane Agnes, see the Special Topic.

HURRICANE THREAT TO THE SOUTHEAST

Today, many atmospheric scientists and others are concerned about the hurricane threat to coastal areas of the southeast United States. Rapid population growth in coastal cities and resorts has necessitated evacuation plans for residents of low-lying areas prone to hurricane storm surges. Florida's coastal population, for example, has doubled in the past 20 years. Many of these residents have been lulled into a false sense of security by the anomalously low frequency of major hurricanes during that same period (Figure 14.14).

* Hurricane names alternate each year between male and female names in alphabetical order. A name is assigned when the storm reaches tropical storm strength.

FIGURE 14.14
A pleasant day on Siesta Key, Sarasota, Florida. Siesta Key is a barrier island just off Florida's Gulf Coast. The relative infrequency of tropical storms and hurricanes has lulled some people in coastal areas of the Southeast into a false sense of security and inspired construction of condominiums within a few meters of high-tide level. [Photograph by J. M. Moran]

Hurricane Agnes

Agnes never exceeded intensity category 1, and yet it was the third most destructive hurricane ever to strike the United States. Property damage topped $2.1 billion, almost twice that of Camille, a category 5 hurricane. Torrential rain rather than strong winds or storm surge was the principal factor in Agnes's destructive rampage.

Agnes began as a weak tropical depression near Cozumel, Yucatan, Mexico, on 14 June 1972 (Figure 1). The depression slowly intensified and drifted eastward along the twentieth parallel. On 16 June, it curved northward and became a tropical storm, with top wind speeds of 62 to 117 km (39 to 73 mi) per

hour. On 17 June, the storm attained hurricane strength as it passed just west of Cuba, and by 18 June, Agnes was southwest of the Florida Keys. That afternoon, peak winds reached 136 km (85 mi) per hour. Agnes shifted to the northeast and moved ashore on the Florida panhandle near Panama City on 19 June and rapidly weakened to a tropical depression. On 18 and 19 June, Agnes spawned 15 tornadoes in Florida and 2 in Georgia.

By 20 June, Agnes's unusually broad circulation was producing heavy rains from northern Florida to southern Virginia, and most of the eastern United

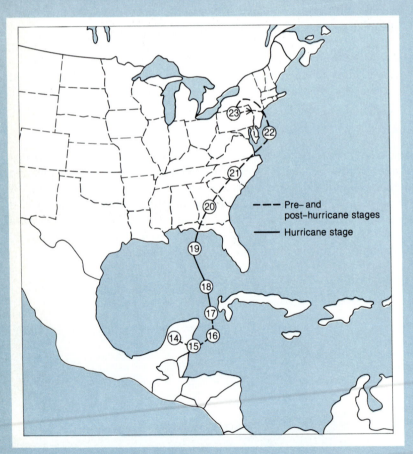

FIGURE 1
The track of Hurricane Agnes, 14–23 June 1972.

FIGURE 2
Waters of the Susquehanna River raged through downtown Wilkes-Barre, Pennsylvania, following heavy rainfall associated with Hurricane Agnes in June 1972. [U.S. Coast Guard photograph]

States was invaded by an unstable tropical air mass. On 21 June, the storm crossed into North Carolina, and the blanket of unstable air and heavy rains extended into eastern Canada. That afternoon, Agnes strengthened to a tropical storm as it tracked offshore near the Virginia–North Carolina border. On 22 June, the storm center was east of Baltimore, but later that day the storm curved toward the northwest and its center tracked inland over New York City. Finally, on 23 June, Agnes was absorbed by an extratropical cyclone over western Pennsylvania and that storm drifted gradually toward the northeast and out over the Atlantic Ocean.

Agnes brought beneficial rains to the Southeast, which had been experiencing a dry spell, but heavy rains in the mid-Atlantic region were anything but beneficial. During the week prior to Agnes's arrival, frontal showers and thunderstorms brought soaking rains from Virginia to New England. Rainfall averaged 3 to 8 cm (1 to 3 in.) and locally topped 15 cm (6 in.).

Agnes's torrential rains falling on already saturated soils produced runoff that triggered many record-breaking river crests and devastating floods.

Heaviest rains fell in a band 350 km (220 mi) wide from western North Carolina northeastward into central New York State. In this hilly terrain, rainfall totals were quite variable but generally ranged from 20 to 30 cm (8 to 12 in.) for the period 18–25 June. Some localities reported total rainfall in excess of 45 cm (18 in.) with more than 25 cm (10 in.) falling in only 24 hours.

Central Pennsylvania was particularly hard hit by flooding. More than 250,000 people were forced from their homes. Crests on the Susquehanna River averaged 4 to 6 m (12 to 18 ft) above flood stage, and at Wilkes-Barre the river crested at more than 6 m (18 ft) above flood stage (Figure 2). Of the 122 fatalities attributed to Agnes, 50 occurred in Pennsylvania, and flood damage in that state accounted for almost two-thirds of total storm damage.

During the 1970s, fewer hurricanes affected the United States than in any prior decade of the century. More than 80 percent of the 40 million people living along the Atlantic and Gulf of Mexico coasts have never experienced a major hurricane.

The danger is particularly acute for residents of the nearly 300 barrier islands that fringe portions of the Atantic and Gulf coasts. Barrier islands protect coastal beaches and wetlands by bearing the brunt of powerful storm-driven sea waves. The energy of storm waves is partially dissipated in shifting the sands that compose barrier islands. However, for decades, many barrier islands have undergone rapid development particularly for resorts and cottages, and some coastal cities, including Miami Beach, are built entirely on barrier islands. The exposed location of these islands makes them vulnerable to the hurricane storm surge. Thus, evacuation of the island population is the most prudent strategy in the event of a major hurricane threat.

The effectiveness of coastal evacuation plans was tested in the late summer of 1985 when Elena (Plate 21) menaced the Gulf of Mexico coast. For four days, Elena followed an erratic path over the Gulf, first taking aim at southern Louisiana, then the Florida Panhandle, and later central Florida before reversing direction and finally coming ashore near Biloxi, Mississippi, on 2 September (Figure 14.15). In probably the greatest evacuation in United States history, nearly a million people from Sarasota, Florida, to New Orleans, Louisiana, were forced to leave beachfront communities and flee to inland shelters.

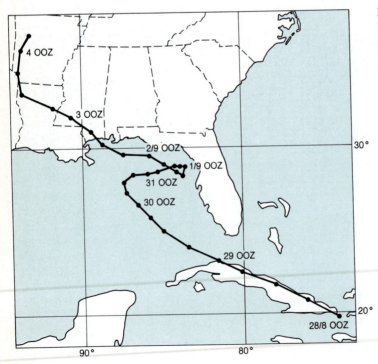

FIGURE 14.15
The track of Hurricane Elena from 28 August 1985 through 4 September 1985. The hurricane's position is reported every six hours. [From the National Hurricane Center, Coral Gables, Florida]

Some returned home only to evacuate again as Elena changed course. Although property damage was considerable because of extensive flooding and winds that exceeded 160 km (100 mi) per hour, only four fatalities were attributed to Elena and none in the area of landfall.

More recently, about midnight on 21 September 1989, Hurricane Hugo (category 4) moved ashore near Charleston, South Carolina. Hugo was a Cape Verde–type hurricane that several days earlier had ripped through the islands of the northeast Caribbean. Driven by peak sustained winds of 215 km (135 mi) per hour, a storm surge of 5 m (16 ft) slammed onto barrier islands off the South Carolina coast, destroying many beachfront homes and structures. Hugo's winds also dealt a severe blow to South Carolina's forests. Over nearly 36 percent of the state's forest area, winds twisted and toppled more than half the trees—a timber loss valued at more than $1 billion. Hugo was the most costly hurricane ever to strike the United States mainland with total damage estimated at $7 billion. Fortunately, advance warning of the hurricane's approach permitted timely evacuation of barrier islands and other low-lying coastal areas. No doubt, the death toll of 21 would have been much higher without evacuation.

HURRICANE MODIFICATION

Spurred by a series of six destructive hurricanes that lashed the East Coast during the mid-1950s, atmospheric scientists stepped up efforts to learn more about hurricanes and to devise hurricane modification techniques. To these ends, **Project STORMFURY** was initiated in 1962. During the next two decades, project scientists advanced our understanding of the formation, structure, and dynamics of hurricanes; improved hurricane forecasting; and upgraded hurricane reconnaissance methods. The original goal of hurricane modification, however, was not realized.

The working hypothesis of Project STORMFURY was that seeding hurricanes with silver iodide (AgI) crystals would reduce wind strength. The argument went as follows. Seeding the band of convective clouds just outside the eye wall would trigger additional latent heat release (as supercooled water droplets converted to ice crystals) and would enhance convection. Artificial invigoration of convection would then dominate the convection in the eye wall, and a new eye wall would form at a greater radius from the storm center. Thereby, the band of strongest wind would be displaced farther from the eye center. By the law of conservation of angular momentum, the circulation would also weaken just as a skater's rotation rate slows when the arms are extended.

In the 1960s and 1970s, there were few opportunities for hurricane-seeding experiments primarily because few hurricanes met the necessary criteria of being intense, having a well-defined eye, and being situated far from land. Some success was reported, however, in reducing winds (between 10 and 30 percent) in four hurricanes that were seeded. Later, these apparent successes were dismissed when new data called into question the original Project STORMFURY hypothesis. For one, it was discovered that convective clouds in hurricanes contain too little supercooled water for seeding to be effective. For another, recent monitoring of unmodified hurricanes shows that changes in eye-wall diameter occur as part of the storm's natural evolution. It appears likely, then, that the Project STORMFURY's "successes" at hurricane modification were by chance rather than be seeding.

Weather Radar

Weather **radar*** is a valuable tool for detecting and tracking severe weather systems. Thunderstorm cells are so small that often they are not sighted directly by the widely spaced network of synoptic weather-observing stations. Weather radar, on the other hand, scans a wide area continuously and can locate small and isolated pockets of precipitation. Conventional weather radar, which has served weather observers for almost 30 years, will soon be replaced by much more sophisticated Doppler radar systems.

CONVENTIONAL RADAR

A conventional weather radar unit operated by the National Weather Service emits short pulses of microwaves having a wavelength of 10 cm. Radar waves are scattered by precipitation but are not scattered by the very small droplets or ice crystals that compose clouds. Thus, weather radar detects ("sees") rain or snow but not the parent clouds. When precipitation is targeted, a portion of the radar waves is scattered back to a receiving unit, which displays the return signal, called a **radar echo,** as electrical pulses on a cathode ray tube similar to a television screen (Figure 14.16A). Because radar signals are sent out and received hundreds of times each second as the radar continuously scans a 360-degree circle (Figure 14.16B), the product is a map of the precipitation pattern surrounding the radar unit. The time interval between emission and reception of the radar signal is calibrated to give the distance to the precipitation.

The intensity of the radar echo depends on the reflectivity of the targeted precipitation and is greatest for large drops and hailstones. To a lesser extent echo intensity is influenced by the concentration of raindrops in the path of the radar beam. Hence, echo intensity is used to distinguish severe thunderstorm cells, which often contain large hail, and as an index of rainfall intensity.

---

* The acronym for RAdio Detection And Ranging.

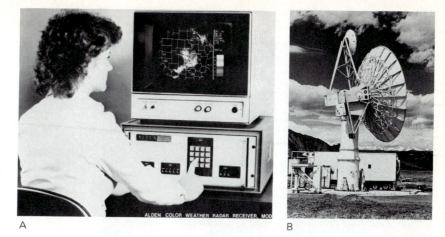

A   B

Some radar units are equipped with electronic devices that show echo intensity on a color scale, so a patch of red indicates very heavy rain and, at the other end of the scale, green indicates very light rain (Plate 22).

Weather radar monitors the development and dissipation of thunderstorm cells, the direction and speed of movement of those cells, and the spiral bands of rainfall associated with hurricanes (Figure 14.17). Conventional radar cannot detect a tornado directly, but in some cases when a tornado is present, a

FIGURE 14.17

Conventional radar display of spiral bands of heavy rainfall associated with a hurricane. [NOAA photograph]

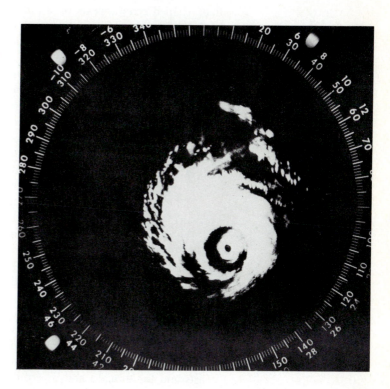

393

FIGURE 14.18
Conventional radar display of a hook echo, which may indicate tornadic circulation of air. [NOAA photograph]

hook-shaped echo appears on a radar screen (Figure 14.18). If present, the hook usually appears on the south side of a severe thunderstorm cell. Apparently, the **hook echo** indicates rainfall being drawn around the mesocyclone within a severe thunderstorm.

To this point, we have described the plan-position indicator (PPI) radar. With this radar unit, a microwave beam sweeps out an almost horizontal circle. The observing circle is limited in size by the curvature of the Earth and may have a radius of up to 400 km (250 mi). A second type of radar display, the range-height indicator (RHI), scans up and down rather than horizontally and is used to determine the altitude of thunderstorm tops by detecting precipitation in the upper reaches of the clouds. As noted in Chapter 13, the altitude of cloud top gives an indication of thunderstorm intensity: the higher the echo, the more intense is the storm. Conventional weather radar units can operate in either the PPI or the RHI mode.

The first reported use of radar for meteorological purposes was on 20 February 1941 when a radar unit on the south coast of England tracked a thunderstorm a distance of 11 km (7 mi). More widespread application of radar for weather analysis began shortly after World War II. The early weather radars were short-range surplus military units. Not until the mid-1950s, following major tornado and hurricane disasters, did the United States Congress allocate funds for the purchase of new long-range radar units designed specifically for meteorological applications. In 1959, these units, called WSR-57 radars, went into service for hurricane detection along the Atlantic and Gulf coasts and for tornado and severe thunderstorm monitoring in the central

United States. In the ensuing years, area coverage by weather radar expanded steadily, and by the late 1960s, radar had become a routine component of televised weathercasts.

DOPPLER RADAR

Over the past two decades, much research and field testing have focused on a new type of weather radar that utilizes the **Doppler effect,** named for Johann Christian Doppler, the Austrian physicist who first explained the phenomenon in 1842. This principle is also the basis for the police radar used to monitor traffic movement and the "gun" used to measure the speed of a pitched ball. The Doppler effect refers to the shift in frequency of sound waves or electromagnetic waves emanating from a moving source. For example, the pitch (frequency) of a train whistle is higher as a train approaches you and drops off as the train moves away. As shown schematically in Figure 14.19A, the crests of sound waves generated by a stationary source form evenly spaced concentric circles about the source, and the wave frequency is uniform. If the source is moving, however, as in Figure 14.19B, the wave crests become more closely spaced in the direction in which the source is moving so that the frequency is greater ahead of the source and less behind the source. You can experience this same change in pitch by moving a buzzing electric razor past your ear.

Doppler radar is a conventional radar that has the added capability of determining the detailed motion of the targeted precipitation toward or away from the radar unit. As a raindrop moves away from or toward a radar unit, the frequency of the radar signal (microwaves) shifts slightly between emission and the return echo. This frequency shift, the Doppler effect, is calibrated in terms of the motion of the target precipitation. With this capability, it is possible to measure the air circulation pattern within a thunderstorm or other storm cloud. Multiple Doppler radar units viewing the same storm simultaneously create a three-dimensional image of air circulation.

FIGURE 14.19
The Doppler effect is the shift in frequency of sound or electromagnetic waves that accompanies the relative motion of the wave source or wave receiver. (A) A sound-wave source (a train whistle, for example) is stationary, and wave frequency is uniform everywhere. (B) The wave source is in motion so that wave frequency is greater ahead of the source than behind the source.

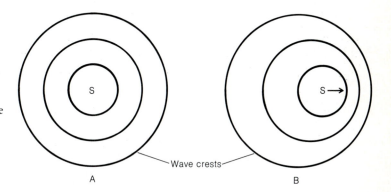

The principal advantage of Doppler radar over conventional radar is that it can determine the detailed circulation *within* a weather system rather than just the general displacement of an area of precipitation. Thus, Doppler radar "sees" inside clouds and severe thunderstorms and detects mesocyclones, developing tornadoes, strong wind shears, and gust fronts. Doppler radar displays, such as the sequence in Plate 23, are color-coded so that greens and blues (cold colors) indicate motion toward the radar and reds and yellows (warm colors) indicate motion away from the radar.

Because Doppler radar monitors a tornado as it develops within a mesocyclone and before it begins descending to the surface, it provides more advanced warning of severe weather than does conventional radar. Field tests of Doppler radar indicate that a warning of up to 20 minutes is quite feasible. Conventional radar, on the other hand, provides an average warning of only about 2 minutes because spotters on the ground must confirm a suspected tornado signature on radar. The greater advanced warning time of Doppler radar will probably mean many more lives saved.

Current radars are scheduled to be replaced by a new multimillion dollar network of Doppler radars by 1993. **NEXRAD,** for NEXt Generation Weather RADar, is a joint effort of the National Weather Service, the Air Weather Service of the United States Air Force, and the Federal Aviation Administration (FAA). It is intended to provide data for civil, military, and aviation needs. Plans call for a network of 113 10-cm microwave units, each capable of completing one 360-degree sweep every 5 minutes, detecting target precipitation at a range of 460 km (286 mi), and measuring velocity of particles as distant as 230 km (143 mi). Data from each NEXRAD unit will be processed by computer into a variety of guidance products such as in Plate 23. Through a telephone link, users may access these products at 157 centers nationwide.

NEXRAD is intended primarily to upgrade the monitoring of severe storm systems. As such, it will probably not significantly improve detection of microbursts and associated wind shear at airports (Chapter 13). Planning is now under way for a separate Doppler radar system based at airport terminals and dedicated to microburst and wind shear detection.

Conclusions

Tornadoes are shortlived, small-scale weather systems that are spawned by severe thunderstorms. Tornado development requires a special combination of atmospheric and terrain conditions such that they are most frequent in spring over the continental interior of the United States. Hurricanes, on the other hand, are longlived, synoptic-scale systems that spend much of their life cycle over tropical oceans. Both tornadoes and hurricanes are capable of causing death and injury and considerable property damage.

With this chapter, we have completed our description of the genesis, life cycle, and characteristics of the various atmospheric circulation systems that affect midlatitude weather. With this background, we are ready to apply principles of meteorology to weather forecasting.

Summary Statements

A tornado is a small mass of air that whirls rapidly about an almost vertical axis and is made visible by clouds, dust, and debris sucked into the system.

When a tornado strikes, damage is caused by very high winds, a strong updraft, an abrupt air pressure drop, and subsidiary vortices. Most tornadoes occur during the spring in a corridor stretching from Texas northward to Nebraska and eastward to Illinois and Indiana.

Synoptic weather conditions favorable for the outbreak of tornadoes progress northward (with the sun) from the Gulf Coast in early spring to southern Canada by early summer.

A tornado develops out of a mesocyclone that forms in the strong updraft of a severe thunderstorm by a process that is not yet well understood. The circulation in a tornado is apparently the consequence of an interaction between a thunderstorm updraft and shear in the horizontal wind.

Based on wind speed and the potential damage to structures, tornadoes are classified as weak, strong, or violent on the F-scale. Most tornadoes are weak, but most fatalities are caused by rare violent tornadoes.

Hurricane damage is a consequence of strong winds, heavy rains, and, in some coastal areas, a storm surge. Based on intensity and potential for damage, a hurricane is rated 1-to-5 on the Saffir–Simpson scale.

Hurricanes develop in a mass of warm, humid air, have no fronts, and are about one-third the size of an extratropical cyclone. Most hurricanes that affect North America originate as convective cloud clusters associated with waves in the trade winds. For reasons unknown, some convective cloud clusters develop into tropical disturbances that may evolve into tropical depressions, tropical storms, and then hurricanes.

A hurricane is an intense oceanic cyclone that originates in tropical latitudes where sea-surface temperatures are 26.5 °C (80 °F) or higher through a depth of at least 60 m (100 ft), the Coriolis effect is significant, and vertical wind shear is weak.

Once a hurricane makes landfall (or passes over relatively cold ocean water), it loses its warm-water energy source, and its circulation rapidly weakens. Heavy rains may continue well inland, however, and can lead to severe flooding.

Weather radar is a valuable tool for detecting the location and movement of areas of rainfall. By using Doppler radar, the detailed circulation within a severe thunderstorm may be monitored.

Key Words

tornado	dust devils	eye wall	radar echo
funnel cloud	hurricanes	typhoons	hook echo
F-scale	storm surge	tropical disturbance	Doppler effect
tornado alley	Saffir-Simpson	tropical depression	Doppler radar
mesocyclone	Hurricane Intensity	tropical storm	NEXRAD
waterspout	Scale	Project STORMFURY	
virga	eye (of a hurricane)	radar	

Review Questions

1. What is a tornado?

2. Distinguish between a tornado and a funnel cloud.

3. Tornadoes usually move in what direction? Speculate on why.

4. What is the principal force operating within a tornado?

5. What is the basis for the F-scale of tornado intensity? Where do most U.S. tornadoes rank on this scale?

6. Why are tornadoes most likely to develop over relatively flat and dry terrain.

7. Where is tornado alley?

8. The locale of principal tornadic activity shifts northward during the spring. Why?

9. Why is our understanding of tornado genesis far from complete?

10. How do tornadoes compare in appearance and intensity with dust devils and waterspouts?

11. How can distant virga be distinguished from a funnel cloud.

12. What is a hurricane?

13. Describe the most destructive effect of a hurricane approaching a low-lying coastal area.

14. What is the basis for the Saffir–Simpson Hurricane Intensity Scale?

15. Describe the isobar pattern about an intense hurricane.

16. Describe the structure of a hurricane and the life cycle of the system.

17. While in tropical latitudes, a hurricane or other tropical storm usually drifts slowly toward the west. Why west?

18. Describe what happens to a hurricane when it reaches middle latitudes.

19. Explain why hurricanes develop only over certain areas of the ocean.

20. Distinguish among (a) tropical disturbance, (b) tropical depression, (c) tropical storm, and (d) hurricane.

Points to Ponder

1. Can a tornado cause a building to explode? Explain your answer.

2. What is understood about the relationship between severe thunderstorms and tornadoes? What is a "mesocyclone"?

3. Describe synoptic weather conditions conducive to tornado development.

4. Why are tornadoes more frequent in spring than in fall?

5. Why do hurricane winds weaken as soon as the system moves over land?

6. Compare and contrast a hurricane with a mature, midlatitude, synoptic-scale cyclone.

7. Distinguish between conventional weather radar and Doppler radar. Why is radar particularly useful for monitoring thunderstorm cells?

8. Refer to Tables 14.5 and 14.6 and speculate on why (a) the ten deadliest hurricanes occurred prior to 1958 and (b) the ten costliest hurricanes occurred since 1953.

Projects

1. From the NOAA Climatic Summary for your area, determine the frequency of tornadoes in your region of the country. What local topographic and atmospheric conditions make your locality more or less prone to tornado development?

2. Find out if the circulation from a hurricane or tropical storm has ever influenced your locality.

Selected Readings

American Meteorological Society. "Is the United States Headed for Hurricane Disaster?" *Bulletin of the American Meteorological Society* 67 (1986):537–538. Includes a statement of concern issued by the AMS.

Cobb, H. "The Siege of New England." *Weatherwise* 42 (1989):262–266. Describes two hurricanes, Carol and Edna, that struck New England within 10 days during the fall of 1954.

Dolan, R., and H. Lins. "Beaches and Barrier Isands." *Scientific American* 257, No. 1 (1987):68–77. Presents a good summary of the vulnerability of barrier islands to coastal storms, especially hurricanes.

Fujita, T. T. "Rugged Rocky Mountain Tornado." *Weatherwise* 41 (1988):80–83. Reports on a rare tornado in the mountainous Teton Wilderness and Yellowstone National Park.

Galway, J. G. "Ten Famous Tornado Outbreaks." *Weatherwise* 34 (1981):100–109. Discusses great tornado outbreaks during the period 1870 to 1979.

Hughes, P. "The Great Galveston Hurricane." *Weatherwise* 32 (1979):148–156. Describes probably the greatest natural disaster in U.S. history.

Ludlum, D. M. "The Great Hurricane of 1938." *Weatherwise* 41 (1988):214–216. Details perhaps the most damaging hurricane to strike New England.

Miller, P. "Tornado!" *National Geographic* 171 (1986):690–715. Presents a well-illustrated discussion of the causes and effects of nature's most violent weather system.

Milner, S. "NEXRAD, The Coming Revolution in Radar Storm Detection and Warning." *Weathewise* 39 (1986):72–85. Provides an excellent summary of the history of development of Doppler radar systems.

Snow, J. T. "The Tornado." *Scientific American* 250, No. 4 (1984):86–96. Gives a well-illustrated review of tornado characteristics and genesis.

Willoughby, H. E., et al. "Project STORMFURY: A Scientific Chronicle 1962–1983." *Bulletin of the American Meteorological Society* 66 (1985):505–514. Traces the historical roots of efforts to modify hurricanes via seeding techniques.

*Probable nor'east to sou'west
    wind, varying to the souhard
    and westard and eastard and
    points between;
high and low barometer,
    sweeping round from place to
    place;
probable areas of rain, snow,
    hail, and drought, succeeded or
    preceded by earthquakes with
    thunder and lightning.*

Mark Twain
NEW ENGLAND WEATHER

15

Weather Analysis and Forecasting

Weather forecasters use sophisticated tools for observing and analyzing the state of the atmosphere,
including satellites, radar, and computers. [Scenes from *The Weather
Channel* Forecast Center and Studio]

401

OST PEOPLE can readily recall the occasions when an erroneous weather forecast upset their plans. It may have been an unexpected thundershower that brought an abrupt end to a softball game, or a raging blizzard that appeared instead of the anticipated clearing skies, or the promised springlike weekend that turned out to be anything but springlike. People seem to remember missed weather forecasts all too clearly and conveniently overlook the many times when the forecast was on target.

Actually, when viewed with the detached objectivity provided by statistical analysis, short-range weather forecasting is found to be surprisingly accurate. For example, the United States **National Weather Service (NWS),** an agency of the **National Oceanic and Atmospheric Administration (NOAA),** issues 24-hour weather forecasts that are correct nearly 85 percent of the time. The popular notion that weather forecasting is seldom accurate probably stems from the simple fact that we notice and remember a missed forecast more readily than an accurate one because of the inconvenience experienced.

How are weather forecasts made? What are the limits of forecast accuracy? On the basis of what you have learned so far, how can you make your own weather forecasts? These are some of the questions considered in this chapter.

World Meteorological Organization

Because the atmosphere is a continuous fluid that envelops the globe, weather observation, analysis, and forecasting require international cooperation. To this end, the **International Meteorological Organization (IMO)** was created in 1878. In 1947, the IMO changed its name to the **World Meteorological Organization (WMO)** and became an agency of the United Nations. Today, the WMO, headquartered in Geneva, Switzerland, coordinates the efforts of more than 145 member nations in a standardized global weather-monitoring network called **World Weather Watch.**

At standard observation times, the state of the atmosphere worldwide is monitored daily by almost 4000 land stations, by more than 7000 ships at sea, by almost 1000 upper-air observing stations, and by reconnaissance aircraft and satellites. These data are transmitted to the three World Meteorological Centers near Washington, D.C.; Moscow; and Melbourne, Australia, where maps and charts are drawn up representing the present state of the atmosphere. From analyses of this information, generalized weather forecasts are prepared. Maps, charts, and forecasts are then sent to National Meteorological Centers (NMCs) located in WMO member nations as well as to 26 Regional Meteorological Centers (RMCs). At NMCs and RMCs, weather information and forecasts are generated and interpreted for each center's area of responsibility and distributed to local weather service offices and then to the public. The United States NMC is located at Camp Springs, Maryland; the Canadian NMC is in Toronto, Ontario.

From the above description we see that weather forecasting entails (1) acquisition of present weather data, (2) depiction of data on weather maps, (3) analysis of data and prediction, and (4) dissemination of weather information and forecasts to users.

Acquisition of Weather Data

Since the invention of the first weather instruments in the seventeenth century, weather monitoring has undergone considerable refinement. Denser monitoring networks, more sophisticated instruments and communications systems, and better trained weather observers have produced an increasingly detailed, reliable, and representative record of weather and climate.

HISTORICAL PERSPECTIVE

The first European explorers to set foot on North American soil were very interested in the weather and climate of the New World. French colonists were the first to spend an entire winter, in 1604–1605, and to establish a settlement on the Atlantic Coast near the present border between Maine and New Brunswick. Samuel de Champlain, geographer for that expedition, commented on the sharp contrast in weather with that of his native France.

> Snow fell on the sixth of October. On the third of December we saw ice passing, which came from some frozen river. The cold was severe and more extreme than in France, and lasted much longer. I believe this is caused by the north-west winds, which pass over mountains continually covered with snow. This we had to a depth of three or four feet up to the end of the month of April; and I believe also that it lasts much longer than it would if the land were under cultivation.*

Almost half of the 79 French colonists failed to survive that first very severe winter.

Some early weather observers used primitive weather instruments, whereas others made qualitative assessments of the weather, jotting down observations in journals or diaries. The first systematic weather observations in North America were made in 1644–45 at Old Swedes Fort (now Wilmington, Delaware). The observer was John Campanius Holm, chaplain of the Swedish military expedition. Other temperature records were begun in Philadelphia in 1731; in Charleston, South Carolina, in 1738; and in Cambridge,

---

* From *The Works of Samuel de Champlain.* H. P. Biggar, ed. Toronto: The Champlain Society, 1922, pp. 302–303.

Massachusetts, in 1753. Thomas Jefferson and James Madison are credited with making the first simultaneous weather observations in the United States, in 1777 and 1778. In the late 1700s and early 1800s, weather recording was generally sparse and sporadic across North America. A notable exception is the New Haven, Connecticut, temperature record, which was begun in 1781 and continues today.

On 2 May 1814, James Tilton, M.D., Surgeon General of the United States Army, issued an order that in retrospect marked the first step in the eventual establishment of a national network of weather-observing stations. Tilton directed the army medical corps to maintain a diary of weather conditions at army posts, with responsibility for weather observations falling to the post's chief medical officer or surgeon. Tilton's objective was to learn more about the climate encountered by troops in the then sparsely populated interior of the continent. He also wanted to assess the relationship between weather and health, for it was a popular notion at the time that weather and climate were important factors in the onset of disease.

It took time for Tilton's order to be implemented. The War of 1812 was still raging, and weather instruments had to be distributed along with directions for proper use. Benjamin Waterhouse, M.D., surgeon at Cambridge, Massachusetts, was the first to submit weather data (for March 1816). By 1818, reports of weather observations at several army posts began trickling in to the Surgeon General's office, and under the direction of Tilton's successor, Joseph Lovell, M.D., the data were compiled, summarized, and eventually published. For this reason, Lovell, rather than Tilton, is sometimes credited with being the founder of the government's system of weather observation.

By 1838, 16 army posts had compiled at least 10 complete, though not aways successive, years of weather data. In ensuing years, the number of military weather-observing stations climbed steadily, reaching 60 by 1843, and by the end of the Civil War, weather records had been assembled for varying periods at 143 locations.

Invention of the telegraph in the mid-1800s spurred Professor Joseph Henry, then Secretary of the new Smithsonian Institution in Washington, D.C., to organize a network of 150 volunteer weather observers. Data were wired to the Smithsonian and displayed on maps. By the 1870s, the Smithsonian's weather network, that operated by the Surgeon General, and another one run by the United States Army Corps of Engineers gradually were incorporated into a single weather observation network within the Army Signal Services (or Corps).

A major impetus for establishing a national weather-observing network was the appalling loss of life and property in shipwrecks caused by surprise storms that swept the Great Lakes. Congress called on the Army to monitor weather conditions and to issue appropriate storm warnings. President Grant signed this resolution into law in early 1870. By 1878, personnel of the United States Army Signal Services were monitoring weather conditions at 284 stations.

In 1891, the military weather network was placed in civilian hands in a new Weather Bureau within the Department of Agriculture, with a special mandate to provide weather and climate guidance to farmers. Later, aviation's need for weather information became increasingly important, and in 1940 the Weather Bureau came under the jurisdiction of the Commerce Department. As the complexity of the physical environment became better understood, the Weather Bureau was reorganized as the National Weather Service (NWS) and was placed under the supervision of the Environmental Science Services Administration (ESSA) in 1965. In 1971, ESSA became the National Oceanic and Atmospheric Administration (NOAA).

In Canada, the earliest weather records date to sporadic observations by British Army officers in the late 1760s. Reverend Alexander Spark, a Presbyterian minister at Quebec City, began the first long-term weather record in December 1798; it ended with his death in March 1819. At Montreal, Thomas McCord and his son, John S. McCord, kept a weather journal from January 1813 to 1842. In addition, the Hudson's Bay Company maintained early nineteenth-century temperature records at its Whale River (1814 to 1816) and Eastmain (1814 to 1821) posts along the east shore of Hudson Bay.

In 1839, the Meteorological Service of Canada was founded and initially placed under military jurisdiction. In 1853, the Service came under civilian control. The first national observation network was established in 1843–44, and the first forecast service began operating in 1876. Cooperation with the United States Weather Bureau dates back to that year, and cooperation increased after World War II with the establishment of joint Arctic weather stations. Today, Canadian weather services are provided by the Atmospheric Environment Service (AES) in the Department of Fisheries and the Environment.

SURFACE WEATHER OBSERVATIONS

Across the United States, nearly 1000 stations routinely monitor surface weather. These stations may be operated by any of the following: (1) National Weather Service personnel, (2) the staff of other government agencies, including the Federal Aviation Administration (FAA), or (3) private citizens or businesses in cooperation with the NWS. At sea and in the Great Lakes, more than 2000 ships also voluntarily gather surface weather data. In addition, the NWS maintains networks of automated weather stations in locations where manned observations are not feasible. This includes, for example, 49 stations on buoys that are moored offshore or in nearshore areas, including the Great Lakes. Observational data from the buoy network are relayed to users via satellite.

Land-based weather-observing stations are part of the **synoptic weather network** or the **basic weather network** or both. Member stations of the synoptic weather network gather weather information primarily to prepare weather maps and forecasts and to exchange data with other nations. These stations observe and report cloud type and sky cover, wind speed and direction, visibility, precipitation, air temperature, dew point, and air pressure. These reports are issued every three hours, beginning at midnight **Greenwich Mean Time (GMT),** which is the time at the Old Royal Observatory, Greenwich, England. (Greenwich is located at 0 degrees longitude, the prime meridian.) Greenwich Mean Time* is the standard used by weather observers all over the world and is sometimes also called Universal Coordinated Time (UCT).

---

* For reference, at 0600 GMT, it is midnight Central Standard Time (CST) in Chicago, and 10 P.M. Pacific Standard Time (PST) in Vancouver, British Columbia.

Special Topic

Aviation Weather Hazards

The *Federal Aviation Administration (FAA)* estimates that 50 percent of all aircraft accidents are weather related. Between 1975 and 1986 weather was the chief causal factor in 26 percent of fatal accidents involving U.S. commercial jetliners. Three major aviation weather hazards are obstructed visibility, turbulence, and icing. In Chapter 13 we described a fourth hazard, the microburst, which can accompany even relatively weak thunderstorms.

In the interest of aircraft safety, pilots obtain preflight and in-fight weather briefings tailored to their individual flight plans. These briefings report current and anticipated weather conditions, including any potential weather hazards. The briefings are given in a special code, which is presented in Appendix III.

Visibility is the maximum horizontal distance at which prominent objects can be seen and identified. The *ceiling*, the altitude of the base of the lowest cloud layer covering more than half the sky, is also important because visibility can be restricted by low stratus clouds. Other visibility-restricting conditions include fog, polluted air, haze, precipitation, and blowing snow, sand, or dust. In general, visibility is good and ceilings are high in anticyclones (except for radiation fog); conversely, visibility can be poor, and ceilings are low in cyclones.

Depending on visibility, pilots fly under one of four flight rules listed in Table 1. When visibility is good, *Visual Flight Rules (VFR)* apply, and a pilot relies on vision to spot landmarks and other aircraft. When visibility is poor, *Instrument Flight Rules (IFR)* apply, and navigation must be aided by instruments.

Because low visibility is the chief reason for flight delays and cancellations, airports should be located where local climate favors good visibility. The ideal site is a moderately elevated area; a low area subject to cold air drainage and frequent radiation fogs is especially unsuitable. The airport should also be upwind of industrial areas to avoid the adverse effects of air pollution on visibility.

As we saw in Chapter 9, *turbulence* is an irregular

Table 1
Flight Rule Criteria

Category	Ceiling (ft)		Visibility (mi)
Low Instrument Flight Rules (LIFR)	<500	and/or	<1
Instrument Flight Rules (IFR)	500 to <1000	and/or	1 to <3
Marginal Visual Flight Rules (MVFR)	1000 to 3000	and/or	3 to 5
Visual Flight Rules (VFR)	>3000	and	>5

Member stations of the basic weather network provide weather observations hourly, mainly for aviation purposes but also to supplement weather forecasting. They measure the same variables as the synoptic weather network stations and, in addition, report cloud height and altimeter settings for aircraft. Pilots obtain weather briefings prior to and during flight and are advised on any weather conditions that may pose a hazard to flight. For a summary description of these hazards, refer to the Special Topic "Aviation Weather Hazards."

Some weather stations are equipped with radar, which tracks the intensity and movement of severe thunderstorms, tornadoes, and hurricanes, as well

flow of air analogous to the white water in the rapids of a swiftly flowing stream. Eddies of varying size, generated by thermal or mechanical influences, develop in the wind. As the mean wind speed increases, eddies generally become more energetic and can adversely affect aircraft in flight.

A shear in wind speed—that is, a change in wind speed over a relatively short distance—most often generates turbulence. The greater the *wind shear,* the more severe is the turbulence. Downbursts, convection currents, the jet stream, and fronts (especially cold fronts) all produce wind shear and turbulence. In convection currents, for example, the wind shear between updrafts and downdrafts produces turbulence that aircraft passengers feel as bumpiness, especially on takeoff and landing. Cumulus clouds are visible evidence of convective turbulence: vertical winds are upward where the sky is cloudy and downward where the sky is clear. Convective turbulence usually poses no problem for aircraft unless the currents surge to high altitudes and trigger thunderstorms and severe wind shear. The likelihood of extreme turbulence and the possibility of damaging hail mean that aircraft should never attempt to fly through a thunderstorm.

Mountain waves can also generate turbulence. Recall from Chapter 12 that a lofty mountain range disturbs large-scale winds so that *standing waves* develop to the lee of the range. Vigorous eddies form below mountain waves and sometimes penetrate to the ground as a rotorlike circulation. Low-flying aircraft can be caught in the rotor and be forced to the ground or into the mountain side. Turbulence may also be encountered at altitudes above the mountain waves, perhaps up to the tropopause.

Often turbulence aloft can be detected by merely observing clouds. In addition to cumulus clouds, the clouds exhibiting a wave pattern (stratocumulus, altocumulus, and cirrocumulus) are associated with turbulent airflow. In other instances, turbulent air occurs without clouds. *Clear air turbulence (CAT)* is difficult to detect except by sophisticated airborne instruments currently under development. Normally, however, CAT is confined to relatively thin layers of the atmosphere, and pilots readily escape this turbulence by changing altitude.

The usual turbulence encountered by commercial airliners is so light as to be merely annoying. On rare occasions when an aircraft flies into an area of severe turbulence, abrupt changes in altitude, attitude, or both occur. For this reason, passengers are best advised to keep seat belts buckled throughout a flight.

Aircraft flown through freezing rain or clouds composed of supercooled water droplets are prone to *icing,* the accumulation of ice on the leading edges of the aircraft. Supercooled raindrops or cloud droplets freeze on contact with the surface of the wings and fuselage. If the aircraft does not have deicing or anti-icing equipment, ice buildup adds weight and interferes with the aircraft's aerodynamics. Icing is most often a hazard for aircraft flying below about 6000 m (20,000 ft) and at air speeds of less than 400 knots. Above 6000 m, air temperatures are so low that clouds are composed almost exclusively of ice crystals, which do not accumulate on the aircraft.

as other areas of precipitation. The National Weather Service operates about 120 radar stations across the United States. Half of these operate continuously; the other half operate only when weather conditions warrant. As noted in Chapter 14, the United States government plans to replace these conventional radar units with a network of Doppler weather radars by the early 1990s.

Besides the nearly 1000 land-based weather stations that provide information of potential use in weather forecasting, an additional 11,590 cooperative weather stations are scattered across the United States. These stations constitute the **National Substation Program,** and their principal function is to record daily precipitation totals and maximum and minimum temperatures for hydrologic, agricultural, and climatic purposes.

UPPER-AIR WEATHER OBSERVATIONS

As noted in Chapter 1, meteorologists monitor the upper atmosphere with radiosondes. A **radiosonde** is a radio-equipped instrument package borne aloft by a balloon. The instruments transmit to a ground station the vertical profiles of air temperature, pressure, and relative humidity up to an altitude of about 30 km (19 mi). In addition, winds at various levels are computed by tracking the balloons with a radio direction-finding antenna (rawinsonde observations). Worldwide readings are made twice each day, at 0000 GMT and 1200 GMT.

In the United States, there are 126 radiosonde stations, 25 of which use special high-altitude balloons that provide data from altitudes above 30 km. Meteorological rockets probe much higher altitudes (to 100 km), but the data obtained from these probes are used primarily for research. Upper-air weather data are also obtained by aircraft, dropwindsondes, radar, and satellites. In some urban areas, low-level soundings (to 3000 m) monitor weather conditions to assess air pollution potential.

Canada's Atmospheric Environment Service operates similar networks of surface and upper-air weather observation stations. The Canadian service has 270 first-order weather stations and more than 3000 climatological stations nationwide.

Meteorology by Satellite

The foremost advantage of weather satellites is that they provide a continuous picture of the state of the atmosphere. The networks of surface and upper-air observation stations detect weather conditions at discrete points only, some of which may be hundreds or thousands of kilometers apart. Satellite imagery fills in the gaps. Hence, satellites "see" subsynoptic weather systems, such as severe thunderstorms, that might escape detection by weather stations. Satellites also enable forecasters to locate and track tropical storms over oceans where weather observations are few and far between.

A variety of weather satellites are either currently in orbit or planned for the future. Weather satellites have been launched by the United States, the Soviet Union, Japan, and the 11-nation European Space Agency. Most are in near-polar orbits at altitudes of 800 to 1000 km (500 to 620 mi), that is, orbits that take the satellites on longitudinal (meridional) trajectories over the polar regions. For example, the U.S. weather satellites NOAA-7 and NOAA-8 observe a strip 1800 km (1100 mi) wide from pole to pole and back in 102 minutes. The satellites then observe the next adjacent meridional strip. With 14 orbits each day, a **polar-orbiting satellite** "sees" a particular locality twice every 24-hour day.

Some weather satellites are in geosynchronous orbits over the equator; that is, they orbit at a rate that matches the Earth's rotation, so they are always above the same spot and scan the same region. From an altitude of about 36,000 km (22,320 mi), each **geosynchronous satellite** monitors almost one-third of the Earth's surface and has a "full-disk" view of the Earth from pole to pole. Because polar-orbiting satellites fly at much lower altitudes, they provide greater resolution (more detail) but over a smaller area and with a longer time interval between images. For more on the orbital characteristics of both polar and geosynchronous satellites, refer to the Mathematical Note at the end of this chapter.

Normally, the Western Hemisphere is monitored by two Geostationary Operational Environmental Satellites, GOES West and GOES East. GOES West, located at approximately 135 degrees W longitude, and GOES East, located at approximately 75 degrees W, provide cloud cover pictures every 30 minutes both night and day. The unexpected failure of GOES East in July 1984 meant that the other geostationary satellite had to do double duty until a new satellite was launched in February 1987. During the interim, the one operational satellite was stationed over 98 degrees W during the Atlantic hurricane season and then moved to 108 degrees W for the remainder of the year.

Sensors on board satellites intercept two types of radiation that emanate from the Earth-atmosphere system: reflected solar radiation (visible) and emitted infrared radiation. Visible satellite imagery is like a black and white photo of Earth taken from space. It reveals cloud patterns, dust storms, and the extent of snow cover over land and ice cover at sea, as well as geographical features such as the Great Lakes and the Black Hills of South Dakota. As a rule, we can discern any object on the Earth's surface larger than 1.0 km (0.6 mi) across. From cloud patterns, we can locate storm centers, fronts, fog, and thunderstorms. The cloud pattern of a hurricane tells meteorologists something about the storm's size, spiral structure, and eye characteristics, and from these it is possible to estimate the stage of the hurricane's development and its strength.

Surveillance of weather systems is made possible by viewing successive satellite images of the same area as a movie loop. Geostationary satellites obtain images every 30 minutes, athough, under special circumstances, the National Weather Service can direct the satellite to scan only a portion of the "full-disk" and so obtain images as frequently as every 3 to 5 minutes. Sequential GOES images are therefore used to monitor the development of thunderstorms and squall lines, and to predict the track of tropical storms by simple extrapolation.

Satellite infrared (IR) sensors measure the heat emitted by land and sea surfaces and the tops of clouds. In Chapter 2, we discussed how emitted radiation depends on the temperature of the radiating body. Based on measured differences in IR emission, sensors can distinguish clouds at low levels, which are relatively warm, from clouds at high levels, which are relatively cold. These measurements can be calibrated on a gray scale in which low clouds appear dark gray and high clouds appear white (Figure 15.1). IR

FIGURE 15.1
GOES full-disk infrared image showing clouds on a grey scale. High clouds are bright white, and low clouds are grey. [NOAA National Environmental Satellite, Data, and Information Service]

sensing of cloud-top temperature is also a means of gauging the depth of convection and the intensity of thunderstorms. As noted in Chapter 13, the altitude of a cloud-top is an indicator of thunderstorm intensity. The higher the top, the lower is the cloud-top temperature, and the more intense is the thunderstorm.

Horizontal winds are inferred from satellite observations of cloud displacement. Clouds move with the wind so that wind speed and direction can be estimated from the drift of clouds in successive satellite images. To determine the actual altitude of those winds, the satellite's IR sensor measures the cloud's radiation temperature, which is then matched to the nearest radiosonde sounding to give cloud (and wind) altitude. Meteorolgists routinely monitor winds in the troposphere by this cloud-track method four times daily and transmit those data to forecast centers.

Vertical winds can also be inferred by satellite even in the absence of convective clouds. IR measurements can distinguish regions of relatively high water vapor concentration from regions of relatively low water vapor concentration. In the satellite image of Figure 15.2, the lighter regions are more humid than the darker regions. We can assume that upward motion characterizes humid regions and subsiding motion characterizes dry regions.

By instrument analysis of the detailed characteristics of the spectrum of radiation emitted by the atmosphere, satellites can obtain soundings.* Now in the final testing phase of development, this technique of measuring vertical profiles of temperature and humidity is expected to be a routine function of geostationary satellites by the early 1990s.

Besides the satellites used for weather observation, some meteorological satellites, such as those in the Nimbus series, are designed mainly for research purposes. Nimbus satellites are in polar orbits and monitor concentrations of water vapor, ozone, and other trace atmospheric gases.

Remote sensing by satellite as well as other advances in monitoring technology are generating a deluge of real-time weather data. This has necessitated development of computerized data management systems. Perhaps foremost among these systems is **McIDAS** (Man–computer Interactive Data Access System), which was designed by scientists at the Space Science and Engineering Center at the University of Wisconsin–Madison. McIDAS receives satellite imagery and soundings, radar displays, and conventional surface and upper-air observations. McIDAS integrates and organizes those data into guidance products that are very useful for potential users. Products include two- and three-dimensional composite images such as that shown in Plate 24. Users access McIDAS on a computer terminal, select some satellite image or other data display, and may choose, for example, to overlay different fields of data on a satellite image.

---

* The specific method used is beyond the scope of this book. For more information, interested readers are urged to consult the article by Smith et al. cited at the end of this chapter.

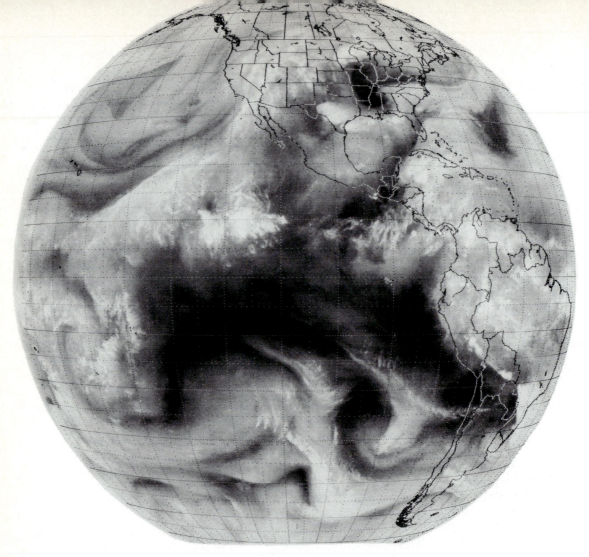

FIGURE 15.2
GOES full-disk infrared image displaying measurements in the water vapor channel.
On this grey scale, the lighter the shade, the more humid the air. [NOAA National
Environmental Satellite, Data, and Information Service]

Data Depiction on Weather Maps

Computers at the National Meteorological Centers use the special symbols
listed in Appendix III to plot weather observations on synoptic and hemispheric weather maps. The weather for each observation station is depicted
on a map by following a conventional **station model.** The station model in
Figure 15.3 shows the symbols for surface weather conditions. By international
agreement, the same symbols and station model are used throughout the
world.

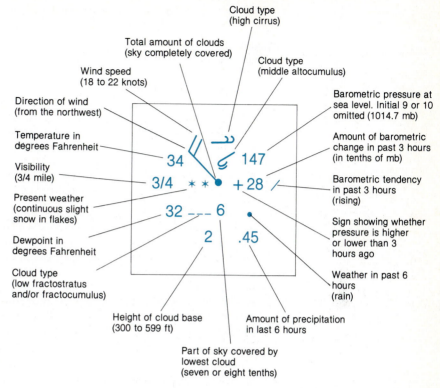

FIGURE 15.3
This conventional weather station model shows the symbols that are used to depict weather observations on surface weather maps.

Because weather systems are three dimensional, both surface and upper-air weather maps are needed. A very different approach is used for the two types of maps, however. Surface weather data are plotted on a constant-altitude (usually sea-level) surface, and upper-air weather data are plotted on constant-pressure (isobaric) surfaces.

SURFACE WEATHER MAPS

It is standard practice for meteorologists to reduce surface air pressure readings to sea level (Chapter 5). Therefore, barometer readings are adjusted so that the air pressure is what it would be if the station were actually at sea level. This procedure is intended to remove the influence of topography on air pressure and thus enable meteorologists to compare surface air pressure readings at weather stations that are situated at different elevations above sea level. The adjusted air pressure readings are plotted on surface weather maps. **Isobars,** lines of equal air pressure, are then constructed at 4-mb intervals to reveal such features as anticyclones and cyclones, troughs and ridges, and horizontal air pressure gradients.

As an illustration, Figure 15.4 presents the surface weather maps compiled for 1–3 April 1982, along with the corresponding GOES East visible satellite imagery. The NOAA *Weekly Weather and Crop Bulletin* reports the following weather summary for each of the three days.

THURSDAY, 1 April 1982—A warm front reached from South Carolina to Arkansas and through central Texas. Warm, moist air from the Gulf of Mexico flowed into Texas and then veered northeastward into Arkansas. Light showers fell in Texas, but moderate to heavy showers and thunderstorms battered Arkansas and northern Mississippi. A complex frontal system through the West produced light to moderate rain and heavy snow in the mountains. The precipitation covered the area west of the Rockies and spread into the northern Plains.

FRIDAY, 2 April 1982—An intense low pressure system deepened in the central Plains. Severe weather moved ahead of the storm through the entire Mississippi Valley and spread slowly eastward to a line from the upper Ohio Valley through Alabama. Light showers and thunderstorms extended to the mid-Atlantic states by the end of the day. The low pressure center deepened to record intensity as it reached the Great Lakes area. Tornadoes, hail, high winds, and heavy rain lashed the area from northeastern Texas into Minnesota and Wisconsin. Rain, wind, and snow from another storm hit the Pacific Northwest.

SATURDAY, 3 April 1982—The intense low pressure system moved slowly into the central lakes area. The cold front, marking the line of severe weather, moved from the central Great Lakes, through the Appalachians to central New York State, and along the east coast to northern Florida. Thunderstorms were not as intense along the front as on Friday, but golf-ball-size hail and high winds were reported from New York to Georgia. Rain showers, with snow in the mountains, spread into central California over the Plateau to the northern Plains.

The major weather event of these three days was the development of an intense cyclone that tracked into the Great Lakes region. (Refer back to Figure 5.2 for the barograph trace during this period at Green Bay, Wisconsin.) On satellite imagery, the mature storm shows up as a huge comma-shaped area of cloudiness. On surface weather maps, the storm center, where air pressure is lowest, is indicated by the symbol L. A pattern of closely spaced isobars surrounds the storm center so that surface winds are strong. Note how fronts originate at the storm center and how isobars bend ('kink") as they cross fronts, indicating that the fronts occupy troughs. Because we can infer surface wind direction from the pattern of isobars, the bending of isobars at fronts tells us that wind shifts direction on either side of a front. Recall from our discussion in Chapter 11 that a wind shift is one characteristic of a frontal passage. Note also how relatively weak air pressure gradients (widely spaced isobars) are associated with the anticyclones (mapped as H).

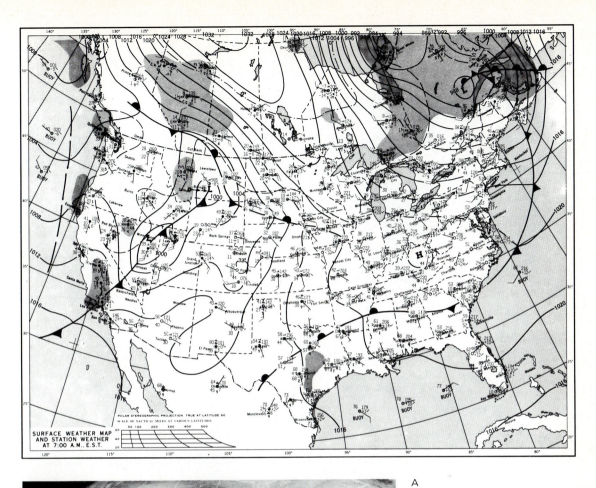

A

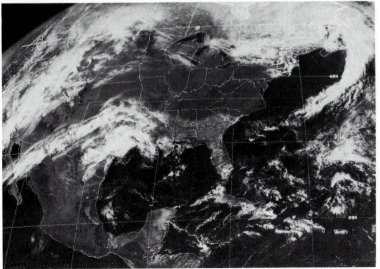

FIGURE 15.4
Surface weather maps and GOES satellite imagery for (A) 1 April 1982, (B) 2 April 1982, and (C) 3 April 1982. [From NOAA]

415

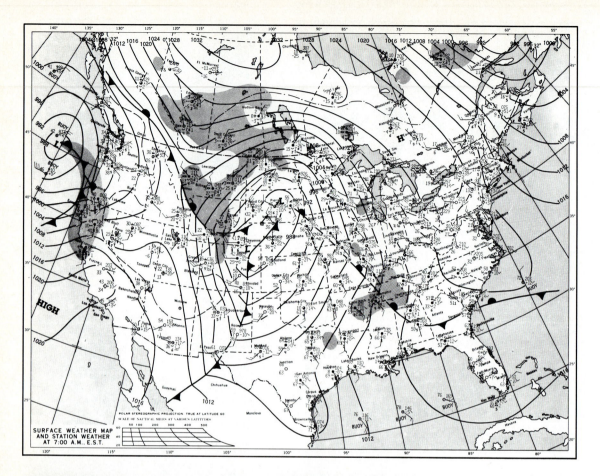

B

FIGURE 15.4 (continued)

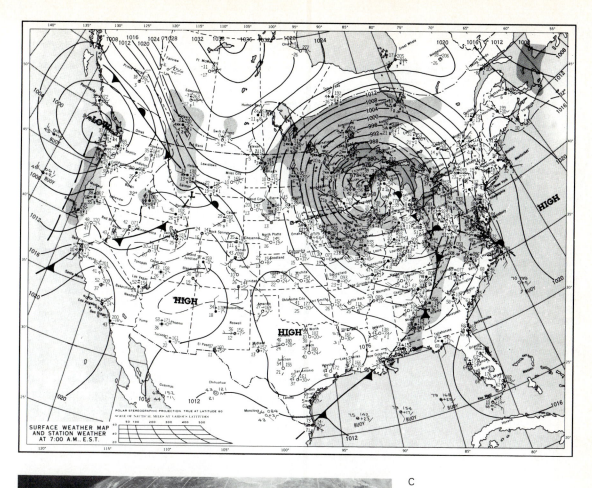

SURFACE WEATHER MAP
AND STATION WEATHER
AT 7:00 A.M. E.S.T.

POLAR STEREOGRAPHIC PROJECTION TRUE AT LATITUDE 60
SCALE OF NAUTICAL MILES AT VARIOUS LATITUDES

C

FIGURE 15.4 (continued)

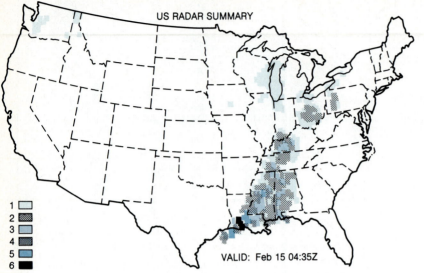

FIGURE 15.5
A sample national weather radar summary on a grey scale. The darkest shading represents the most intense echoes (heaviest rain). [Courtesy of Universal Weather and Aviation, Inc.]

Surface synoptic weather maps are drawn every three hours for North America and at six-hour intervals for the Northern Hemisphere. Special charts are also constructed that summarize a variety of weather elements including, for example, (1) maximum and minimum temperatures for 24-hour periods, (2) precipitation amounts for 6 hours and for 24 hours, and (3) observed snow cover. A radar summary chart is also issued hourly (Figure 15.5).

UPPER-AIR WEATHER MAPS

Upper-air weather data acquired by radiosondes are plotted on constant-pressure surfaces. Applying the basic laws of atmospheric physics, meteorologists compute the actual altitudes corresponding to these measurements. For example, we can determine the altitude of the 500-mb surface, that is, the altitude at which the air pressure drops to about one-half its standard sea-level value. By plotting altitudes of the 500-mb level as monitored simultaneously by all radiosonde stations across North America, meteorologists can then construct a map representing the "topography" of the 500-mb surface. They simply draw contours through localities where the 500-mb level is at the same altitude, a procedure that requires some interpolation between stations.

The altitude of a pressure surface (such as the 500-mb surface) varies from place to place, primarily because of differences in temperature of air below the pressure surface. The drop of air pressure with altitude is more rapid in cold air masses than in warm air masses. Because raising the temperature of air reduces its density, greater altitudes are required for warm air to exhibit the same drop in pressure as cold air. This means, for example, that the 500-mb level is at a low altitude when the air below is relatively cold and at a high altitude when the air below is relatively warm. The contours of an isobaric surface therefore always show a gradual slope downward from the warm tropics to the colder polar latitudes.

Upper-air observations indicate that horizontal winds parallel height contour lines at the 500-mb level. Where contours are closely spaced (a steep height gradient), winds are strong, and where contours are far apart (a weak height gradient), winds are light. Why are winds associated with a height gradient? A height gradient develops in response to a horizontal air temperature gradient, and where there is a horizontal air temperature gradient, there are also horizontal gradients in air density and air pressure. As we saw in Chapter 9, a horizontal air pressure gradient generates wind.

The 500-mb surface is so far above the friction layer that the synoptic-scale and global-scale winds are essentially geostrophic where contours are straight, and gradient where contours are curved. The large-scale wind at the 500-mb level is thus the product of interactions among a horizontal height gradient, the Coriolis effect, and the centripetal force.

On upper-air weather maps, contours exhibit both cyclonic (counterclockwise) and anticyclonic (clockwise) curvature. These are the ridges and troughs in the westerlies that we described in Chapter 10. Contours may also define a series of concentric circles, perhaps indicating a cutoff or blocking circulation pattern. At the center of a ridge, the air column is relatively warm and contour heights are high, so we label the ridge with an **H**. An upper-air ridge may be linked to a warm-core anticyclone at the surface. In contrast, at the center of a trough, the air column is relatively cold and contour heights are low, so the trough is labeled with an **L**. An upper-air trough may be linked to a cold-core cyclone at the surface.

From the preceding discussion, it follows that warm-core cyclones (thermal lows) and cold-core anticyclones (polar and arctic highs) do not appear on 500-mb weather maps. These pressure systems are simply too shallow to influence the air circulation pattern at the 500-mb level.

The 500-mb maps for 1–3 April 1982 are presented as Figure 15.6. The 500-mb weather chart is particularly useful because winds at that level closely approximate the upper-level steering winds and the trough and ridge patterns responsible for the development and displacement of surface weather systems. Solid lines are height contours of the 500-mb surface labeled in feet above sea level; the contour interval (difference between successive contours) is 200 ft. Dashed lines are isotherms labeled in degrees Celsius. Flags indicate wind direction and speed at the 500-mb level (Appendix III). Winds paralleling contour lines describe cyclonic flow in troughs and anticyclonic flow in ridges. In line with our discussion in Chapter 11, note how the Great Lakes storm in Figure 15.4C is linked to a cold trough aloft in Figure 15.6C.

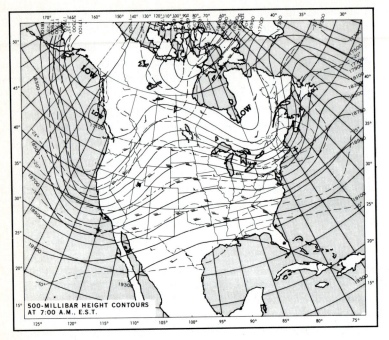

A

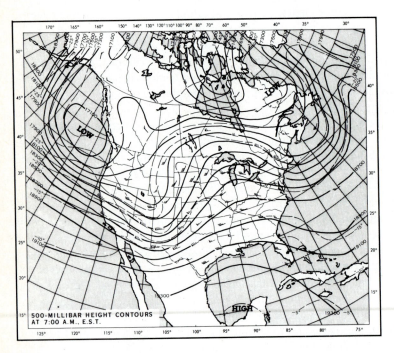

B

FIGURE 15.6
The 500-mb analyses for (A) 1 April 1982, (B) 2 April 1982, and (C) 3 April 1982. [From NOAA]

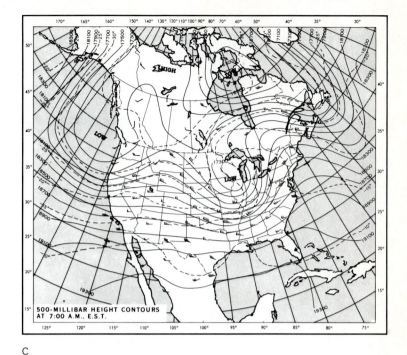

500-MILLIBAR HEIGHT CONTOURS AT 7:00 A.M., E.S.T.

FIGURE 15.6 (continued)   C

Winds flowing across isotherms produce cold or warm air advection. **Cold air advection** takes place when winds blow from colder localities toward warmer localities. **Warm air advection** occurs when winds blow in the opposite sense. Warm air advection causes the 500-mb surface to rise, and cold air advection causes the 500-mb surface to lower. Cold air advection at 500-mb thus deepens troughs and weakens ridges, whereas warm air advection strengthens ridges and weakens troughs.

When cold air advects into troughs at the same time that warm air advects into ridges, the circulation becomes more meridional. On the other hand, if warm air advects into troughs while cold air advects into ridges, the circulation becomes more zonal. As we saw in Chapter 10, shifts in the upper-air circulation pattern between meridional and zonal have important implications for air mass exchange and storm development.

The National Meteorological Centers (NMCs) issue 500-mb maps twice each day based on upper-air observations at 0000 GMT and 1200 GMT. One set of maps covers North America and another the Northern Hemisphere. Although we have focused our discussion of upper-air weather maps on the 500-mb level, similar analyses are routinely constructed twice daily for the 850-, 700-, 300-, 250-, 200-, and 100-mb levels. The 300- and 250-mb analyses are quite important because they are near the strongest winds of the midlatitude jet stream.

Weather Prediction

Meteorologists at National Meteorological Centers analyze satellite observations and weather data plotted on surface and upper-air weather maps. From these analyses, weather forecasts are then prepared. Forecasting the weather is extremely challenging, primarily because it involves many variables and a vast quantity of weather data. For these reasons, computerized numerical models of the atmosphere have been developed to assist forecasters.

NUMERICAL WEATHER FORECASTING

The first experimental use of an electronic computer as a guidance tool in weather forecasting was conducted at the Institute for Advanced Study at Princeton University in April 1950. A primitive computer, less powerful than today's personal computer, successfully forecast the horizontal air pressure pattern at 5 km (16,000 ft) altitude over North America 24 hours in advance. By 1955, computers were routinely generating weather forecasts from surface and upper-air weather observations.

An electronic computer is programmed with a numerical model of the atmosphere, that is, a model consisting of mathematical equations that relate winds, temperature, pressure, and water vapor concentration. (Many of these basic equations are presented in the Mathematical Notes of this book.) Beginning with present (real-time) weather data, numerical models predict pressure surface altitudes at some future time, say, 10 minutes hence. With these predicted conditions as a new starting point, another forecast is then computed for, say, the subsequent 10 minutes. The computer repeats this process again and again until a weather map is generated for the next 12, 24, 36, and 48 hours. In this process, tens of millions of computations must be performed each second on a vast array of observational data; hence, there is a need for a high-speed electronic computer that can accommodate huge quantities of data.

NMC computers apply different numerical models to specific portions of the atmosphere. As of this writing the LFM (Limited-area Fine-mesh Model) is being replaced by the Nested Grid Model (NGM), which was developed in 1984 by staff scientists at the National Meteorological Center. Already the NGM has proved its value in providing better predictions of East Coast winter storms. Table 15.1 lists a sample of the guidance materials produced by numerical models.

Some computerized numerical models of the atmosphere are designed to operate over different spatial scales depending on the forecast range. For medium-range forecasts (up to 10 days), observational data are fed into the computer from all over the globe, since within that forecast range a weather

Table 15.1

Sample Computer-Generated Guidance Maps
Used in Weather Forecasting for Periods of
12, 24, 36, and 48 Hours

Surface air pressure
1000-mb to 500-mb thickness (a measure of temperature)
700-mb vertical velocity and 12-hour precipitation totals
700-mb height contours
Surface to 500-mb average relative humidity
500-mb height contours
700-mb vorticity

---

system may travel long distances. On the other hand, for short-range forecasts
(up to 3 days), the model utilizes data drawn from a more restricted region
of the globe. Compared to a global model, a regional model offers the advantage of greater resolution of data over a smaller area of interest.

SPECIAL FORECAST CENTERS

Not all weather forecasts are prepared by the National Meteorological Centers.
Responsibility for forecasting tropical storms and hurricanes is divided among
two centers: the **National Hurricane Center (NHC)** in Coral Gables, Florida, and the Central Pacific Hurricane Center in Honolulu, Hawaii. Local and
regional weather service offices transmit information from these centers to the
public as advisories, warnings, or statements. The goal is to provide at least
12 hours of daylight warning so that coastal residents may prepare for a
hurricane.

Specially instrumented reconnaissance aircraft complement satellite and
radar surveillance of hurricanes and tropical storms. Aircraft fly directly into
and through the storms and determine the precise location of the eye, measure
wind speeds, and obtain soundings by dropwindsonde. By extrapolating from
air pressure readings at flight altitude (usually 1500 to 3000 m), scientists also
determine the sea-level air pressure at the storm center. These data are immediately radioed back to the NHC.

The principal challenge to forecasters at the National Hurricane Center is
prediction of the track of a hurricane. Normally, such forecasts are issued every
6 hours and cover periods up to 72 hours in the future. The basis for hurricane
track forecasts is a blend of climatology (records of tracks of similar hurricanes
in the past), numerical models, and the experience of the forecaster. Although
the advent of satellite monitoring of hurricanes (and their precursors) led to
a significant improvement in the accuracy of track forecasts in the early 1960s,
track forecasting skill has improved very little since then. Beyond 24 hours,

accuracy declines rapidly and approaches zero for 72-hour forecasts. Apparently, the major problem is the poor quality of observational data initially fed into the numerical model.

If, on the other hand, the hurricane track forecast is correct, another numerical model can predict quite accurately the location and height of the storm surge along coastal areas. Forecasters report good success with the NHC numerical model SLOSH (Sea, Lake and Overland Surges from Hurricanes). By consulting local topographic maps, officials can use SLOSH predictions to identify areas that are likely to be inundated by flood waters and to plan evacuation routes accordingly.

Beginning with the 1983 hurricane season, the NHC has been including a probability forecast as part of its advisory statements. Probability is defined as the percent chance that the center of a hurricane or tropical storm will pass within 105 km (65 mi) of any of 46 designated Gulf and East Coast communities from Brownsville, Texas, to Eastport, Maine. The first probability forecast is usually issued 72 hours in advance of the storm's anticipated landfall. At that time, by convention, the probability is set no higher than 10 percent for any community. Probabilities increase to 13 to 18 percent at 48 hours, 20 to 25 percent at 36 hours, 35 to 50 percent at 24 hours, and 60 to 80 percent at 12 hours before the storm's expected landfall.

Meteorologists at the **National Severe Storms Forecast Center (NSSFC)** at Kansas City, Missouri, monitor atmospheric conditions for the potential development of severe local storms (including winter blizzards) and issue watches for severe thunderstorms and tornadoes. Watches go out 2 to 6 hours in advance of expected severe weather and usually cover an area of about 65,000 km^2 (25,000 mi^2). Actual severe weather warnings are made by local or regional weather service offices.

FORECAST SKILL

Just how accurate are today's computer-guided weather forecasts? This question can be answered by comparing the accuracy of modern weather forecasting with predictions based on persistence (forecasting no change in present weather) or with predictions based on climatology (forecasts derived from past weather records). Table 15.2 rates the accuracy of computerized weather forecasting according to whether or not it exceeds the accuracy of persistence or climate forecasting. Results show that the accuracy of all types of weather forecasting deteriorates rapidly for periods longer than 48 hours and that accuracy is minimal beyond 10 days.

Several factors contribute to the rapid decline in forecasting skill. Computerized forecasts are only as accurate as the input (observational) data and predictive equations (numerical model) allow them to be. There is nothing magical about a computer. Errors are introduced by (1) missing or inaccurate

Table 15.2

Accuracy of Weather Forecasting Compared
with Accuracy of Persistence
and Climatological Forecasting[a]

Forecast period	Forecast accuracy
To 12 hours	Considerable
12–48 hours	Considerable
2–5 days	Moderate for temperature
	Slight for precipitation
6–10 days	Low for average temperature
	Slight for precipitation
10–30 days	Minimal
More than 30 days	Minimal

[a] In the middle latitudes of the Northern Hemisphere.
Source: "Policy Statement of the American Meteorological Society on
Weather Forecasting." *Bulletin of the American Meteorological Society* 60
(1979):1453–1454.

observations, (2) failure of weather stations to detect all mesoscale and microscale circulation systems, and (3) imprecise predictive equations that include assumptions and first approximations. Unfortunately, the adverse effects of these errors grow as the forecast period lengthens. In today's numerical models of the atmosphere, the impact of even small errors doubles about every 2 to 2.5 days over the forecast period. When computerized numerical models are run out to 10 days, some elements of daily forecasts are useful for some localities to about 6 or 7 days. Beyond that, chaos sets in.

Nevertheless, the accuracy of numerical weather forecasting has shown slow but steady improvement over the past three decades, and prospects for continued progress are good. This trend is largely due to (1) better understanding of atmospheric processes that makes possible the development of more realistic numerical models of the atmosphere, (2) larger and faster computers, (3) more reliable and sophisticated observational tools, including Doppler radar and remote soundings by satellite, and (4) denser observational networks worldwide.

The computer is never likely to replace human weather forecasters. The products of computerized numerical models function as guidance materials for the forecaster. He or she must understand the characteristic errors and biases of the various numerical models as well as atmospheric processes and weather-observing systems. As much as numerical weather prediction has advanced in recent years, the best forecasters still rely heavily on their personal experience and intuition, tempered by their knowledge of how the atmosphere works. A good forecaster does not start with the model output in preparing a

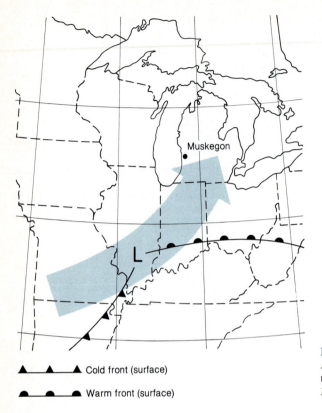

Muskegon

FIGURE 15.7

An early winter storm tracks northeastward through the Midwest and causes heavy snowfall at Muskegon, Michigan.

▲ ▲ ▲ Cold front (surface)

● ● ● Warm front (surface)

forecast, but rather starts with previous and current observations, forming a view of what the atmosphere has been doing and ideas on what is likely to happen in the future.

Special local or regional conditions may call for significant modification of a computer-generated weather forecast. Forecasters must analyze and interpret computerized predictions and adapt those forecasts as necessary to regional and local circumstances. Consider an example.

Suppose that an intense early winter storm tracks northeastward through the Midwest, across central Illinois, and into eastern lower Michigan, as shown in Figure 15.7. Muskegon, Michigan, is on the cold, snowy side of the storm's path, so residents experience strong northeast winds accompanied by blowing snow. After the storm passes, winds shift to the north and northwest, and cold air advection begins. A computerized numerical model might predict clearing skies for Muskegon, but local conditions dictate a different result. Strong winds from the northwest advect cold, dry air across the relatively warm waters of Lake Michigan, giving rise to **lake-effect snows** on the lake's eastern shore (Chapter 12). Because of this local effect, northwest winds may bring more snow to Muskegon after the storm has passed than northeast winds did when the storm was nearby.

LONG-RANGE FORECASTING

Recently, researchers reported success in using numerical models to predict the ENSO (El Niño/Southern Oscillation) event of 1986–87 some three to nine months in advance. As noted earlier, however, the reliability of detailed weather forecasts for periods longer than 10 days is minimal. The Long Range Prediction Branch of the Climate Analysis Center in Camp Springs, Maryland, prepares 30-day (monthly) and 90-day (seasonal) generalized weather "outlooks" that identify areas of expected positive and negative **anomalies** (departures from long-term averages) in temperature and precipitation. An example is shown in Figure 15.8.

Forecasting the prevailing circulation pattern at the 700-mb level is the first step in forecasting monthly temperature and precipitation anomalies at the surface. Basically, the present circulation pattern is extrapolated into the future, although an effort is also made, based on historical data, to identify features of the present pattern that are most likely to persist. The prevailing westerly flow at the 700-mb level permits identification of areas of strong warm and cold air advection as well as principal storm tracks. Predictions of surface temperature and precipitation anomalies are derived from these data.

A somewhat different approach is utilized for 90-day outlooks. Forecasters rely more on long-term trends and on recurring events and attempt to isolate persistent circulation features from prior months and seasons. In the analog technique, for example, a computer searches its 40-year archive (memory) of past seasonal weather patterns for the closest match with the present season's

FIGURE 15.8
Winter weather outlook for 1987–88 as issued in late fall 1987 by the National Weather Service. (A) The probability of the mean temperature being above or below the long-term average. (B) The probability of the total precipitation being above or below the long-term median. Winter encompasses the 90-day period of December 1987 through February 1988. [From NOAA]

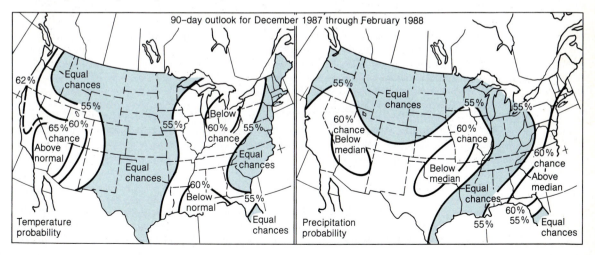

weather pattern. The 90-day outlook is then based on whatever followed the best historical match.

Statistics on seasonal forecast skill can be a bit confusing. For the past 23 years, the National Weather Service has reported average skills of only 8 percent for temperature outlooks and 4 percent for precipitation outlooks. To put these statistics in perspective, forecast skill would be 0 percent if based on chance alone and 100 percent for a perfect forecast. The greatest success so far has been with winter temperature outlooks, with a forecast skill of 16 to 18 percent.

A promising area of research in long-range weather forecasting concerns atmospheric teleconnections. A **teleconnection** is a linkage between weather changes occurring in widely separated regions of the globe, often many thousands of kilometers apart. An example is the linkage between the ENSO phenomenon and midlatitude weather extremes (Chapter 10).

Jerome Namias and his colleagues at the Scripps Oceanographic Institution have been studying teleconnections that may aid in seasonal weather forecasting for North America. They are investigating the relationship between anomalies in sea-surface temperature over the Pacific and the westerly circulation pattern that prevails over North America during the following season. That the ocean somehow influences atmospheric circulation follows from our discussion of global radiation in Chapters 2 and 4. After all, the ocean is in contact with the atmosphere over about 72 percent of the Earth's surface, and the ocean is the principal sink for solar radiation. Sea-surface temperatures thus influence air pressure gradients over the ocean and thereby help to shape the downwind planetary circulation pattern.

Namias and other researchers argue that just such a coupling between ocean and atmosphere triggered the 1988 drought over the North Central United States. Anomalously cold water in the central and eastern tropical Pacific (La Niña) plus unusually high sea-surface temperatures to the north (near Hawaii) interacted with the atmosphere to produce a long-wave pattern that was more meridional than usual. Recall from Chapter 10 that this circulation pattern featured a warm anticyclone over the central contiguous United States that persisted through much of the spring and summer of 1988.

SINGLE-STATION FORECASTING

Short-term weather forecasts based on weather observations at one location, known as **single-station forecasts,** may be derived from the principles of weather behavior discussed in the previous chapters of this book. Because such forecasts are based on rules applied at only one location, they tend to be quite generalized and tentative, and complications may crop up as local conditions are modified by changes elsewhere. Table 15.3 is a sample list of rules of thumb applicable to midlatitude weather. You may wish to add to this list. Sometimes, but not usually, weather proverbs are useful in forecasting. This is the subject of the Special Topic, ''Weather Proverbs: Fact or Fiction?''

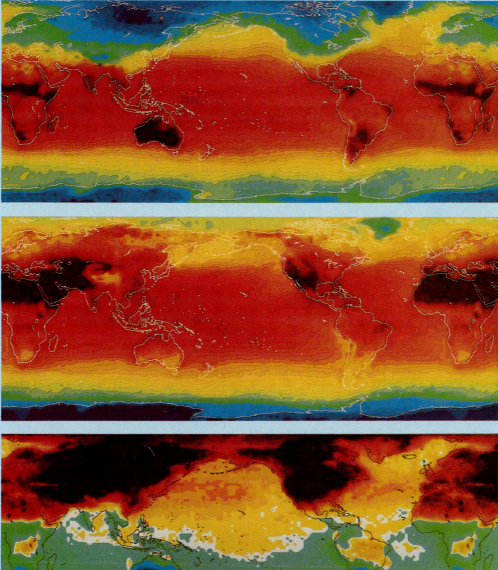

PLATE 17
Infrared sensors on board weather satellites measure surface temperature patterns over the globe. *Top*: January 1979; *middle*: July 1979; *bottom*: difference between January and July. Subfreezing temperatures are indicated by green and blue; warmer temperatures are red and brown. Note that the greatest seasonal temperature contrast occurred over the continents. [Courtesy of NASA]

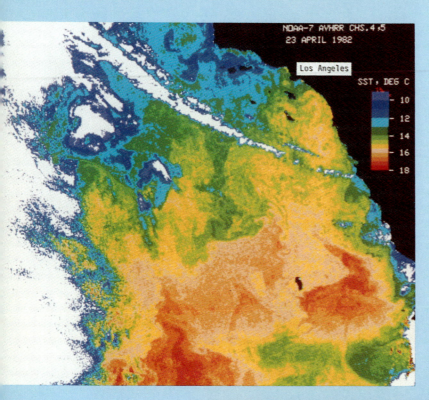

PLATE 18
Sea-surface temperature patterns as determined by infrared sensors on board the NOAA-7 satellite. *Top*: 23 April 1982. The black area in the upper right is the coast of southern California, and the white patches are clouds. Note the red area where sea-surface temperatures are near 18 °C (65 °F). *Bottom*: The same area nearly one year later (1 April 1983) shows a greatly expanded area of surface waters having temperatures near or above 18 °C (65 °F). [Courtesy of NOAA]

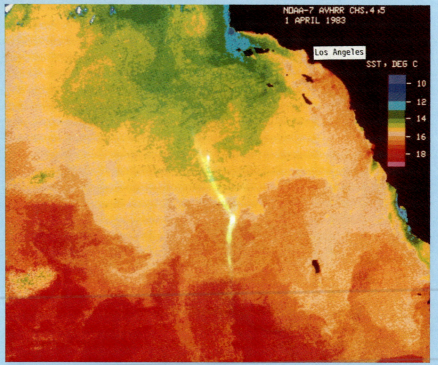

PLATE 19
Color-enhanced IR satellite image shows 17 urban heat islands in southeastern New England. Number 7 is Metropolitan Boston. Temperature–color calibration: orange–yellow, 25.5–31.0 °C; light–dark green, 16.5–25.0 °C; light–dark blue, 6.0–16.0 °C. This image was taken by the NOAA-5 satellite on 23 May 1978 at 1400 GMT. Relatively warm areas on Cape Cod and adjacent islands are probably due to land-use patterns or soil types. [Courtesy of Michael Matson, National Environmental Satellite Service, NOAA]

PLATE 20
Pouchlike mammatus clouds occur on the underside of a thunderstorm anvil and sometimes indicate severe weather. [Photograph by J. M. Moran]

PLATE 21
Hurricane Elena over the Gulf of Mexico as photographed from the Space Shuttle
Discovery on 2 September 1985. Note the spiral cloud bands surrounding the central eye
of the storm. The Earth's curvature is visible in the background. [NASA photograph
courtesy of the Johnson Space Center, Houston, Texas]

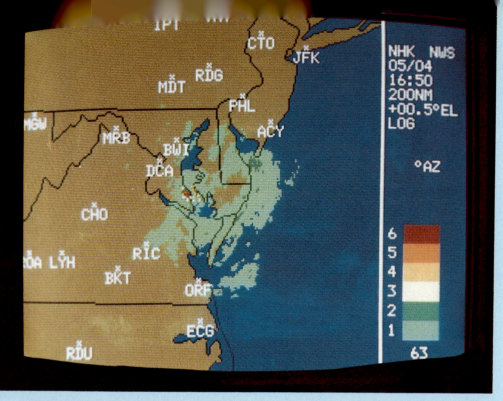

PLATE 22

Conventional weather radar displays featuring echo intensity graduated by color. The most intense echo (heavy rain) is indicated by dark red, and the weakest echo (light rain) by light green. The radar image below was taken 11 minutes after the radar image above and indicates a rapid intensification of rain showers over the mid-Atlantic coast. [Courtesy of Alden Electronics]

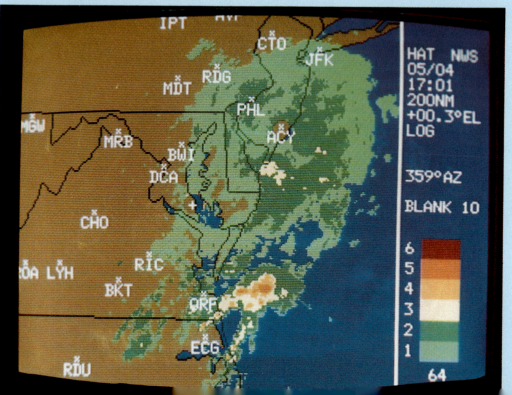

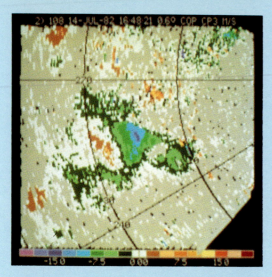
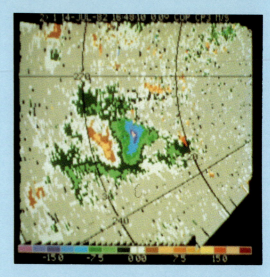

PLATE 23

Life cycle of a microburst as documented in a sequence of Doppler radar images. The radar unit is located to the northeast, and range rings are drawn at 5-mile intervals. Orange and yellow (warm colors) indicate air motion away from the radar; blue and green (cold colors) indicate air motion toward the radar. A starburst circulation pattern—characteristic of a microburst—appears at the center of the first three images. A microburst develops in the first two images, reaches maximum intensity in the third image, and rapidly dissipates in the fourth image. The time span between the first and final images is about 7 minutes. [National Center for Atmospheric Research/ National Science Foundation]

PLATE 24

Cloud image generated from GOES observations over the Gulf of Mexico, Louisiana, Mississippi, and Alabama at 1800 GMT, 10 September 1985. This technique of animated display was developed in conjunction with McIDAS. [Photo courtesy of W. L. Hibbard, Space Science and Engineering Center, University of Wisconsin–Madison; and the American Meteorological Society]

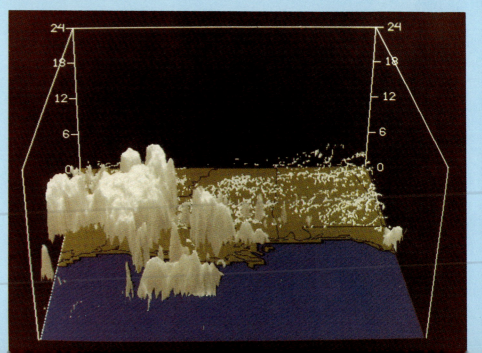

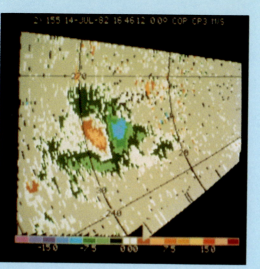
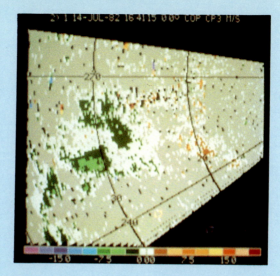

PLATE 25
Computer-enhanced GOES satellite IR image of the ash cloud ejected by Mount St. Helens. The image was taken at 4:45 P.M. Pacific Daylight Time on the day of the eruption, 18 May 1980. The ash cloud shows up as a dark blotch covering southern Washington, northern and eastern Idaho, southwestern Montana, and western Wyoming. The dark area to the north and northwest is high clouds. [Courtesy of Michael Matson, National Environmental Satellite Service, NOAA]

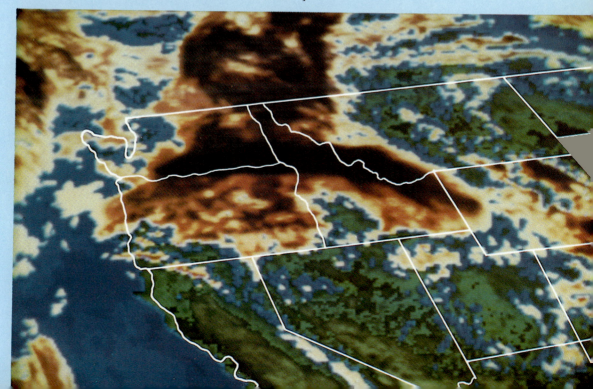

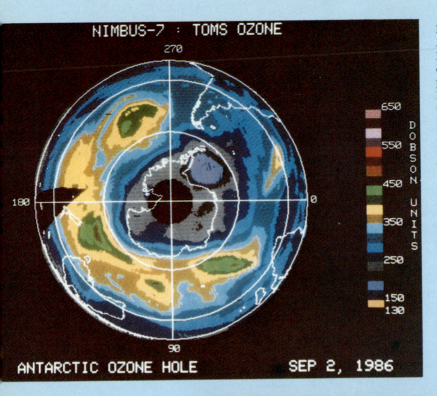

NIMBUS-7 : TOMS OZONE

ANTARCTIC OZONE HOLE   SEP 2, 1986

DOBSON UNITS

650
550
450
350
250
150
130

PLATE 26

Maps of total ozone in the Southern Hemisphere illustrating the Antarctic ozone hole. These images were produced at NASA's Goddard Space Flight Center using data obtained by the TOMS (Total Ozone Mapping Spectrometer) on board the Nimbus-7 satellite. The ozone hole is portrayed in gray and violet colors and is surrounded by a ring of high total ozone (yellow, green, and brown) at middle latitudes. As illustrated in the color bar at the right, ozone concentrations are expressed in Dobson Units. (One Dobson Unit is a hundredth of a millimeter and refers to the depth of ozone produced if all the ozone in a column of the atmosphere were brought to sea-level temperature and pressure.) Between 2 September 1986 (*left*) and 2 October 1986 (*below*), the region of very low total ozone (violet color) expanded and covered an area the size of the United States. [Courtesy of NASA]

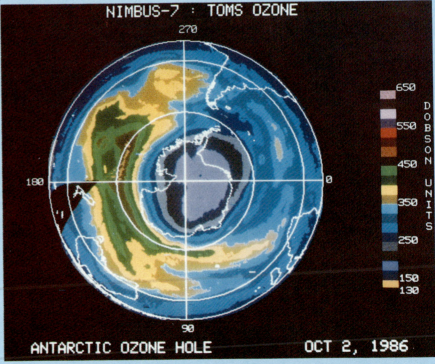

NIMBUS-7 : TOMS OZONE

ANTARCTIC OZONE HOLE   OCT 2, 1986

DOBSON UNITS

650
550
450
350
250
150
130

Table 15.3
Rules of Thumb for Single-Station Weather Forecasting

1. At night, air temperatures will be lower if the sky is clear than if the sky is cloud covered.
2. Clear skies, light winds, and a fresh snow cover favor extreme nocturnal radiational cooling and very low air temperatures by dawn.
3. Falling air pressure may indicate the approach of stormy weather, whereas rising air pressure suggests that fair weather is in the offing.
4. The appearance of cirrus, cirrostratus, and altostratus clouds, in that order, indicates overrunning ahead of a warm front and the possibility of precipitation.
5. A counterclockwise wind shift from northeast to north to northwest (called *backing*) is usually accompanied by clearing skies and cold air advection.
6. A clockwise wind shift from east to southeast to south (called *veering*) is usually accompanied by clearing skies and warm air advection.
7. A wind shift from northwest to west to southwest is usually accompanied by warm air advection.
8. If radiation fog lifts by late morning, a fair afternoon is likely.
9. With west or northwest winds, a steady or rising barometer, and scattered cumulus clouds, fair weather is likely to persist.
10. Towering cumulus clouds by midmorning may indicate afternoon showers or thunderstorms.

Observations derived from studying records of past weather events may aid single-station weather forecasting. Such records exhibit a **fair-weather bias.** That is, fair-weather days outnumber stormy days almost everywhere. In fact, if we boldly predict that all days will be fair, we probably will be correct more than half the time. The only merit of this exercise would be to establish a baseline for evaluating the accuracy of more sophisticated weather-forecasting techniques. That is, we would expect traditional forecasting methods to score higher than forecasting based solely on a fair-weather bias.

Another characteristic of past weather records is **persistence,** or the tendency for weather episodes to persist for some period of time. For example, if the weather has been cold and stormy for several days, the weather may well continue that way for many more days. The same weather records also show, however, that an episode of one weather type typically gives way to another weather type very abruptly, usually in a day or less. Weather forecasts based on persistence alone are therefore prone to serious error.

A third approach is based on climate; that is, we prepare a weather forecast for a particular day based on the type of weather that occurred on the same day in years past. Suppose, for example, that climate records of your town indicate that it has rained on 7 August only 12 times in the past 100 years. Accordingly, you predict that the probability of rain next 7 August is only 12 percent, and you confidently plan a picnic. The problem with this approach is that, statistics aside, there is no guarantee that it will not rain

Weather Proverbs: Fact or Fiction?

Over the centuries, people have developed weather proverbs to serve as guides for predicting weather. These proverbs are based on casual observations, usually of a weather event or a change in animal behavior, that are linked to a subsequent weather event. When weather proverbs are subjected to close scientific scrutiny, a few prove to be substantially correct, but most turn out to be merely picturesque myths with little or no scientific basis.

The Zuni tribe of New Mexico has a weather proverb that often proves to be true.

> When the sun is in his house, it will rain soon.

The phrase "in his house" refers to a halo surrounding the sun. Recall from Chapter 8 that a halo forms when the sun's rays are refracted by the ice crystals that compose high, thin cirrus clouds. Recall also from Chapter 11 that cirrus clouds signal the beginnings of overrunning as warm air is advected aloft. The appearance of a halo around the sun thus indicates the approach of a storm, but a halo does not ensure that precipitation will follow. Because cirrus clouds may spread as much as 1000 km (620 mi) ahead of the storm center, the storm may well change direction or die out before reaching your area.

Another scientifically sound proverb uses rainbows to predict the day's weather.

> A rainbow in the morning
> Is the sailor's warning.
> A rainbow at night
> Is the sailor's delight.

The basis for this proverb's validity (as described in Chapter 8) is that (1) in midlatitudes, weather systems usually progress from west to east and (2) an observer of a rainbow must face the rainshower with the sun at his or her back.

Most weather proverbs, however, have no validity. For example, many cultures have proverbs that deal with the ability of animals to forecast weather changes. Probably the most famous of these is the myth that a groundhog can determine on the second day of February whether winter is about to end. If the sun is shining and the groundhog can see its shadow, winter will supposedly last another six weeks. If it is cloudy, winter will soon end. Interestingly, the proverb originated in medieval Europe, and Groundhog Day was

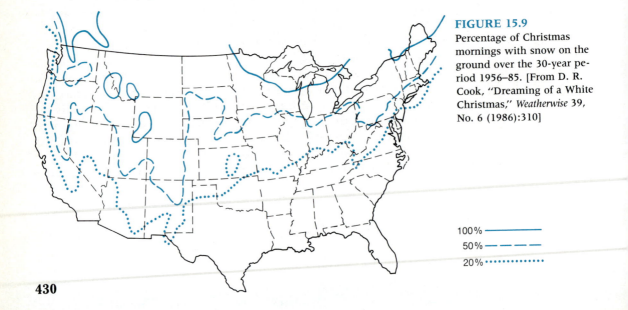

FIGURE 15.9
Percentage of Christmas mornings with snow on the ground over the 30-year period 1956–85. [From D. R. Cook, "Dreaming of a White Christmas," *Weatherwise* 39, No. 6 (1986):310]

100% ——————
50% – – – – –
20% ···········

known as Candlemas Day. At that time, no definite period was assigned for winter to stay. Then the proverb read:

If Candlemas Day be fair and bright
Winter will have another fight;
But if Candlemas Day brings clouds and rain
Winter is gone and won't come again.

Common sense shows this proverb to be false. Many parts of the United States and Canada are snow covered on the second of February, and it is unlikely that the groundhog even makes an appearance. What if the day dawns clear and then turns cloudy? Would the weather for the next six weeks depend on whether or not the groundhog was an early riser? Ridiculous, isn't it? Moreover, no evidence exists that the presence or absence of sunshine on one particular day is an indicator of the weather for weeks to come. Weather is too variable and too complex to support such a simple cause/effect relationship.

Other weather proverbs deal with the ability of animals such as squirrels, deer, rabbits, and caterpillars to predict the severity of the upcoming winter. For example: "When squirrels lay in a large store of nuts, expect a cold winter." However, no evidence of any linkage exists between the two. Squirrels are more likely to store more nuts if more nuts are available! The squirrels are simply responding to a good acorn-growing season, rather than exhibiting an innate ability to predict future weather.

Another familiar, but false, weather proverb concerns lightning: "Lightning never strikes twice in the same place." Recall our discussion of cloud-to-ground lightning in Chapter 13. An electrical discharge usually follows the shortest path between the ground and the cloud, so that lightning often strikes a tree, an electrical pole, a tower, or other tall object. As long as this object is not destroyed by the first lightning stroke, there is no physical reason why lightning cannot strike the object again.

This concluding verse, attributed to E. V. Lucas (1863–1938), illustrates one universal truth about weather—no matter what the weather, not everybody will be happy.

The Duke of Rutland urged the folk to pray
For rain; the rain came down the following
  day,
The pious marvelled, the skeptics murmured
  fluke,
The farmers, late with hay, said, "Damn the
  Duke."

next 7 August. Another example of climate-derived weather forecasting is statistics on the likelihood of a white Christmas, as shown in Figure 15.9.

PRIVATE FORECASTING

In our description of weather forecasting, we have focused mainly on the role of government agencies. In addition, numerous private weather forecasters and forecast services analyze weather maps and other guidance materials supplied by the National Meteorological Centers and tailor them for the private sector's special needs. For example, a private weather forecaster retained by an appliance store chain might warn store executives of a pending heat wave so that stores might be stocked with an adequate supply of fans and air conditioners. Another private forecaster might advise an electric power company of expected winter temperatures so the energy supplier could better anticipate public fuel demands. In this way, private forecasters supplement the efforts of government weather forecasting.

Communication and Dissemination

Weather maps, charts, forecasts, and outlooks issued by the United States National Meteorological Center are transmitted to regional Weather Service Forecast Offices (WSFOs).

When hazardous weather appears either possible or probable, the National Weather Service issues weather watches and weather warnings covering the affected geographical area for a specified time period. In general, a **weather watch** is indicated when hazardous weather is considered possible on the basis of current or anticipated atmospheric conditions. People in the designated area need not interrupt their normal activities except to remain alert for threatening weather and to keep the television or radio on for further advisories. A **weather warning** is issued when hazardous weather is likely or is occurring somewhere in the region. People are then advised to take all necessary safety precautions.

Watches and warnings are issued for tornadoes, severe thunderstorms, hurricanes, floods, and winter storms. Normally, a tornado warning is issued only after a tornado has actually been sighted. If conventional radar is used, a suspected tornado signature (a hook echo, for example) must be confirmed by ground observation. On the other hand, if the more reliable Doppler radar is used (Chapter 14), ground confirmation may not be required before a tornado warning is issued. The warning bulletin specifies the location of the tornado, its anticipated path, and the time when the tornado is expected in the warning area.

FIGURE 15.10
A conventional weather facsimile machine used to relay and reproduce weather maps, satellite imagery, and other weather guidance materials. [Courtesy of Alden Electronics]

Winter storm warnings may specify heavy snow, blizzard conditions, or an ice storm. Usually, a heavy snow warning is issued if snowfall is expected to total at least 4 in. over 12 hours or 6 in. over 24 hours. A **blizzard warning** means that falling or blowing snow is expected to be accompanied by winds in excess of 55 km (35 mi) per hour, reducing visibility to 150 m (450 ft) or less. A severe blizzard features winds stronger than about 70 km (45 mi) per hour, visibility near zero, air temperature below −12 °C (10 °F), and dangerously low windchill equivalent temperatures. Ice storm warnings mean that potentially dangerous accumulations of freezing rain or sleet are expected on the ground and other exposed surfaces.

Because weather is highly changeable, weather observations and guidance information such as weather maps, forecasts, watches, and warnings must be communicated as rapidly as possible both nationally and internationally. For this reason, the World Meteorological Organization and its member nations maintain elaborate communications networks consisting of a variety of systems. Weather information is relayed by satellite, teletypewriter, radio, and facsimile systems like the one shown in Figure 15.10, which reproduce maps, charts, and satellite images.

Conversion to the **Automation of Field Operations and Service (AFOS)** systems has made the United States weather communications network more efficient and reliable. Each AFOS system has two main compo-

FIGURE 15.11

The console unit of the National Weather Service weather data communications system, called AFOS for Automation of Field Operations and Service. [NOAA photograph]

FIGURE 15.12

At the push of a button, this portable weather radio issues NOAA weather reports and forecasts. [Photograph by Laura Carlson]

nents: (1) a minicomputer that collects weather data and transmits them to the forecaster and other AFOS stations and (2) a console of television-type screens for displaying data and maps and a keyboard for data and message input (Figure 15.11). The AFOS system has replaced the slower and less dependable teletypewriters and facsimile machines at most NWS offices. Although facsimile transmissions take 5 to 10 minutes and teletypewriters process about 100 words per minute, the AFOS system can reproduce data and maps in only 15 seconds and can transmit 3000 words per minute. In addition, the AFOS system is less prone than the old system to service interruptions during severe weather, when the system is most needed. Each AFOS system ties into a master computer in Maryland.

The public receives regular weather reports and forecasts on the radio, on commercial television and cable-TV weather channels, and in the newspapers. In many cities, weather information is also available via recorded telephone announcement systems. In addition, NOAA sponsors low-power VHF-FM radio transmitters that broadcast continuous weather information 24 hours a day. A taped message is repeated every 4 to 6 minutes and is revised routinely every 2 to 3 hours. Should hazardous weather threaten, watches, warnings, and advisories are issued. Although the range of these broadcasts is only 50 to 65 km (30 to 40 mi), an estimated 90 percent of the United States population is within range of a transmitter. Broadcasts can be picked up on commercially available weather radios (Figure 15.12), some of which are equipped with an alarm that sounds when a weather warning is issued.

In addition, during hurricane season, the National Hurricane Center (in cooperation with NOAA, NBC News, and *USA Today*) operates a 24-hour telephone hotline. A recorded message provides late information on an approaching hurricane or tropical storm. The telephone number (in the United States) is (900) 410–NOAA; there is a small per-minute fee.

Conclusions

Weather forecasting is a complex and challenging science that depends on the efficient interplay of weather observation, data analysis by meteorologists and computers, and rapid communications. Meteorologists have achieved a very respectable level of accuracy for short-range weather forecasting. Further improvement is expected with denser surface and upper-air observation networks, more precise numerical models of the atmosphere, larger and faster computers, and more sophisticated techniques of remote sensing by satellite. If these advances are to be realized, however, continued international cooperation is essential, for the atmosphere is a continuous fluid that knows no political boundaries.

What, then, are the prospects for climate forecasting? So far, accuracy in long-range forecasting has been minimal, and the prospects for improvement in the near future are not promising. The reasons will become evident in Chapters 17 and 18, where we discuss principles of climate, climatic trends, and climatic variability. Before that, however, we turn our attention to the topic of air pollution meteorology.

Mathematical Note

Some Orbital Characteristics of Weather Satellites

Both types of weather satellites, geosynchronous and polar orbiting, move through space at constant speeds in orbits that are very nearly circular. Because friction is negligibly small, the only forces acting on an orbiting satellite are gravitation and the centripetal force.

Gravitation (F), the attractive force between two objects, is directly proportional to the product of the masses of the two objects (m_1, m_2) and inversely proportional to the square of the distance (r) between the two objects. That is,

$$F \propto \frac{m_1 \times m_2}{r^2}$$

The constant of proportionality is G, the gravitational constant. Hence,

$$F = \frac{G(m_1 \times m_2)}{r^2}$$

Assume that a weather satellite has a unit mass and that the distance between the satellite and the Earth's center of mass is equal to the Earth's radius (R) plus the satellite's altitude (h). Then the gravitational attraction between the Earth and the satellite is given by

$$F = \frac{Gm_E}{(R + h)^2}$$

where m_E is the mass of the Earth.

The centripetal force (C_e) acting on a satellite of unit mass is computed as

$$C_e = \frac{V^2}{R + h}$$

where V is the orbital speed of the satellite.

For computational purposes, we can set the grav-

Summary Statements

The fluid nature of the atmosphere requires international cooperation in the gathering and interpretation of surface and upper-air weather data. To this end, the World Meteorological Organization coordinates an international effort of weather observation, analysis, and forecasting.

Surface weather is monitored at land stations and by automated weather stations and ships at sea. Upper-air weather data are profiled by rawinsonde measurements, and additional weather observations are supplied by radar, aircraft, and satellites.

Weather satellites have the advantage of monitoring a continuous broad field of view. Polar-orbiting satellites follow relatively low meridional orbits. Geosynchronous satellites are positioned over the equator and orbit at the Earth's rotational rate.

Upper-air weather data are plotted on constant-pressure surfaces, and surface weather data are plotted on a constant-altitude (sea-level) surface.

Because many variables are involved in the atmospheric system and because huge quantities of weather observations are generated, weather data are analyzed and forecasts are prepared by high-speed electronic computers using numerical models of the atmosphere.

Although the accuracy of short- and medium-range weather forecasting has improved steadily over the past three decades, forecasting skill deteriorates rapidly for periods longer than 48 hours and is minimal beyond 10 days.

One-station weather forecasting may be based on principles of weather behavior, fair-weather bias, climatic records, or persistence of weather episodes.

Weather information and forecasts are communicated to users via teletypewriters, radio, facsimile systems, television, and newspapers.

---

itational force equal to the centripetal force since the satellite's acceleration is zero. But recall from earlier discussions (Chapter 9) that this is *not* a case of balanced forces because the satellite continually changes direction as it describes a circular orbit.
Hence, with this qualification,

$$F = C_e$$

or

$$\frac{Gm_E}{(R + h)^2} = \frac{V^2}{R + h}$$

Solving for V and simplifying,

$$V = \left[\frac{Gm_E}{R + h} \right]^{1/2}$$

From this equation, the orbital speed (V) of a satellite is inversely proportional to its altitude (h). Thus, in order to remain in orbit, a satellite at low altitudes

must move faster than a satellite at high altitudes. For a geosynchronous satellite, an orbital altitude is selected so that the corresponding orbital speed means that the satellite always views the same portion of the Earth. For today's polar-orbiting weather satellites, on the other hand, an orbital altitude is selected so that the corresponding orbital speed takes the satellite over all points on the Earth's surface twice daily at the same local time, at 7:30 A.M. and 2:30 P.M. Such an orbit is described as sun-synchronous.

A geostationary satellite at an altitude of 36,000 km (22,320 mi) has an orbital speed of about 11,000 km (6820 mi) per hour and completes one orbit in 24 hours. A polar-orbiting satellite at an altitude of 855 km (530 mi) has an orbital speed of almost 27,000 km (16,740 mi) per hour and completes one orbit every 102 minutes.

Key Words

National Weather
  Service (NWS)
National Oceanic and
  Atmospheric
  Administration
  (NOAA)
International
  Meteorological
  Organization (IMO)
World Meteorological
  Organization
  (WMO)
World Weather Watch

synoptic weather
  network
basic weather
  network
Greenwich Mean
  Time (GMT)
National Substation
  Program
radiosonde
polar-orbiting satellite
geosynchronous
  satellite

McIDAS
station model
isobars
cold air advection
warm air advection
National Hurricane
  Center (NHC)
National Severe
  Storms Forecast
  Center (NSSFC)
lake-effect snows
anomalies

teleconnection
single-station
  forecasts
fair-weather bias
persistence
weather watch
weather warning
blizzard warning
Automation of Field
  Operations and
  Service (AFOS)

Review Questions

1. Why does weather forecasting require international cooperation?

2. Describe the major steps involved in the preparation of weather forecasts.

3. What events provided the impetus for the establishment of a national weather observational network?

4. Distinguish between the activities of the synoptic weather network and the activities of the basic weather network.

5. What is Greenwich Mean Time (GMT)?

6. What time is it at your locality at 0100 GMT?

7. What is the chief purpose of the National Substation Program?

8. Distinguish between a satellite in polar orbit and a satellite in geosynchronous orbit.

9. List the advantages of satellites in weather observation.

10. What is the basic difference between surface weather maps and upper-air weather maps?

11. What is meant by a numerical model of the atmosphere?

12. Why does the reliability of numerical weather prediction deteriorate rapidly for forecast periods beyond 48 hours?

13. Why are computerized weather forecasts not likely to replace human weather forecasters? Provide some illustrations.

14. Describe the basis for long-range (30-, 60-, 90-day) weather forecasts.

15. What is a teleconnection? Describe how teleconnections might aid long-range weather forecasting.

16. Speculate on how long-term weather records could be used to formulate seasonal weather forecasts.

17. Explain why there might be a linkage between sea-surface temperatures and planetary-scale atmospheric circulation patterns.

18. An episode of one weather type usually gives way to another weather type very abruptly—in a single day or less. Explain why.

19. What is meant by a "fair-weather bias?"

20. Distinguish between weather watches and weather warnings.

Points to Ponder

1. What is the advantage of simultaneous weather observations?

2. Explain how weather satellites are able to provide nighttime observations of cloud cover.

3. For a sea-level location, determine the fraction of the atmosphere that lies below (a) the 100-mb level, (b) the 700-mb level, and (c) the 850-mb level.

4. Explain why warm air advection causes the 500-mb surface to rise and why cold air advection causes the 500-mb surface to lower.

5. Explain why cold-core anticyclones and warm-core cyclones do not appear on 500-mb maps.

6. Why are winds associated with contour gradients on upper-air weather maps?

7. Describe how computers are used to forecast the weather.

8. On average in midlatitutdes in summer, a 500-mb ridge develops over the continents and a 500-mb trough occurs over the oceans. In winter, it's just the reverse: a 500-mb trough develops over the continents and a 500-mb ridge occurs over the oceans. Please give an explanation.

Projects

1. Design an experiment in which you test the validity of (a) persistence weather forecasting and (b) weather forecasting based on past climate.

2. Speculate on why there is a "fair-weather bias" and test the notion that there really is such a bias in your locality.

Selected Readings

FAA and NOAA. *Aviation Weather Services*. Washington, DC: U.S. Government Printing Office, 1979. 123 pp. Describes the weather maps and forecast charts available for pilots.

Gilman, D. "Predicting the Weather for the Long Term." *Weatherwise* 36 (1983):290–297. Includes a discussion on how long-range weather outlooks are prepared.

Glahn, H. R. "Yes, Precipitation Forecasts Have Improved." *Bulletin of the American Meteorological Society* 66 (1985):820–830. Demonstrates an improvement in precipitation probability forecasts for the period 1967–1982.

Hughes, P. "American Weather Services." *Weatherwise* 33 (1980):100–111. Discusses the principal people and events in the history of the National Weather Service.

Kierein, T. "The Hi-Tech World of TV Weathercasting." *Weatherwise* 41 (1988):150–154. Gives an illustrated tour of the making of a television weathercast.

Pool, R. "Is Something Strange About the Weather?" *Science* 243 (1989):1290–1293. Presents a good summary of the recent interest in applying chaos theory to long-range weather forecasting.

Smith, W. L., et al. "The Meteorological Satellite: Overview of 25 Years of Operation." *Science* 231 (1986):455–462. Thoroughly summarizes the capabilities of today's weather satellites along with projected developments for the future.

Spencer, R. W., and J. R. Christy. "Precise Monitoring of Global Temperature Trends from Satellites." *Science* 247 (1990):1558–1562. Describes how satellites are used to monitor Earth's surface temperature patterns.

Thaler, J. S. "West Point—152 Years of Weather Records." *Weatherwise* 32 (1979):112–115. Discusses the history of one of the longest weather records in the nation.

Tribbia, J. J., and R. A. Anthes. "Scientific Basis of Modern Weather Prediction." *Science* 237 (1987): 493–499. Provides a relatively sophisticated summary of the historical roots and future directions of numerical weather prediction.

*This goodly frame, the earth,
seems to me a sterile
promontory; this most excellent
canopy the air, look you, this
brave o'erhanging firmament,
this majestical roof fretted with
golden fire—why, it appeareth
no other thing to me than a
foul and pestilent congregation
of vapours.*

William Shakespeare
HAMLET

16

Air Pollution Meteorology

The quality of the air we breathe is
strongly influenced by wind speed
and atmospheric stability. [Photo-
graph by Mike Brisson]

AIR POLLUTION is at least as old as civilization itself. The first air pollution episode probably took place when early humans tried to build a fire in a poorly ventilated cave. Reference to polluted air appears as early as Genesis (19:28): "Abraham beheld the smoke of the country go up as the smoke of a furnace." About 400 B.C., Hippocrates noted the pollution of city air. And in 1170 A.D., Maimonides wrote of Rome: "The relation between city air and country air may be compared to the relation between grossly contaminated, filthy air, and its clear, lucid counterpart."

The Industrial Revolution was the single greatest contributor to air pollution as a chronic problem in Europe and North America. As industrial innovations spread from one nation to another, so too did air pollution. In the United States, in post–Civil War days, cities swelled with new industries and new immigrants to work in those industries. By the turn of the century, the urban environment was becoming increasingly fouled by the fumes of foundries and steel mills. In those days, a city took pride in smokestacks like those shown in Figure 16.1; they signaled a prosperous economy. Efforts to regulate air quality were meager, and little was known about the health effects

FIGURE 16.1
Industrial smokestacks in 1906 Pittsburgh meant a prosperous economy. [Carnegie Library of Pittsburgh]

of polluted air. In an attempt to placate a wheezing and coughing populace, some physicians even argued that polluted air had medicinal value.

Concern over polluted air probably does not stem from disenchantment with the fruits of industrialism. Belching smokestacks still signal a healthy economy, but many people are troubled by the possibility that polluted air is adversely affecting their health, agricultural productivity, and the weather. In this chapter, we examine two questions related to these concerns: (1) How do weather conditions influence air pollution levels? (2) What is the impact of air pollution on weather?

Air Pollutants

As we pointed out in Chapter 1, many gases and aerosols that can be air pollutants are actually normal constituents of the atmosphere. These substances become pollutants when their concentrations increase to levels that threaten the well-being of living things or disrupt physical or biological processes. In the Special Topic, "Principal Air Pollutants," we survey the major air pollutants, noting their natural cycling within the environment, and the contributions to pollution made by human activity.

Air Pollution Episodes

On the morning of 26 October 1948, a fog blanket reeking of pungent sulfur dioxide fumes spread over the town of Donora in Pennsylvania's Monongahela Valley. Before the fog lifted five days later, almost half the area's 14,000 inhabitants had fallen ill and 20 had died. This killer fog resulted from the combination of mountainous topography and stable atmospheric conditions that trapped and concentrated deadly effluents from the community's steel mill, zinc smelter, and sulfuric acid plant.

Air pollutants are especially dangerous when atmospheric conditions favor their concentration. Once pollutants are emitted into the atmosphere, their concentrations usually begin to decline. The rate of concentration decrease, or dilution, is determined in part by the extent to which pollutants mix with cleaner air: the more thorough the mixing, the more rapid is the rate of dilution. When conditions in the atmosphere favor rapid dilution, the impact of air pollution is usually minor. On other occasions, termed **air pollution episodes,** atmospheric conditions minimize dilution, and the impact of air pollution can then be severe, particularly on human health. The two weather conditions that most influence the rate of dilution are wind speed and atmospheric stability.

Principal Air Pollutants

Here we summarize the sources, cycling, and some of the impacts of the major air pollutants.

Oxides of Carbon. Through respiration, organisms release carbon dioxide (CO_2) to the atmosphere, and through photosynthesis, plants take up carbon dioxide. Other natural sources of CO_2 include forest and brush fires and volcanic activity. Combustion of fossil fuels, coal, oil, and natural gas for electric power generation and space heating releases carbon dioxide into the atmosphere. About 55 percent of this carbon dioxide remains in the atmosphere, and most of the remainder dissolves in ocean water. Today the concentration of carbon dioxide in the atmosphere is about 352 parts per million (ppm) and is rising at a rate of about 16 ppm per decade. As we will see in Chapter 18, this upward trend in atmospheric CO_2 may affect the global climate.

By far the most important natural source of atmospheric carbon monoxide (CO) is the combination of oxygen with methane or other volatile organic compounds (described below). Carbon monoxide is removed from the atmosphere by the activity of certain soil microorganisms and by chemical reactions that convert CO to CO_2. The net result is a harmless Northern Hemisphere average concentration of less than 0.15 ppm.

In developed nations of the Northern Hemisphere, the principal source of CO derived from human activity is combustion of fossil fuels, especially by motor vehicles. Breathing CO causes drowsiness, slows reflexes, and impairs judgment; at high concentrations death ensues. Because the gas is odorless and tasteless, carbon monoxide defies detection by human senses and constitutes a serious health hazard, especially where the concentration is high, as it can be in highway tunnels and underground parking garages.

Recently, scientists discovered that the major source of CO in the Southern Hemisphere and the tropics is burning of forests and savannas for land clearing. Instrument measurements onboard aircraft and the space shuttle indicate that this contribution may be comparable in magnitude to global fossil fuel combustion.

Volatile Organic Compounds. Volatile organic compounds, commonly called *hydrocarbons*, include a wide variety of chemicals that contain only hydrogen and carbon. Of the hydrocarbons that occur naturally in the atmosphere, methane (CH_4) has the highest concentration (1.684 ppm, on average). Methane is produced when biological material decays in the absence of oxygen, for example, in swamps, rice paddies, and the rumen of cattle. At normal background levels, methane is nonreactive; that is, it does not chemically interact with other substances and causes no detrimental health effects. However, methane is the chief component of natural gas, and concentrated methane may be explosive. As we will see in Chapter 18, some scientists are concerned about the potential effect of methane on global climate. Methane is an infrared-absorbing gas—that is, it contributes to the greenhouse effect—and the mean global concentration of methane has been rising at a rate of 0.016 ppm per year.

Vegetation emits various hydrocarbons. Terpenes are an example. Occurring in concentrations of less than 0.1 ppm, they are chemically reactive and responsible for the aromas of pine, sandalwood, and eucalyptus trees. Recent studies suggest that natural biogenic hydrocarbons (other than methane) may play an important role in the formation of smog.

In addition to biogenic sources, reactive hydrocarbons are emitted during the incomplete combustion of gasoline by motor vehicles. This source is, in fact, responsible for hundreds of different hydrocarbons. Because gasoline is very volatile, some hydrocarbons (perhaps as much as 15 percent of the total in some cities) also escape to the air during gasoline delivery and refueling operations at service stations.

Although our understanding of the natural cycling of hydrocarbons in the atmosphere is as yet minimal, we do know that the typically low concentrations of most hydrocarbons found in city air appear to pose no direct environmental threat. However, serious health hazards arise from the products of complex chemical reactions involving hydrocarbons and other air pollutants (particularly nitrogen oxides) in the presence of sunlight. These photochemical reactions produce smog. In addition, some hydrocarbons, such as benzene (C_6H_6), an industrial solvent, and benzopyrene, a product of fossil fuel and tobacco combustion, are also carcinogenic.

Oxides of Nitrogen. The action of soil bacteria is responsible for most of the nitric oxide (NO) that is produced naturally and released to the atmosphere. Within the atmosphere, NO combines readily with oxygen to form nitrogen dioxide (NO_2). Together, these two oxides of nitrogen are usually referred to as NO_x.

Although human activities contribute only about 10 percent of the atmosphere's total load of NO_x, our contributions tend to be much more concentrated than the natural atmospheric average. Based on the mode of formation, scientists distinguish between *thermal NO_x* and *fuel NO_x*. Thermal NO_x forms when high combustion temperatures, such as occur inside an automobile engine, cause nitrogen (N_2) and oxygen (O_2) in the air to combine. Fuel NO_x forms when nitrogen contained in a fuel, such as coal, is oxidized. (N_2 in the fuel combines with O_2 in the air.) For both modes of formation, NO is generated initially, and then when vented and cooled, some of the NO converts to NO_2. About half of fuel NO_x comes from stationary sources (power plants, primarily), and the other half comes from motor vehicle emissions.

Nitrogen dioxide (NO_2) is a much more serious air pollutant than its precursor NO; the toxicity of NO_2 is about four times that of NO. Nitrogen dioxide at high levels is believed to contribute to heart, lung, liver, and kidney damage, and it is linked to the incidence of bronchitis and pneumonia. Moreover, because nitrogen dioxide occurs as a brownish haze, it reduces visibility. When NO_2 combines with moisture in the air, nitric acid (HNO_3), a corrosive substance, is formed. Oxides of nitrogen are also precursors of smog.

Compounds of Sulfur. Sulfur enters the atmosphere naturally as sulfur dioxide (SO_2) from volcanic eruptions, as sulfate particles from sea spray, and as hydrogen sulfide (H_2S) produced when organic matter decays anaerobically (in the absence of oxygen). These sulfur compounds are washed from the air by precipitation and are taken up by soil, vegetation, and surface waters.

A variety of human activities also releases sulfur compounds into the atmosphere, at about one-third the rate of emission by natural sources. Most of our contribution comes from the combustion of fossil fuels (chiefly coal and oil) that contain sulfur as an impurity and emit sulfur dioxide when burned. The principal sources of sulfur dioxide are coal-fired electric-generating plants (69 percent) and industrial boilers (8 percent). In addition, the smelting of sulfur-bearing ores—lead, zinc, and copper sulfides—is a source of sulfur dioxide (8 percent), as are petroleum refining (5 percent), motor vehicles (5 percent), and oil burning for residential and commercial space heating (5 percent).

In the atmosphere, sulfur dioxide is converted to sulfur trioxide (SO_3) and sulfate aerosols. Sulfate aerosols restrict visibility and, in the presence of water, form droplets of sulfuric acid (H_2SO_4), a highly corrosive substance that also lowers visibility. Both SO_2 and SO_3 irritate respiratory passages and can aggravate asthma, emphysema, and bronchitis. Sulfate particles and sulfuric acid droplets are thought to increase human vulnerability to respiratory infection.

Certain industrial activities, including paper and pulp processing, emit hydrogen sulfide (H_2S) as well as a family of organic sulfur-containing gases called mercaptans. Even at extremely low concentrations, these compounds are foul smelling. Hydrogen sulfide tarnishes silverware and copper facings and discolors lead-based paints.

Smog. When vehicular traffic is congested (during morning and evening rush hours, for example), *photochemical smog* is likely to form. Oxides of nitrogen in motor vehicle exhaust and hydrocarbons (from various anthropogenic and biogenic sources) react in the presence of sunlight to produce a noxious, hazy mixture of suspended particles and gases. Products include ozone (O_3), formaldehyde, ketones, and PAN (peroxyacetyl nitrates), substances that irritate the eyes and damage the respiratory system. Although smog is most common in urban areas, winds can transport auto exhaust into suburban and rural areas, where the sun's rays trigger smog development.

Whereas levels of ozone at the Earth's surface average only about 0.02 ppm, in thick smog ozone concentrations may exceed 0.5 ppm. At these relatively high concentrations, ozone irritates the eyes and the mucous membranes of the nose and throat. It also degrades rubber and fabrics, retards tree growth, and damages crops.

Suspended Particulates. To this point, our survey of the major air pollutants has focused primarily on gases. Now we turn our attention to the multitude of tiny solid particles and liquid droplets that are suspended in the atmosphere. Collectively, these particles and droplets are termed suspended particulates, or *aerosols*. Sea-salt spray, soil erosion, volcanic activity, and various industrial emissions account for about one half the atmosphere's total aerosol load. The other half is largely the consequence of the atmospheric reactions among various gases.

Perhaps the most common particulates are dust and soot. Most dust is produced when wind erodes soil; this erosion is often accelerated by agricultural activity. Soot, tiny solid particles of carbon, is emitted during the incomplete combustion of fossil fuels and refuse. In urban-industrial air, suspended particulates may consist of a wide variety of materials depending on the specific types of mining, milling, or manufacturing. Urban-industrial particulates usually include a diverse array of trace metals such as lead, nickel, iron, zinc, copper, magnesium, and cadmium. These particulates pose a significant health hazard because their typically small size allows them to be inhaled deeply into the lungs. In addition, air may contain asbestos fibers, pesticides, and fertilizer dust.

Air normally also contains fungal spores and pollen. Disturbance of the land by farming and construction promotes the abundant growth of ragweed and other weeds, the pollen of which evokes allergic reactions, such as hayfever, in roughly 1 out of every 20 people.

WIND SPEED

Intuitively, we know that air is likely to mix more thoroughly on a windy day than on a calm day. When it is windy, turbulent eddies mix polluted air and cleaner air and thereby accelerate dilution. But when the wind is calm, dilution is by the very slow process of **molecular diffusion,** that is, dispersal at the molecular scale. As a general rule, with every doubling of wind speed, the concentration of air pollutants is cut in half (Figure 16.2).

Certain weather patterns favor light winds and thus inhibit the dispersal of contaminants, Over a broad area at the center of an anticyclone, for example, horizontal air pressure gradients are weak. Consequently, winds are very light or calm, and pollutants do not disperse readily. On the other hand, within a cyclone, steeper air pressure gradients mean stronger winds and more rapid dilution of air pollution. In addition, the rain or snow associated with a cyclone cleans the air by washing pollutants to the ground.

Wind speed is influenced not only by horizontal air pressure gradients but also by surface roughness (friction). In a city, winds are slowed by the rough surface created by the canyonlike topography of tall buildings and narrow streets. Average wind speeds may be 25 percent slower within a city than in the surrounding countryside. When regional (synoptic-scale) winds are light (less than 15 km per hour), the contrast between city and country winds is even more pronounced, amounting to a wind speed reduction of up

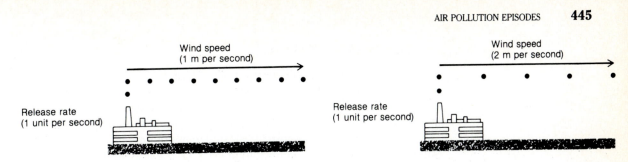

FIGURE 16.2
Doubling the wind speed from 1 to 2 m per second increases the spacing between puffs of smoke by a factor of two, thereby reducing pollution concentrations by one half.

to 30 percent. The dilution of air pollutants by wind is thus impeded in urban localities, the very places where most of the contaminants are generated.

Another consequence of the frictional interaction of the wind with obstacles on the Earth's surface is the formation of zones of light and irregular winds that can trap air pollutants. As we saw in Chapter 9, the horizontal wind breaks into turbulent eddies to the lee of an obstacle such as a building. Immediately downwind from the obstacle, the irregular eddy motion forms a closed circulation, known as a **wake.** If a smoke plume enters the wake of a building, as in Figure 16.3A, the smoke is trapped and may circulate into the building's ventilation system. Hence, smokestacks must be constructed to a height such that the plume clears the wake of nearby buildings (Figure 16.3B). This stack height is generally taken to be 2.5 times the height of the nearest obstacle.

FIGURE 16.3
(A) If a smokestack is too low, effluents may be trapped within the wake of nearby buildings or the chimney itself. (B) If a smokestack is constructed to the height of good engineering practice (2.5 times the height of the nearest obstacle), effluents clear the wake and downwash and trapping are avoided.

A

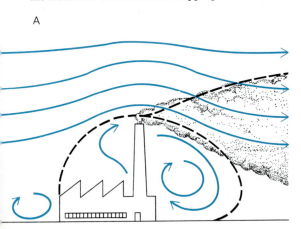

B

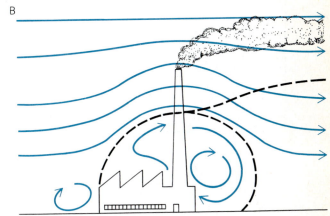

ATMOSPHERIC STABILITY

As we saw in Chapter 6, stability affects vertical motion within the atmosphere. Convection and turbulence are enhanced when the air is unstable and inhibited when the air is stable. The stability of air thus influences the rate at which polluted air mixes with clean air. A parcel of polluted air (a puff of smoke, for example) emitted into unstable air undergoes more mixing than does the same parcel of polluted air emitted into stable air. Stable air inhibits the upward transport of air pollutants, and a layer of stable air aloft may act as a lid over the lower troposphere that traps air pollutants. The continual emission of contaminants into stable air results in the accumulation and concentration of pollutants.

Mixing depth is the vertical distance between the Earth's surface and the altitude to which convection currents (that is, the mixing) reach. When mixing depths are great (many kilometers, for example), the relative abundance of clean air allows pollutants to mix and dilute rapidly. When mixing depths are shallow (less than 1 km, for example), air pollutants are restricted to a smaller volume of air, and concentrations may approach hazardous levels. When air is stable, convection is suppressed and mixing depths are low. When air is unstable, convection is enhanced and mixing depths increase. Because solar heating triggers convection, mixing depths tend to be greater in the afternoon than in the morning, greater during the day than at night, and greater in summer than in winter.

We can sometimes estimate the stability of air layers by observing the behavior of a plume of smoke belching from a smokestack. If the smoke is entering an unstable air layer, the plume undulates, as in Figure 16.4. In general, this plume behavior indicates that polluted air is mixing readily with the surrounding cleaner air, thereby facilitating dilution. The net effect is improved air quality (except where the plume loops to the ground). On the

FIGURE 16.4
Smoke plumes entering unstable air exhibit looping that is indicative of strong mixing and dilution of air pollutants.

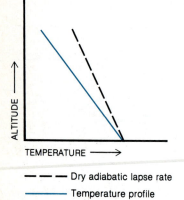

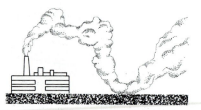

- - - - Dry adiabatic lapse rate

——— Temperature profile

ALTITUDE ——→

TEMPERATURE ——→

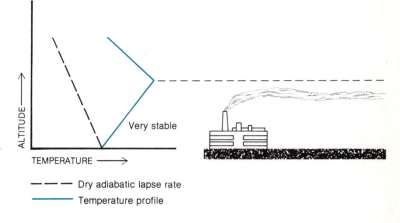

FIGURE 16.5
A temperature inversion within a surface air layer indicates very stable conditions. A smoke plume entering such stable air forms a thin ribbon extending downwind from the stack, and dilution of pollutants is minimal.

other hand, a smoke plume that flattens and spreads slowly downwind, as in Figure 16.5, indicates very stable conditions and minimal dilution.

In summary, air stability influences the rate at which polluted air and cleaner air mix. If air layers are stable, dilution is inhibited, but if air layers are unstable, dilution is enhanced.

TEMPERATURE INVERSIONS

An air pollution episode is most likely when a persistent **temperature inversion** develops. Recall from Chapter 6 that within an air layer characterized by a temperature inversion, air temperatures increase with altitude (the *inverse* of the normal situation in the troposphere). Warm, light air thus overlies cooler, denser air. This is an extremely stable stratification that strongly inhibits mixing and dilution. A temperature inversion can form by (1) subsidence of air, (2) extreme radiational cooling, or (3) advection of air masses. The resulting inversion may occur aloft or at the surface.

A **subsidence temperature inversion** forms a lid over a wide area, often encompassing several states at one time. It develops during a period of fair weather when the hemispheric circulation pattern causes a warm anticyclone to stall. An anticyclone is characterized by air that subsides and is thereby warmed by adiabatic compression. Subsiding, warm air is prevented from reaching the Earth's surface by the **mixing layer** in which air is thoroughly mixed by convection (Figure 16.6). Air temperatures within the mixing layer decrease with altitude, but air just above the mixing layer, having been warmed by adiabatic compression, is significantly warmer than air at the top of the mixing layer. An elevated temperature inversion thus separates the mixing layer from the compressionally warmed air above. Under these conditions, air pollutants are distributed throughout the mixing layer up to the

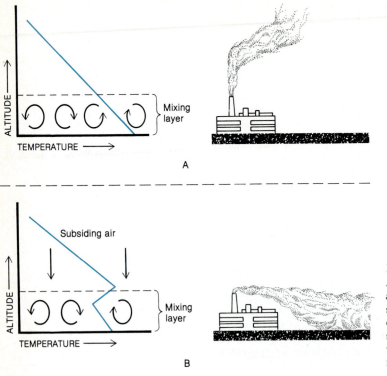

FIGURE 16.6

A temperature inversion can develop aloft through subsidence of air. A sounding prior to subsidence (A) is compared with a sounding during subsidence (B). The temperature inversion acts as a lid over the lower atmosphere, trapping air pollutants.

altitude of the temperature inversion. This situation is sometimes referred to as **fumigation.**

Radiational temperature inversions are perhaps more common and are often more localized than subsidence temperature inversions. At night, under clear skies, the Earth's heat is rapidly lost to space through emission of infrared radiation, a process called **nocturnal radiational cooling.** The surface air layer is then chilled by contact with the cooler ground. Because the air at the surface is coldest, a surface temperature inversion develops. After sunrise, as solar radiation is absorbed by the ground and heat is conducted and convected to the overlying air, the inversion gradually disappears and a normal temperature lapse rate is restored. In winter, however, when the sun's rays are weak and where snow covers the ground, a radiational temperature inversion may persist for several days, or even weeks at a time and may severely inhibit the dispersal of air pollutants.

Advecting air masses can also give rise to temperature inversions, which sometimes develop at the base of the Rocky Mountains. As shown in Figure 16.7, a westerly airflow is compressionally warmed as it is drawn down the leeward slopes of the mountain range. Along the foot of the mountains, however, surface winds advect cold air southward. Hence, a temperature inversion aloft separates the warm air above from the surface layer of cold air.

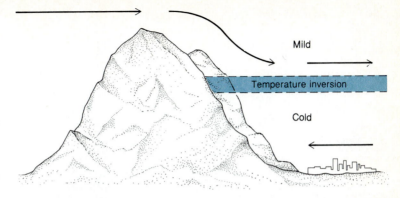

Although temperature inversions also characterize warm and cold fronts, these inversions have little adverse effect on air quality because the fronts are moving and the accompanying precipitation washes pollutants from the air.

AIR POLLUTION POTENTIAL

Weather conditions that favor shallow mixing depths, air stagnation, and air pollution episodes occur with varying frequency in different places and at different times of the year. Areas with particularly high air pollution potential include southern and coastal California, portions of the Rocky Mountain states, and the Appalachian Mountain valleys. In general, in much of the West, air quality is lowest in winter; in the East, air quality is lowest in autumn. In southern California, air pollution is highest in summer. These seasonal changes in air quality are due to normal seasonal shifts in circulation patterns.

Many regions with high air pollution potential are also locales of great topographic relief, since hills and mountain ranges can block horizontal winds that would disperse polluted air. In addition, radiational temperature inversions that form in lowlands, such as river valleys, are often strengthened by an accumulation of cold, dense air that drains downward from nearby highlands. As illustrated in Figure 16.8, the result is a persistent stable stratification of mild, light ir over cool, dense air.

Los Angeles is particularly susceptible to air pollution episodes because of its topographic setting, a high concentration of pollutant sources (millions of

FIGURE 16.8
Under the influence of gravity, cold dense air drains downslope and strengthens the temperature inversion in the valleys.

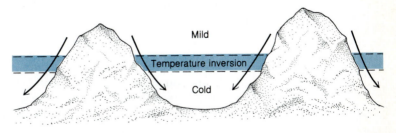

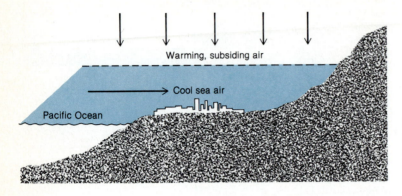

FIGURE 16.9
Prevailing atmospheric circulation patterns and topographic features combine to give Los Angeles an unusually high air pollution potential.

cars), and frequent periods of air stability. Figure 16.9 shows the air circulation and topographic features that influence air quality in that city. Weather in the Los Angeles area, like the weather throughout much of California, is strongly influenced by the eastern edge of the semipermanent Pacific anticyclone. This subtropical high is responsible not only for California's famous fair weather, but also for air that gently subsides over the city. Subsiding air is prevented from reaching the ground by a shallow onshore flow of cool maritime air from over the adjacent Pacific Ocean. Because of adiabatic compression, the subsiding air aloft is warmer than the underlying maritime layer. Consequently, a subsidence temperature inversion typically develops at about 700 m (2300 ft) over Los Angeles, and convection is confined to the shallow maritime layer. As a result, mixing depths are low on perhaps two-thirds of the days of the year.

The exceptionally high incidence of extremely stable atmospheric conditions in Los Angeles is aggravated by topographic barriers. The city is situated on a plain that opens to the Pacific and is rimmed on three sides by mountains. Because of this topography, cool breezes sweeping inland from the ocean are unable to flush pollutants out of the city. The mountains and a temperature inversion aloft thus encase the city in its own fumes, and within this crucible a complex photochemistry takes place that produces smog.

Because air pollutants can be a serious threat to human health, weather reports in many cities regularly include air quality bulletins. These advisories sometimes refer to the Pollutant Standards Index (PSI) as a measure of air quality. In the Special Topic, "Pollutant Standards Index," we describe how the PSI is determined and some effects of air pollutants on human health.

NATURAL CLEANSING PROCESSES

Conditions favoring the accumulation and concentration of pollutants in the air are countered to some extent by natural removal mechanisms. Some particulates are removed from the air when they strike and adhere to buildings and other structures, a process called **impaction.** Aerosols are also subject to

Pollutant Standards Index

The U.S. Environmental Protection Agency (EPA) has devised an index of air quality that gauges the relative health hazard of polluted air. Individuals with heart or lung ailments, people planning strenuous physical activity, and especially the elderly are well advised to consult this index to judge the safety of the air on any given day.

In response to federal law, the EPA established uniform air quality standards for six air pollutants (Table 1). These standards apply to the ambient (outdoor) air and are of two types, primary and secondary. A *primary air quality standard* is the maximum exposure level of an air pollutant that can be tolerated by human beings without ill effects. A *secondary air quality standard* is the maximum allowable concentration of an air pollutant when considering its potential harmful impact on vegetation, visibility, personal comfort, and climate.

The *Pollutant Standards Index (PSI)* is computed for a particular location using the air pollutant that exhibits the highest concentration when compared with its primary air quality standard. If the concentration is at the primary standard, the PSI is assigned a value of 100; if it is twice the standard, the PSI value is 200, and so on. As shown in Table 2, PSI values are divided into six ranges that correspond to the risk to human health. PSI values higher than 100 are considered unhealthy, particularly for elderly persons and individuals with heart or respiratory problems. A PSI value of 300 or higher is considered hazardous; during such episodes, everyone should avoid physical exertion outdoors.

When PSI values exceed 100, what is the potential impact on human health? Although many parts of the human body can be eventual targets, the initial attack of air pollutants is via the respiratory system. Inhaled air follows a long pathway before finally reaching the alveoli (air sacs) of the lungs. Within air sacs, oxygen is removed from the air by red blood cells, which then carry it to all parts of the body. If the blood is deprived of its oxygen supply, the person succumbs to *asphyxiation,* or oxygen starvation. Because pollutants are inhaled into the respiratory tract along with air, they follow the same route to the air sacs. In sufficiently high concentrations, some air pollutants are asphyxiating agents, whereas others irritate the tissues of the respiratory tract.

Carbon monoxide (CO), a constituent of motor vehicle exhaust, is the most common asphyxiating air pollutant. Within the air sacs, CO displaces the oxygen on the hemoglobin molecules (the transporters of oxygen in red blood cells). As increasing concentrations of carbon monoxide are inhaled, the quantity of life-sustaining oxygen that is transported by the bloodstream decreases. Initially, the victim experiences dizziness, headache, and impaired perception; higher concentra-

Table 1

National Ambient Air Quality Standards for Criteria Pollutants

Pollutant	Averaging time[a]	Primary standard	Secondary standard
Total suspended particulates (micrograms[b] per m³)	1 yr	75	60
	24 hr	260[c]	150
Sulfur oxides (ppm)	1 yr	0.03	—
	24 hr	0.14[c]	—
	3 hr	—	0.5
Carbon monoxide (ppm)	8 hr	9[c]	9
	1 hr	35[c]	35
Nitrogen dioxide (ppm)	1 yr	0.053	0.053
Ozone (ppm)	1 hr	0.12[c]	0.12[c]
Lead (micrograms[b] per m³)	3 mos	1.5	—

[a] Averaging time is the period over which concentrations are measured and averaged.
[b] A microgram is one millionth of a gram.
[c] Concentration not to be exceeded more than once (on separate days) per year.
Source: United States Environmental Protection Agency.

Table 2
Pollutant Standards Index (PSI) of Air Quality

Pollutant Standards Index value statement	Description	Episode level	General health effects and cautions
0–50	Good		
51–100	Moderate		
101–200	Unhealthy	Primary standard	Symptoms are slightly aggravated in susceptible persons. Symptoms of irritation occur in healthy population. Persons with heart or respiratory ailments should reduce physical exertion and outdoor activity.
201–300	Very unhealthy	Alert	Symptoms are significantly aggravated, and exercise tolerance is decreased in persons with heart or lung disease. Symptoms are widespread in healthy population. Elderly persons and those with heart or lung disease should stay indoors and reduce physical activity.
301–400	Hazardous	Warning	Premature onset of some diseases is noted, in addition to significant incidence of symptoms and decreased exercise tolerance in healthy persons. Elderly persons and those with heart or lung disease should stay indoors and avoid physical exertion. General population should avoid outdoor activity.
401 or more	Hazardous	Emergency	Premature death occurs in ill and elderly persons. Healthy persons experience adverse symptoms affecting normal activity. All persons should minimize physical exertion, avoid traffic, and remain indoors, keeping windows and doors closed.

Source: Wisconsin Department of Natural Resources, Bureau of Air Quality Management.

tions result in nausea, visual problems, and, ultimately, death.

Gases that act mainly as respiratory tract irritants include ozone, sulfur dioxide, and nitrogen dioxide. Each causes somewhat different reactions, but the effects generally include persistent cough and heavy secretion of mucus. The mucus clogs smaller pathways leading to the lungs' air sacs, thereby decreasing the amount of oxygen available to the blood.

Particulates that reach the lungs may cause illness and, eventually, even death. The most severe health problems result from lengthy exposure to relatively high concentration levels, as in such occupations as mining, metal grinding, and manufacturing of abrasives. Resultant lung diseases, which are usually named for the type of particulates involved, include *black lung* (from coal dust), *silicosis* (from quartz dust generated during mining), *asbestosis* (from asbestos fibers), and *brown lung* or *byssinosis* (from cotton dust). Depending on the type of particulate and the concentrations inhaled, the lungs sustain irritation, allergic reactions, or scarring of tissue, which becomes a potential site for tumor development. Victims typically experience coughing and shortness of breath and, in the long run, may develop pneumonia, chronic bronchitis, emphysema, or lung cancer.

gravitational settling, which is most effective for aerosols with radii greater than a tenth of a micrometer. The heavier and larger aerosols have greater terminal velocities and settle more rapidly than do the smaller aerosols. For this reason, larger aerosols tend to settle nearer their source, whereas smaller aerosols may be transported by wind many kilometers and to great altitudes before they finally settle to the ground. The combined processes of impaction and gravitational settling are sometimes referred to as **dry deposition.**

The most effective natural pollution removal mechanism, however, is scavenging by rain and snow. In localities with moderate precipitation, as much as 90 percent of suspended aerosols is removed by **precipitation scavenging.** Gaseous pollutants are somewhat less susceptible to scavenging than are aerosols, but they do dissolve to some extent in raindrops and in cloud droplets. Although scavenging of air pollutants enhances air quality, the process degrades the quality of the rain and snow, sometimes to the degree that surface waters are polluted and aquatic life is threatened. We discuss this problem later in the chapter.

Air Pollution's Impact on Weather

We have seen how weather conditions influence air pollution potential. Air pollution also impacts weather. Air pollution affects the amount of cloudiness, and the quantity and quality of precipitation, especially downwind from large urban-industrial areas. In Chapter 18, we discuss how air pollution may be affecting global climate.

URBAN WEATHER

Certain air pollutants usually found in urban air, including a variety of dust particles and acid droplets, can influence the development of clouds and precipitation within and downwind from a city. These pollutants, many of which are hygroscopic, serve as nuclei for cloud droplets and thus accelerate condensation. In addition, the heat island effect of large urban areas (Chapter 12) spurs uplift of air and consequent cloud formation.

The influence of urban air pollution on condensation and precipitation is illustrated by the typical climatic contrasts between urban and rural areas. Winter fogs occur about twice as frequently in cities as in the surrounding countryside. Downwind from cities, rainfall may be enhanced by 5 to 10 percent. The greater contrasts tend to occur on weekdays, when urban-industrial activity is at its peak, suggesting that increased precipitation is at least partially due to urban-industrial air pollution. Data from the **Metropolitan Meteorological Experiment (METROMEX)** indicate significantly greater precipitation enhancement downwind of St. Louis. METROMEX scientists analyzed weather observations during an intensive field study over five years (1971–1975) and concluded that summer rainfall was up to 30 percent

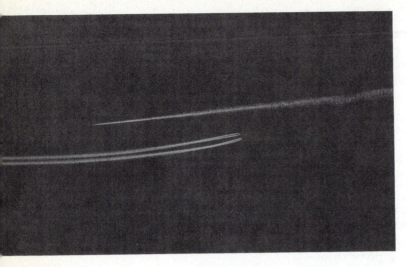

FIGURE 16.10
Contrails produced by jet aircraft rapidly spread laterally to form cirrus clouds. [Photograph by Mike Brisson]

greater downwind of St. Louis than upwind of the city. This rainfall anomaly was attributed to the combined effect of urban contributions of heat and "giant" cloud condensation nuclei.

Because precipitation, fog, and cloudiness in urban areas often have adverse effects on both surface and air transportation, any artificial increase in these conditions is potentially troublesome. Reduced visibility, for example, slows surface traffic, contributes to auto accidents, and curtails air travel. In the last decade, significant improvement in local urban visibilities has been reported and is apparently the consequence of enforcement of stricter air quality standards.

Jet aircraft traffic is modifying the cloud cover, especially along heavily traveled air corridors between major cities. The visible jet contrails etching the sky, like those shown in Figure 16.10, are composed of ice crystals that are traceable to the water vapor and condensation nuclei produced by jet engines as combustion products (refer back to the Special Topic, "Clouds by Mixing," in Chapter 6). Contrails sometimes dissipate rapidly and sometimes spread laterally as a thin cirrus overcast. Increased cloudiness, in turn, reduces sunshine penetration and may enhance local precipitation by serving as a source of ice-crystal nuclei for lower clouds.

ACID DEPOSITION

As we saw in Chapter 6, the atmospheric subcycle of the hydrologic cycle purifies water. That is, through evaporation or sublimation and subsequent condensation or deposition, dissolved and suspended substances are removed from water. As raindrops and snowflakes fall from clouds to the ground, however, they wash pollutants from the air, thereby altering the chemistry of the precipitation.

Rain is normally slightly acidic because it dissolves some atmospheric carbon dioxide, producing a weak acid solution. Where the air is polluted with oxides of sulfur and oxides of nitrogen, these gases interact with moisture in the atmosphere to produce tiny droplets of sulfuric acid (H_2SO_4) and nitric acid (HNO_3). These substances dissolve in precipitation and thereby increase its acidity. Precipitation that falls through such contaminated air may become 200 times more acidic than normal. Furthermore, in the absence of precipitation, sulfuric acid droplets convert to acidic particles that settle to the ground as dry deposition (Figure 16.11). The combination of acidic precipitation with dry deposition is often referred to as **acid deposition.**

The range of acidity and alkalinity, called the **pH scale,** is shown in Figure 16.12, which compares the normal acidity of rainwater with the pH

FIGURE 16.11

An instrument for monitoring precipitation quality. Separate containers collect dry and wet deposition for chemical analysis. One container is open only during fair weather and collects dry deposition. Rainfall activates a switch that covers the dry deposition container and opens the other to collect precipitation. [Photograph by J. M. Moran]

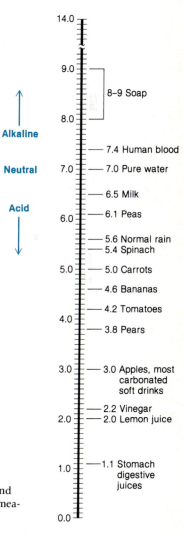

FIGURE 16.12

The scale of acidity and alkalinity, the pH scale, and pH values for some common substances. pH is a measure of hydrogen ion concentration.

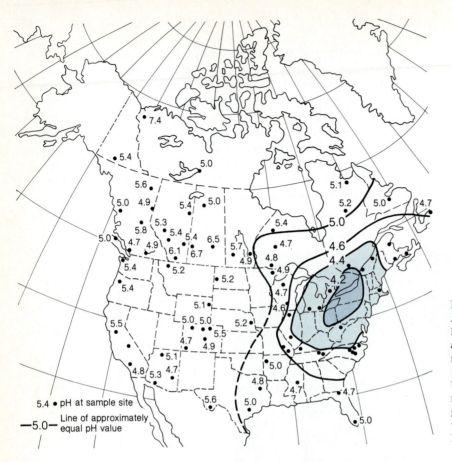

FIGURE 16.13

Acidity of precipitation over North America. Figures are average annual pH values of rain and snow for 1982. [Data supplied by the United States National Atmospheric Deposition Program and the Canadian Network for Sampling of Precipitation]

Legend on map:

5.4 • pH at sample site

—5.0— Line of approximately equal pH value

values of some other familiar substances. The normal pH of rainwater is 5.6; rain that is more acidic than normal is called **acid rain.*** Note that the pH scale is logarithmic; that is, each unit increment corresponds to a tenfold change in acidity. Hence, a drop in pH from 5.6 down to a value of 3.6 represents a hundredfold (10×10) increase in acidity.

Gene E. Likens (now of the New York Botanical Garden) and his associates reported an increase in the acidity of rainfall over the eastern United States during the 1955 to 1973 period. Their findings were later confirmed and updated by measurements made by the National Atmospheric Deposition Program in the United States and by the Canadian Network for Sampling Precipitation (Figure 16.13). Much of this upswing in precipitation acidity can be attributed to acid rain precursors emitted during fuel combustion for power,

---

* Distilled water that is saturated with carbon dioxide has a pH of 5.6. This is why a pH of 5.6 is taken as the threshold for acid precipitation; that is, rain or snow having a pH under 5.6 is described as *acidic*. However, recent studies demonstrate that small amounts of naturally occurring acids (other than carbonic acid) lower the normal pH of precipitation closer to 5.0. This argues for a revision of the current pH criterion for acid precipitation.

industry, and motor vehicles. Coal burning for electric power generation is the principal source of sulfur oxides, whereas high-temperature industrial processes and internal combustion engines produce nitrogen oxides.

Where acid rains fall on soils or noncarbonate bedrock that cannot neutralize the acidity, lakes and streams become more acidic. Acid-sensitive areas in North America are shown in Figure 16.14. In such areas, spring snowmelt can be a serious problem because of the sudden pulse of acidic meltwater entering waterways. A 1985 survey by the U.S. Environmental Protection Agency (EPA) found that pH levels were 5 or lower in about 10 percent of lakes in upper Michigan and in the Adirondack Mountains of New York. Excessively acidic lake or stream water disrupts the reproductive cycles of fish, and acid rains leach metals (such as aluminum) from the soil, washing them into lakes and streams where they may harm fish, microorganisms, and aquatic plants. As a consequence of increased acidity of aquatic habitats, fish populations in some lakes and streams in Norway, Sweden, eastern Canada, and the northeastern United States have declined or been eliminated.

FIGURE 16.14

Surface waters (lakes and rivers) in many areas of North America are sensitive to acidification. These areas typically lack natural buffers (substances that neutralize acids) in the soil or bedrock. [From J. N. Galoway and E. B. Cowling, "The Effects of Precipitation on Aquatic and Terrestrial Ecosystems: A Proposed Precipitation Chemistry Network," *Journal of Air Pollution Control Association* 28 (1978):233]

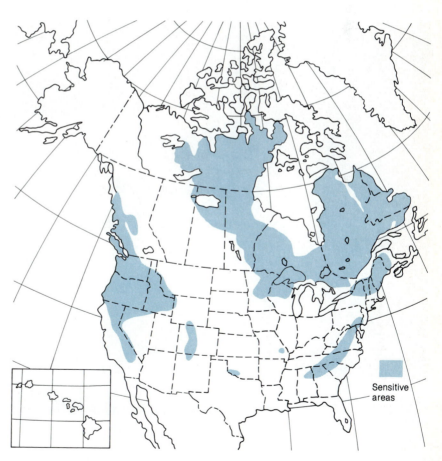

Sensitive areas

As part of the EPA's National Acid Precipitation Assessment Program, scientists artificially acidified a portion of a small lake in northern Wisconsin to determine the impact on the lake's aquatic life under controlled conditions. The 8-hectare (20-acre) Little Rock Lake is typical of the thousands of small acid-sensitive lakes that dot northern Wisconsin, Minnesota, and Upper Michigan. Scientists anticipate that the experiment will enable them to predict the probable long-term response of similar lakes to acid precipitation.

Little Rock Lake is shaped like an hourglass with two basins of about equal area joined by a narrows. Following an intensive field study of the lake's properties, scientists installed a 75-m (230-ft) plastic barrier to separate the two basins at the narrows (Figure 16.15). In the ensuing three years (1986 to

FIGURE 16.15
Little Rock Lake in northern Wisconsin is the subject of an ongoing experiment on the effects of acidified waters on aquatic life. The basin to the left of the barrier has been artificially acidified, while the basin to the right is unaltered and serves as a standard for comparison with the acidified basin. [Photograph by J. M. Moran]

1988), sulfuric acid was added to one of the lake basins so that the pH was gradually lowered from its initial value of 6.0 to 4.5. The other lake basin was not disturbed and served as a standard for comparison with the acidified basin. Scientists are now studying the effects of acidification on water chemistry and aquatic species diversity and abundance. Eventually, either natural processes or human intervention will restore the acidified lake basin to its original state.

Recent studies suggest that, in addition to threatening aquatic systems, acid rains and fogs may be implicated along with other air pollutants (especially ozone) in the decline and dieback of coniferous forests in West Germany, upstate New York, and northern New England. Another costly impact of acid rains is the accelerated weathering of building materials, especially limestone, marble, and concrete. Metals, too, corrode faster than normal when exposed to acidic moisture.

Because winds aloft can transport oxides of sulfur and nitrogen many thousands of kilometers from tall stack sources, acid rain is becoming a global problem. Acid rains have been reported from such isolated localities as the Hawaiian Islands and the central Indian Ocean. Long-range transport of acid-rain precursors has even strained the traditionally amiable relationship between the United States and Canada.

Acid rains threaten Canada's primary industries, lumber and fisheries. Perhaps half the acidic rainfall in Canada can be traced to industrial emissions from Ohio Valley industries and power plants. Canadian government officials are pressing their Washington counterparts to enact stricter controls on polluting industries and power plants in order to ease the problem, which some Canadians feel will become severe within the next decade. In response, in 1987, the Reagan administration proposed to spend $2.5 billion on a five-year effort to demonstrate methods to cut sulfur oxide and nitrogen oxide emissions from power plants. More recently, the Bush administration has set goals to reduce sulfur dioxide and nitrogen oxide emissions at 107 coal-fired power plants in 18 states by the year 2000.

The Ozone Shield

In general, residents of urban-industrial areas run the greatest risk of adverse health effects due to polluted air, but some air pollutants threaten the well-being of all living things everywhere. These pollutants threaten the delicate stratospheric ozone layer that shields organisms from dangerous levels of solar ultraviolet radiation.

Within the troposphere, ozone (O_3) is a component of smog and a serious air pollutant. Ironically, its presence in minute concentrations (10 ppm) in the stratosphere is essential for the continuation of life on Earth. As noted in

Chapter 2, two sets of UV-absorbing chemical reactions take place within the stratosphere: one set generates and the other destroys ozone. These absorption processes shield organisms from exposure to potentially lethal intensities of UV radiation, hence the term **ozone shield.** A portion of the UV radiation that reaches the Earth's surface is responsible for sunburn and causes or contributes to skin cancer. This biologically effective radiation, called **UVB,** spans the wavelengths from 0.28 to 0.32 micrometer.

UVB absorption is very sensitive to changes in stratospheric ozone concentration. As a rule, every 1 percent decrease in ozone concentration results in a 2 percent increase in UVB that passes through the ozone shield. The amount of increased UVB that would actually reach the Earth's surface hinges on atmospheric factors such as cloudiness and dustiness. Various studies suggest that a 2.5 percent thinning of the ozone shield could boost the rate of human skin cancers by 10 percent. In the Spring of 1990, a report from an observatory in the Swiss Alps indicated that the flux of UVB had increased by about 1 percent per year since 1981.

CHLOROFLUOROCARBONS (CFCs)

Probably the most serious threat to the ozone shield is from certain chlorofluorocarbons (CFCs). Most of us have encountered several CFCs, knowingly or unknowingly, in our daily lives (Table 16.1). CFCs are widely used as chilling agents in refrigerators and air conditioners, for cleaning electronic components, and in the manufacture of foams used for insulation and fast-food packaging. Use of CFCs as a propellant in common household aerosol sprays, such as deodorants and hairsprays, was banned from the United States and Canada in the late 1970s.

F. S. Rowland and M. J. Molina of the University of California at Irvine first warned of the danger of CFCs to the ozone shield in 1974. They theorized a disturbing scenario. Certain CFCs are inert (chemically nonreactive) in the

Table 16.1
The Most Commonly Used Chlorofluorocarbons (CFCs)

Type	Formula	Primary use	Residence time,[a] yr
CFC-11	CCl_3F	Foam blowing agent; aerosol propellant	76
CFC-12	CCl_2F_2	Refrigeration and air conditioning	139
CFC-113	$C_2Cl_3F_3$	Solvent for cleaning electronic microcircuits	92

[a] Time required for 63 percent of the CFC to be washed from the atmosphere.

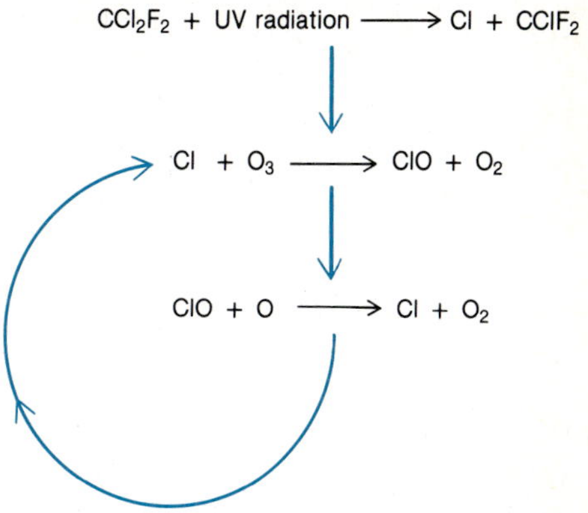

$$CCl_2F_2 + UV\ radiation \longrightarrow Cl + CClF_2$$

$$Cl + O_3 \longrightarrow ClO + O_2$$

$$ClO + O \longrightarrow Cl + O_2$$

FIGURE 16.16
A chain reaction involving CCl_2F_2, a chlorofluorocarbon (CFC) that is widely used as a refrigerant. Through this reaction, chlorine repeatedly reacts with and destroys ozone molecules.

troposphere, where they have been accumulating since production began back in the 1930s. The inert CFCs gradually migrate upward into the stratosphere. At altitudes above about 25 km (15 mi), intense UV radiation breaks down the CFCs, causing them to release chlorine, a gas that readily reacts with and destroys ozone (Figure 16.16). By a chain reaction, each chlorine atom destroys perhaps a hundred thousand ozone molecules.

THE ANTARCTIC OZONE HOLE

During the Southern Hemisphere spring (mainly September and October), the ozone layer in the Antarctic stratosphere thins drastically. And every November the ozone level recovers. Although the British Antarctic Survey team first reported this phenomenon in 1985, massive ozone depletions had been measured during the prior eight Antarctic springs but had been promptly dismissed as the product of instrument error. The area of Antarctic ozone depletion, dubbed an **ozone hole,** is about the size of the continental United States (Plate 26).

Analysis of measurements by ozone monitors* onboard NASA's Nimbus 7 satellite indicated that the Antarctic ozone hole deepened from the late 1970s into the mid-1980s. This prompted speculation of a possible link to CFCs and led to an intensive field investigation by specially instrumented aircraft during the Antarctic spring of 1987. That investigation not only measured a record ozone depletion of 50 percent but also revealed exceptionally high concentrations of chlorine monoxide (ClO), a product of chemical reactions known to destroy ozone.

---

* Satellite sensors measure trends in stratospheric ozone concentrations by comparing absorption at different wavelengths of solar radiation. Ozone absorbs some wavelengths strongly and other wavelengths very little. If radiation at the absorbed wavelengths increases relative to that at the unabsorbed wavelengths, then the ozone concentration has declined.

Although the 1987 findings strongly implicated CFCs as the cause of the deepening of the Antarctic ozone hole, events of 1988 reminded investigators that the matter is far from being totally understood. During the Antarctic spring of 1988, ozone loss amounted to only about 15 percent. Apparently, changes in atmospheric circulation patterns played a pivotal role in reversing the deepening trend of the ozone hole.

A belt of westerly winds, known as the **circumpolar vortex,** encircles and contains the Antarctic ozone hole. When this circulation is vigorous (as it was during the winter and spring of 1987), milder air from lower latitudes is kept out of the Antarctic stratosphere and temperatures plunge. At temperatures below -78 °C (-108 °F), ice clouds form in the polar stratosphere and their component ice crystals play an essential (catalytic) role in ozone-destroying chemical reactions. When the circumpolar vortex weakens, however, as it did in 1988, milder air is transported into the polar stratosphere, temperatures rise, few ice clouds develop, and ozone depletion is less. In 1989, the vortex again strengthened and ozone depletion matched the record levels of 1987.

Discovery of the deepening Antarctic ozone hole and speculation of a possible ink to CFCs spurred a massive reanalysis of global ozone measurements made by ground-based and satellite instruments. In March 1988, NASA's Ozone Trends Panel reported that the mean annual ozone concentration in the Northern Hemisphere had declined by about 2 percent between 1969 and 1986. For the latitude belt between 30 and 64 degrees N, the decline ranged from 1.7 to 3.0 percent, with the greatest drop at high latitudes in winter. Whether CFCs are actually responsible for the downward trend in ozone is not known. In addition, scientists investigating stratospheric chemistry in the Arctic during early 1989 discovered ozone-destroying chlorine compounds and a slight thinning of Arctic ozone.

Currently, CFC production is partially limited in the United States, Canada, Sweden, and Norway. Perhaps two-thirds of world production continues, although growing international acceptance of the CFC threat to the ozone shield is beginning to change that. In September 1987, under United Nations auspices, representatives of 23 nations met in Montreal and negotiated a treaty to cut by 50 percent the global output of CFCs by June 1998. To date, 53 nations—accounting for about 90 percent of global CFC output—have ratified the so-called Montreal Protocol. This action together with growing public awareness of the threat to the ozone shield has spurred the chemical industry to accelerate efforts to develop CFC substitutes.

Yet, even if the Montreal Protocol achieves its goal, the atmospheric content of ozone-destroying chlorine will continue to grow because of the long residence times of CFCs in the atmosphere (Table 16.1). This has prompted some scientists to argue that the Montreal Protocol is inadequate and to push for a complete worldwide phaseout of CFCs by the end of the century. Opposing this plan are developing nations such as China who need CFCs for continued economic growth and cannot afford costly substitutes.

Conclusions

Weather influences, and is influenced by, air pollutants. Atmospheric stability and circulation patterns affect the rate of dilution of air pollutants. Some air pollutants, in turn, affect cloud development and the quality and quantity of precipitation, especially downwind of large urban-industrial areas. In Chapter 18 we consider the possible impact of air pollution on large-scale trends in climate, but first we turn to a new subject, the nature of climate and climatic behavior.

Summary Statements

Most air pollutants are cycled naturally within the atmosphere, but concentrations may reach levels that threaten human health or disrupt physical and biological processes.

Strong winds and unstable air enhance the rate of dilution of air pollutants, whereas weak winds and stable air suppress dilution.

A temperature inversion consists of an extremely stable stratification of light, mild air over denser, cooler air. An inversion may develop at the Earth's surface or aloft through subsidence of air, extreme radiational cooling, or air mass advection. Temperature inversions greatly inhibit the dispersal of air pollutants.

Many regions with high air pollution potential are also locales of great topographic relief, because hills and mountain ranges can block horizontal winds that disperse polluted air. Furthermore, cold air drainage into river and mountain valleys can reinforce radiational temperature inversions.

Conditions that favor the accumulation and concentration of pollutants in the atmosphere are countered to some extent by natural cleansing processes including dry deposition and scavenging by rain and snow.

Urban heat and air pollutants affect precipitation, cloudiness, and fog development within and downwind from large metropolitan areas. Some urban air pollutants are hygroscopic and thereby function as efficient cloud condensation nuclei.

In localities where the air is polluted by oxides of sulfur or nitrogen, precipitation becomes strongly acidic. Acid deposition threatens aquatic life and corrodes structures.

The formation of ozone in the stratosphere protects life on Earth by filtering out harmful intensities of solar ultraviolet radiation. This ozone shield is threatened by chlorofluorocarbons (CFCs) accumulating in the atmosphere. CFCs are inert in the troposphere, but they drift into the stratosphere where intense UV causes them to break down and release chlorine, a gas that reacts with and destroys ozone.

At this point, the cause of the deepening Antarctic ozone hole is not known, although CFCs and variations in atmospheric circulation are likely implicated.

Key Words

air pollution episodes
molecular diffusion
wake
mixing depth
temperature inversion
subsidence temperature
 inversion

mixing layer
fumigation
radiational temperature
 inversions
nocturnal radiational
 cooling
impaction

gravitational settling
dry deposition
precipitation scavenging
Metropolitan Meteoro-
 logical Experiment
 (METROMEX)
acid deposition

pH scale
acid rain
ozone shield
UVB
ozone hole
circumpolar vortex

Review Questions

1. Under what conditions are natural components of the atmosphere considered to be pollutants?

2. As a general rule, how does wind speed affect the concentration of air pollutants?

3. Why is the dilution of air pollutants particularly impeded in urban areas?

4. Define mixing depth. Explain how and why the mixing depth varies with (a) time of day and (b) season.

5. How does the stability of air influence the rate of dilution of air pollutants?

6. Describe the behavior of smoke plumes in (a) stable air and (b) unstable air.

7. Why do subsidence temperature inversions generally cover a larger area than radiational temperature inversions?

8. Why are radiational temperature inversions often short-lived? Under what conditions might a radiational temperature inversion persist for many days?

9. In areas of great topographic relief, cold air drainage strengthens radiational temperature inversions. Explain this statement.

10. What factors contribute to the relatively high air pollution potential of Los Angeles?

11. Compare the natural atmospheric cleansing processes for the troposphere and the stratosphere.

12. What is the most effective natural air pollution removal mechanism?

13. How do air pollutants influence urban weather?

14. Describe the potential climatic influence of jet plane contrails.

15. Why is rainfall normally slightly acidic?

16. What air pollutants are acid rain precursors? Identify their sources.

17. What is the ozone shield?

18. What is thought to be the most serious threat to the ozone shield? How might these chemicals affect stratospheric ozone?

19. How might thinning of the stratospheric ozone shield affect the amount of biologically effective ultraviolent radiation (UVB) that reaches the Earth's surface? How might this impact human health?

20. Speculate on the major sources and types of air pollutants in nonindustrialized nations.

Points to Ponder

1. Midlatitude cyclones are beneficial to air quality. Explain this statement.

2. Some electric power utilities have reduced ground-level concentrations of air pollutants by building taller smokestacks. How do taller smokestacks contribute to the acid rain problem?

3. Comment on the notion that at least some air pollution is an inevitable consequence of our way of life.

4. Why is a temperature inversion a case of "extreme" atmospheric stability?

5. What types of weather patterns favor (a) low air quality and (b) high air quality?

6. Describe how a temperature inversion develops by (a) subsidence of air, (b) extreme radiational cooling, and (c) air mass advection.

7. Many aerosols in urban air are hygroscopic. What does this imply about the development of clouds downwind from a city?

8. It is possible for temperature inversions to form simultaneously at the surface and aloft in the atmosphere. Speculate on the sequence of weather events that would give rise to such a situation.

Projects

1. Evaluate the air pollution potential of your community in terms of (a) frequency of stable atmospheric conditions, (b) topographic influences, and (c) locations of major pollutant sources.

2. The types of pollutants that foul the air of a particular locality depend on the specific kinds of industrial, domestic, and agricultural activities that take place there. Prepare a list of pollution sources and types of air pollutants for your community. Is information available on the amounts of air pollutants emitted by each source?

3. Maintain a daily log describing the appearance of local industrial smoke plumes. Determine whether the shape and behavior of the plumes indicate unstable or stable conditions. If a temperature inversion is evident, does it usually form at the surface or aloft? Taking daily photographs of plumes may aid your analysis.

4. Has your community ever experienced a severe air pollution episode? If so, was there an increased incidence of respiratory illness? You may wish to refer to past issues of local newspapers or consult with your local public health agency.

5. Collect a sample of snow, melt it, and filter the meltwater. Examine the residue on the filter paper under a microscope. Describe what you see and speculate on its origins. Compare the appearance of samples taken from different locations in your community.

Selected Readings

Changnon, S. A., Jr. "More on the LaPorte Anomaly: A Review." *Bulletin of the American Meteorological Society* 61 (1980):702–711. Concludes that anomalous precipitation at LaPorte, Indiana, between the late 1930s and the 1960s was linked to urban-industrial air pollution.

Likens, G. E., et al. "Acid Rain." *Scientific American* 241 (1979):43–51. Discusses trends in acid deposition over North America and Western Europe.

Postel, S. "Air Pollution, Acid Rain, and the Future of Forests." *Worldwatch Paper,* No. 58 (1984):1–54. Reviews possible link between acid deposition and the dieback of coniferous forests in portions of Europe and northeastern North America.

Rowland, F. S. "Can We Close the Ozone Hole?" *Technology Review* 90, No. 6 (1987):50–58. Gives an excellent summary of the CFC threat to the ozone shield authored by one of the scientists who first proposed the concept.

Schindler, D. W. "Effects of Acid Rain on Freshwater Ecosystems." *Science* 239 (1988):149–156. Presents an excellent summary of what is currently understood about the impact of acidic precipitation.

Shaw, R. W. "Air Pollution by Particles." *Scientific American* 257, No. 2 (1987):96–103. Discusses the role of acidic sulfate particles in the development of haze.

Stolarski, R. S. "The Antarctic Ozone Hole." *Scientific American* 258, No. 1 (1988):30–36. Reviews the thinning of ozone during the Antarctic spring and possible causes for the recent acceleration of that thinning.

17

World Climates

Planet Earth is a mosaic of many different types of climate. The variety of the Earth's climates is best viewed in mountainous regions where, over several thousand meters of altitude, the climate exhibits the same variations as in thousands of kilometers of latitude. [Photograph by J. M. Moran]

H

AVING EXAMINED the properties of the atmosphere, the governing principles of weather, and the various atmospheric circulation systems, we now turn to climate. In this chapter we examine some of the basics of climatology, the classification of climate, and global patterns of climate. Climatology is a complex discipline, however, and we cannot present more than a cursory summary of major concepts in this chapter. The interested reader is referred to the Selected Readings at the close of this chapter for more on the subject. In the next chapter, we review the climatic record, how climate changes, and the possible causes of climatic variability.

Describing Climate

Climate is defined as the weather of some locality averaged over some time period, the seasonal distribution of weather, plus the extremes in weather behavior. Climate must be specified for a place and time period because, like weather, climate varies both spatially and temporally. Thus, for example, the climate of Minneapolis is different from that of Miami, and winters in Minneapolis were somewhat milder in the 1940s and 1950s than in the 1840s and 1850s.

Extremes in weather are important aspects of the climatic record. Hence, daily weather reports usually include the highest and lowest temperatures ever recorded for that date. Climatic summaries typically identify such extremes as the coldest, warmest, driest, wettest, snowiest, or cloudiest month or year on record. Farmers, for example, are interested in knowing not only the long-term average rainfall during the growing season but also the frequency of drought. Electric utilities are interested in the hottest summers and coldest winters on record so that they might anticipate extremes in residential energy demands.

Climate is usually described in terms of normals, means, and extremes of a variety of weather elements including temperature, precipitation, and wind. Weather extremes for the world and for North America are listed in Appendix IV. Climatic summaries are available in tabular form for climatic divisions of each of the states and Canadian provinces as well as for major cities. A sample climatic summary is shown in Table 17.1 (pages 470–71), and climatic data for selected North American cities are presented in Appendix V. A narrative description of local or regional climate usually accompanies these data. In addition, a vast variety of climatic maps have been constructed for individual continents, countries, and regions.

In the United States, the National Weather Service is responsible for gathering the basic weather data used in climatological summaries. Data are processed, entered into archives, and made available to the public at the **National Climate Data Center (NCDC)** in Asheville, North Carolina. The NCDC also disseminates climatic information and prepares analyses for spe-

1816, The Year Without a Summer?

The summer of 1816 is famous for its anomalous cold. In fact, 1816 is sometimes described as "the year without a summer." Unseasonable snowfall and freezes damaged many crops in the then-agrarian northeastern United States and adjoining portions of Canada. More than 90 percent of the corn crop, the prime food staple, was lost. Some superficial accounts of events of that summer give the impression that unusual cold persisted throughout the summer and affected the entire civilized world. Such was not the case.

In the Northeast, the summer of 1816 was punctuated by several outbreaks of unusually cold weather. Killing frosts occurred in northern and interior southern New England as well as in Quebec in early June and July and again in late August. The June cold snap was accompanied by moderate to heavy snowfalls in the highlands. No sooner had farmers replanted their frozen crops than a killing freeze would strike again. The miserably short and disastrous growing season came to an end with a general hard freeze on 27 September.

These brief but unusually cold episodes interrupted longer spells of seasonably warm weather. Mean June and July temperatures were 1.6 to 3.3 C°(3 to 6 F°) below average. Individually, these monthly anomalies fall within the expected range of climatic variability. Some previous and subsequent Junes, Julys, and Augusts have been at least as cold as the summer months of 1816, if not colder. What is most notable about the summer of 1816 is the persistence of a strong negative temperature anomaly through all three summer months. Nevertheless, the summer in the Northeast as a whole fell within the range of expected climatic variability.

What about the weather elsewhere during the summer of 1816? We know that it was probably not the same everywhere as in the Northeast, for climatic anomalies are usually geographically nonuniform in both magnitude and sign. Unfortunately, little weather information is available from the central and western United States and Canada because few settlements existed there at the time. France, Germany, and Great Britain did experience an unusually cold summer, but this is not surprising. The westerly long-wave pattern that advects unseasonably cold air into the Northeast would also favor western Europe with the same anomalous weather. The two localities are typically one wavelength apart. In contrast, available data indicate that east-central and eastern Europe experienced a warmer than usual summer.

The purpose of this discussion is not to diminish the climatic significance of the summer of 1816 or the hardships that people suffered then. It is, rather, to show that weather extremes typically fall within the expected range of climatic variability. We have more to say about the summer of 1816 in Chapter 18, where the controversy surrounding the impact of volcanic eruptions on climate is discussed.

cialized uses. In Canada, comparable services are provided by the Atmospheric Environment Service, headquartered at Downsview, Ontario.

THE CLIMATIC NORM

A potentially misleading term used in climatic summaries is "normal" or "norm." It does not indicate just average conditions, for the **climatic norm** encompasses the total variation in the climatic record, that is, both averages and extremes. Hence, for example, the occurrence of an exceptionally cold winter may not be "abnormal" because it may fall well within the expected range of variability of winter temperature. Even the cold summer of 1816, described in the Special Topic as "1816, The Year Without a Summer?," may have been a normal, albeit extreme, event.

Table 17.1
Normals, Means, and Extremes, Green Bay, Wisconsin

Latitude: 44° 29′N   Longitude: 88° 08′W   Elevation: 682 ft above sea level   Time zone: Central

	(a)	Jan	Feb	Mar	Apr	May	Jun	Jul	Aug	Sep	Oct	Nov	Dec	Year
Temperature (°F)														
Normals														
Daily maximum		22.5	26.9	37.0	53.7	66.6	76.2	80.9	78.7	69.8	58.5	42.0	28.5	53.4
Daily minimum		5.4	8.7	20.1	33.6	43.5	53.1	58.1	56.3	47.9	38.2	26.3	13.0	33.7
Monthly		13.9	17.8	28.6	43.7	55.0	64.7	69.5	67.5	58.9	48.4	34.2	20.8	43.6
Extremes														
Record highest	36	50	55	73	89	91	97	99	99	95	88	72	62	99
Year		1961	1981	1967	1980	1959	1971	1977	1955	1955	1963	1953	1970	Jul 1977
Record lowest	36	−31	−26	−29	7	21	32	40	38	24	15	−9	−27	−31
Year		1951	1971	1962	1954	1966	1958	1965	1967	1949	1966	1976	1983	Jan 1951
Normal degree days														
Heating (base 65 °F)		1581	1322	1128	639	325	91	17	39	192	515	924	1370	8143
Cooling (base 65 °F)		0	0	0	0	18	82	156	116	9	0	0	0	381
% of possible sunshine	36	48	52	53	52	60	64	66	63	55	48	37	39	53
Mean sky cover (tenths), sunrise to sunset	36	6.6	6.5	6.6	6.8	6.4	6.0	5.6	5.8	6.0	6.4	7.4	7.1	6.4
Mean number of days, sunrise to sunset														
Clear	36	8.0	7.2	7.2	6.3	6.9	7.4	8.5	8.8	8.1	7.3	5.0	6.4	87.2
Partly cloudy	36	6.3	6.5	7.6	7.7	9.6	11.1	12.1	10.5	9.7	8.4	6.4	6.0	101.9
Cloudy	36	16.7	14.6	16.2	16.0	14.5	11.5	10.4	11.7	12.2	15.3	18.6	18.6	176.2
Precipitation, 0.01 in. or more	36	10.3	8.3	10.9	11.0	11.2	10.6	9.7	10.4	10.2	8.8	9.3	10.9	121.7
Snow, ice pellets, 1.0 in. or more	36	3.7	2.7	2.8	0.8	0.*	0.0	0.0	0.0	0.0	0.1	1.4	3.2	14.7
Thunderstorms	36	0.1	0.2	1.2	2.4	4.1	6.9	6.6	5.8	4.1	1.9	0.6	0.2	34.0
Heavy fog, visibility ¼ mi. or less	36	1.7	2.6	2.6	2.2	1.5	1.4	1.1	2.5	2.0	2.8	2.4	2.4	25.2
Temperature (°F)														
Maximum														
90 and above	24	0.0	0.0	0.0	0.0	0.*	1.4	2.9	1.3	0.2	0.0	0.0	0.0	5.8
32 and below	24	24.5	19.2	8.4	0.5	0.0	0.0	0.0	0.0	0.0	0.0	4.7	20.3	77.5
Minimum														
32 and below	24	30.7	27.5	26.8	14.2	2.8	0.0	0.0	0.0	0.6	7.8	21.9	29.4	161.5
0 and below	24	12.7	7.8	1.4	0.0	0.0	0.0	0.0	0.0	0.0	0.0	0.3	6.5	28.6
Avg. station press. (mb)	13	991.3	992.0	989.3	989.3	988.4	988.2	989.8	991.0	991.4	991.5	990.7	991.0	990.3
Relative humidity (%)														
Hour 00	24	75	76	77	75	75	80	82	86	86	81	81	79	79
Hour 06	24	76	78	81	79	79	81	85	89	89	85	83	80	82
Hour 12 (local time)	24	69	68	66	59	55	58	57	61	63	62	70	73	67
Hour 18	24	72	70	68	60	57	59	60	65	70	71	76	76	67

(a) Length of record in years.   * Less than 0.05.   T Trace amount.

Table 17.1 (continued)

	(a)	Jan	Feb	Mar	Apr	May	Jun	Jul	Aug	Sep	Oct	Nov	Dec	Year
Precipitation (in.)														
Water equivalent														
Normal		1.19	1.05	1.90	2.70	3.13	3.17	3.25	3.16	3.17	2.10	1.76	1.42	28.00
Maximum monthly	36	2.64	3.56	4.68	5.52	8.21	8.47	6.50	9.04	7.80	5.00	4.96	3.15	9.04
Year		1950	1953	1977	1953	1973	1967	1950	1975	1965	1954	1985	1971	Aug 1975
Minimum monthly	36	0.12	0.04	0.31	0.98	0.56	0.31	0.83	0.90	0.28	T	0.16	0.10	T
Year		1981	1969	1978	1963	1981	1976	1981	1955	1976	1952	1976	1960	Oct 1952
Maximum in 24 hr	36	1.14	1.78	1.44	2.00	3.28	2.65	2.95	4.60	2.99	3.68	2.30	1.55	4.60
Year		1980	1966	1979	1981	1973	1969	1959	1975	1964	1954	1985	1959	Aug 1975
Snow, ice pellets														
Maximum monthly	36	28.0	20.6	22.2	11.8	2.6				T	1.7	16.5	27.0	28.0
Year		1982	1962	1972	1977	1960				1965	1959	1985	1977	Jan 1982
Maximum in 24 hr	36	8.8	9.2	10.1	10.2	2.2				T	1.6	8.2	11.1	11.1
Year		1982	1959	1964	1977	1960				1965	1959	1977	1985	Dec 1985
Wind														
Mean speed (mph)	36	11.1	10.7	11.1	11.6	10.4	9.3	8.3	8.0	9.1	10.0	11.1	10.8	10.1
Prevailing direction through 1963		SW	SW	NE	NE	NE	SW	SW	SW	SW	SW	SW	SW	SW
Fastest mile														
Direction	35	W	W	W	SW	SW	SW	NE	SW	W	SW	W	W	SW
Speed (mph)	35	61	66	68	59	109	73	70	50	66	65	67	52	109
Year		1950	1951	1951	1964	1950	1953	1957	1950	1951	1951	1955	1957	May 1950
Peak gust														
Direction	2	NW	NW	E	NW	SW	NW	NW	W	SW	SW	W	SW	W
Speed (mph)	2	38	46	46	49	52	49	52	53	47	44	41	52	53
Date		1985	1985	1985	1984	1985	1984	1985	1985	1985	1984	1985	1984	Aug 1985

By international convention,* climatic norms are computed from averages and extremes of weather elements compiled over a 30-year period. Current climatic summaries are based on weather records from 1961 to 1990. The average July rainfall, for example, is the simple average of the total rainfall during each of 30 consecutive Julys from 1961 through 1990. The 30-year period is adjusted every 10 years to add the latest decade and drop the earliest one.

Selection of a 30-year period for summarizing climatic data may be inappropriate for many applications because climate can change significantly in periods shorter than 30 years. For other purposes, a 30-year period provides a short-sighted view of the climatic record. Compared with the longer-term

---

* In 1933, the International Meteorological Organization recommended summarization of climate data as 30-year normals. Previously, averages typically covered the entire period of record.

climatic record, the current 1961–1990 "norm," for example, was an unusually mild period over the eastern two-thirds of the United States. Nonetheless, people use the averages and extremes of past climate as a general guide to future expectations.

CLIMATIC ANOMALIES

Climatologists often compare the weather of a specific week, month, or year with the past climatic record. Such comparisons carried out over wide geographical areas show that departures from long-term climatic averages, called **anomalies,** never occur with the same sign or magnitude everywhere. As an illustration, consider Figure 17.1, which shows the temperature anomaly pattern for the winter of 1980–81 (December through February) across the United States, excluding Alaska and Hawaii. Note that average winter temperatures were above the long-term averages (positive anomaly) in the western two-thirds of the nation and below the long-term averages (negative anomaly) in the eastern third of the nation. Furthermore, the magnitude of the anomaly, positive or negative, varied from one place to another. For example, the winter of 1980–81 was up to 4 F° (2 C°) milder than the long-term average over most of Texas and 2 to 4 F° (1 to 2 C°) colder than the long-term average over the bordering state of Louisiana.

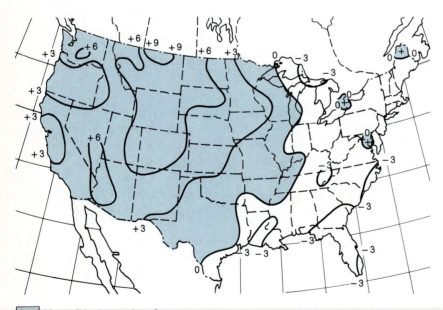

FIGURE 17.1
Temperature anomaly pattern across the United States during the winter of 1980–81 (December through February). Numbers are departures from the long-term average in F°. [NOAA data]

Mean winter temperature above the long-term average

Mean winter temperature below the long-term average

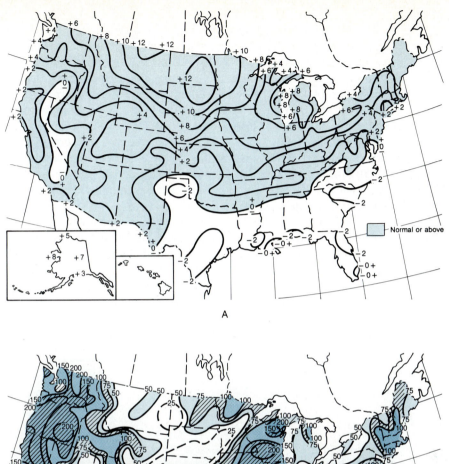

A

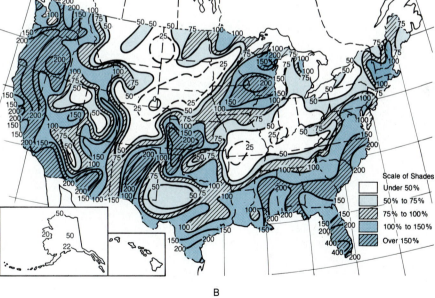

B

FIGURE 17.2

A precipitation anomaly pattern typically is much more complex than a temperature anomaly pattern. (A) Temperature anomalies across the United States for February 1983 are departures from 30-year averages and are expressed in F°. (B) Precipitation anomalies for the same month are expressed as percentages of long-term averages. [NOAA data]

The geographic nonuniformity of climatic anomalies is linked to the prevailing westerly wave pattern. Average temperatures, for example, mirror the air mass advection that is controlled by the dominating westerly flow. Hence, during the winter of 1980–81, the prevailing westerly wave pattern favored more than the usual cold air advection into the East and anomalous warm air advection into the West. The airflow aloft (the westerlies) determines the location of weather extremes such as drought or very cold temperatures. In view of the number of westerly waves that typically encircle the hemisphere (Chapter 10), a single weather extreme never occurs over an area as large as the United States or Canada; that is, severe cold or drought never grips the entire nation at the same time.

Rainfall typically forms considerably more complex anomaly mosaics than does temperature, as illustrated by Figure 17.2. This is due to the greater spatial differences in rainfall arising from the variability of storm tracks and the almost random distribution of convective showers. For these reasons, in spring in midlatitudes, for instance, even adjoining counties may experience opposite rainfall anomalies (one having above-average rainfall and the other below-average rainfall).

Special Topic

Agroclimatic Compensation: The Benefits and Limitations

Because of the geographic nonuniformity of climatic anomalies, crop yields in any one season are likely to vary from one place to another. As a consequence, lower crop yields in one area may be compensated for to some extent by higher crop yields in another area. For example, in the spring of 1982, corn planting was delayed for nearly a month in Iowa and Kansas by wet fields. As a consequence, per-hectare yields in these states were 5 and 9 percent lower, respectively, than in the prior year. Meanwhile, eastern corn belt states such as Ohio and Indiana experienced quite favorable weather throughout the growing season, and their per-hectare yields jumped 22 and 18 percent, respectively, from 1981 levels. Interestingly, these yields resulted in a record high average yield per hectare for the corn belt as a whole.

The impact of a weather extreme on crop production typically hinges on its time of occurrence during the growing season. For example, by the time drought and excess heat afflicted the Midwest and Plains states in 1983, much of the wheat crop had already matured. Although the yields per hectare of the later-maturing corn and soybeans declined by over 28 percent and 19 percent, respectively, from the previous year, the 1983 wheat harvest broke a record for yield per hectare. The impact of weather extremes can also vary on a larger geographical scale. Although the United States experienced very favorable growing conditions for corn production from May to September 1982, the Republic of South Africa received less than half its normal summer rainfall (December 1981 through March 1982). As a consequence, the United States enjoyed a record corn harvest, whereas the Republic of South Africa suffered a reduction in corn yield of nearly 30 percent. Yields declined over much of Africa and South America that year but increased in North America and Europe. For

From an agricultural perspective, the geographic nonuniformity of climatic anomalies may be advantageous in that some compensation is implied. That is, poor growing weather and consequent low yields in one area may be compensated for to some extent by better growing weather and increased yields elsewhere. This is known as **agroclimatic compensation** and is discussed further in the Special Topic "Agroclimatic Compensation: The Benefits and Limitations."

Geographic nonuniformity also characterizes trends in climate. A trend in the average annual temperature of the Northern Hemisphere is therefore not necessarily representative of all localities within the Northern Hemisphere. Over the same period, some locations experience cooling trends, whereas others experience warming trends, regardless of the direction of the hemispheric temperature trend. Not only is it misleading to assume that the direction of large-scale climatic trends applies to all localities, but it is also erroneous to assume that the magnitude of climatic trends is the same everywhere. A small change in the average hemispheric temperature typically translates into a much larger change in some areas and into little or no change in other areas.

1982, the gains more than compensated for the declines as the worldwide average yield per hectare increased by 3 percent.

Agroclimatic compensation has its limitations, however. More favorable weather conditions in a region where low soil fertility already limits crop yields would probably not compensate for a yield decline caused by unfavorable weather in a region with highly fertile soils. For example, many of the soils in Illinois and Iowa developed from loess, a dusty, wind-borne sediment that was deposited during the Ice Age. As a rule, the thicker the deposit of loess, the more fertile is the soil. Even with fertilizer applications, corn yields on the shallower loess soils are often only half those of the deeper loess soils. Hence, favorable corn-growing weather in areas with thin loess soils would probably not compensate for the decline in yields resulting from poor growing weather in areas with deep loess soils.

Another limitation of agroclimatic compensation is that intensive cultivation of most vegetable and fruit crops generally is confined to specific regions. In the United States, nearly half the vegetables for fresh market and food processing are grown in California. In March and April 1983, cold winds and heavy rainfall delayed the planting and harvesting of vegetables, thereby reducing significantly the nation's supply of these vital foodstuffs. Although the return of good weather spurred vegetable production, the inclement episode contributed to a 4 percent production decline for that year and a nearly 6.5 percent increase in market price, a rate twice that of the inflation rate.

For nations as large as the United States and Canada, which have diverse climatic regimes that support a variety of crops, agroclimatic compensation provides some flexibility in feeding the populace, but it may mean switching to more abundant foodstuffs and paying more for food. For smaller nations, agroclimatic compensation may not be possible within their borders, and weather extremes may require the importation of foodstuffs, perhaps at elevated prices.

Air Mass Climatology

Although the climate of a locality is traditionally described by normals, means, and extremes of various weather elements, an alternative approach, known as **air mass climatology,** has some interesting implications. In this approach, the frequency with which various types of air masses develop over a locality or are advected into a locality is used to describe the climate of that locality. For example, during January a northern United States city on average might have cold, dry air 60 percent of the time; mild, humid air 30 percent of the time; and mild, dry air 10 percent of the time.

Earlier we identified the major source regions for North American air masses (Chapter 11). Air masses actually originate in the clockwise divergent flow that characterizes surface winds in anticyclones. As air streams away from an anticyclone, the temperature and humidity of the air change to some extent, depending on the nature of the surface over which it travels (air mass modification). In this way, a single anticyclone can be the source of many different types of air masses, depending on the specific modification that takes place. Recall, for example, that winds spiraling outward from subtropical anticyclones are dry on the eastern flank and humid on the western flank. Air masses to the east of subtropical highs are therefore dry, whereas those to the west are humid.

In a study of air mass climatology, W. M. Wendland of the Illinois State Water Survey and R. A. Bryson of the University of Wisconsin-Madison identified the anticyclonic sources for Northern Hemisphere air masses. They did this by determining the average surface streamline pattern across the hemisphere for each month of the year. Average **streamlines** represent the mean paths of air moving horizontally. Figure 17.3 is an example of the Wendland–Bryson analysis. Anticyclonic flow is indicated by streamlines that turn into a clockwise and outward spiral. Of the 19 anticyclonic source regions, some were over the oceans, some were over the continents, and a few were in the Southern Hemisphere. Five sources persisted through the entire year, and three persisted for about 11 months. The others were prominent for 1 to 9 months.

Where streamlines from different anticyclones meet, a zone of confluence forms. A **confluence zone** is a boundary between air masses and is equivalent to an average frontal position. Confluence zones are indicated by dashed lines in Figure 17.3. As a rule, these zones separate distinctly different types of climate. Climate is quite uniform within airstreams but changes significantly across confluence zones.

Using air mass frequency to describe climate appears to be a valid approach in light of the apparent air mass control of the location of certain vegetational communities. For example, R. A. Bryson has demonstrated a close correspondence between the region dominated by cold, dry arctic air and the location of the coniferous boreal forest of Canada. The southern

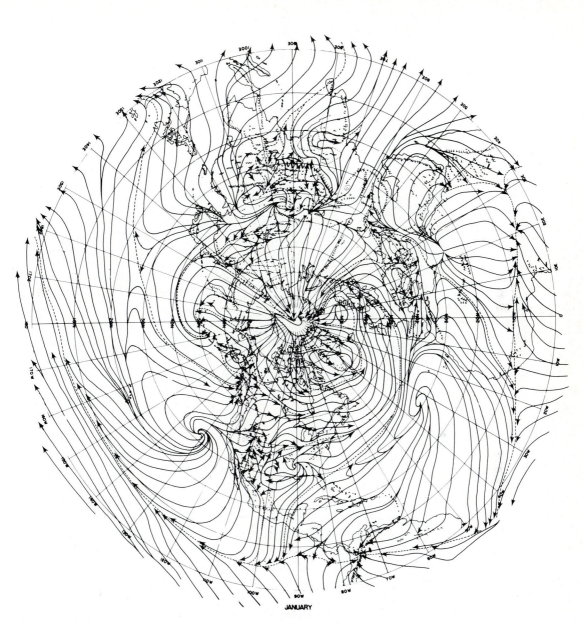

FIGURE 17.3

In this streamline analysis of mean January winds over the Northern Hemisphere, the anticyclonic spirals are air mass source regions. Dashed lines mark confluence zones that are climatic air mass boundaries. [From W. M. Wendland and R. A. Bryson, "Northern Hemisphere Airstream Regions," *Monthly Weather Review* 109 (1981):257. Graphics by Raymond Steventon, Center for Climatic Research, University of Wisconsin-Madison. *Monthly Weather Review* is a publication of the American Meteorological Society.]

Northern boreal forest border

Summer position of arctic frontal zone

Southern boreal forest border

Winter position of arctic frontal zone

FIGURE 17.4
Northern and southern borders of Canada's boreal forest correspond closely to the average positions of the leading edge of arctic air in summer and winter, respectively. [From R. A. Bryson. "Air Masses, Streamlines and the Boreal Forest." *Geographical Bulletin* 8 (1966):266. Reproduced by permission of the Minister of Supply and Services, Canada.]

boundary of the boreal forest nearly coincides with the average position of the leading edge of arctic air (the arctic front) during winter, and the northern border of the forest closely corresponds to the average position of the leading edge of arctic air during summer (Figure 17.4).

Climate Controls

Many factors, working together, shape the climate of any locality. Climatic controls include (1) latitude, (2) elevation, (3) topography, (4) proximity to large bodies of water, and (5) atmospheric circulation. For all practical purposes, the first four controls are fixed and so exert a regular and predictable influence on climate. In Chapter 2, we saw that seasonal changes in incoming solar radiation as well as length of day vary with latitude. Recall from Chapter 4 our discussion of how air temperature responds to these regular variations of solar energy input. Elevation also influences air temperature and whether precipitation falls in the form of rain or snow. Topography can affect the distribution of cloud and precipitation patterns. For example, in Chapter 6, we learned why the windward slopes of high mountain barriers are wetter than the leeward slopes. The great thermal stability of large bodies of water moderates the temperature of downwind localities, reducing the seasonal temperature contrast and lengthening the growing season.

The fifth climatic control, atmospheric circulation, is considerably less regular and, consequently, less predictable than the others. This control encompasses the combined influence of all the weather systems described in Chapters 10–14. This variability is especially characteristic of synoptic- and subsynoptic-scale weather systems. Global-scale circulation systems, such as the prevailing wind belts, the subtropical anticyclones, and the intertropical convergence zone, exert a somewhat more systematic impact on climate. How these and the other controls interact to shape the climates of the continents will become clearer as we describe the Earth's major climate groups.

Global Patterns of Climate

Viewed globally, climate exhibits some regular patterns. As we describe these patterns, keep in mind that they may be significantly modified by local and regional climate controls.

TEMPERATURE

If we ignore the influence of mountainous terrain on air temperature, then mean annual isotherms roughly parallel latitude circles (Figure 17.5), underscoring the importance of solar radiation and solar altitude as climate controls.

FIGURE 17.5
Mean annual global sea-level temperatures in °F.

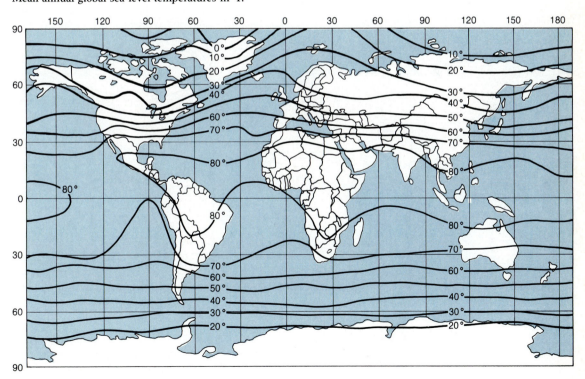

Interestingly, the latitude of highest mean annual temperature, the so-called **heat equator,** is located about 10 degrees north of the geographical equator. Mean annual isotherms are symmetrical with respect to the heat equator and decline in magnitude toward the poles.

The heat equator is in the Northern Hemisphere because, overall, that hemisphere is warmer than the Southern Hemisphere. Several factors contribute to this hemispheric temperature contrast. For one, the polar regions of the two hemispheres have different radiational characteristics. Most of the Antarctic continent is submerged under a massive glacial ice sheet, so the surface has a very high albedo for solar radiation and is the site of intense radiational cooling, especially during the long polar night. In contrast, the Northern Hemisphere polar region is mostly ocean. Although the Arctic Ocean is usually ice covered, patches of open water develop in summer and lower the overall surface albedo. Hence, differences in radiational characteristics mean that the Arctic is warmer than the Antarctic.

A second factor contributing to the relative warmth of the Northern Hemisphere is that hemisphere's greater fraction of land in tropical latitudes. Because land surfaces warm up more than water surfaces in response to the same insolation, tropical latitudes are warmer in the Northern Hemisphere

FIGURE 17.6
Mean global sea-level temperatures for January in °F.

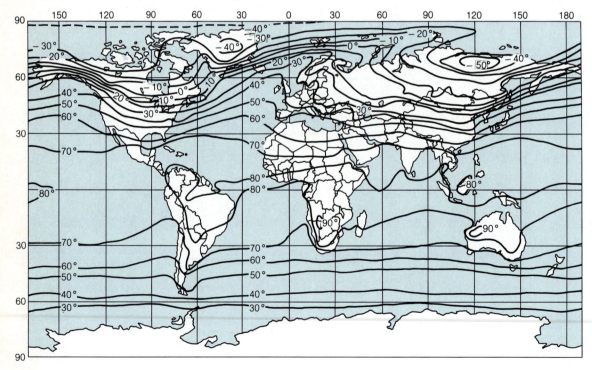

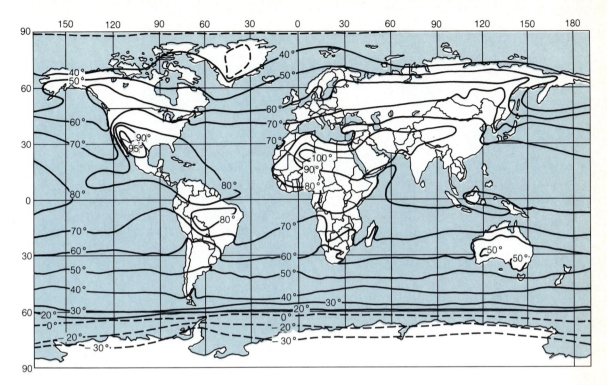

FIGURE 17.7
Mean global sea-level temperatures for July in °F.

than in the Southern Hemisphere. A third contributing factor is ocean circulation. Apparently, ocean currents transport more warm water to the Northern than to the Southern Hemisphere.

Systematic isotherm patterns also appear when we consider the worldwide distribution of mean January temperature (Figure 17.6) and mean July temerature (Figure 17.7); January and July are usually the coldest and warmest months of the year. If we neglect the influence of topography on air temperature, isotherms tend to parallel latitude circles. However, monthly isotherms exhibit some notable north-to-south bends primarily because of land/sea contrasts and the influence of ocean currents. As the year progresses from January to July and to the next January, isotherms in both hemispheres shift north and south in tandem, but the greater annual range in air temperature over land than over water produces a more pronounced latitudinal shift of isotherms over the continents than over the oceans. The meridional temperature gradient, however, is steeper in the winter hemisphere than in the summer hemisphere. From earlier discussions (Chapter 10), we know that this steeper temperature gradient means a more vigorous circulation and stormier weather in the winter hemisphere.

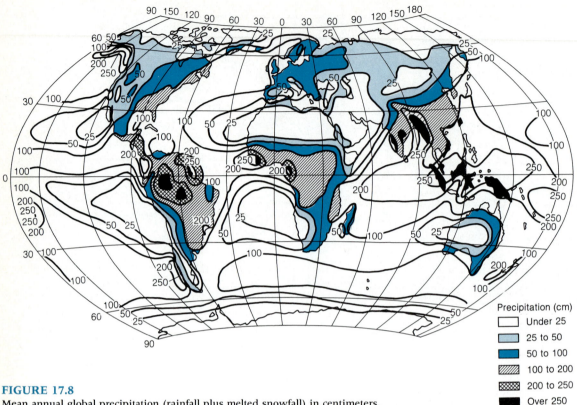

FIGURE 17.8
Mean annual global precipitation (rainfall plus melted snowfall) in centimeters.

PRECIPITATION

The global pattern of mean annual precipitation (rain plus melted snow) exhibits great spatial variability (Figure 17.8). Some of this variability can be attributed to the influence of topography and the distribution of land and sea, but of greater importance is the global-scale circulation. The intertropical convergence zone (ITCZ), subtropical anticyclones, and the prevailing wind belts give a roughly zonal (west–east) pattern to precipitation distribution. In addition, the regular shifts of these circulation features through the year cause the seasonality of precipitation that is typical of many localities.

At low latitudes near the equator, convective activity associated with the trade wind convergence triggers abundant rainfall year-round. In the adjacent belt poleward to about 20 degrees latitude, rainfall depends on seasonal shifts of the ITCZ and the subtropical anticyclones. Poleward shifts of the ITCZ cause summer rains, whereas the equatorward shift of the subtropical highs brings winter drought. This is the belt of tropical monsoon circulation described in Chapter 12.

Poleward of this belt, from about 20 to 35 degrees N and S, the subtropical anticyclones dominate the climate all year. Subsiding dry air on the anticyclones' eastern flanks is responsible for the Earth's major subtropical deserts. On the other hand, unstable humid air that characterizes the western flanks of the subtropical anticyclones produces relatively moist conditions.

Between about 35 and 40 degrees latitude, precipitation is governed by the prevailing westerlies and subtropical anticyclones. Typically, on the western side of continents, winter cyclones migrating with the westerlies bring moist weather, but in summer, westerlies shift poleward and the area lies under the dry eastern flank of a subtropical anticyclone. Hence, summers are dry. At the same latitudes, but on the eastern side of continents, the climate is dominated by westerlies in winter and the moist airflow on the western flank of the subtropical anticyclone in summer. Thus, rainfall is triggered by cyclonic activity in winter and by convection in summer and shows little seasonal variability.

Poleward of about 40 degrees latitude, precipitation generally declines as lower temperatures reduce the mean saturation vapor pressure (Chapter 6). Although precipitation is generally not seasonal, the tendency in the continental interiors is for more rainfall in summer. This is a result of higher air temperatures and greater convective activity in summer and of lower air temperatures and more frequent anticyclones in winter.

Our description of annual global precipitation is somewhat idealistic and requires some qualification. Land/sea distribution and topography add complexity to the zonal distribution of precipitation. More rain falls over the oceans than over the continents, and north-to-south-oriented mountain ranges induce wet windward slopes and extensive leeward rain shadows. Furthermore, statistics on annual precipitation totals fail to convey some other important aspects of precipitation, including the average amount of rainfall per day and the season-to-season and year-to-year reliability of precipitation. As a rule, rainfall is most reliable in maritime climates, less reliable in continental localities, and least reliable in arid regions. However, drought is possible anywhere, even in maritime climates.

Climate Classification

In response to the interaction of many controls, the world's climates form a complex mosaic. Climatologists have attempted to simplify and organize the myriad of climate types by devising classification schemes that group together climates with common characteristics. Classification schemes typically group climates according to (1) the meteorological basis of climate or (2) the environmental effects of climate. The first is a genetic climate classification, and the second is an empirical climate classification. One of the most popular classification systems, designed by Wladimir Köppen, combines these two approaches.

Recognizing that vegetation indigenous to a region is a natural indicator of regional climate, Köppen delineated climatic boundaries throughout the world based on the limits of vegetational communities and the monthly means of temperature and precipitation. Köppen's climate classification scheme uses letters to symbolize five main groups of world climates: **(A)** tropical rainy; **(B)** dry; **(C)** midlatitude rainy, mild winter; **(D)** midlatitude rainy, cold winter; and **(E)** polar. Each of these climates is further subdivided into principal climate types, designated by the addition of another letter symbol to form a pair. Even further qualification of the climate may be specified by adding a third letter symbol. With a combination of only two or three letters, the general climate characteristics of a region can thus be described and classified.

Since its introduction in 1918, Köppen's climate classification has undergone numerous and substantial revisions by Köppen himself and by other climatologists. One version of the Köppen-based climate classification is pre-

Table 17.2

Climate Classification

Climate groups	Climate types	Precipitation
A Tropical humid	**Ar,** tropical wet	Not over 2 dry months
	Aw, tropical wet-and-dry	High-sun wet (zenithal rains), low-sun dry
B Dry	**BS,** semiarid (steppe)	
	BSh (hot), tropical– subtropical	Short moist season
	BSk (cold), temperate– boreal	Meager rainfall, most in summer
	BW, arid (desert)	
	BWh (hot), tropical– subtropical	Constantly dry
	BWk (cold), temperate– boreal	Constantly dry
C Subtropical	**Cs,** subtropical dry summer	Summer drought, winter rain
	Cf, subtropical humid	Rain in all seasons
D Temperate	**Do,** oceanic	Rain in all seasons
	Dc, continental	Rain in all seasons, accent on summer; winter snow cover
E Boreal	**E,** boreal	Meager precipitation through-out year
F Polar	**Ft,** tundra	Meager precipitation through-out year
	Fi, ice cap	Meager precipitation through-out year
H Highland	**H,** variable	Variable

Source: G. T. Trewartha and L. H. Horn. *An Introduction to Climate.* 5th ed. New York: McGraw–Hill, 1980, p. 227.

sented by G. T. Trewartha and L. H. Horn in their text, *An Introduction to Climate* (1980). Trewartha and Horn identify seven main climate groups (Table 17.2); five are based on temperature, one is based on precipitation, and one refers to mountainous regions. The worldwide distribution of these climate groups is shown in Figure 17.9.

TROPICAL HUMID CLIMATES (A)

Tropical humid climates constitute a discontinuous belt straddling the equator and extending poleward to near the Tropic of Cancer in the Northern Hemisphere and the Tropic of Capricorn in the Southern Hemisphere. Air temperatures are high and show little month-to-month variability throughout the year. The coolest monthly mean temperature is no lower than 18 °C (64 °F), and there is no frost. The temperature contrast between the warmest and coolest month is typically less than 10 C° (18 F°). In fact, the diurnal temperature range exceeds the annual temperature range. This monotonous air temperature regime is the consequence of consistently intense insolation and nearly uniform day length throughout the year.

Although tropical humid climate types are not readily distinguishable on the basis of temperature, there are important differences in precipitation regimes. Tropical humid climates are therefore subdivided into tropical wet climates **(Ar)** and tropical wet-and-dry climates **(Aw)**. Although both climate types feature abundant annual rainfall, more than 100 cm (40 in.), differences occur in length of rainy season.

In tropical wet climates, the yearly rainfall of 175 to 250 cm (70 to 100 in.) supports the world's most luxuriant vegetation (Figure 17.10). Tropical rainforests occupy the Amazon Basin of Brazil, the Congo Basin of Africa, and the islands of Micronesia. For the most part, rainfall is distributed uniformly throughout the year, although in some areas there is a brief (1 to 2 months) relatively dry season. Rainfall occurs as heavy downpours in frequent thunderstorms triggered by local convection and by surges of the ITCZ. Convective rainfall, controlled by insolation, typically peaks in midafternoon, the warmest time of day. Because water vapor concentrations are very high, even the slightest cooling during the early morning hours results in dew or fog, which gives the region a sultry, steamy appearance.

For the most part, tropical wet-and-dry climates **(Aw)** border tropical wet climates and are transitional poleward to the subtropical dry climates. **Aw** climates support the savanna, tropical grasslands with scattered deciduous trees (Figure 17.11). Summers are wet and winters are dry, with the dry season lengthening poleward. This marked seasonality of rainfall is linked to shifts of the ITCZ and to subtropical anticyclones, which follow the seasonal excursions of the sun. In summer, surges of the ITCZ trigger convective rainfall; in winter, the weather is dominated by the dry eastern flank of the subtropical anticyclones.

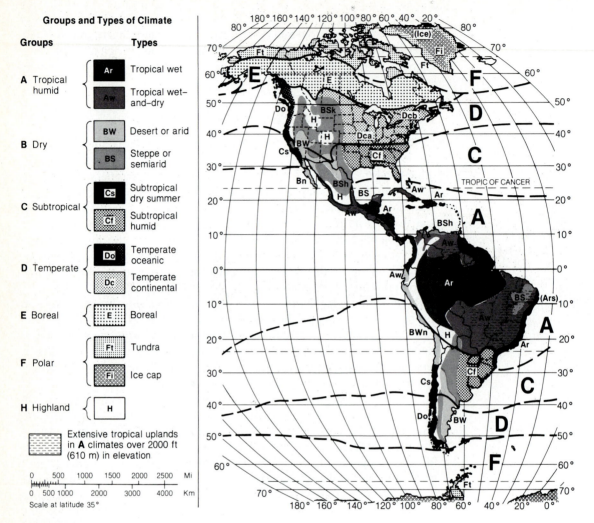

Groups and Types of Climate

Groups | **Types**

A Tropical humid
- Ar — Tropical wet
- Aw — Tropical wet-and-dry

B Dry
- BW — Desert or arid
- BS — Steppe or semiarid

C Subtropical
- Cs — Subtropical dry summer
- Cf — Subtropical humid

D Temperate
- Do — Temperate oceanic
- Dc — Temperate continental

E Boreal
- E — Boreal

F Polar
- Ft — Tundra
- Fi — Ice cap

H Highland
- H

Extensive tropical uplands in **A** climates over 2000 ft (610 m) in elevation

Scale at latitude 35°

FIGURE 17.9

Distribution of major climate groups across the globe. [Modified after G. T. Trewartha and L. H. Horn. *An Introduction to Climate,* 5th ed. New York: McGraw-Hill, 1980. Used with permission.]

Annual mean temperatures in **Aw** climates are only slightly lower, and the annual temperature range is only slightly greater, than in the tropical wet climates **(Ar).** Diurnal temperature variations are noticeably greater, however. In summer, frequent cloudy skies and high humidities suppress the diurnal temperature range by reducing solar heating during the day and radiational cooling at night. In winter, on the other hand, persistent fair skies have the opposite effect and widen the diurnal temperature range. Cloudy, rainy summers also mean that the year's highest temperatures typically occur toward the close of the dry season in late spring.

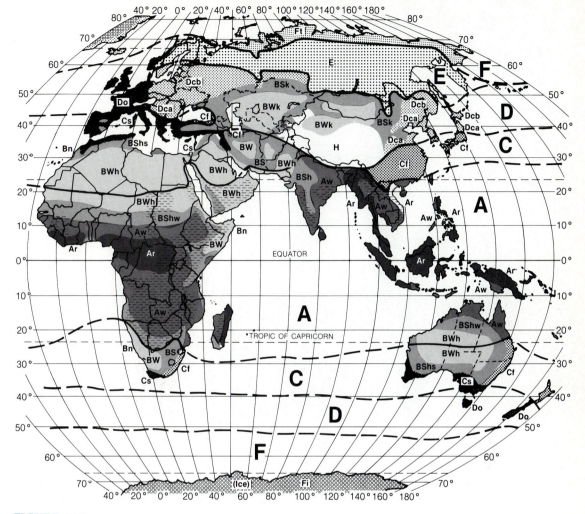

FIGURE 17.9 (continued)

DRY CLIMATES (B)

Dry climates characterize those regions where annual potential evaporation exceeds annual precipitation. Because evaporation is controlled by air temperature, it is not possible to specify some rainfall amount as the criterion for dry climates. Rainfall is not only limited in **B** climates; it is also highly variable and unreliable. As a general rule, the lower the mean annual rainfall, the greater is its variability.

The world's dry climates encompass a larger land area than any other single climate grouping. Perhaps 30 percent of the Earth's land surface, stretch-

FIGURE 17.10

Luxuriant tropical rainforests of the Amazon Basin of Brazil. [FAO photograph by Peyton Johnson]

FIGURE 17.11

Grassland savanna of Tanzania. [Visuals Unlimited photograph by E. C. Williams]

FIGURE 17.12
Sparsely vegetated desert of the American Southwest. [Photograph by John D. Cunningham]

ing from the tropics to midlatitudes, is characterized by a moisture deficit of varying degree. These are the climates of the world's deserts and steppes where vegetation is sparse and equipped with special adaptations that permit survival under conditions of severe moisture stress (Figure 17.12). Dryness is the consequence of either subtropical anticyclones or the rain-shadow effect of high mountain barriers. Mean annual temperatures are latitude dependent, as is the variation of mean monthly temperatures through the year.

On the basis of degree of dryness, we distinguish two climate types: steppe or semiarid **(BS)** and arid or desert **(BW).** The steppe or semiarid climates are transitional to more humid climates and usually border arid climates. We further distinguish warm, dry climates of tropical latitudes **(BSh** and **BWh)** from cold, dry climates of midlatitudes **(BSk** and **BWk).**

Descending stable air on the eastern flanks of the subtropical anticyclones gives rise to the tropical dry climates **(BSh** and **BWh).** These huge semipermanent pressure cells dominate the weather year-round near the tropics of Cancer and Capricorn. Consequently, dry climates characterize North Africa eastward to northwest India, the southwestern United States and northern Mexico, coastal Chile and Peru, southwest Africa, and much of the interior of Australia.

Although persistent and abundant sunshine is the general rule in dry tropical climates, some important exceptions occur. In coastal deserts bordered by cold ocean waters, a shallow layer of stable maritime air drifts inland. The desert air thus features high relative humidity, persistent low stratus clouds and fog, and considerable dew formation. Examples are the Atacama Desert of Peru and Chile, the Namib Desert of southwest Africa, and portions of the coastal Sonoran Desert of Baja California and the coastal Sahara Desert of northwest Africa. These anomalous desert climates are designated **Bn** or **BWn.**

Cold, dry climates of midlatitude **(BWk** and **BSk)** lie in the rain shadows of great mountain ranges. They occur primarily in the Northern Hemisphere and are found to the lee of the Sierra Nevada and Cascade ranges in North America, and the Himalayan chain in Asia. Because these dry climates are at higher latitudes than their tropical counterparts, mean annual temperatures are lower and the seasonal temperature contrast is greater. Anticyclones dominate winter weather, resulting in cold, dry conditions, whereas summers are hot and generally dry. The meager precipitation is produced by scattered convective showers in summer.

SUBTROPICAL CLIMATES (C)

Subtropical climates are situated just poleward of the tropics of Cancer and Capricorn and are dominated by seasonal shifts of the subtropical anticyclones. There are two basic climate types: subtropical dry summer (or Mediterranean) climates **(Cs)** and subtropical humid climates **(Cf).**

Mediterranean climates occur on the west side of continents between about 30 and 45 degrees latitude. In North America, mountain ranges confine this climate type to a narrow coastal strip of California. Elsewhere, **Cs** climates rim the Mediterranean Sea and occur in extreme southwestern and southeastern Australia. Summers are dry because at that time of year **Cs** regions are under the stable subsiding air on the eastern flanks of the subtropical highs. The equatorward shift of the subtropical highs in autumn allows oceanic cyclones to migrate inland, bringing moderate winter rainfall. Annual precipitation totals are in the range of 40 to 80 cm (16 to 32 in.).

Although Mediterranean climates exhibit a pronounced seasonality in precipitation (dry summer, wet winters), the temperature regime is quite variable. In coastal areas, cool onshore breezes prevail, lowering mean annual temperatures and reducing seasonal temperature contrasts. Well inland, however, away from the ocean's moderating influence, summers are considerably warmer, resulting in higher mean annual temperatures and greater seasonal temperature differences than in coastal **Cs** localities. An illustration of the contrast in temperature regime within **Cs** regions is provided by the climatic records of coastal San Francisco and inland Sacramento, California (Figure 17.13). Although the two cities are separated by only about 145 km (90 mi), the climate of Sacramento is much more continental than that of San Francisco.

Subtropical humid climates **(Cf)** occur on the eastern sides of continents between about 25 and 40 degrees latitude. **Cf** climates are situated primarily in the southeastern United States, southeastern South America, eastern China, and southern Japan and on the southeastern coasts of South Africa and Australia. These climates feature abundant precipitation (76 to 165 cm, or 30 to 65 in.), which occurs throughout the year. In summer, **Cf** regions are dominated by a flow of sultry maritime tropical air on the western flanks of

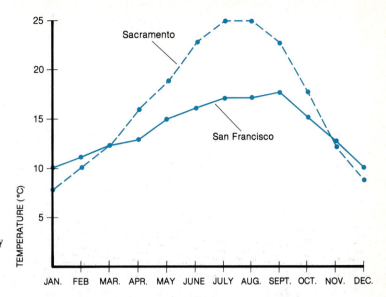

FIGURE 17.13

Note the contrast in the march of monthly mean temperatures between Sacramento, California, and more maritime San Francisco. [NOAA data]

the subtropical anticyclones. Consequently, summers are hot and humid with frequent thunderstorms. As noted in Chapter 14, tropical storms contribute significant rainfall to some North American and Asian **Cf** regions. In winter, after the subtropical highs shift toward the equator, **Cf** regions come under the influence of migrating midlatitude cyclones and anticyclones.

A noteworthy exception to the nonseasonality of precipitation in **Cf** regions is the interior of southern China, where winters are relatively dry. This is because prevailing winter winds are from the cold and dry north and northwest. This anomalous **Cf** climate is designated **Cw.**

In **Cf** localities, summers are hot and winters are mild. Mean temperatures of the warmest month are typically in the range of 24 to 27 °C (75 to 80 °F). For the coolest month, average temperatures are typically in the 4 to 13 °C (40 to 55 °F) range. Freezes and snowfall are infrequent.

TEMPERATE CLIMATES (D)

The primary distinction among temperate climates is between temperate oceanic **(Do)** and temperate continental **(Dc).** As the names imply, the first features a strong maritime influence, whereas the second is highly continental. These are midlatitude climates.

Temperate oceanic climates are found on the windward side of continents, mainly poleward of 40 degrees latitude. These climates characterize a narrow coastal strip in the Pacific Northwest from northern California northwestward into Alaska, along the southwestern coastal plain of South America, throughout most of western Europe, and in small portions of Australia and New

Zealand. In these regions, a strong maritime influence prevails year-round. Consequently, temperatures are relatively mild for the latitude, and seasonal and diurnal temperature contrasts are reduced. Cold waves and heat waves are rare, and the growing season is long (up to 210 freeze-free days).

The precipitation regime also reflects the maritime influence in **Do** climates. Maritime polar (mP) air masses dominate to produce persistent episodes of low clouds and light-to-moderate rainfall, with rainfall amounts dependent on topography. Droughts are infrequent. The mP air is stable, so convective showers are unusual.

Temperate continental climates **(Dc)** occur only in the Northern Hemisphere from about 40 to 50 degrees N. These **Dc** climates are inland and on the leeward side of continents: the northeastern third of the United States, Eurasia, and extreme eastern Asia. A marked continentality increases inland with maximum temperature contrasts between the coldest and warmest months as great as 25 to 35 C° (45 to 63 F°). **Dc** climates are divided into the southerly **Dca** climates with cool winters and warm-to-hot summers and the northerly **Dcb** climates with cold winters and mild summers. The freeze-free season varies in length from 7 months in the south to 3 months in the north.

Table 17.3

The Twenty Snowiest Major Cities in the United States Based on Cumulative Totals for the Period 1975–1985

Buffalo	2710.9 cm	1067.3 in.
Rochester, N.Y.	2635.8	1037.7
Salt Lake City	1622.6	638.8
Minneapolis–St. Paul	1605.5	632.1
Albany	1604.3	631.6
Cleveland	1592.6	627.0
Denver	1483.4	584.0
Milwaukee	1400.3	551.3
Detroit	1176.8	463.3
Chicago	1114.8	438.9
Pittsburgh	1069.6	421.1
Hartford, Conn.	1062.0	418.1
Boston	1058.7	416.8
Providence	872.7	343.6
Dayton, Ohio	849.9	334.6
Indianapolis	824.2	324.5
Columbus, Ohio	804.9	316.9
St. Louis	700.3	275.7
Cincinnati	685.5	269.9
Philadelphia	678.2	267.0

Source: P. R. Chaston, "1975–1985 Snowiest Major Cities," *Weatherwise* 39, No. 1 (1986):44–45.

The weather in **Dc** regions is very changeable and dynamic because these areas are swept by cyclones and anticyclones and by surges of contrasting air masses. Polar front cyclones dominate winter, bringing episodes of light-to-moderate frontal precipitation. These storms alternate with surges of dry polar and arctic air masses. In summer, cyclones are weak and infrequent because the principal storm track shifts poleward. Summer rainfall is mostly convective, and amounts can be very heavy locally in severe thunderstorms. Although precipitation is distributed fairly uniformly throughout the year, in most places there is a summer maximum.

In the northern portions of **Dc** climates, winter snowfall becomes an important factor. The amount of snow and the persistence of snow cover increase northward (Table 17.3). Because of its high albedo for solar radiation and its excellent emission of infrared, a snow cover chills and stabilizes the overlying air. For these reasons, a snow cover is self-preserving, so that once an extensive snow cover is established in early winter, it tends to persist.

BOREAL CLIMATE (E)

The boreal climate occurs only in the Northern Hemisphere as an east-to-west band between 50 or 55 degrees N and 65 degrees N latitude. It is a region of extreme continentality and very low mean annual temperatures. Summers are short and cool, and winters are long and bitterly cold. Both continental polar (cP) and arctic (A) air masses originate here, and this area is the site of an extensive coniferous forest. Because midsummer freezes are possible, the growing season is precariously short.

Weak cyclonic activity occurs throughout the year and yields meager precipitation (typically less than 50 cm, or 20 in.). Convective activity is rare. A summer precipitation maximum is due to the winter dominance of cold, dry air masses. Snow cover is persistent in winter.

POLAR CLIMATES (F)

Polar climates occur poleward of the Arctic and Antarctic circles at 66 degrees 30 minutes latitude. These boundaries correspond roughly to localities where the mean temperature for the warmest month is 10 °C (50 °F). These limits also approximate the tree line, the poleward limit of tree growth. Poleward are tundra (Figure 17.14) and the Greenland and Antarctic ice sheets. A distinction is made between tundra **(Ft)** and ice cap **(Fi)** climates, with the dividing criterion being 0 °C (32 °F) for the mean temperature of the warmest month. Vegetation is sparse in **Ft** regions and nonexistent in **Fi** areas.

Polar **(F)** climates are characterized by extreme cold and slight precipitation, which is mostly in the form of snow (less than 25 cm, or 10 in., melted, per year). Although summers are cold, the winters are so extremely cold that **F** climates feature a marked seasonal temperature range. Mean annual temperatures are the lowest in the world.

FIGURE 17.14
The tundra of northern Canada. [Photograph by Steve McCutcheon]

HIGHLAND CLIMATES (H)

Highland climates encompass a wide variety of climate types that characterize mountainous terrain. Altitude, latitude, and exposure are among the factors that shape a complexity of climate types. Climate-ecological zones are telescoped in mountainous areas. That is, in ascending several thousand meters of altitude, we encounter the same biotic-climatic zones that we would experience in traveling several thousand kilometers of latitude. As a general rule, every 300 m (1000 ft) of elevation corresponds roughly to a northward advance of 500 km (310 mi).

Conclusions

In this chapter, we described and classified climate with a focus on present climates. We see that the world is a mosaic of many climates that are shaped by many interacting controls. Changes in these controls mean that climate also changes. How and why climate changes is the subject of the next chapter.

Summary Statements

Climate is defined as weather conditions averaged over some time period, the seasonal distribution, plus extremes in weather. Climate is usually specified for a particular location and for some time interval.

The climatic norm encompasses both means and extremes of weather elements. By convention, norms are computed for a 30-year period. The current climatic norm is based on the period 1961–1990.

Anomalies in climate—departures from long-term averages—are geographically nonuniform in both sign (positive or negative) and magnitude. The same geographic nonuniformity also characterizes large-scale climatic trends.

An alternate way of describing the climate of a locality is by frequency of occurrence of the various types of air masses that develop over or are advected into that locality.

Climatic controls include the various weather systems operating at all spatial and temporal scales, along with insolation, land/water distribution, and topography.

The globe is a mosaic of many different types of climate that may be grouped and classified by meteorological causes or environmental effects. The Köppen climate classification system combines the two approaches and has undergone considerable refinement since its introduction.

The globe's major climate groups are tropical humid, dry, subtropical, temperate, boreal, polar, and highland.

Key Words

climate
National Climate Data
    Center (NCDC)
climatic norm

anomalies
agroclimatic
    compensation

air mass climatology
streamlines

confluence zone
heat equator

Review Questions

1. Define climate.

2. What is meant by the climatic "norm?"

3. By international convention, what period of time is used as the basis for computation of climatic norms? Is this a reasonable time frame?

4. What is a climatic anomaly?

5. How is air mass frequency used to describe the climate of a locality?

6. Explain how a single anticyclone can be the source of many different types of air masses.

7. What is the purpose of a streamline analysis?

8. How does arctic air mass frequency correspond to the location of Canada's boreal forest?

9. List and describe briefly the principal controls of climate.

10. What are the bases for climate classification?

11. How does Köppen's classification scheme group global climates?

12. In tropical humid climates, how does the diurnal temperature range compare with the annual temperature range?

13. On the continents, which climate grouping covers the largest area?

14. How does the variability (and hence, reliability) of rainfall change as annual rainfall totals decrease?

15. How is a desert climate defined? What are the principal controls of desert climates?

16. How do seasonal shifts of the subtropical anticyclones induce seasonal changes in rainfall in subtropical latitudes?

17. Explain the seasonality of rainfall in (a) India and (b) coastal California.

18. Contrast temperate oceanic climates with temperate continental climates.

19. How does a seasonal snow cover influence climate?

20. Why is convective activity rare in boreal climates and polar climates?

Points to Ponder

1. Explain why a 30-year period provides a short-sighted view of the climatic record.

2. Why are climatic anomalies geographically nonuniform in sign (direction) and magnitude?

3. Explain why a precipitation anomaly map typically is considerably more complex than a temperature anomaly map.

4. What is the significance of the geographic nonuniformity of climatic anomalies for (a) agricultural productivity and (b) home heating and cooling demands?

5. Speculate on how long-term changes in climate controls might influence climate.

6. Climatic controls are really not independent of one another. Provide some examples of linkages.

Projects

1. In what climate grouping is your locality? How closely does the climate of your area match the general description given in this chapter?

2. Consult the climatic records for your area.

When did records begin? During the record period, were there any changes in the location where weather was monitored? Speculate on the possible implications of such changes.

Selected Readings

Bryson, R. A., and F. K. Hare (Eds.). *World Survey of Climatology 11, Climates of North America.* New York: Elsevier Scientific, 1974. 420 pp. Includes data summaries and reviews of the major features of the climates of the United States, Canada, and Mexico.

Guttman, N. B. "Statistical Descriptors of Climate." *Bulletin of the American Meteorological Society* 70 (1989):602–607. Discusses how climatic data may be made more useful for predictive purposes.

Hamilton, K. "Early Canadian Weather Observers and the 'Year Without a Summer'." *Bulletin of the American Meteorological Society* 67 (1986):524–532. Presents a description based on newspaper accounts and records of amateur weather observers.

Hughes, P. "1816—The Year Without a Summer."

Weatherwise 32 (1979):108–111. Speculates on possible causes of the anomalously cold summer.

Kunkel, K. E., and A. Court. "Climatic Means and Normals—A Statement of the American Association of State Climatologists." *Bulletin of the American Meteorological Society* 71 (1990):201–204. Discusses the limitations of 30-year climate averages.

Oliver, J. E., and R. W. Fairbridge (Eds.). *The Encyclopedia of Climatology.* New York: Van Nostrand Reinhold, 1987. 986 pp. Provides a comprehensive and detailed treatise on climatology.

Oliver, J. E., and J. J. Hidore. *Climatology, An Introduction.* Columbus, OH: C. E. Merrill, 1984. 381 pp. Provides a well-illustrated survey of the principles of climate and the climates of the globe.

Trewartha, G. T., and L. H. Horn. *An Introduction to Climate*, 5th ed. New York: McGraw–Hill, 1980. 416 pp. Serves as a basic textbook on the meteorological basis of climate plus descriptions of primary climate belts.

Wendland, W. M., and R. A. Bryson. "Northern Hemisphere Airstream Regions." *Monthly Weather Review* 109 (1981):255–270. Determines air mass source regions on the basis of streamline analysis.

*Now from the smooth deep
ocean-stream the sun
Began to climb the heavens,
and with new rays
Smote the surrounding fields*

Homer
ILIAD

18

Climatic Record and Climatic Variability

The climatic future, like the climatic past, will be shaped by many interacting factors including changes in solar radiation, cloudiness, atmospheric chemistry, and sea-surface temperatures. [Photograph by Mike Brisson]

499

$\Large{C}$LIMATE IS inherently variable. As we saw in the previous chapter, climate differs from place to place; climate also varies with time. In this closing chapter, we describe what is understood about past variations in climate, factors that may be responsible for climate variability, and prospects for the climatic future.

The Climatic Past

As we saw in Chapter 15, the reliable instrument-based record of past weather and climate in most places is limited to only a century or so. For information on earlier variations in climate, we must rely on indirect evidence. In this chapter's first Special Topic, "Reconstructing Past Climates," we describe some of the methods whereby scientists have succeeded in extending the climatic record back in time. In this section, we summarize what is currently understood about the Earth's climatic record.

GEOLOGIC TIME AND CONTINENTAL DRIFT

As we journey back through the millions and millions of years that constitute **geologic time,** the climatic record becomes extremely fragmented and unreliable. From various lines of evidence, including fossils and sedimentary rock layers, some generalized sketches of past climate emerge (Table 18.1).

Table 18.1

Summary of Earth's Climatic History over the Past 350 Million Years

Years before present	Climatic event
350–250 million	Pangaean Ice Age
250–50 million	Relatively mild episode
50 million	Onset of cooling
15–10 million	Glaciation of Antarctica
2 million	Beginning of major glacial epoch in North America
1 million	Major interglacial episode
25,000	Last major glacial advance
18,000	Last glacial maximum Onset of warming
10,000	Glacier retreats from northern United States
7000–5000	Peak mild episode

Reconstructing Past Climates

For times and places where there is no instrument record of weather, it may be possible to infer past climatic information from various sensors that substitute for actual weather instruments. These climatic sensors, sometimes called *proxy climate data sources,* include historical records, pollen, tree growth-rings, deep-sea sediment cores, and glacial ice cores.

Under cautious scrutiny certain historical documents can yield a wealth of information on the climatic past. Almanacs, personal diaries, old newspapers, and ships' logs may contain qualitative and some quantitative references to weather and climate. Other types of documents refer only indirectly to weather and climate but can be useful nonetheless. Records of success of grain harvests, quality of wine, or various phenological events (such as dates of blooming of plants) provide indirect indications of the growing-season weather. For example, in his fascinating book, *Times of Feast, Times of Famine,* Emmanuel Le Roy Ladurie relies heavily on vineyard records to reconstruct the climate of western Europe in the Middle Ages.

We must be careful in interpreting climatic information from historical documents chiefly because many factors other than climate usually influence such records. For example, in addition to the growing-season weather, vineyard harvest dates are affected by fluctuations in the wine market as well as by human judgment as to when the grapes are ripe and ready for harvest. It is also well to bear in mind that people have always applied ingenuity to ameliorate the impact of climate—especially extremes in climate. Hence, climatic information derived from records of human activity is not always reliable, and corroborative data from other sources are needed to support any climatic inferences.

Ponds, peat bogs, and swamps are favorable sites for the accumulation and preservation of wind-borne pollen, the tiny fertilizing component of a seed plant. Pollen grains mix with other sediments (clay, silt, and other organic debris) that also settle and accumulate in these low-lying depositional areas. Upward of 20,000 pollen grains may be mixed in a cubic centimeter of mud. If we assume that accumulated pollen is the product of nearby vegetation and that climate largely determines vegetation, then climate may be inferred from pollen. When the climate changes significantly, the vegetation changes and so does the pollen. Thus, changes in abundance of pollen of different species at various depths within accumulated sediment may reflect changes in climate. Scientists extract sediment cores from swamps and pond bottoms (Figure 1), separate pollen from its host sediment, and reconstruct the sequence of past vegetational changes. From the climatic requirements of the reconstructed vegeta-

FIGURE 1

Investigators extract a sediment core from the bottom of a frozen pond. Identification of pollen separated from the sediment allows scientists to reconstruct past vegetation and climate. [Photograph courtesy of the Center for Climatic Research, University of Wisconsin-Madison]

tion (based on modern species distribution and modern climate), scientists decipher the sequence of past climatic shifts.

Much of the pollen/climate research conducted in North America has focused on vegetation and climate variations over the past 15,000 years, that is, since the last glacial maximum. Using sophisticated statistical techniques, scientists have reconstructed remarkably detailed quantitative climatic data. For example, the pollen record from Kirchner Marsh near Minneapolis, Minnesota, yielded a record of variations in July mean temperature and annual precipitation back to 12,000 years ago. Unfortunately, very few sites favor the accumulation and preservation of continuous long-term pollen/climate records. Most such records come from a few geographical areas, including the Great Lakes region, interior New England, and western mountain valleys.

Analysis of variations in the thickness of annual growth-rings of certain trees also can yield detailed information on the climatic past. The study of tree growth-rings for climatic data is known as *dendroclimatology*. Andrew E. Douglass, who was a solar astronomer at the University of Arizona, pioneered this work in the American Southwest early in this century. Today, his successors at the University of Arizona's Laboratory of Tree-Ring Research are attempting to reconstruct past climates using special statistical approaches and computers. At the onset of each growing season, plant tissue located immediately beneath tree bark produces relatively large thin-walled wood cells that give the wood a relatively light appearance. Wood cells produced in summer are thick-walled and give the wood a darker appearance. A year's growth of spring wood plus summer wood constitutes an annual growth-ring. Hence, counting the number of growth-rings gives the tree's age in years. Because tree growth-ring widths normally decrease as the tree ages, widths are usually expressed in terms of a *tree-ring index*, which is the ratio of the actual tree-ring width to the width expected based on tree age. The index is relatively low in stressful growing seasons and high in favorable growing seasons.

Trees growing under harsh conditions at climatically marginal sites are the most sensitive to climatic fluctuations, and their growth-rings are the most reliable sensors of climate. A simple, hollow drill is used to extract cores from living trees or cut wood. Usually, cores are taken from many trees at one site, and tree-ring indexes are averaged. In western and southwestern North America, the primary locales of dendroclimatic research, scientists sample ponderosa pine, Douglas fir, or the exceptionally longlived bristlecone pine. By assiduous matching of tree growth-ring records from living trees with those from timbers in prehistoric dwellings, detailed tree-ring chronologies can be extended back thousands of years. This matching technique is known as *cross-dating*.

Deep-sea sediment cores are also sources of past climatic data. Tiny clay particles and the shell and skeletal remains of marine organisms gradually settle out of ocean water and accumulate as sediment on the sea floor. In the open ocean, far from shoreline turbulence that might disturb sedimentation, an orderly record of environmental conditions accumulates. Scientists aboard specially outfitted ships extract cores from ocean bottom sediments and analyze layers of sediment for their climatic implications.

Much of what we know today about the large-scale climatic shifts that took place over the past 500,000 years came out of the *Climate/Long Range Investigation, Mapping and Prediction (CLIMAP)* project. Sponsored by the National Science Foundation and initiated in 1971, this ambitious research effort was carried out by a cadre of scientists and support personnel from a number of universities and representing a variety of disciplines. CLIMAP's primary objective was to reconstruct global climate for various periods over the past 2 million years, relying mostly on analyses of shell remains of marine organisms contained in deep-sea sediment cores. One important finding of CLIMAP was the quasi-regularity of shifts between glacial climates and interglacial climates. Glacial-interglacial climatic shifts were reconstructed by measuring the ratio of two isotopes* of oxygen (^{16}O and ^{18}O) contained in the shells of plankton, tiny free-floating marine organisms.

Water is composed of both ^{16}O and ^{18}O. When water evaporates from the ocean, however, much of the ^{18}O is left behind. This means that snow is enriched in ^{16}O relative to ^{18}O. Hence, as ice sheets thicken and spread across the land during major glacial climatic episodes, the lighter oxygen isotope (^{16}O) becomes locked up in glacial ice preferentially over the heavier oxygen isotope (^{18}O). And, ocean waters become enriched in ^{18}O relative to ^{16}O Plankton living during these times incorporate into their shells more ^{18}O than do plankton living during interglacial episodes. By analyzing oxygen isotope ratios in plankton shells found at various horizons within deep-sea sediment cores, CLIMAP investigators were able to reconstruct the major glacial-interglacial climatic shifts of the past 500,000 years.

Glacial ice cores also can be useful in climate reconstruction. Where the climate favors more annual snowfall than snowmelt, a glacier may develop. Under the mounting weight of accumulating snow, annual layers of snow compact and are transformed into solid ice that flows under the influence of gravity—that is, a glacier forms. Vertical cores extracted from a glacier thus contain continuous records of past seasonal snowfalls preserved as discrete layers of ice. By studying oxygen isotope ratios of ice samples taken from various layers within an ice core, scientists can decipher the record of past air temperature. During relatively cold

episodes, snowfall is enriched in ^{16}O relative to ^{18}O (whereas ocean water is enriched in ^{18}O relative to ^{16}O). Thus, the lower the concentration of heavy oxygen in ice core samples, the lower is the indicated temperature.

Layers of ice in glacial ice cores also contain bubbles of trapped air that provide clues to changes in atmospheric chemistry. In 1988, Soviet and French scientists reported on their analysis of air trapped in a 2200-m (7150-ft) ice core extracted from the Antarctic ice sheet at Vostok station. The ice core spanned 160,000 years. Chemical analysis indicated that atmospheric CO_2 concentration varied between 200 ppm during glacial climatic episodes and 270 ppm during interglacials. This finding suggests that variations in CO_2 may amplify the Milankovitch cycles (discussed elsewhere in this chapter) in explaining major climatic shifts during the Ice Age.

In summary, studies of historical documents, pollen, tree growth-rings, deep-sea sediments, and glacial ice cores have produced much information on the climatic past. Many other proxy climatic data sources exist, but space does not permit a description of them; interested readers are urged to consult the references at the close of this chapter for details.

* Isotopes of a given element differ from one another on the basis of atomic mass.

Dealing with time frames of hundreds of millions of years is complicated by geologic processes such as mountain building and continental drift. Today we consider topography and land/water distribution to be fixed climatic controls, but in the perspective of geologic time, they are variable. Mountain ranges have risen and eroded away; seas have invaded and withdrawn from the land, continually altering the shapes of continents; and land masses have drifted slowly across the face of the globe.

Continental drift, an idea first proposed in 1912 by Alfred Wegener, a German geophysicist, but not widely accepted until the 1960s, might explain such seemingly anomalous finds as 200-million-year-old glacial deposits in the Sahara Desert, fossils of tropical plants in Greenland, and fossil coral reefs in Wisconsin. These materials were most likely formed when the continents were situated at different latitudes than at present. The Earth's crust is divided into a dozen gigantic, rigid plates that drift slowly (about 2 to 10 cm per year).

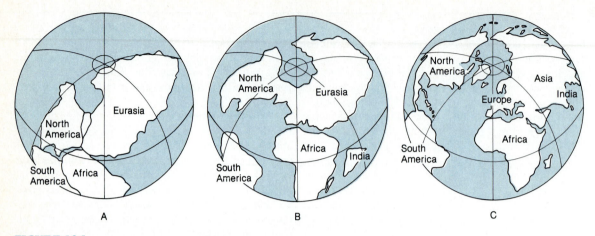

A   B   C

FIGURE 18.1
Originally, today's continents formed a single supercontinent called Pangaea. (A) Perhaps 200 million years ago the supercontinent began to break up and the fragments drifted apart. (B) The probable configuration of the drifting continents approximately 65 million years ago. (C) Today the continents have this arrangement, but they continue to slowly drift. [From P. J. Wyllie, *The Way the Earth Works*. Copyright © 1976 John Wiley & Sons. Reprinted by permission of John Wiley & Sons, Inc.]

The continents are part of the plates and drift with them over the globe. Geologic evidence suggests that about 200 million years ago there was just one supercontinent, called Pangaea (for "all land"), which subsequently split into smaller land masses. Its constituent land masses, the continents we now know, drifted slowly apart (Figure 18.1), eventually reaching their present locations. What preceded Pangaea is the subject of much speculation.

THE PAST MILLION YEARS

As we focus on the climatic record of the past million years or so, we do not have to be concerned with the effects of continental drift and mountain building. For all practical purposes, mountain ranges and continents were essentially as they are today. Because climate varies over a wide range of time scales, it is, however, useful to view the climatic record of the past million years in the perspective provided by progressively narrower time frames. Such an approach helps resolve the complex oscillations of climate into somewhat simpler fluctuations.

When compared with climatic conditions that prevailed over most of geologic time, the climate of the last million years has been anomalous in favoring the development of huge glacial ice sheets. During most of Earth's history the average global temperature may have been 10 C° (18 F°) warmer than it was in the last million years. The trend toward colder conditions began perhaps 50 million years ago and culminated in the Ice Age. Reconstructions based on deep-sea sediment cores indicate that during the Ice Age the climate shifted numerous times between conditions conducive to the expansion of glaciers, a **glacial climate,** and conditions conducive to the decay of glaciers, an **interglacial climate.** Because climatic trends are nonuniform geographically, temperature changes were greater in some latitude belts and less in others. A variety of evidence indicates that temperature fluctuations between major glacial and interglacial climatic episodes typically amounted to about 2 C° (3.6 F°) in the tropics, 6 to 8 C° (11 to 14.4 F°) at midlatitudes, and 10 C° (18 F°) or more at high latitudes.

During major glacial climatic episodes, the Laurentide ice sheet developed over central Canada and spread westward almost to the Rocky Mountains, eastward to the ocean, and southward over the northern tier states of the United States (Figure 18.2). At the same time, a much smaller ice sheet formed over northwestern Europe. Because a vast quantity of water was locked up

FIGURE 18.2

The extent of glaciation over North America 18,000 years ago, the time of the last glacial maximum.

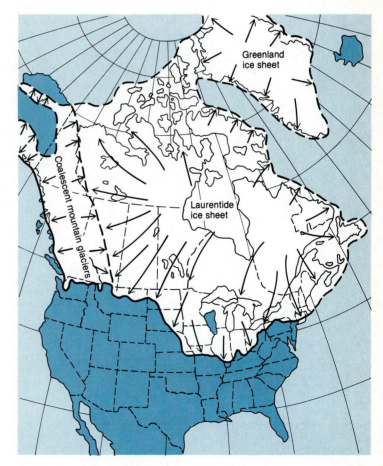

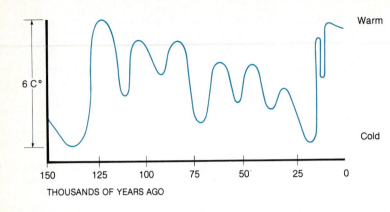

FIGURE 18.3

Generalized curve of midlatitude air temperature over the past 150,000 years based on pollen records, sea-level fluctuations, and reconstructed sea-surface temperatures. [Reprinted from W. L. Gates and Y. Mintz, *Understanding Climatic Change,* 1975, with permission from the National Academy of Sciences, National Academy Press, Washington, DC.]

in these ice sheets, sea level fell by perhaps 130 m (400 ft), exposing portions of the continental shelf, including a land bridge linking Siberia and North America. The Laurentide and European ice sheets thinned and retreated, and may even have disappeared entirely, during the relatively mild interglacial episodes, which typically lasted about 10,000 years. Throughout the interglacials, however, glacial ice cover persisted over most of Greenland and Antarctica, as it still does today.

Resolution of the past climatic record improves somewhat when we shift focus to the last 150,000 years. The temperature curve in Figure 18.3 was derived from a combination of midlatitude sea-surface temperature indicators, fossil pollen data, and reconstructed sea-level fluctuations. Many shifts between glacial and interglacial climatic episodes are visible.

In the perspective of the past 22,000 years, even more climatic detail appears. The midlatitude temperature curve in Figure 18.4 is based primarily on reconstructed glacial fluctuations and pollen studies. The last major glacial climatic episode began about 25,000 years ago and reached its peak about

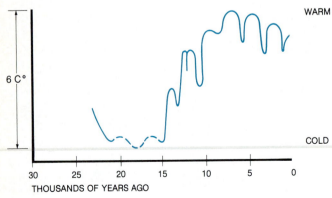

FIGURE 18.4

Generalized curve of midlatitude air temperature over the past 22,000 years based on pollen records, tree-line changes, and glacial ice fluctuations. [Reprinted from W. L. Gates and Y. Mintz, *Understanding Climatic Change,* 1975, with permission from the National Academy of Sciences, National Academy Press, Washington, DC.]

18,000 years ago when global temperatures averaged 4 to 6 C° (7.2 to 10.8 F°) lower than at present. The last glacial maximum was followed by a warming trend that triggered oscillatory glacial retreat. The Laurentide ice sheet withdrew from the Great Lakes region about 10,000 years ago and finally melted away by about 6000 years ago.

The postglacial warming trend culminated in the so-called **climatic optimum** about 5000 to 7000 years ago, a time when global temperatures were somewhat warmer than at present. A recent pollen-based climate reconstruction indicates that 6000 years ago, July temperatures were about 2 C° (3.6 F°) warmer than at present over most of Europe.

By examining documentary records of old farming communities in eastern Europe, the British meteorologist and historian Hubert H. Lamb derived a generalized temperature curve for the past 1000 years, as shown in Figure 18.5. The most notable feature of this record is the relatively mild conditions of the Middle Ages and the sharp cooling that followed from about 1400 to 1850—a period that has come to be known as the **Little Ice Age.** Other independent lines of evidence have confirmed that this was indeed a relatively cool period throughout the world, with mean annual global temperatures perhaps 1 to 2 C° (1.8 to 3.6 F°) lower than at present. Sea ice expanded equatorward, mountain glaciers advanced, and growing seasons shortened—bringing much hardship to many people.

With the invention of weather instruments and the establishment of weather observational networks throughout the world, the climatic record becomes much more detailed (Chapter 15). The most reliable temperature records date from the birth of the World Meteorological Organization (and the National Weather Service) in the late 1800s because by then weather observations were made under standardized conditions. Examination of temperature trends over the past 100 years is instructive as to the basic variability of climate.

FIGURE 18.5
Generalized curve of air temperatures in eastern Europe over the past 1000 years based on a winter severity index developed by H. H. Lamb. [Adapted from H. H. Lamb, "Climatic Fluctuations." In H. Flohn (Ed.), *World Survey of Climatology,* Vol. 2, *General Climatology.* New York: Elsevier, 1969, p. 236.]

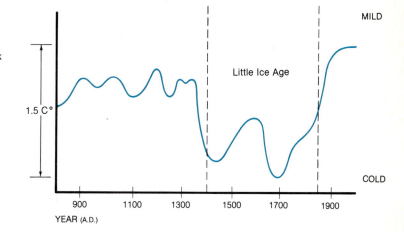

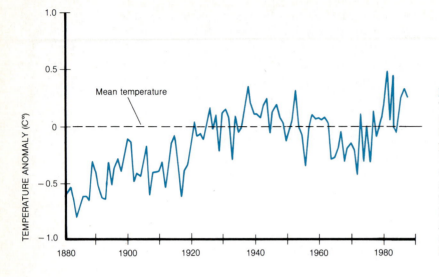

FIGURE 18.6
The variation of mean annual temperature of the Northern Hemisphere, 1881–1983, is expressed as departures from a 100-year mean temperature (in C°). [Data from P. D. Jones et al., "Variations in Surface Air Temperatures: Part 1. Northern Hemisphere, 1881–1980," *Monthly Weather Review* 110 (1982):67, and subsequent publications by the same authors]

In Figure 18.6, the mean annual Northern Hemisphere temperature is expressed as a departure from the long-term average. Note that the temperature trend is upward from the 1890s through the late 1930s, downward into the 1970s, and then upward again into the 1980s. The total temperature fluctuation amounts to only ±0.6 C° (±1.1 F°) about the century average. Remember, however, that we are looking at a hemispheric mean temperature, so the trend was amplified or reversed or both in specific locations. For example, Figure 18.7 shows that trends in winter (December through February) mean temperature for much of the same period varied considerably in different regions of the United States.

We gain a somewhat different perspective on temperature variations over the past 100 years when we consider the global rather than hemispheric scale. J. E. Hansen and his colleagues at NASA found that the post-1940 cooling trend, though quite pronounced poleward of the Northern Hemisphere tropics, was minor elsewhere. Mean global temperatures increased about 0.5 C° (0.9 F°) between 1880 and 1985, with the greatest warming at midlatitudes of both hemispheres. By 1980, the global mean temperature was almost as high as it was in 1940. As of this writing, 1987 and 1988 were the two warmest years on record globally.

Critics sometimes question the validity of mean hemispheric and global temperature trends. They cite as potential sources of error: (1) improved sophistication and reliability of weather instruments through the period of record, (2) changes in location and exposure of instruments at most long-term weather stations, and (3) huge gaps in monitoring networks, especially over the oceans. By careful statistical treatment of available data, however, large-scale temperature trends have been confirmed with a reasonable degree of certainty.

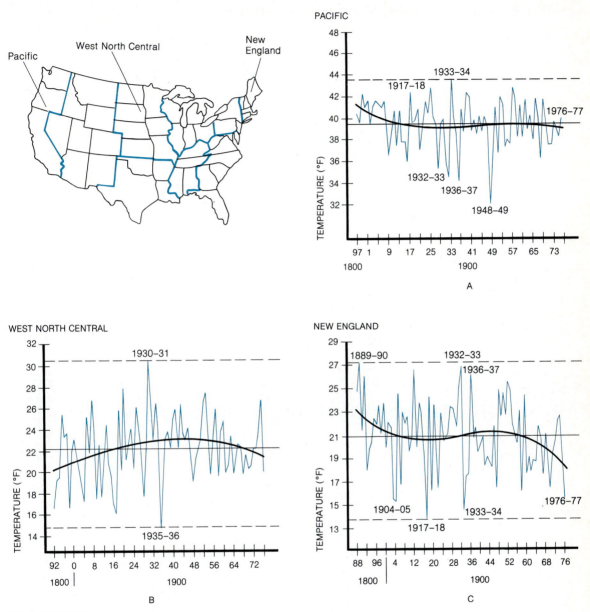

FIGURE 18.7
The geographic nonuniformity in climatic trends is illustrated by differences in trends of mean winter (December through February) temperatures in three regions of the United States: (A) Pacific, (B) West North Central, and (C) New England. Data are from the late 1800s to 1977. [From H. F. Diaz and R. G. Quayle, "The 1976–77 Winter in the Contiguous United States in Comparison with Past Records," *Monthly Weather Review* 106 (1978):1402–1405]

Lessons of the Climatic Record

What does the climatic record tell us about climatic behavior? After all, that is a major reason for journeying into the climatic past. Perhaps the first and foremost lesson of the climatic past is that climate is inherently variable and it varies over a wide range of time scales, from years to millennia. Another conclusion is that variations in climate impact society.

CLIMATE AND SOCIETY

There is little doubt that climate and climatic variations have played a role in the course of history since humans emerged during the Ice Age. Sometimes climate was probably the key factor.

In their book, *Climates of Hunger,* R. A. Bryson and T. J. Murray describe several cases in the distant past when climatic change severely affected societies. They argue convincingly, for example, that prolonged drought contributed to the decline and fall of (1) the Harappan civilization of the Indus Valley region of northwest India about 1700 B.C., (2) the Mycenaean civilization of Greece in 1200 B.C., and (3) the Mill Creek culture of northwest Iowa about 1200 A.D. Other scientists propose that a succession of severe droughts forced the Pueblo Indians of Mesa Verde in southwestern Colorado to abandon their homes around 1300 A.D.

Special Topic

Climatic Change and the Norse Greenland Tragedy

The late ninth century saw the beginning of a lengthy episode of unusually mild conditions throughout much of the Northern Hemisphere. A relatively mild climate enabled Viking explorers to probe the northerly reaches of the Atlantic Ocean. Previously, severe cold and extensive drift ice had proved insurmountable obstacles to European navigators. By 930 A.D. the Vikings established the first permanent settlement in Iceland, some 970 km (600 mi) west of Norway and just south of the Arctic Circle.

Among Iceland's early inhabitants was Eric the Red, a troublesome individual whose exploits eventually caused his banishment from Iceland in 982 A.D. He sailed west and discovered a new land, which he named Greenland. Then as now, much of Greenland was buried under a massive glacial ice sheet. The only habitable lands were small patches scattered along the coast, hemmed in by the sea and ice sheet and separated by treacherous mountain ridges and deep fjords. Some historians speculate that Eric called the new land Greenland to entice others to follow him. At the time, however, the climate was so mild that some sheltered valleys were probably greener than they are today.

In such a place, on Greenland's southwest shore, Eric founded the first of two Norse settlements (Figure 1). Although never prosperous, Norse colonization of Greenland persevered for nearly five centuries. The Norse subsistence economy was primarily agrarian and dependent on raising cattle, sheep, and goats. In addition, the Norse hunted migratory harp seals in spring

In these and other studies of the impact of climatic change on society, one message is very clear. The people most vulnerable to climatic change are those living in areas where the climate is marginal for survival. These are typically climatic zones where barely enough rain falls or where the growing season is just long enough for crops to mature. Even a small change in these critical parameters in the wrong direction can spell disaster. This is apparently what happened to the early Greenland settlements, as discussed in the Special Topic "Climatic Change and the Norse Greenland Tragedy."

Interestingly, the fluctuations of climate that seriously impacted past civilizations likely fall within the range of climatic variability that appears in the modern climatic record. Even the very cold summer of 1816, described in a Special Topic in Chapter 17, falls within the range of expected climatic variability, albeit at an extreme end. Theoretically, then, the atmospheric system could shift to any anomalous weather pattern at any time. In effect, whatever happened in the climatic past can happen again. This conclusion also underscores the notion that past climatic episodes differ from present climatic episodes in frequency rather than in type of episode. For example, an exceptionally cold winter is the consequence of a usual westerly wave pattern that occurs with unusually high frequency or persistence (Chapter 10). A climatic shift to drier conditions results from more frequent dry weather circulation patterns. The individual weather patterns are not themselves unusual, but their more frequent occurrence may be anomalous.

FIGURE 1

Locations of Ancient Norse settlements on coastal Greenland. [From R. A. Bryson and T. J. Murray. *Climates of Hunger.* Madison, WI: The University of Wisconsin Press, 1977, p. 48.]

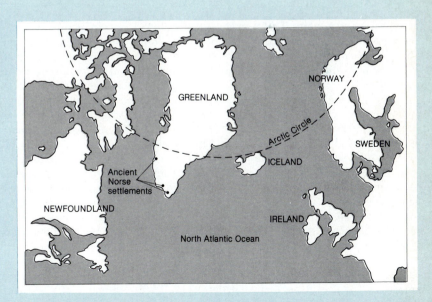

How rapidly do climatic variations take place? Are climatic shifts abrupt or gradual? A climatic shift that takes 100 years is abrupt in the context of the long-term climatic record, which stretches back a million years, but the same shift is gradual when viewed in the limited human perspective of historical climate. In terms of this rate criterion, the climatic record indicates that significant climatic variations tend to occur abruptly rather than gradually. Now, as in the past, a significant variation in climate can occur so rapidly that humankind may be unable to adjust to the shift.

We must also be careful to distinguish between a shortlived meandering from the climatic norm and a semipermanent climatic change. The Great Plains drought in the 1930s, the three severe midwestern winters in the late 1970s, and the tragic Sahelian droughts were all temporary climatic fluctuations. In contrast, the climatic cooling of the late fourteenth century that heralded the Little Ice Age constituted a longer-term climatic change.

RHYTHMS IN THE RECORD

The recorded and reconstructed climatic record has long been the target of painstaking analyses for possible trends or regular rhythms. One formidable challenge in this search is to separate signal from noise. Some climatic elements are so variable (the noise) through time that detection of any cycles or trends (the signal) requires close scrutiny and meticulous dissection of the climatic

and caribou in autumn. They even embarked on long, dangerous hunting expeditions northward along Greenland's west coast in search of walrus and polar bear, whose valuable tusks and hides they traded to Europeans for durable goods.

Initially, the colonists fared relatively well; the population of the Eastern Settlement climbed to an estimated 4000 to 5000, and that of the Western Settlement peaked at perhaps 1000 to 1500. By about 1350 A.D., however, the Western Settlement was vacant, apparently the victim of some sudden calamity. The larger Eastern Settlement also succumbed, but more gradually, so that by 1500 A.D., Norse society in Greenland had been erased. What happened to the Norse settlements in Greenland can only be inferred from their graves, the ruins of their homes and barns, and a few chronicles. There were no survivors. In 1921, an expedition from Denmark examined the Norse remains and found evidence that the settlers had suffered

a painful annihilation. Grazing land was buried under surging lobes of glacial ice, and most farmland was made useless by permafrost. Near the end, the descendants of a robust and hardy people were ravaged by famine; they were crippled, dwarflike, and diseased.

Many explanations have been proposed for the extinction of Norse society in Greenland, and there were probably many contributing factors. Climate change and the inability (or unwillingness) of the Norse to adapt to an increasingly stressful climate and unreliable food sources, however, appear to have been the major causes of disaster. By the 1300s, the climate was cooling rapidly, heralding the Little Ice Age. Drift ice expanded over the North Atlantic, hampering and eventually halting navigation between Greenland and Iceland. All contact between the Norse settlements and the outside world ended shortly after 1400 A.D.

Deteriorating climate in Greenland meant wetter summers with poor haying conditions that caused ma-

record. In recent years, use of high-speed electronic computers programmed with sophisticated statistical routines has greatly facilitated the search for climatic rhythms and trends. The motivation behind all this effort is obvious: identification of any statistically real periodicities or trends in the climatic record would be a powerful tool in both weather and climate forecasting. What are the results of such investigations?

So far, few cycles have been identified in the climatic record that are significant in a rigorous statistical sense, and none of the cycles has much practical value for weather or climate forecasting. Cycles established as statistically real are (1) the familiar annual and diurnal radiation/temperature cycles and (2) a less familiar quasibiennial (almost every 2 years) cycle in various midlatitude climatic elements. The first means only that winters are cooler than summers and nights are cooler than days. Examples of the second type of cycle include an approximate 2-year fluctuation in Midwestern rainfall and a 25.5-month oscillation in a lengthy temperature record (1659 to present) from central England. A 20- to 22-year recurrence of drought on the western High Plains, and the major glacial-interglacial climatic fluctuations are actually only quasiperiodic (not quite regular) and need to be more thoroughly understood before they can be used to reliably forecast future weather and climate. Trends may be visible in the climatic record, but unless a trend is demonstrated to be part of a statistically significant cycle, there is no guarantee that the trend will not end abruptly or reverse direction at any time.

jor livestock losses. Snowier and longer winters probably claimed an even greater toll of livestock—especially among newborn animals. Climatic change also disrupted the harp seal migration and decimated caribou herds, further reducing food sources. About the same time, the Norse had to compete for shrinking food sources with the Inuit people who had migrated southward along Greenland's west coast. Competition for food evidently led to hostilities between the two groups.

Unable to survive the climatic stress, the Norse succumbed to famine; yet the neighboring Inuit people survived. The Inuits succeeded probably because their hunting skills and techniques were more suited to the hostile climatic conditions. On the other hand, the Norse refused to adopt the hunting practices of the Inuits and clung tenaciously to their traditional subsistence methods. Unfortunately, those methods were no longer suited to the new climatic regime.

The Norse tragedy in Greenland may be one of only a very few historical examples of the extinction of a European society in North America. What lesson does it teach contemporary society? The lesson is that climate changes and that it can change rapidly—sometimes with serious, even disastrous, consequences. Nowhere are people more vulnerable to climatic shifts than in regions where the climate is just marginal for their survival. These are regions where barely enough rain falls to sustain crops and livestock or where mean temperatures are so low and the growing season so short that only a few hardy crops can be cultivated successfully. In such regions, even a small change in climate in the wrong direction can make agriculture impossible. If inhabitants cannot locate and utilize new food sources, or if they cannot migrate to more hospitable lands, their fate may be similar to that of the Greenland Norse.

Explaining Climatic Variability

There is no simple explanation for why climate varies. The complex spectrum of climatic variability is a response to the interactions of many processes both internal and external to the Earth-atmosphere system. These processes are shown schematically in Figure 18.8. Internal processes are indicated by open arrows, and external processes are represented by solid arrows.

There are perhaps as many hypotheses about the causes of climatic variations as there are scientists who have seriously investigated the question. Many of these ideas evolved from efforts to explain the great Ice Age, but some stemmed from attempts to explain short-term fluctuations of climate. One way to organize the many hypotheses on the causes of climatic variability is to match a possible cause (or forcing) with a specific climatic oscillation on the basis of the period of that oscillation. This approach is displayed sche-

FIGURE 18.8

Climatic variability is influenced by many processes, both internal (white arrows) and external (black arrows) to the Earth-atmosphere system. [Reprinted from W. L. Gates and Y. Mintz, *Understanding Climatic Change,* 1975, with permission from the National Academy of Sciences, National Academy Press, Washington, DC.]

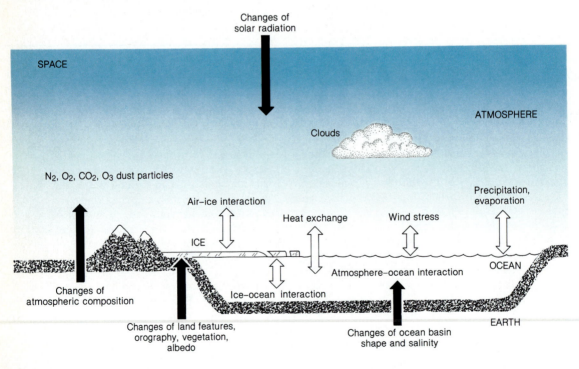

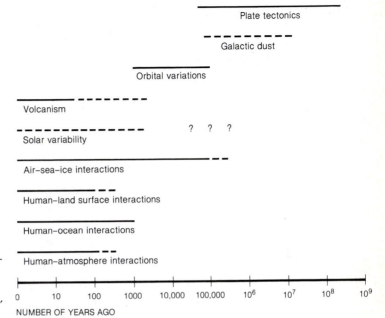

Air–sea evolution

Plate tectonics

Galactic dust

Orbital variations

Volcanism

? ? ?

Solar variability

Air–sea–ice interactions

Human–land surface interactions

Human–ocean interactions

Human–atmosphere interactions

| 0 | 10 | 100 | 1000 | 10,000 | 100,000 | 10^6 | 10^7 | 10^8 | 10^9 |

NUMBER OF YEARS AGO

FIGURE 18.9
One way to speculate on the possible cause of a climatic oscillation is to match the period of that oscillation with the period of an appropriate forcing phenomenon. Lines are dashed where there is great uncertainty. [Reprinted from *Geological Perspectives on Climatic Change,* 1978, with permission from the National Academy of Sciences, National Academy Press, Washington, D.C.]

matically in Figure 18.9. For example, mountain building and continental drift might explain climatic changes over periods of hundreds of millions of years. Systematic changes in the Earth's orbit about the sun may account for climatic shifts of the order of 10,000 to 100,000 years. Sunspots and variations in the sun's energy output may be associated with climatic fluctuations of decades to centuries, and volcanic eruptions may induce climatic fluctuations lasting a few months or years. Note, however, that matching a possible cause and effect is no guarantee of a real physical relationship; we could well be dealing with mere coincidence.

Hypotheses concerning climatic variability can also be organized in the context of the global radiation balance. Energy that goes into the Earth-atmosphere system must ultimately equal energy that leaves the system (Chapter 2). That is, the net input of solar radiation must balance the output of infrared radiation. Any change in either energy input or energy output will shift the Earth-atmosphere system to a new equilibrium state and the planet's climate thereby changes. The present state of global radiation balance could be altered by fluctuations in (1) the solar constant, (2) the planetary albedo, or (3) the gas and aerosol composition of the atmosphere. We examine this approach quantitatively in the Mathematical Note at the end of this chapter.

Climate and Solar Variability

Changes in the sun's energy output or variations in the Earth's orbit about the sun could alter the solar constant or the distribution of solar radiation over the Earth's surface. Any of these changes could in turn alter climate.

THE SOLAR CONSTANT

Is the solar constant* really constant? As demonstrated in this chapter's Mathematical Note, even a 1 percent change in the solar constant could significantly alter the radiative equilibrium temperature of the Earth-atmosphere system. In fact, only a 0.5 percent fluctuation in the solar constant could account for the total variation of the mean annual hemispheric temperature over the past century. Until recently, scientists have been unable to monitor small-scale changes in the solar constant. Older ground-based instruments lacked the necessary sensitivity, and high-resolution instruments onboard satellites had not been operating long enough to provide reliable records. This situation began to change on 14 February 1980 with the launch of NASA's Solar Maximum Mission (SMM) satellite (Figure 18.10) into orbit 550 km (340 mi) above the Earth's surface.

Extremely sensitive radiation sensors on the SMM satellite could detect changes as small as 0.001 percent in the sun's total radiative output, called **solar irradiance.** Analysis of measurements by the SMM and other satellites indicates a decline in solar irradiance between 1981 and 1986 amounting to about 0.018 percent per year. This was a total reduction of about 0.10 percent in almost six years. Had this trend continued, the dimming of the sun probably would have become climatically significant by 1990. However, solar irradiance began to increase after the Fall of 1986, about the time of the last sunspot minimum, and continued to increase into 1990. This correspondence suggests that solar irradiance may vary with the 11-year sunspot cycle.

Skepticism regarding the reliability of SMM irradiance data appears to be ill founded. The satellite experienced technical difficulties in November 1980 and was repaired by NASA's space shuttle in April 1984. However, SMM findings compare favorably with data from other solar-irradiance-monitoring experiments. Radiation sensors aboard the Nimbus 7 satellite, launched in 1978, detected a 0.015 percent per year decline in solar irradiance. In addition, probes by rockets and high-altitude balloons found a downward trend of about 0.016 percent per year. Unfortunately, the SMM satellite finally went out of service and plunged to Earth in 1989.

---

* In Chapter 2, the Earth's solar constant was defined as the flux of solar radiation on a surface oriented perpendicular to the solar beam at the top of the atmosphere when the Earth is at its mean distance from the sun.

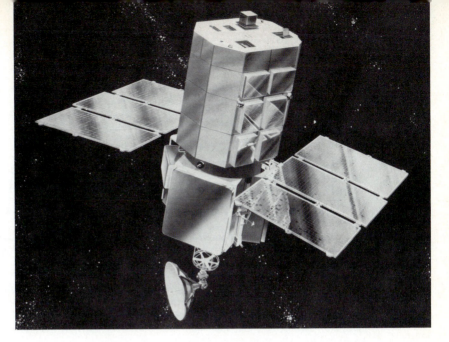

SUNSPOTS

Both popular and technical literature contain much speculation on the possible link between the Earth's weather and climate and sunspot activity. **Sunspots** are relatively large (typically thousands of kilometers in diameter) dark blotches that appear on the face of the sun. As shown in Figure 18.11, a sunspot consists of a dark central area, called an **umbra,** which is ringed by an outer, lighter area, termed a **penumbra.** The umbra radiates at about 4000 K, the penumbra at 5400 K, and the surrounding surface of the sun, the **photosphere,** at 5800 K.

FIGURE 18.11

This sequence of images, taken by sensors aboard NASA's Solar Maximum Mission spacecraft, shows an active solar region containing four large sunspots, which are moving toward the edge of the sun. Image widths in these photographs represent a distance on the sun's surface of about 360,000 km (224,000 mi). [Courtesy of NASA]

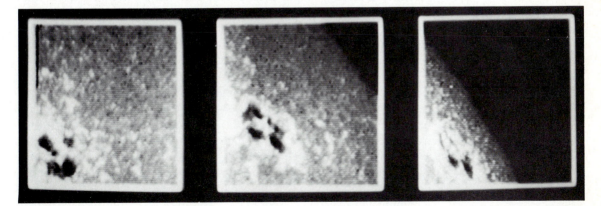

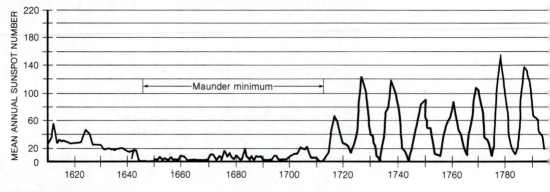

FIGURE 18.12

On this graph of mean annual sunspot number, note the Maunder minimum from 1645 to 1715. Note also that the time between sunspot maxima is not quite regular, but varies between about 10 and 12 years. [From J. A. Eddy. "The Case of the Missing Sunspots," *Scientific American* 236, No. 5 (1977):82–83. Copyright © 1977 by Scientific American, Inc. All rights reserved.]

As early as 28 B.C., Chinese astronomers observed sunspots with the unaided eye by viewing the sun's reflection in a quiet pond. Galileo is credited with being the first to study sunspots telescopically in 1610, and thereafter sunspots became objects of considerable scientific curiosity. Speculation on a possible sunspot–climate link was spurred by Heinrich Schwabe's discovery in 1843 of the regularity of sunspot activity. As shown in Figure 18.12, the number of sunspots varies systematically with an average 11-year period between sunspot maxima. Also an approximate 22-year oscillation ("double" sunspot cycle) occurs in the strong magnetic field that is associated with sunspots. Note, however, that the sunspot record is not precisely periodic and that variations of 10 to 12 years or more occur between maxima or minima. As of this writing, sunspots are nearing a peak.

As noted above, recent satellite measurements of solar irradiance indicate that the sun's energy output varies directly with sunspot number; that is, more sunspots mean a brighter sun (perhaps because of a concurrent increase in faculae activity). Thus more sunspots may translate to a warmer Earth, assuming that all other climate controls are constant. Past records of sunspot activity and climate reveal some intriguing correspondences.

In 1893, while searching old records at the Old Royal Observatory at Greenwich, England, E. Walter Maunder discovered that sunspot activity was greatly reduced in the 70 years between 1645 and 1715. The total number of sunspots observed during this period, now called the **Maunder minimum,** was fewer than is typical in a single year today. Strangely, the scientific community largely ignored Maunder's finding until the 1970s, when Maunder's work was reinvestigated and confirmed by John A. Eddy of the University Corporation for Atmospheric Research. Eddy pointed out that the Maunder minimum and an earlier episode of reduced sunspot activity, called the **Spörer**

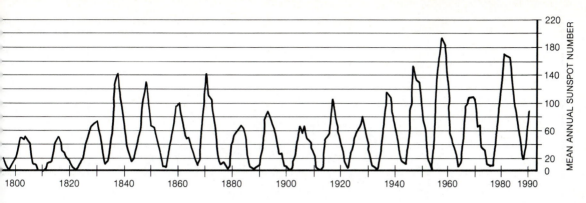

MEAN ANNUAL SUNSPOT NUMBER

minimum (1450–1550), happened to coincide with two relatively cold pulses of the Little Ice Age in Western Europe. However, skeptics dismiss the climatic significance of this correspondence, arguing that anomalous cold did not prevail throughout the sunspot minima and was not global in extent. The late climatologist Helmut Landsberg of the University of Maryland noted that globally the coldest episode of the Little Ice Age occurred 100 years *after* the Maunder minimum. On the other hand, relatively mild weather coincided with a period of considerable sunspot activity between about 1100 and 1250.

There is also a proposed match between the 22-year "double" sunspot cycle and the frequency of drought on the western High Plains. By analyzing tree growth-ring records from 40 sites in the western United States, Charles Stockton of the University of Arizona's Laboratory of Tree-Ring Research was able to extend the region's drought chronology back to the seventeenth century. He discovered that drought recurred every 20 to 22 years.

Are these correlations actual cause–effect relationships or are they merely coincidences? So far, investigators have not come up with an adequate explanation of the mechanism by which sunspot activity might influence climate, but the structure of sunspots is an interesting avenue of inquiry. In 1979, Douglas V. Hoyt, then a solar researcher for NOAA, reported that the ratio of the area of the sunspot umbra (U) to the area of its penumbra (P) may be proportional to the brightness of the sun, or the solar constant. As shown in Figure 18.13, the U/P ratio trended upward from 1874 to the 1930s and then downward to 1970, closely paralleling the variation in the mean annual Northern Hemisphere temperature during the same period. It is therefore possible that the brightness of the sun is influenced by sunspot structure. If that is the case, then some possible linkages exist between solar and climatic variability.

Currently, however, there is no scientific consensus on a definitive solar–climate link. A 1982 report by the National Research Council* summarizes a

---

* National Research Council. *Solar Variability, Weather, and Climate.* Washington, D.C.: National Academy Press, 1982, p. 5.

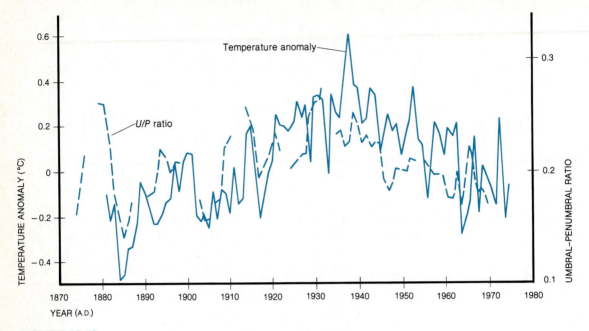

FIGURE 18.13

Over the past hundred years, a close parallel has existed between the variation in mean annual temperature of the Northern Hemisphere (solid line) and the ratio of sunspot umbra (*U*) area to penumbra (*P*) area (dashed line). This suggests a possible cause-effect relationship. [From D. V. Hoyt, "Climatic Change and Solar Variability," *Weatherwise* 33, No. 2 (1980):69]

century of research on this issue by stating ". . . in our view, none of these endeavors, nor the combined weight of all of them, has proved sufficient to establish unequivocal connections between solar variability and meteorological response." The report goes on to urge a shift in research emphasis away from a search for correlations and toward a better understanding of the physical relationship between the sun and the Earth-atmosphere system.

ASTRONOMICAL CHANGES

The regular oscillations of glacial and interglacial climates as revealed by analyses of deep-sea sediment cores may be related to systematic changes in the geometry of the Earth's orbit about the sun. These changes, first identified in the middle of the nineteenth century, were extensively investigated in the 1920s and 1930s by the Serbian mathematician Milutin Milankovitch.

Milankovitch studied cyclical variations in three elements of Earth–sun geometry: (1) the precession (wobble) of the Earth's axis of rotation such that the spin axis describes a circle; (2) the tilt of the Earth's axis (varying between 22.1 and 24.5 degrees); and (3) the eccentricity (departure from a circle) of the Earth's orbit. The precession cycle alters the dates of perihelion and aphelion and thereby increases the seasonal contrast in one hemisphere and decreases the seasonal contrast in the other hemisphere. As the axial tilt increases, the seasonal contrast increases so that winters are colder and summers are warmer in both hemispheres. Changes in orbital eccentricity affect the Earth–sun distance and, hence, the seasonal variation in solar constant. These cycles are illustrated schematically in Figure 18.14.

Although the **Milankovitch cycles** do not alter the solar constant appreciably, they do change significantly the latitudinal and seasonal distribution

FIGURE 18.14

The Milankovitch cycles in Earth-sun geometry that may determine the timing of major glacial-interglacial climatic shifts. Note that the diagrams greatly exaggerate changes in geometry.

Milankovitch Cycles

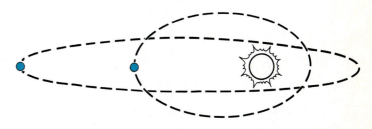

A. Changes in eccentricity: ~100,000 years

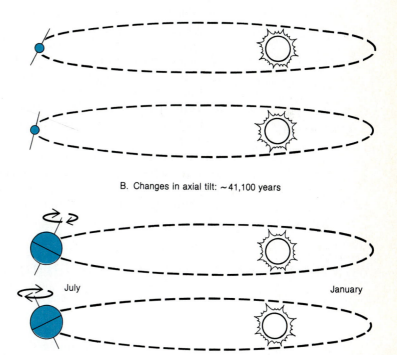

July   January

B. Changes in axial tilt: ~41,100 years

C. Changes in axial precession: ~23,000 years

of solar radiation received by Earth. Milankovitch proposed that glacial climatic episodes were initiated at times when the Earth–sun geometry favored minimum summer insolation at high northern latitudes. During these times, some of the winter snows falling in Canada would presumably survive the summer, and a repetition of many such years woud initiate glaciation on a continental scale.

Milankovitch hand-calculated latitudinal variations in insolation for 600,000 years prior to the year 1800. More recently, with some refinement in methodology and use of high-speed electronic computers, these calculations were repeated and extended over a longer time period. This research demonstrated that systematic changes in precession, axial tilt, and eccentricity would induce climatic cycles having periods of about 23,000 years, 41,000 years, and 100,000 years, respectively. Discovery of climatic cycles of similar periodicities in deep-sea sediment cores strongly argues for variations in Earth–sun geometry as the cause of the regular large-scale fluctuations of the planet's glacial ice cover.

Although the close correspondence between Milankovitch cycles and reconstructed climatic fluctuations of the Ice Age is impressive, the precise physical linkage between glacial-interglacial climates and periodic changes in Earth–sun geometry is not yet understood. Furthermore, the Milankovitch cycles do not account for the lengthy periods of climatic quiescence during the millions of years prior to the Ice Age.

Discovery of a possible relationship between systematic changes in Earth–sun geometry and major glacial-interglacial climate shifts has prompted climatologists to search for influences of Milankovitch cycles on other smaller scale climatic fluctuations. For example, J. E. Kutzbach of the University of Wisconsin–Madison has linked the climatic optimum (described earlier) to the Earth's precession cycle. Between 5000 and 8000 years ago, Earth was closest to the sun (perihelion) in June rather than January. In the Northern Hemisphere, this would have meant about 5 percent more solar radiation during summer and about 5 percent less solar radiation during winter. The consequent increase in seasonal air temperature contrast probably altered the planetary circulation. For one, evidence suggests that the monsoon circulation of subtropical latitudes was stronger during the climatic optimum than it is today.

Volcanoes and Climatic Variability

The spectacular eruption of Mount St. Helens on 18 May 1980 (Figure 18.15) spurred much speculation on the possible climatic impact of volcanic eruptions. An estimated 5 billion metric tons of volcanic ash and gases were blasted 10 km (6 mi) into the stratosphere.

The notion that volcanoes somehow influence weather and climate has been around for more than two centuries. Benjamin Franklin proposed that

FIGURE 18.15
The spectacular eruption of Mount St. Helens on 18 May 1980 apparently had a negligible impact on hemispheric temperatures. [Photograph courtesy of the U.S. Geological Survey, EROS Data Center]

the eruption of Iceland's Laki volcano in the summer of 1783 was responsible for the severe winter of 1783–84. The unusually cool summer of 1816 followed the violent eruption of Tambora, an Indonesian volcano, in the spring of 1815. Several relatively cold years also occurred after Krakatoa, also in Indonesia, blew its top in 1883. Is the relationship between volcanic eruptions and atmospheric cooling real or merely coincidental?

Until recently, climatologists were concerned primarily with the potential climatic impact of the relatively fine ash particles thrown high into the stratosphere during violent volcanic eruptions. Theoretically, volcanic ash should raise the planetary reflectivity (albedo), and a higher albedo means less solar radiation reaching the Earth's surface and lower air temperatures. Today, however, most climatologists agree that ash particles, being relatively large, quickly settle out of the stratosphere to the ground with little or no long-term effect on the radiation balance. The large-scale climatic impact of violent volcanic eruptions now appears to depend more on the volume of sulfur oxide gases ejected.

Once in the stratosphere, sulfur oxide gases convert to tiny droplets of sulfuric acid. The extremely small size of the acid droplets (less than 1 micrometer in diameter), coupled with the stratosphere's extreme stability and absence of precipitation, means that the sulfuric acid droplets remain suspended in the stratosphere for many months to perhaps several years before finally settling to the Earth's surface. While in the stratosphere, the veil of

523

sulfuric acid droplets interacts with solar radiation, so that a portion of the radiation is absorbed, causing stratospheric warming and reducing the amount of radiation available to the troposphere. Furthermore, some solar radiation is scattered by the sulfuric acid droplets back to space, also contributing to cooling at the Earth's surface. Over the years, a succession of volcanic eruptions has produced a permanent sulfurous layer in the lower stratosphere, and sulfur oxide emissions from an occasional violent volcanic eruption temporarily thickens the layer.

Apparently, only those volcanic eruptions rich in sulfur oxide gases and sufficiently explosive to send ejecta well into the stratosphere can disturb hemispheric or global climate. Even then, surface hemispheric mean temperatures are not likely to be lowered by more than 1 C° (1.8 F°). A volcanic eruption in this category was the 1963 eruption of Agung in Bali, which, according to one estimate, lowered mean global air temperatures by about 0.2 C° (0.4 F°) for a year or two. Although the 1980 eruption of Mount St. Helens produced about as much ash as the Agung explosion, the Mount St. Helens ejecta were low in sulfur oxides and had no detectable influence on large-scale climate.

Although the Mount St. Helens eruption did not influence hemispheric or global climate, some short-term localized effects on surface temperatures did occur immediately downwind of the volcano over eastern Washington, Idaho, and western Montana. Over these areas, the ash plume, clearly visible in the satellite photograph in Plate 25, was sufficiently thick to alter the local radiation balance for 12 to 24 hours after the eruption. C. Mass and A. Robock of the University of Maryland report that on the day of the eruption, blockage of the sun by volcanic ash lowered surface air temperatures over eastern Washington by up to 8 C° (14.4 F°). That night, over Idaho and western Montana, low-level ash impeded infrared cooling, thereby elevating temperatures by as much as 8 C°.

The violent eruption of the Mexican volcano El Chichón on 4 April 1982 probably had a much greater impact on large-scale climate than the Mount St. Helens eruption. In fact, the eruption of El Chichón may turn out to be a more significant perturber of climate than any volcanic eruption of the past two centuries. Although many volcanic eruptions have been more violent and have spewed out much more ejecta, the El Chichón eruption was unusually rich in sulfur oxides. Conversion of these gases to sulfuric acid droplets and the gradual spreading of the aerosol veil through the stratosphere of the Northern Hemisphere probably caused cooling on a large scale. Scientists employed several computerized global circulation models to assess the effect of the El Chichón veil on surface temperatures; these studies generally agree on a possible hemispheric cooling of about 0.5 C° (0.9 F°). Precisely when this cooling took place is, however, a point of disagreement. Even now, many years after the El Chichón eruption, we cannot pinpoint the peak cooling period because the volcanic contribution cannot be distinguished from other influences on climate.

Air Pollution and Climatic Variability

Among the many contemporary environmental issues is the possibility that certain human activities may disrupt the Earth-atmosphere radiation balance and thereby contribute to variations in climate. From our study of the fundamentals of weather, we can deduce that human activity may influence climate in at least three ways: (1) by altering the radiational properties of the Earth's surface (changing the albedo, for example), (2) by venting waste heat into the atmosphere, and (3) by changing the concentrations of certain key gaseous or aerosol components of the atmosphere. Although many questions remain unanswered, the first two activities seem to be important only on a local scale, that is, in large metropolitan areas. We examined these influences in Chapter 16. In this chapter, we consider the third potential influence on climate: the impact of air pollution on climatic variations at the hemispheric or global scale. We focus specifically on the climatic implications of elevated levels of atmospheric carbon dioxide and aerosols.

THE CARBON DIOXIDE QUESTION

In recent years, many atmospheric scientists have expressed concern about the possible climatic ramifications of the steady rise in concentration of atmospheric carbon dioxide (CO_2). Higher carbon dioxide levels likely will enhance the greenhouse effect—that is, increase absorption and reradiation of infrared radiation and consequently warm the lower atmosphere worldwide. Actually, the possibility of CO_2-induced global warming was first proposed in 1827 by the French mathematician Baron Joseph Fourier.

Although G. S. Callendar, a British engineer, reported an upward trend in atmospheric carbon dioxide in 1939, the increase likely began during the Industrial Revolution when reliance on fossil fuels began to increase. Fossil fuel combustion accounts for about 80 percent of the increase in carbon dioxide concentration, and deforestation is likely responsible for the balance. The burning of coal, oil, and natural gas produces carbon dioxide as a by-product. Clearing of tropical forests also contributes CO_2 via burning, organic decay of wood residue, and reduced photosynthetic removal of carbon dioxide from the atmosphere.

Systematic measurements of atmospheric carbon dioxide levels began in 1957 at NOAA's Mauna Loa Observatory in Hawaii under the direction of Charles D. Keeling of Scripps Institution of Oceanography. That record is shown in Figure 18.16. At 3400 m (10,200 ft) above sea level, Mauna Loa Observatory is far enough away from major industrial sources of air pollution that carbon dioxide levels monitored there are considered representative of at least the Northern Hemisphere. Also beginning in 1957, atmospheric CO_2 has been monitored at the South Pole station of the U.S. Antarctic Program. Interestingly, the South Pole record closely parallels that at Mauna Loa.

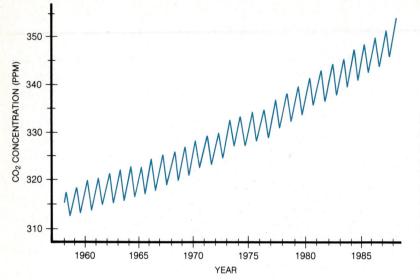

FIGURE 18.16
Upward trend in atmospheric carbon dioxide levels as measured at Mauna Loa Observatory, Hawaii. Annual cycles of photosynthesis and respiration account for the yearly oscillation in atmospheric CO_2. [Data from C. D. Keeling et al., Scripps Institution of Oceanography, as reported in *Geophysical Monitoring for Climate Change*, NOAA]

The Mauna Loa and the South Pole records show an annual carbon dioxide cycle due to seasonal changes in Northern Hemisphere vegetation; carbon dioxide levels fall during the growing season (when photosynthetic removal exceeds respiration) and recover in winter (when respiration exceeds photosynthesis). This annual cycle is superimposed on a nearly exponential rate of growth. The annual rate of increase of carbon dioxide rose from about 0.7 ppm (parts per million) in 1958 to 1.0 to 1.5 ppm during the 1980s. Carbon dioxide concentration increased from 315 ppm in 1958 to 352 ppm in 1989. Sketchy data suggest that atmospheric CO_2 concentration may have climbed by 24 percent since its estimated preindustrial level of 280 ppm.

What might this rapid rise in atmospheric carbon dioxide mean for the climatic future? Researchers have tried to answer this question by using numerical global climate models and high-speed electronic computers. In such experiments, the typical procedure is to test the sensitivity of the model by changing one of the variables—in this case, carbon dioxide concentration. Initially, the model is run to a state of equilibrium using current atmospheric conditions. The model's carbon dioxide concentration is then raised (usually doubled), and the numerical model is run to a new equilibrium state. The difference between the initial and final equilibrium states is assumed to be the consequence of elevated carbon dioxide concentration.

Depending on the specific global climate model used, these experiments predict that the Earth's average surface temperature could rise between 2 and 5 C° (4 and 10 F°) with a doubling of atmospheric CO_2—possible by the middle of the next century. Warming could match the total post–Ice Age temperature rise, albeit at a rate 10 to 100 times faster. Predictive models provide little detail on regional or local climatic change, but the warming is likely to be geographically nonuniform in magnitude. Models predict that polar warming will be up to three times the global average change. In addition, models generally agree that annual rain and snowfall will increase poleward of about 30 degrees latitude and within about 5 degrees of the equator, and decrease elsewhere. However, significant summer drying is predicted for the continental interiors of midlatitudes, where much of the world's food is grown.

CO_2-induced global warming may be exacerbated by rising levels of other IR-absorbing trace gases, the concentrations of which are typically expressed in parts per billion (ppb). These gases include methane (CH_4) at 1684 ppb, nitrous oxide (N_2O) at 304 ppb, and chlorofluorocarbons (CFCs) at less than 0.4 ppb. The climatic significance of these trace gases lies in their strong absorption of outgoing infrared radiation within the atmospheric windows, especially in the 7- to 13-micrometer wavelength range. (Recall from Chapter 2 that within atmospheric windows, infrared absorption by water vapor, carbon dioxide, and ozone is minimal.) Trace gas absorption is directly proportional to concentration, so that doubling the concentration doubles the absorption. The combined climatic impact of rising levels of these gases could match that of increasing carbon dioxide.

The concentration of methane in the atmosphere is increasing by about 1 percent per year. Recent studies of air trapped in ancient glacial ice cores suggest that the methane increase began more than a century ago and has doubled since then. Nitrous oxide is increasing at about 0.25 percent per year at least since 1976 and, based on ice-core analysis, is now about 9 percent higher than it was at the turn of the century. CFCs are currently the fastest growing of the greenhouse trace gases. The two most common CFCs (CFC-11 and CFC-12) are increasing at about 5 percent per year.

The cause of the methane increase is not known but may be linked to biological activity in termites and the stomachs of cattle and sheep, or as effluvium of marshes, rice paddies, and landfills. Increases in the other two gases are likely linked to industrial activity (Chapter 16). (Recall that CFCs also appear to be the culprits in the erosion of the ozone shield.)

What would be the societal impact of CO_2-induced warming? If the carbon dioxide level actually doubles by the middle of the next century, as some studies predict, the consequent global climatic change could be greater than any experienced thus far in the 10,000 years of civilization. Virtually every sector of society would be affected to some degree, with agriculture likely to be the most seriously disrupted. Heat and moisture stress would likely cut crop yields, and traditional farming practices would have to change. For the North American grain belt, for example, higher temperatures and more frequent drought might necessitate a switch from corn to wheat and require more irrigation.

Predicted amplification of a global warming trend in polar latitudes has prompted much speculation on the fate of the Greenland and Antarctic ice sheets. Some studies contend that the warming would melt enough ice that sea level would rise and inundate densely populated coastal areas. In addition, warming causes ocean water to expand, and this would also contribute to a sea-level rise. The combined effect of melting glaciers together with expansion of ocean water would raise sea level an estimated 0.2 to 1.5 m (1 to 5 ft) in 60 years.

Some researchers hasten to point out that CO_2-induced warming may actually yield some benefits. Warmer winters would reduce fuel demand for space heating in middle and high latitudes. In their 1988 summary of potential socioeconomic impacts in the Great Lakes region (due to a CO_2-doubling), S. J. Cohen and T. R. Allsopp of the Atmospheric Environment Service of Canada reported that in Ontario milder winters would reduce energy demand for space heating by 45 percent (more than offsetting an estimated 7 percent increase in summer air conditioning demand). Warmer winters would also lengthen the navigation season on lakes, rivers, and harbors where ice cover is a problem.

Has the warming begun? At many long-term weather stations in the Midwest and Great Plains, the summer of 1988 was one of the driest and hottest on record (Chapter 10). The extreme heat triggered much speculation that a major climatic change was underway and that the much heralded CO_2-induced global warming had at last begun. Although acknowledging that one long hot summer does not constitute a climatic change, some atmospheric scientists were quick to point out recent trends in climate that are similar to changes that would attend a greenhouse warming as predicted by global climate models. These trends include:

1. Occurrence of the six warmest years of the past 100 years during the 1980s, with 1988 the warmest.
2. Measurement of the lowest global mean lower-stratospheric temperatures on record during 1985, 1986, and 1987.
3. Since 1940, in the Northern Hemisphere, an increase in average precipitation at high latitudes (35 to 70 degrees N) and a decrease at low latitudes (5 to 35 degrees N).

But before residents of Boston and New Orleans quit the city and head for high ground to avoid rising ocean water, a note of caution is warranted. Even though some recent climatic trends are consistent with the changes that are predicted to accompany CO_2-induced warming, there is as yet no evidence of a direct cause–effect relationship. Climate controls other than carbon dioxide may well be responsible for these observed trends. Furthermore, the atmosphere is a complex and highly interactive system, so that other processes may compensate for any CO_2-induced warming. For instance, the ocean has great thermal stability, which for a time may slow the warming. In addition, warmer conditions in high latitudes may mean more snowfall and eventually more, not less, glacial ice!

As pointed out earlier in this chapter, serious questions surround the validity of the global mean temperature record that shows a 0.5 C° (0.9 F°) warming between the 1880s and 1980s. In an effort to get away from the problem of large gaps in the worldwide network of thermometers and obtain a more reliable estimate of global mean temperature, R. W. Spencer of the Marshall Flight Center and J. R. Christy of the University of Alabama, Huntsville, analyzed lower atmospheric temperatures derived from radiation sensors in satellites. In early 1990, Spencer and Christy reported no obvious trend in global mean temperature between 1979 and 1988.

Criticism is also directed at the computerized numerical models that have been used to predict the climatic impact of rising levels of CO_2. For example, models vary in sophistication and in their ability to simulate the climatic influence of ocean currents, cloud cover, and atmospheric water vapor. Thus, contrary to model predictions, the Arctic has been cooling—not warming—since the late 1970s.

In spite of uncertainties, some scientists argue that so much is at stake that action must be taken now to head off enhanced greenhouse warming. They call for (1) a 50 percent reduction in global fossil-fuel consumption, (2) greater reliance on nonfossil fuels (solar power, for example), (3) higher energy efficiencies (more miles per gallon, for example), (4) massive reforestation and a halt to deforestation, and (5) phasing out of CFCs. They hasten to point out that these actions are advisable even if greenhouse warming fails to materialize because they will help alleviate other serious environmental problems. For example, phasing down fossil fuels will also reduce the problem of acid deposition and air pollution. Similarly, eliminating CFCs will lessen the threat to the ozone shield.

THE DUST QUESTION

Some atmospheric scientists believe that increased dustiness, called **turbidity,** of the atmosphere has the opposite effect of elevated carbon dioxide levels. Instead of warming the Earth, an increase in turbidity adds to the reflectivity of the Earth-atmosphere system, reducing insolation and thereby cooling the lower atmosphere. Other atmospheric scientists disagree, arguing that increased turbidity resulting from human activity, such as industry and agriculture, promotes warming at the Earth's surface. They contend that these aerosols tend to scatter solar radiation back toward the surface, where it is absorbed, and that the larger dust particles absorb and reemit infrared radiation to enhance the greenhouse effect.

Resolution of this disagreement hinges on a better understanding of the net effect of aerosols on the global radiation balance. It is known that the percentage of insolation that is scattered downward toward the Earth's surface versus the percentage that is scattered back into space varies with the optical properties of the aerosol. In addition, the albedo of the Earth's surface underlying an aerosol layer may help determine whether the net effect is heating or cooling.

The impact of aerosols on climate is thus a complex and unresolved problem. Although stratospheric aerosols of volcanic origin probably trigger cooling at the Earth's surface, the impact of aerosols produced by human activity, most of which do not reach the stratosphere, is uncertain.

Earth's Surface and Climatic Variability

In Chapter 2, we saw that the Earth's surface, which is mostly water, is the prime absorber of solar radiation. Any change in the characteristics of the land and water surface or in the relative distribution of land and sea may affect the radiation balance and hence the climate.

On land, variations in regional snow cover may trigger important climatic fluctuations. This is because an extensive heavy snow cover has a refrigerating effect on the atmosphere (Chapter 4). Fresh-fallen snow typically reflects 80 percent or more of the incident solar radiation, thereby substantially reducing the amount of solar heating and lowering the daily maximum temperature. Snow is also an excellent emitter of infrared radiation, so heat is quickly radiated off to space at night, especially on nights when skies are clear. Because of this radiative feedback, a snow cover tends to be self-sustaining. This effect may be further enhanced by storm tracks that often follow the periphery of a regional snow cover, where horizontal air temperature gradients are steep. This places the snow-covered area on the cold, snowy side of migrating cyclones, thereby enforcing the chill. Hence, an unusually extensive snow cover favors persistence of an anomalously cold episode.

Although large-scale changes in surface characteristics of the continents may affect climate, changes in ocean characteristics and circulation may be much more important. This follows from our discussion of radiative transfer within the Earth-atmosphere system (Chapter 2). Because ocean waters cover nearly three quarters of the global surface and exhibit a very low albedo for solar radiation, the ocean is the principal absorber of insolation. Anything that alters this strong absorption, such as fluctuations in sea ice cover, is likely to affect radiative equilibrium and climate. As noted in Chapter 15, a connection appears to exist between anomalies in sea-surface temperatures and atmospheric long-wave circulation patterns. Shifts in sea-surface temperature anomaly patterns may therefore alter the prevailing circulation and climate. The atmosphere-ocean system involves a two-way interaction, however, and determining which dominates is difficult: the ocean's influence on the atmosphere or the atmosphere's influence on the ocean. Atmospheric impacts on the ocean are relatively short term, whereas the ocean's impacts on the atmosphere are relatively long term because of the great thermal stability of the ocean.

Factor Interaction

We have now identified and elaborated on many of the potential causes of climatic variability. Note, however, that this is not an exhaustive list of all possible controls of climatic variability. For example, subtle astronomical shifts may contribute to short-term climatic fluctuations. We have also treated the

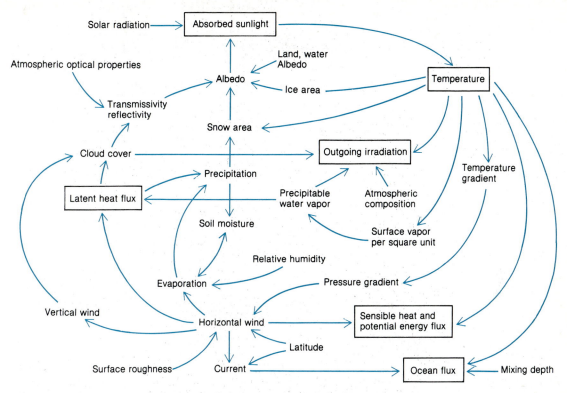

FIGURE 18.17
Climatic variability is influenced by the complex interaction of many processes operating within the Earth-atmosphere system. [From W. W. Kellogg and S. H. Schneider,
"Climate Stabilization: For Better or for Worse?" *Science* 186 (1974):1164. Copyright
1974 by AAAS.]

various climatic controls as if each acted independently of the others. This is
clearly not the case. Within the Earth-atmosphere system, many factors are
linked in complex cause–effect chains, as illustrated in Figure 18.17. Internal
and external forces probably work together to induce climatic shifts.

An illustration of how several factors might interact to shape large-scale
variations in climate is provided by J. E. Hansen and his colleagues at NASA.
Using a numerical model of the atmosphere, they predicted how rising carbon
dioxide levels would affect mean global temperature from 1880 to 1980. In
Figure 18.18A, the predicted temperature is compared with the actual variation in global temperature for the same period. The fit between the predicted
and actual temperature variation improves substantially when the influence
of volcanic aerosols on temperature is added (Figure 18.18B). The similarity
between the two curves becomes even closer when the influence of solar
irradiance is entered into the model (Figure 18.18C). Even with these three
climatic controls considered together, however, the two curves do not match,
suggesting that additional factors are affecting global climate variations.

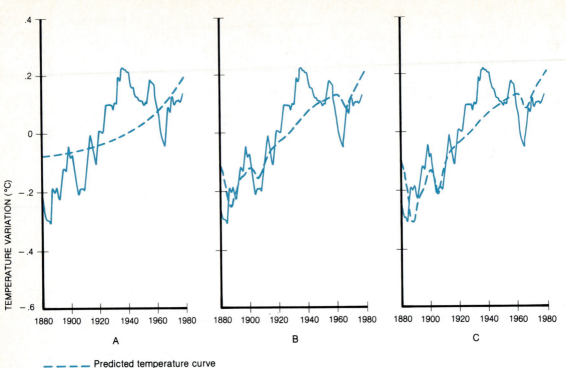

Factor interactions involve feedback loops that may at one extreme amplify (positive feedback) and at the other extreme weaken (negative feedback) climatic fluctuations. Consider an illustration. If rising carbon dioxide level accentuates the greenhouse effect, tropospheric temperatures will rise, and the pack ice cover in polar regions will be reduced. Less pack ice, in turn, will lower the albedo of polar seas, further warming the troposphere (positive feedback). On the other hand, higher temperatures at the Earth's surface mean more evaporation, which may result in thicker and more persistent cloud cover. Recall from Chapter 2 that clouds have both cooling and warming effects on the lower atmosphere. The relatively high albedo of cloud tops causes cooling, and cloud absorption and reradiation of terrestrial IR contributes to the greenhouse effect. Recent satellite measurements indicate that the cooling effect dominates. Thus, a thicker and more persistent cloud cover would cool the Earth's surface (negative feedback).

In part because of factor interaction, it is difficult to isolate simple cause–effect relationships. Even when a near-perfect match in periodicities occurs between some internal or external forcing factor (such as a volcanic eruption) and a climatic response, there is no guarantee that the relationship is actually one of cause and effect. In the absence of a demonstrated physical linkage, the relationship may well be coincidental.

The Climatic Future

What does the climatic future hold for us? Theoretically, we could take two approaches in attempting to answer this question. The first is numerical modeling of the atmosphere. Numerical models have been used to forecast both the climatic future and the climatic past. Typically, these predictions identify broad regions of expected positive and negative climatic anomalies. Earlier, we discussed the use of numerical models to predict the impact of rising levels of atmospheric CO_2. Numerical models have also been used to "predict" global climate 18,000 years ago, the last glacial maximum. However, most modelers agree that global climate models are in need of considerable refinement. For one, today's models may not adequately simulate the role of oceans in climate variability. Second, they may not accurately portray local and regional conditions and thereby may miss important feedback processes such as produced by clouds. The latter problem stems from the fact that climate models use data grids that partition the atmosphere into cells, each covering an area about the size of Colorado. A mean state of the atmosphere (temperature, humidity, cloud cover, for example) is computed for each cell. This problem of low resolution can only be solved by faster and larger supercomputers.

A second approach to climatic forecasting is to analyze the various factors that may have contributed to past climatic fluctuations and to extrapolate their influence into the future. This is an empirical method. Atmospheric scientists have long probed climatic records in search of cycles that might be extrapolated into the future. As noted earlier, however, none of the statistically significant oscillations that appear in the climatic record has much practical value for climate (or weather) forecasting.

If all other controls of climate remain the same, then increasing concentration of atmospheric carbon dioxide and other infrared-absorbing gases is likely to cause global warming. To what extent all other controls remain the same, however, is open to question. The climatic future will evidently remain a mystery until we have more accurate numerical models of the atmosphere, denser weather observation networks, and a better understanding of how climatic controls interact. It is likely that the climatic future, like the climatic past, will be shaped by many interacting factors both external and internal to the Earth-atmosphere system. At present, we are confident only that our climate will vary (as it has in the past), but precisely how it will vary is simply not known.

Conclusions

The interactions of many factors causes climate to vary on many time scales. Although we can isolate specific climatic controls that are internal or external to the Earth-atmosphere system, our understanding of how these controls interact is far from complete. This state of the art limits the ability of atmospheric scientists to forecast the climatic future. We should therefore be wary of simplistic scenarios of the climatic future, for not enough is known about the causes of climatic variations.

Continued research on climate is needed. It is reasonable to assume that climatic change is governed by physical laws; that is, variations in climate are not arbitrary, random events. As scientists more fully comprehend the laws governing climatic variability, their ability to predict the climatic future will improve. Meanwhile, trends in climate must be monitored closely, especially in view of the strong dependence of our food supply and energy demands on climate. In spite of the scientific uncertainty, some experts argue that we should plan now for the worst possible future climate scenarios. What is your opinion?

Summary Statements

Because of the rapid deterioration of climatic detail as we go back in time, the most reasonable view of ancient climates is in terms of average climates prevailing over long periods of time.

Continental drift and mountain building complicate our investigation of the Earth's climatic record over the hundreds of millions of years that constitute geologic time.

The climatic record becomes more detailed and reliable as we approach the present, especially since the beginning of instrument-based records and standardized weather observation practices.

The people most vulnerable to climatic fluctuations are those who live in areas where the climate is just marginal for survival. In such areas even small changes in climate can be disastrous.

How we view the rate of climatic change depends on our time frame of reference, but climate can change so rapidly that humankind may be unable to adjust.

Few statistically significant cycles have been identified in the climatic record, and none of these has much practical value for weather or climate forecasting.

Climatic variability may be explained by forces both internal and external to the Earth-atmosphere system that affect the global radiation balance.

One approach to explaining climatic variability is to match an appropriate cause with a specific climatic oscillation on the basis of the period of that oscillation. Such a match is, however, no guarantee of a real physical relationship.

Measurements by instruments aboard the SMM and other satellites indicate that the solar constant is not constant. Changes in solar irradiance may be linked to sunspot activity.

The exact linkage between sunspots and climate variability is not known, although correspondences may exist between prolonged sunspot minima and relatively cool periods and between the "double" sunspot cycle and drought on the western High Plains.

Regular long-term changes in Earth–sun geometry (precession, axial tilt, eccentricity) may explain

Radiative Equilibrium and Climatic Change

We can express the condition of radiative equilibrium within the Earth-atmosphere system in a mathematical expression that equates energy input to energy output. Then, by perturbing the variables in the equation, we can simulate climatic change.

Radiation incident on the Earth-atmosphere system is given by the solar constant, S. A portion of this energy is reflected or scattered back into space, the amount depending on the planetary albedo, α. The energy actually available to drive the atmosphere is

$$S(1 - \alpha)$$

Planet Earth intercepts this energy as a disk having an area of πR^2, where R is the radius of the Earth. The net energy input is then given by

$$(\pi R^2)S(1 - \alpha)$$

Energy is emitted by the Earth-atmosphere system in the form of infrared radiation. If we assume the Earth-atmosphere system to be a perfect radiator, then we can describe the energy output by the *Stefan–Boltzmann law* (see the Mathematical Note at the end of Chapter 2). This law states that the total radiational energy output (at all wavelengths) of a blackbody is proportional to the radiating temperature, T (in kelvins), raised to the fourth power, that is, T^4. In this relationship, the constant of proportionality is σ, the Stefan–Boltzmann constant. However, the Earth-atmosphere system is not a blackbody, so we must introduce a correction factor to account for its actual radiating properties. The correction factor, ϵ, is called the *effective emissivity*. The energy output of the Earth-atmosphere system is therefore

$$\epsilon \sigma T^4$$

This energy is emitted from the entire surface area $(4\pi R^2)$ of the spherical Earth. The total energy output is thus given by

$$(4\pi R^2)\epsilon \sigma T^4$$

At radiative equilibrium, what comes in (energy input) must equal what goes out (energy output), so

$$(\pi R^2)S(1 - \alpha) = (4\pi R^2)\epsilon \sigma T^4$$

This simplifies to

$$S(1 - \alpha) = 4\epsilon \sigma T^4$$

Solving for T, the temperature of the Earth-atmosphere system at radiative equilibrium, we have

$$T = [S(1 - \alpha)/4\epsilon \sigma]^{1/4}$$

The radiative equilibrium temperature thus depends on (1) the solar constant, (2) the planetary albedo, and (3) the effective emissivity. A change in any one or any combination of these variables will change the value of T and hence the climate. Consider some illustrations.

First, keep α and ϵ constant. By mathematical manipulation, we can show that a 1 percent change in the solar constant translates into a 0.6 C° (1.1 F°) change in radiative equilibrium temperature. This does not appear to be a major temperature change, but recall that climatic trends are geographically nonuniform in magnitude (and direction). In some localities, therefore, this temperature change will be amplified considerably. If we keep S and ϵ constant and increase the planetary albedo from its present value of 30 percent up to 35 percent (perhaps by increasing the ocean's ice cover), the radiative equilibrium temperature will drop by about 4.5 C° (8.1 F°).

As a third illustration, we could hold S and α constant and vary the effective emissivity. As noted earlier, the effective emissivity depends on the radiative properties of the Earth-atmosphere system, so any change in the chemistry of the Earth-atmosphere system could alter ϵ. Such chemical changes might include, for example, alterations in levels of atmospheric carbon dioxide or aerosols.

large-scale glacial-interglacial climatic shifts as recorded in deep-sea sediment cores.

The potential impact of violent volcanic eruptions on large-scale climate is negligible unless the eruptive cloud is relatively rich in sulfur oxide gases. In the stratosphere, these gases convert to sulfuric acid droplets that increase the planetary albedo.

Fossil fuel combustion is raising atmospheric carbon dioxide levels. Uncompensated, this trend will likely intensify the greenhouse effect. CO_2-induced warming may be enhanced by rising levels of other infrared-absorbing trace gases that are active within atmospheric windows.

The potential global climatic impact of elevated aerosol levels in the troposphere is not known.

Changes in the radiative characteristics of the Earth's land and water surfaces or in the relative distribution of land and sea may influence the global radiation balance and, hence, the climate.

Factor interactions involve feedback loops that might at one extreme amplify and at the other extreme weaken climatic oscillations.

With regard to the climatic future, atmospheric scientists are confident only that the climate will vary (as it has in the past), but precisely how it will vary is simply not known.

Key Words

geologic time	climatic optimum	umbra	Spörer minimum
continental drift	Little Ice Age	penumbra	Milankovitch cycles
glacial climate	solar irradiance	photosphere	turbidity
interglacial climate	sunspots	Maunder minimum	

Review Questions

1. How do we know that climate is inherently variable?

2. What happens to the reliability of the climatic record as we go back in time? Why?

3. How might continental drift and mountain building influence the climate of some locality over millions of years?

4. What is the difference between a glacial climate and an interglacial climate?

5. Characterize the climatic optimum and the Little Ice Age.

6. What considerations have led some scientists to question the validity of global and hemispheric temperature trends compiled from measurements made over the past 100 years?

7. Provide some examples of the societal impact of climatic variability.

8. The people most vulnerable to climatic change are those living in areas where the climate is just marginal for survival. What is meant by this statement? Give some illustrations.

9. Are there cycles in the climatic record that are useful for forecasting the climatic future? Explain.

10. What is meant by the globe's radiative equilibrium?

11. What is a sunspot? Describe its structure.

12. Is the sunspot record really periodic in a rigorous statistical sense? Explain why or why not.

13. What is the Maunder minimum, and what might have been its influence on climate?

14. How does the Milankovitch theory seek to explain the major climatic shifts of the Ice Age?

15. How might violent volcanic eruptions influence global climate?

16. Identify those climatic forcing phenomena that might contribute to short-term variations in climate, that is, variations over years and decades.

17. Evaluate humankind's contributions thus far to climatic variability. How might this change in the future?

18. What human activities are contributing to the upward trend in atmospheric carbon dioxide concentration?

19. How is the ocean involved in the controversy now surrounding the potential climatic impacts of rising carbon dioxide levels?

20. Explain why aerosols injected into the stratosphere may have a more prolonged impact on climate than aerosols that are confined to the troposphere.

Points to Ponder

1. A high statistical correlation between two factors does not necessarily indicate a cause–effect relationship. Give an example of how this statement applies to the search for the causes of climatic variability.

2. If the Milankovitch theory is correct, then the next ice age will peak about 23,000 years from now. Does this imply that each successive year from now on will get progressively colder? Explain your response.

3. How does the geographic nonuniformity of climatic trends complicate the search for causes of climatic variability?

4. In general terms, describe how sensitivity to climatic change varies geographically. Speculate on

why polar regions appear to be more sensitive than other areas of the globe to large-scale climatic change.

5. In attempting to explain climatic variability, we cannot focus simply on changes that occur in the atmosphere alone. We must also consider changes in the oceans and in the Earth's land surface characteristics. Explain.

6. How might ocean currents play an important role in climatic variability?

7. From what you have learned about conditions required for hurricane development (Chapter 14), speculate on the possible effects of CO_2-induced warming on the intensity and frequency of hurricanes.

Projects

1. Is there any indication that recent climatic trends have had any impact on your region's economy?

2. Speculate on the possible economic impact in your area of future climatic (a) cooling and (b) warming.

Selected Readings

Brooks, C.E.P. *Climate Through the Ages.* New York: Dover Publications, 1970, 395 pp. Represents a classic treatment of Earth's climatic history first published in 1926 and revised in 1949.

Bryson, R. A., and T. J. Murray. *Climates of Hunger.* Madison, Wis.: University of Wisconsin Press, 1977. 171 pp. Describes how past variations in climate have influenced people living in agriculturally marginal regions of the world.

Cohen, S. J., and T. R. Allsopp. "The Potential Impacts of a Scenario of CO_2-Induced Climatic Change on Ontario, Canada." *Journal of Climate* 1 (1988): 669–681. Provides a fascinating summary of the potential economic impacts of climatic change that may accompany a doubling of atmospheric carbon dioxide.

COHMAP Members. "Climatic Changes of the Last 18,000 Years: Observations and Model Simulations." *Science* 241 (1988):1043–1052. Reports on the appli-

cation of global atmospheric circulation models in an extensive investigation of past climates by a multi-institutional consortium of scientists.

Covey, C. "The Earth's Orbit and the Ice Ages." *Scientific American* 250, No. 2 (1984):58–66. Describes the Milankovitch theory of the link between Earth–sun geometry and major fluctuations in continental-scale glaciers.

Jones, P. D., and T. M. L. Wigley. "Global Warming Trends." *Scientific American* 263, No. 2 (1990):84–91. Reviews problems involved in computing global temperature trends.

Le Roy Ladurie, E. *Times of Feast, Times of Famine: A History of Climate Since the Year 1000.* New York: Doubleday, 1971. 426 pp. Includes a fascinating reconstruction of the climatic past based primarily on documentary sources.

McGovern, T. H. "The Economics of Extinction in Norse Greenland." in T.M.L. Wigley et al. *Climate and*

History. New York: Cambridge University Press, 1981, pp. 404–443. Describes the various factors that probably contributed to the decline of the Norse settlements in Greenland.

Ramanathan, V. "The Greenhouse Theory of Climate Change: A Test by an Inadvertent Global Experiment." *Science* 240 (1988):293–299. Thoroughly reviews the possible climatic effects of rising levels of atmospheric CO_2 and other IR-absorbing trace gases.

Rampino, M. R., and S. Self. "The Atmospheric Effects of El Chichón." *Scientific American* 250, No. 1 (1984):48–57. Reviews the possible global climatic significance of this 1982 volcanic eruption.

Revelle, R. "Carbon Dioxide and World Climate." *Scientific American* 247, No. 2 (1982):35–43. Presents an excellent overview of the atmospheric carbon dioxide problem and its potential implications.

Rotberg, R. I., and T. K. Rabb. *Climate and History.* Princeton, N.J.: Princeton University Press, 1981, 280 pp. Contains a collection of informative articles primarily on the methodology of climatic reconstruction.

Schneider, S. H. "Climate Modeling." *Scientific American* 256, No. 5 (1987):72–80. Presents an interesting critique of climate modeling as applied to projected trends in atmospheric carbon dioxide level.

Schneider, S. H. *Global Warming.* San Francisco: Sierra Club Books, 1989, 317 pp. Discusses the causes and possible consequences of an enhanced greenhouse effect.

Turco, R. P., et al. "Climate and Smoke: An Appraisal of Nuclear Winter." *Science* 247 (1990):166–175. Updates what is understood about the potential global climatic impact of nuclear war.

White, R. M. "Greenhouse Policy and Climate Uncertainty." *Bulletin of the American Meteorological Society* 70 (1989):1123–1127. Provides a concise summary of the problems involved when public policy-making decisions are based on uncertain climate forecasts.

Milestones in the History of Meteorology (to 1950)

ca.	500	B.C.	Classification of world climates by latitude as torrid, temperate, and frigid (Parmenides).
ca.	400		Rainfall measured in India.
	334		Aristotle's *Meteorologica* was the first work on the atmospheric sciences.
	61	A.D.	Seneca complained of the air pollution of Rome.
	1442		Rain gauges in use in Korea.
ca.	1500		Leonardo Da Vinci invented the hygrometer.
	1643		Torricelli invented the barometer.
	1644		First weather records in America were made by Rev. John Campanius near present site of Wilmington, DE.
	1658		Perier ascended the Puy-de-Dome, France, and demonstrated the decline of pressure with altitude.
	1660		Von Guericke noted that a severe storm follows a sudden drop in barometric pressure.
ca.	1670		Mercury used in thermometers for first time.
	1683		Edmund Halley published first comprehensive map of winds.
	1687		Isaac Newton developed his three laws of motion.
	1743		Benjamin Franklin deduced the progressive movement of a hurricane.
	1752		Benjamin Franklin demonstrated that lightning is an electrical phenomenon.
	1760		Joseph Black formulated the concept of latent heat.
	1781		Systematic weather records began at New Haven, CT.
	1800		Sir William Herschel discovered energy transfer in the infrared.
	1802		Classification of cloud types by Luke Howard.
	1806		Sir Francis Beaufort developed a wind scale.
	1825		E. F. August developed the psychrometer.
	1835		Gustav Gaspard de Coriolis demonstrated quantitatively the effects of rotational forces.
	1843		Discovery of the sunspot "cycle."
	1844		Invention of the aneroid barometer.

1856	William Ferrel proposed a scheme of the general circulation of the atmosphere using three cells.
1857	Lorin Blodget published *Climatology of the United States.*
1869	Cleveland Abbe in Cincinnati prepared first regular weather maps for part of the United States.
1878	Establishment of the International Meteorological Organization.
1884	Ludwig Boltzmann derived the Stefan–Boltzmann law.
1894	Wilhelm Wien developed the displacement law of radiation.
1916	Pyranometer for measuring global radiation was developed.
1917	Vilhelm Bjerknes formulated polar front theory.
1918	Wladimir Köppen introduced his climatic classification scheme.
ca. 1928	The first radiosonde developed.
1933	Tor Bergeron published paper on "Physics of Cloud and Precipitation."
1937	Carl-Gustaf Rossby introduced techniques for forecasting upper westerly waves.
1941	Development of weather radar.
1944	Hurd Curtis Willett produced atmospheric cross sections showing the jet stream.
1946	Vincent J. Schaefer and Irving Langmuir performed the first cloud-seeding experiments.
1946	John von Neumann began mathematical modeling of weather.
1947	Establishment of World Meteorological Organization.
1948	Beginning of rocket probes of upper atmosphere.

Modified from J. F. Griffiths. "A Chronology of Items of Meteorological Interest." *Bulletin of the American Meteorological Society* 58 (1977):1058–1067.

II

The Standard Atmosphere

Altitude (km)	Temperature (°C)	Pressure (mb)	P/P_0*	Density (kg/m^3)	D/D_0*
30.00	−46.60	11.97	0.01	0.02	0.02
25.00	−51.60	25.49	0.03	0.04	0.03
20.00	−56.50	55.29	0.05	0.09	0.07
19.00	−56.50	64.67	0.06	0.10	0.08
18.00	−56.50	75.65	0.07	0.12	0.09
17.00	−56.50	88.49	0.09	0.14	0.12
16.00	−56.50	103.52	0.10	0.17	0.14
15.00	−56.50	121.11	0.12	0.20	0.16
14.00	−56.50	141.70	0.14	0.23	0.19
13.00	−56.50	165.79	0.16	0.27	0.22
12.00	−56.50	193.99	0.19	0.31	0.25
11.00	−56.40	226.99	0.22	0.37	0.30
10.00	−49.90	264.99	0.26	0.41	0.34
9.50	−46.70	285.84	0.28	0.44	0.36
9.00	−43.40	308.00	0.30	0.47	0.38
8.50	−40.20	331.54	0.33	0.50	0.40
8.00	−36.90	356.51	0.35	0.53	0.43
7.50	−33.70	382.99	0.38	0.56	0.45
7.00	−30.50	411.05	0.41	0.59	0.48
6.50	−27.20	440.75	0.43	0.62	0.50
6.00	−23.90	472.17	0.47	0.66	0.54
5.50	−20.70	505.39	0.50	0.70	0.57
5.00	−17.50	540.48	0.53	0.74	0.60
4.50	−14.20	577.52	0.57	0.78	0.63
4.00	−11.00	616.60	0.61	0.82	0.67
3.50	−7.70	657.80	0.65	0.86	0.70
3.00	−4.50	701.21	0.69	0.91	0.74
2.50	−1.20	746.91	0.74	0.96	0.78
2.00	2.00	795.01	0.78	1.01	0.82
1.50	5.30	845.59	0.83	1.06	0.86
1.00	8.50	898.76	0.89	1.11	0.91
0.50	11.80	954.61	0.94	1.17	0.95
0.00	15.00	1013.25	1.00	1.23	1.00

*P/P_0 = ratio of air pressure to sea-level value; D/D_0 = ratio of air density to sea-level value.

Weather Map Symbols

BY INTERNATIONAL agreement, a standard set of symbols is plotted on weather maps to represent weather conditions. This standardization facilitates the international exchange of weather information. Presented here is an abridged listing of weather symbols. We also include in this appendix a hurricane tracking chart.

AIR PRESSURE TENDENCY

Rising, then falling; same as or higher than 3 hours ago

Barometric pressure now higher than 3 hours ago

Rising, then steady; or rising, then rising more slowly

Rising steadily, or unsteadily

Falling or steady, then rising; or rising, then rising more rapidly

Steady; same as 3 hours ago

Falling, then rising; same as or lower than 3 hours ago

Barometric pressure now lower than 3 hours ago

Falling, then steady; or falling, then falling more slowly

Falling steadily, or unsteadily

Steady or rising, then falling; or falling, then falling more rapidly

CLOUD ABBREVIATIONS

St stratus

Fra fractus

Sc stratocumulus

Ns nimbostratus

As altostratus

Ac altocumulus

Ci cirrus

Cs cirrostratus

Cc cirrocumulus

Cu cumulus

Cb cumulonimbus

543

CLOUD TYPES

Cu of fair weather, little vertical development and seemingly flattened

Cu of considerable development, generally towering, with or without other Cu or Sc bases all at same level

Cb with tops lacking clear-cut outlines, but distinctly not cirriform or anvil shaped; with or without Cu, Sc, or St

Sc formed by spreading out of Cu; Cu often present also

Sc not formed by spreading out of Cu

St or StFra, but no StFra of bad weather

StFra and/or CuFra of bad weather (scud)

Cu and Sc (not formed by spreading out of Cu) with bases at different levels

Cb having a clearly fibrous (cirriform) top, often anvil shaped, with or without Cu, Sc, St, or scud

Thin As (most of cloud layer semitransparent)

Thick As, greater part sufficiently dense to hide sun (or moon), or Ns

Thin Ac, mostly semitransparent; cloud elements not changing much and at a single level

Thin Ac in patches; cloud elements continually changing and/or occurring at more than one level

Thin Ac in bands or in a layer gradually spreading over sky and usually thickening as a whole

Ac formed by the spreading out of Cu or Cb

Double-layered Ac, or a thick layer of Ac, not increasing; or Ac with As and/or Ns

Ac in the form of Cu-shaped tufts or Ac with turrets

Ac of a chaotic sky, usually at different levels; patches of dense Ci usually present also

Filaments of Ci, or "mares tails," scattered and not increasing

Dense Ci in patches or twisted sheaves, usually not increasing, sometimes like remains of Cb; or towers or tufts

Dense Ci, often anvil shaped, derived from or associated with Cb

Ci, often hook shaped, gradually spreading over the sky and usually thickening as a whole

Ci and Cs, often in converging bands, or Cs alone; generally overspreading and growing denser; the continuous layer not reaching 45° altitude

Ci and Cs, often in converging bands, or Cs alone; generally overspreading and growing denser; the continuous layer exceeding 45° altitude

Veil of Cs covering the entire sky

Cs not increasing and not covering entire sky

Cc alone or Cc with some Ci or Cs, but the Cc being the main cirriform cloud

CLOUD COVER

◯	No clouds
	One tenth or less
	Two tenths or three tenths
	Four tenths
	Five tenths
	Six tenths
	Seven tenths or eight tenths
	Nine tenths or overcast with openings
●	Completely overcast (ten tenths)
⊗	Sky obscured

WIND SPEED

	Knots	Miles per hour	Kilometers per hour
◎	Calm	Calm	Calm
	1–2	1–2	1–3
	3–7	3–8	4–13
	8–12	9–14	14–19
	13–17	15–20	20–32
	18–22	21–25	33–40
	23–27	26–31	41–50
	28–32	32–37	51–60
	33–37	38–43	61–69
	38–42	44–49	70–79
	43–47	50–54	80–87
	48–52	55–60	88–96
	53–57	61–66	97–106
	58–62	67–71	107–114
	63–67	72–77	115–124
	68–72	78–83	125–134
	73–77	84–89	135–143
	103–107	119–123	192–198

FRONTS

Fronts are shown on surface weather maps by the symbols below. (Arrows—not shown on maps—indicate direction of motion of front.)

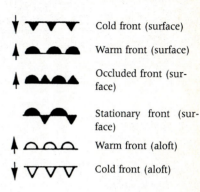

Cold front (surface)

Warm front (surface)

Occluded front (surface)

Stationary front (surface)

Warm front (aloft)

Cold front (aloft)

WEATHER CONDITIONS

Cloud development NOT observed or NOT observable during past hour

Clouds generally dissolving or becoming less developed during past hour

State of sky on the whole unchanged during past hour

Clouds generally forming or developing during past hour

Visibility reduced by smoke

Light fog (mist)

Patches of shallow fog at station, NOT deeper than 6 feet on land

More or less continuous shallow fog at station, NOT deeper than 6 feet on land

Lightning visible, no thunder heard

Precipitation within sight, but NOT reaching the ground

Drizzle (NOT freezing) or snow grains (NOT falling as showers) during past hour, but NOT at time of observation

Rain (NOT freezing and NOT falling as showers) during past hour, but NOT at time of observation

Snow (NOT falling as showers) during past hour, but NOT at time of observation

Rain and snow or ice pellets (NOT falling as showers) during past hour, but NOT at time of observation

Freezing drizzle or freezing rain (NOT falling as showers) during past hour, but NOT at time of observation

Slight or moderate dust storm or sandstorm, has decreased during past hour

Slight or moderate dust storm or sandstorm, no appreciable change during past hour

Slight or moderate dust storm or sandstorm has begun or increased during past hour

Severe dust storm or sandstorm, has decreased during past hour

Severe dust storm or sandstorm, no appreciable change during past hour

Fog or ice fog at distance at time of observation, but NOT at station during past hour

Fog or ice fog in patches

Fog or ice fog, sky discernible, has become thinner during past hour

Fog or ice fog, sky NOT discernible, has become thinner during past hour

Fog or ice fog, sky discernible, no appreciable change during past hour

Intermittent drizzle (NOT freezing), slight at time of observation

Continuous drizzle (NOT freezing), slight at time of observation

Intermittent drizzle (NOT freezing), moderate at time of observation

Continuous drizzle (NOT freezing), moderate at time of observation

Intermittent drizzle (NOT freezing), heavy at time of observation

Intermittent rain (NOT freezing), slight at time of observation

Continuous rain (NOT freezing), slight at time of observation

Intermittent rain (NOT freezing), moderate at time of observation

Continuous rain (NOT freezing), moderate at time of observation

Intermittent rain (NOT freezing), heavy at time of observation

Intermittent fall of snowflakes, slight at time of observation

Continuous fall of snowflakes, slight at time of observation

Intermittent fall of snowflakes, moderate at time of observation

Continuous fall of snowflakes, moderate at time of observation

Intermittent fall of snowflakes, heavy at time of observation

Slight rain shower(s)

Moderate or heavy rain shower(s)

Violent rain shower(s)

Slight shower(s) of rain and snow mixed

Moderate or heavy shower(s) of rain and snow mixed

Moderate or heavy shower(s) of hail, with or without rain, or rain and snow mixed, not associated with thunder

Slight rain at time of observation; thunderstorm during past hour, but NOT at time of observation

Moderate or heavy rain at time of observation; thunderstorm during past hour, but NOT at time of observation

Slight snow, or rain and snow mixed, or hail at time of observation; thunderstorm during past hour, but NOT at time of observation

Moderate or heavy snow, or rain and snow mixed, or hail at time of observation; thunderstorm during past hour, but NOT at time of observation

Haze | Widespread dust in suspension in the air, NOT raised by wind, at time of observation | Dust or sand raised by wind at time of observation | Well-developed dust whirl(s) within past hour | Dust storm or sandstorm within sight of or at station during past hour

Precipitation within sight, reaching the ground but distant from station | Precipitation within sight, reaching the ground, near to but NOT at station | Thunderstorm, but no precipitation at the station | Squall(s) within sight during past hour or at time of observation | Funnel cloud(s) within sight of station at time of observation

Showers of rain during past hour, but NOT at time of observation | Showers of snow, or of rain and snow, during past hour, but NOT at time of observation | Showers of hail, or of hail and rain, during past hour, but NOT at time of observation | Fog during past hour, but NOT at time of observation | Thunderstorm (with or without precipitation) during past hour, but NOT at time of observation

Severe dust storm or sandstorm has begun or increased during past hour | Slight or moderate drifting snow, generally low (less than 6 ft) | Heavy drifting snow, generally low | Slight or moderate blowing snow, generally high (more than 6 ft) | Heavy blowing snow, generally high

Fog or ice fog, sky NOT discernible, no appreciable change during past hour | Fog or ice fog, sky discernible, has begun or become thicker during past hour | Fog or ice fog, sky NOT discernible, has begun or become thicker during past hour | Fog depositing rime, sky discernible | Fog depositing rime, sky NOT discernible

Continuous drizzle (NOT freezing), heavy at time of observation | Slight freezing drizzle | Moderate or heavy freezing drizzle | Drizzle and rain, slight | Drizzle and rain, moderate or heavy

Continuous rain (NOT freezing), heavy at time of observation | Slight freezing rain | Moderate or heavy freezing rain | Rain or drizzle and snow, slight | Rain or drizzle and snow, moderate or heavy

Continuous fall of snowflakes, heavy at time of observation | Ice prisms (with or without fog) | Snow grains (with or without fog) | Isolated starlike snow crystals (with or without fog) | Ice pellets or snow pellets

Slight snow shower(s) | Moderate or heavy snow shower(s) | Slight shower(s) of snow pellets, or ice pellets with or without rain, or rain and snow mixed | Moderate or heavy shower(s) of snow pellets, or ice pellets, or ice pellets with or without rain or rain and snow mixed | Slight shower(s) of hail, with or without rain or rain and snow mixed, not associated with thunder

Slight or moderate thunderstorm without hail, but with rain and/or snow at time of observation | Slight or moderate thunderstorm, with hail at time of observation | Heavy thunderstorm, without hail, but with rain and/or snow at time of observation | Thunderstorm combined with dust storm or sandstorm at time of observation | Heavy thunderstorm with hail at time of observation

KEY TO AVIATION WEATHER OBSERVATIONS

LOCATION IDENTIFIER TYPE AND TIME OF REPORT *	SKY AND CEILING	VISIBILITY WEATHER AND OBSTRUCTION TO VISION	SEA-LEVEL PRESSURE	TEMPERATURE AND DEW POINT	WIND	ALTIMETER SETTING	REMARKS AND CODED DATA
MCI SA Ø758	15 SCT M25 OVC	1R-F	132	/58/56	/1897	/993/	RØ1VR2ØV4Ø

SKY AND CEILING

Sky cover contractions are for each layer in ascending order. Figures preceding contractions are base heights in hundreds of feet above station elevation. Sky cover contractions used are:

CLR = Clear: Less than Ø.1 sky cover.
SCT = Scattered: Ø.1 to Ø.5 sky cover.
BKN = Broken: Ø.6 to Ø.9 sky cover.
OVC = Overcast: More than Ø.9 sky cover.

— = Thin (When prefixed to SCT, BKN, OVC).
—X = Partly obscured: Ø.9 or less of sky hidden by precipitation or obstruction to vision (bases at surface).
X = Obscured: 1.Ø sky hidden by precipitation or obstruction to vision (bases at surface).

A letter preceding the height of a base identifies a ceiling layer and indicates how ceiling height was determined. Thus:

E = Estimated
M = Measured
W = Vertical visibility into obscured sky

V = Immediately following the height of a base indicates a variable ceiling.

VISIBILITY

Reported in statute miles and fractions (V = Variable)

WEATHER AND OBSTRUCTION TO VISION SYMBOLS

A	Hail	IC	Ice crystals
BD	Blowing dust	IF	Ice-fog
BN	Blowing sand	IP	Ice pellets
BS	Blowing snow	IPW	Ice pellet showers
D	Dust	K	Smoke
F	Fog	L	Drizzle
GF	Ground fog	R	Rain
H	Haze	RW	Rain showers

S	Snow		
SG	Snow grams		
SP	Snow pellets		
SW	Snow showers		
T	Thunderstorms		
T+	Severe thunderstorm		
ZL	Freezing drizzle		
ZR	Freezing rain		

Precipitation intensities are indicated thus: - Light; (no sign) Moderate; + Heavy

WIND

Direction in tens of degrees from true north, speed in knots; ØØØØ indicates calm. G indicates gusty. Q indicates Squalls. Peak wind speed in the past 1Ø minutes follows G or Q when gusts or squalls are reported. The contraction WSHFT, followed by GMT time group in remarks, indicates windshift and its time of occurrence. (Knots x 1.15 = statute mi/hr.)

EXAMPLES: 3627 = wind from 36Ø Degrees at 27 knots;
3627G4Ø = wind from 36Ø Degrees at 27 knots, peak speed in gusts 4Ø knots

ALTIMETER SETTING

The first figure of the actual altimeter setting is always omitted from the report.

RUNWAY VISUAL RANGE (RVR)

RVR is reported from some stations. For planning purposes, the value range during 1Ø minutes prior to observations and based on runway light setting 5 are reported in hundreds of feet. Runway identification precedes RVR report.

PILOT REPORTS (PIREPs)

When available, PIREPs in fixed-format may be appended to weather observations. PIREPs are designated by UA or UUA for urgent PIREPs.

DECODED REPORT

Kansas City International: Record observation completed at Ø758 GMT 15ØØ feet scattered clouds, measured ceiling 25ØØ feet overcast. visibility 1 mile, light rain, fog, sea-level pressure 1Ø13.2 millibars, temperature 58ºF, dewpoint 56ºF, wind from 18Øº, at 7 knots, altimeter setting 29.93 inches. Runway Ø1, visual range 2ØØØ feet lowest 4ØØØ feet highest in the past 1Ø minutes.

* TYPE OF REPORT

SA = a scheduled record observation
SP = an unscheduled special observation indicating a significant change in one or more elements
RS = a scheduled record observation that also qualifies as a special observation

The designator for all three types of observations (SA, SP, RS) is followed by a 24 hour-clock-time-group in Greenwich Mean Time (GMT or Z).

U.S. DEPARTMENT OF COMMERCE—NATIONAL OCEANIC AND ATMOSPHERIC ADMINISTRATION—NATIONAL WEATHER SERVICE

KEY TO AVIATION WEATHER FORECASTS

TERMINAL FORECASTS contain information for specific airports on expected ceiling, cloud heights, cloud amounts, visibility, weather and obstructions to vision, and surface wind. They are issued 3 times/day, amended as needed, and are valid for up to 24 hours. The last six hours of each forecast period are covered by a categorical statement indicating whether VFR, MVFR, IFR or LIFR conditions are expected (L in LIFR and M in MVFR indicate "low" and "marginal"). Terminal forecasts are written in the following form:

CEILING: Identified by the letter "C" (for lowest layer with cumulative sky cover greater than 5/10).
CLOUD HEIGHTS: In hundreds of feet above the station (ground)
SKY COVER AMOUNT (including any obscuration)
CLOUD LAYERS: Stated in ascending order of height
VISIBILITY: In statute miles (omitted if over 6 miles)
WEATHER AND OBSTRUCTION TO VISION: Standard weather and obstruction to vision symbols are used
SURFACE WIND: In tens of degrees and knots (omitted when less than 6 knots)

EXAMPLE OF TERMINAL FORECAST

DCA 221010: DCA Forecast 22nd day of month - valid time 10Z-10Z.
10SCT C18 BKN 5SW—3415G25 OCNL C8 X 1/2SW: Scattered clouds at 1000 feet, ceiling 1800 feet broken, visibility 5 miles, light snow showers, surface wind from 340 degrees at 15 knots. Gusts to 25 knots, occasional ceiling 8 hundred feet sky totally obscured, visibility 1/2 mile in moderate snow showers.
12Z C50 BKN 3312G22: By 12Z becoming ceiling 5000 feet broken, surface wind 330 degrees at 12 knots. Gusts to 22.
04Z MVFR CIG: Last 6 hours of FT after 04Z marginal VFR due to ceiling.

AREA FORECASTS are 12-hour aviation forecasts plus a 6-hour categorical outlook prepared 3 times/day, with each section amended as needed, giving general descriptions of potential hazards, airmass and frontal conditions, icing and freezing level turbulence and low-level windshear and significant clouds and weather for an area the size of several states. Heights of cloud bases and tops, turbulence and icing are referenced ABOVE MEAN SEA LEVEL (MSL); unless indicated by Ceiling (CIG) or ABOVE GROUND LEVEL (AGL). Each SIGMET OR AIRMET affecting an FA area will also serve to amend the Area Forecast.

SIGMET, AIRMET and CWA messages (in-flight advisories) broadcast by FAA on NAVAID voice channels warn pilots of potentially hazardous weather. SIGMET's concern severe and extreme conditions of importance to all aircraft (i.e., icing, turbulence and dust storms/sandstorms or volcanic ash). Convective SIGMET's are issued for thunderstorms if they are sufficiently strong, wide spread or embedded. AIRMET's concern less severe conditions which may be hazardous to aircraft particularly smaller aircraft and less experienced or VFR only pilots. CWA's (Center Weather Advisories) concern both SIGMET and AIRMET type conditions described in greater detail and relating to a specific ARTCC area.

WINDS AND TEMPERATURES ALOFT (FD) FORECASTS are 6, 12, and 24-hour forecasts of wind direction (nearest 10° true N) and speed (knots) for selected flight levels. Forecast Temperatures Aloft (°C) are included for all but the 3000-foot level.

EXAMPLES OF WINDS AND TEMPERATURES ALOFT (FD) FORECASTS:
```
FD WBC 121645
BASED ON 121200Z DATA
VALID 130000Z FOR USE 2100-0600Z. TEMPS NEG ABV 24000 FT

       3000    6000    9000    12000   18000   24000   30000   34000   39000
BOS    3127    3425 17  342011  3421-11 3516-27 3512-38 311649  292451  281451
JFK    3026    3327 28  3324-12 3322-16 3122-27 2923-38 284248  285150  285749
```
At 6000 feet MSL over JFK wind from 330° at 27 knots and temperature minus 8°C.

TWEB (CONTINUOUS TRANSCRIBED WEATHER BROADCAST) - Individual route forecasts covering a 25-nautical-mile zone either side of the route. By requesting a specific route number, detailed en route weather for a 12- or 18-hour period (depending on forecast issuance) plus a synopsis can be obtained.

PILOTS. . . . report in-flight weather to nearest FSS. The latest surface weather reports are available by phone at the nearest pilot weather briefing office by calling at H+10.

U.S. DEPARTMENT OF COMMERCE—NATIONAL OCEANIC AND ATMOSPHERIC ADMINISTRATION—NATIONAL WEATHER SERVICE—REVISED JANUARY 1984

NOAA PA 73029

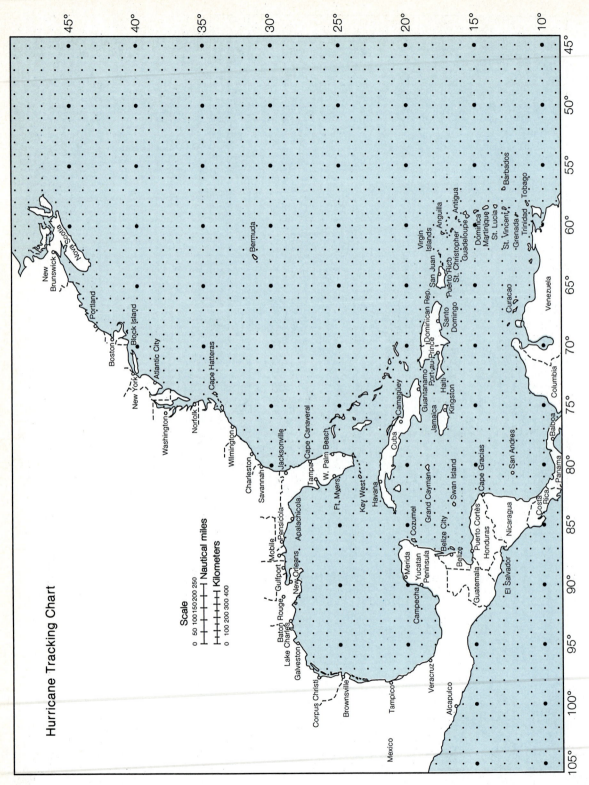

Hurricane Tracking Chart

Scale

0 50 100 150 200 250 Nautical miles

0 100 200 300 400 Kilometers

Weather Extremes

Temperature	
Maximum	
United States	57 °C (134 °F), Death Valley, CA, 10 July 1913
Canada	45 °C (113 °F), Midale and Yellow Grass, Sask., 5 July 1937
World	58 °C (136 °F), El Azizia, Libya, 13 September 1922
Minimum	
United States	−62.1 °C (−79.8 °F), Prospect Creek, AK, 23 January 1971
	−56.5 °C (−70 °F), Rogers Pass, MT, 20 January 1954
Canada	−63 °C (−81.4 °F), Snag, Yukon, 3 February 1947
World	−89 °C (−129 °F), Vostok, Antarctica, 21 July 1983
Precipitation	
24-hour maximum	
United States	1092 mm (43 in.), Alvin, Tx, 25–26 July 1979
Canada	489 mm (19.3 in.), Ucluelet, B.C., 6 October 1967
World	1880 mm (73.62 in.), Cilaos, La Reunion, Indian Ocean, 15–16 March 1952
One-month maximum	
United States	1817 mm (71.54 in.), Helin Mine, CA, January 1909
	2718 mm (107 in.), Kuki, Maui, HI, March 1942
Canada	2235.5 mm (88.01 in.), Swanson Bay, B.C., November 1917
World	9300 mm (366.14 in.), Cherrapunji, Assam, India, July 1861
One-year maximum	
United States	1878 cm (739 in.), Kuki, Maui, HI, December 1981–December 1982
Canada	8122 mm (319.8 in.), Henderson Lake, B.C., 1931
World	26461 mm (1041.78 in.), Cherrapunji, Assam, India, August 1860–August 1861
One-year minimum	
United States	0.0 mm (0.0 in.), Death Valley, CA, 1919
	0.0 mm (0.0 in.), Bagdad, CA, 3 October 1912–8 November 1914
Canada	81.2 mm (1.23 in.), Eureka, NWT, 1956
World	0.0 mm (0.0 in.), Arica, Chile, October 1903–December 1917

Snowfall
24-hour maximum
United States   192.5 cm (75.8 in.), Silver Lake, Boulder County, CO, 14
                                 −15 April 1921

Canada   118 cm (46 in.), Lakelse Lake, B.C., 17 January 1974
Single storm maximum
United States   480 cm (189 in.), Mt. Shasta Ski Bowl, CA, 13−19
                                 February 1959

One-month maximum
United States   991 cm (390 in.), Tamarack, CA, January 1911
Season maximum
United States   2850 cm (1122 in.), Paradise Ranger Station, Mt. Rainier,
                                 WA, 1971−72

Canada   2446.5 cm (964 in.), Revelstoke, Mt. Copeland, B.C.,
                                 1971−72

Atmospheric pressure
Maximum
United States   1068 mb (31.43 in.), Barrow, AK, 3 January 1970
                                 1063 mb (31.40 in.), Helena, MT, 9 January 1962
Canada   1068 mb (31.43 in.), Mayo, Yukon, 1 January 1974
World   1083.8 mb (32.005 in.), Agata, Siberia, USSR, 31
                                 December 1968

Minimum
United States   892 mb (26.35 in.), Matecumbe Key, FL, 3 September
                                 1935
Canada   940.2 mb (27.77 in.), St. Anthony, Nfld., 20 January 1977
World   870 mb (25.69 in.), Typhoon Tip, Pacific Ocean, 12
                                 October 1979

V

Selected Climatic Data for United States and Canada

T = daily average temperature (°C); P = monthly precipitation (millimeters).

		Jan	Feb	Mar	Apr	May	Jun	Jul	Aug	Sep	Oct	Nov	Dec	Annual
Montgomery, AL	T	8.3	9.9	13.9	18.3	22.2	26.1	27.8	27.2	24.9	18.3	12.8	9.4	18.3
	P	107	114	150	112	102	89	122	81	119	58	76	122	1250
Fairbanks, AK	T	−25	−20	−13.3	−1.1	8.9	15.0	16.7	13.9	7.2	−3.9	−15.6	−23.3	−3.3
	P	13	10	10	8	15	33	46	48	28	18	18	18	264
Phoenix, AZ	T	11.1	13.3	16.1	20.0	25.0	30.0	33.3	32.2	29.4	22.8	16.1	11.7	21.7
	P	18	15	20	8	3	5	18	25	15	15	13	20	180
Little Rock, AR	T	4.4	6.7	11.1	16.7	21.1	25.6	27.8	27.2	23.3	17.2	10.6	6.1	16.7
	P	99	97	119	137	135	94	91	79	109	71	112	107	1250
Los Angeles, CA	T	13.9	15.0	15.6	16.7	18.3	20.6	23.3	23.9	22.8	20.6	17.2	14.4	18.3
	P	94	76	61	30	5	1.0	0.0	3	8	5	46	51	376
San Francisco, CA	T	8.9	11.1	11.7	12.8	14.4	16.1	16.7	17.2	17.8	16.1	12.2	9.4	13.9
	P	117	81	66	38	8	3	1.0	1.0	5	28	61	91	500
Denver, CO	T	−1.1	1.1	3.3	8.3	13.9	19.4	22.9	21.7	17.2	11.1	3.9	0.6	10.0
	P	13	18	30	46	64	41	48	38	30	25	20	15	389
Hartford, CT	T	−3.9	−2.2	2.8	9.4	15.0	20.6	22.8	21.7	17.2	11.1	5.6	−1.7	10.0
	P	89	81	107	102	86	86	79	102	99	89	102	107	1128
Wilmington, DE	T	−0.6	0.6	5.6	11.1	16.7	21.7	24.4	23.9	20.0	13.3	7.8	2.2	12.2
	P	79	76	99	86	81	89	99	102	91	74	84	89	1052
Washington, DC	T	1.7	2.8	7.8	13.3	18.3	22.8	25.6	25.0	21.1	14.4	8.9	3.3	13.9
	P	81	71	97	86	99	102	107	130	91	81	79	86	1105
Miami, FL	T	19.4	20.0	22.2	23.9	25.6	27.2	27.8	28.3	27.8	25.6	22.8	20.0	24.4
	P	53	51	48	79	165	234	152	178	206	180	69	48	1463
Atlanta, GA	T	5.6	7.2	11.1	16.7	20.6	24.4	26.1	25.6	22.8	16.7	11.1	6.7	16.1
	P	124	112	150	112	102	86	119	86	81	64	86	107	1234
Honolulu, HI	T	22.8	22.8	23.3	24.4	25.6	26.1	26.7	27.2	27.2	26.7	25.0	23.3	25.0
	P	97	69	89	38	30	13	13	15	15	48	81	86	597
Boise, ID	T	−1.1	2.2	5.0	9.4	13.9	18.9	23.9	22.2	17.2	11.1	4.4	0.0	10.6
	P	41	28	25	30	30	25	8	10	15	20	33	33	297
Chicago, IL	T	−6.1	−3.3	2.2	9.4	15.0	20.6	22.8	22.2	18.3	12.2	4.4	−2.2	9.4
	P	41	33	66	94	81	104	91	89	86	58	53	53	848
Indianapolis, IN	T	−3.3	−1.1	4.4	11.1	16.7	22.2	23.9	22.8	19.4	12.8	5.6	0.0	11.1
	P	66	64	91	94	94	102	109	89	69	64	76	76	996
Des Moines, IA	T	−7.2	−4.4	1.7	9.9	16.7	22.2	24.4	23.3	18.3	12.2	3.9	−3.3	10.0
	P	25	28	56	81	102	108	81	104	79	56	38	25	782

		Jan	Feb	Mar	Apr	May	Jun	Jul	Aug	Sep	Oct	Nov	Dec	Annual
Topeka, KS	T	−3.3	0.0	5.5	12.8	18.3	23.3	26.1	25.0	20.0	13.9	6.1	0.0	12.2
	P	23	25	56	79	102	130	102	94	86	71	46	33	848
Louisville, KY	T	0.0	2.2	7.2	13.9	18.3	23.3	25.6	24.4	21.1	14.4	7.8	2.8	13.3
	P	86	81	119	104	107	91	104	84	86	66	89	89	1107
Baton Rouge, LA	T	11.1	12.2	15.6	20.0	23.9	26.7	27.8	27.2	25.6	20.0	15.0	11.7	20.0
	P	117	127	117	142	122	79	180	127	112	66	102	127	1417
Portland, ME	T	−5.6	−5.0	0.0	6.1	11.7	16.7	20.0	19.4	15.0	8.9	3.3	−3.3	7.2
	P	97	91	102	99	84	79	71	71	84	97	119	114	1105
Baltimore, MD	T	0.6	1.7	6.1	12.2	17.2	22.2	25.0	24.4	20.6	13.9	7.8	2.2	12.8
	P	74	74	94	86	86	99	99	117	89	79	79	86	1062
Boston, MA	T	−1.1	−0.6	3.3	9.4	15.0	20.0	23.3	22.2	18.3	12.8	7.2	1.1	11.1
	P	99	94	104	94	89	74	69	94	86	86	107	114	1113
Detroit, MI	T	−5.0	−3.3	1.7	8.9	14.4	20.0	22.2	21.1	17.2	11.1	4.4	−2.2	9.4
	P	48	43	64	81	71	86	79	81	56	53	58	64	1016
Minneapolis, MN	T	−11.7	−7.8	−1.7	7.8	14.4	20.0	22.8	21.7	16.1	10.0	0.6	−7.2	7.2
	P	20	20	43	51	81	104	89	91	64	46	33	23	671
Jackson, MS	T	7.8	9.4	13.3	18.3	22.8	26.1	27.8	27.2	24.4	18.3	12.8	9.4	18.3
	P	127	114	150	147	122	74	112	94	91	66	107	137	1341
Springfield, MO	T	0.0	2.2	7.2	13.3	18.3	22.8	25.6	25.0	21.1	14.4	7.2	2.2	13.3
	P	41	53	86	102	109	119	91	71	107	81	74	66	1003
Helena, MT	T	−7.8	−3.3	0.0	5.6	11.1	15.6	20.0	18.9	13.3	7.2	−0.6	−4.4	6.1
	P	15	10	18	28	43	51	25	30	46	15	13	15	290
Omaha, NE	T	−6.1	−2.7	2.8	11.1	17.2	22.8	25.6	23.9	18.9	13.3	4.4	−2.8	10.6
	P	20	48	48	74	109	102	91	104	89	53	33	20	770
Las Vegas, NV	T	7.2	10.0	12.8	17.8	22.8	28.9	32.2	31.1	26.6	20.0	12.2	7.2	18.9
	P	13	13	10	5	5	3	10	13	8	5	10	8	107
Concord, NH	T	−6.7	−5.6	0.0	6.7	12.8	18.3	21.1	19.4	15.0	8.9	2.7	−4.4	7.2
	P	71	64	74	76	74	74	74	84	79	79	94	86	927
Newark, NJ	T	−0.6	0.6	5.0	11.1	16.7	22.2	25.0	24.4	20.0	13.9	7.8	2.2	12.2
	P	79	76	107	91	91	74	97	109	94	79	91	86	1074
Albuquerque, NM	T	1.7	3.9	7.2	12.8	17.8	22.8	25.0	23.9	20.0	14.4	6.7	3.3	13.3
	P	10	10	13	10	13	13	33	38	20	23	10	13	206
New York, NY	T	0.0	0.6	5.0	11.1	16.7	21.7	25.0	23.9	20.0	14.4	8.3	2.2	12.2
	P	81	79	107	97	97	81	97	102	94	86	104	97	1120
Raleigh, NC	T	4.4	5.6	9.4	15.6	19.4	23.3	25.6	25.0	21.7	15.6	10.0	5.6	15.0
	P	91	86	94	74	94	94	112	112	84	69	74	79	2062
Bismarck, ND	T	−13.9	−10.0	−3.3	5.6	12.8	17.8	21.1	20.6	13.9	7.8	−1.7	−9.4	5.0
	P	13	10	18	38	56	76	51	51	43	20	13	13	391
Cleveland, OH	T	−3.3	−2.8	2.8	8.9	14.4	20.0	22.2	21.1	17.8	11.7	5.6	−0.6	10.0
	P	64	56	76	84	84	89	86	86	74	61	71	71	894
Oklahoma City, OK	T	2.2	5.0	9.4	15.6	20.0	25.0	27.8	27.2	22.8	16.7	9.4	4.4	15.6
	P	25	33	53	74	140	99	76	61	86	69	38	30	785
Portland, OR	T	3.9	6.1	7.8	10.0	13.9	16.7	20.0	19.4	17.2	12.2	7.8	5.0	11.7
	P	157	99	91	58	53	38	13	33	41	76	132	163	950
Pittsburgh, PA	T	−2.8	−1.7	3.3	10.0	15.6	20.0	22.2	21.7	17.8	11.1	5.6	−0.6	10.0
	P	74	61	91	84	89	84	97	84	71	64	58	66	922
Providence, RI	T	−2.2	−1.7	2.8	8.9	14.4	19.4	22.2	21.7	17.8	11.7	6.1	0.0	10.0
	P	104	94	109	102	89	71	76	102	89	97	107	114	1151
Columbia, SC	T	7.2	8.3	12.2	17.8	22.2	25.6	27.2	26.7	23.9	17.2	12.2	8.3	17.2
	P	109	102	132	91	97	112	137	142	107	66	64	89	1247
Rapid City, SD	T	−6.1	−3.3	0.6	7.2	13.3	18.3	22.8	21.7	16.1	10.0	1.7	−3.3	8.3
	P	10	15	25	51	66	84	53	36	25	20	13	10	414

		Jan	Feb	Mar	Apr	May	Jun	Jul	Aug	Sep	Oct	Nov	Dec	Annual
Nashville, TN	T	2.8	4.4	9.4	15.6	20.0	24.4	26.1	25.6	22.2	15.6	9.4	5.0	15.0
	P	114	102	142	114	117	94	97	86	94	66	89	117	1232
Dallas–Fort Worth,	T	6.7	8.9	13.3	18.9	23.3	27.8	30.0	30.0	26.1	20.0	13.3	8.9	18.9
TX	P	41	48	61	91	109	66	51	46	84	64	46	43	747
Houston, TX	T	11.1	12.8	16.1	20.6	23.9	27.2	28.3	28.3	25.6	21.1	15.6	12.2	20.0
	P	81	81	69	107	119	102	84	94	124	94	86	94	1138
Salt Lake City, UT	T	−1.7	1.1	5.0	9.4	15.0	14.4	25.0	23.9	18.3	11.7	4.4	−1.1	11.1
	P	36	33	43	56	38	25	18	23	23	28	30	36	388
Burlington, VT	T	−8.3	−7.8	−1.7	6.1	12.8	18.3	21.1	19.4	15.0	8.9	2.8	−5.0	6.7
	P	46	43	56	69	76	91	86	99	81	71	71	61	856
Richmond, VA	T	2.8	3.9	8.3	14.4	18.9	23.3	25.6	25.0	21.1	15.0	9.4	4.4	14.4
	P	81	79	91	74	91	91	130	127	89	94	84	86	1120
Seattle, WA	T	5.0	6.7	7.8	10.0	13.3	16.1	18.3	18.3	16.1	12.2	7.8	6.1	11.7
	P	150	107	94	64	43	38	23	36	51	86	137	160	1013
Charleston, WV	T	0.6	2.2	7.2	12.8	17.8	21.7	23.9	23.3	20.0	13.3	7.2	2.7	12.8
	P	89	79	102	89	94	84	137	107	76	66	74	84	1077
Madison, WI	T	−8.9	−6.7	−0.6	7.8	13.9	18.9	21.1	20.0	15.6	10.0	1.7	−5.5	7.2
	P	28	25	56	79	84	109	99	97	79	56	46	38	782
Cheyenne, WY	T	−3.3	−1.7	0.0	5.6	11.1	16.7	20.6	19.4	14.4	8.9	1.7	−1.7	7.8
	P	10	10	25	30	61	51	48	36	25	18	13	10	338
Calgary, Alb.	T	−9.9	−8.8	−4.4	3.6	9.8	13.0	16.7	15.1	10.9	5.4	−2.2	−6.6	3.6
	P	17	20	26	35	52	88	58	59	35	23	16	15	444
Prince George,	T	−11.3	−7.5	−2.3	4.3	9.7	12.9	14.9	13.7	10.1	4.8	−2.5	−6.6	3.3
B.C.	P	56	44	36	28	43	62	64	65	56	59	57	56	626
Winnipeg, Man.	T	−17.7	−15.5	−7.9	3.3	11.3	16.5	20.2	18.9	12.8	6.2	−4.8	−12.9	2.5
	P	26	21	27	30	50	81	69	70	55	37	29	22	517
Saint John, N.B.	T	−6.9	−6.4	−2.0	3.8	9.4	13.9	17.2	17.1	13.4	8.3	2.7	−4.3	5.4
	P	144	122	106	105	98	94	87	108	104	108	153	133	1362
Goose Bay, Nfld.	T	−16.6	−14.9	−8.4	−1.6	5.1	11.9	16.3	14.7	10.1	3.2	−4.4	−12.9	0.2
	P	72	63	68	62	56	72	84	91	76	63	67	63	837
Frobisher Bay,	T	−26.5	−25.5	−21.5	−13.7	−3.1	3.6	7.9	6.9	2.2	−4.7	−12.3	−20.5	−8.9
NWT	P	22	26	19	21	19	33	53	53	43	34	33	24	380
Halifax, N.S.	T	−3.3	−3.6	−0.1	4.8	9.9	14.4	18.5	18.8	15.5	10.3	5.3	−0.9	7.4
	P	141	119	113	112	109	94	94	96	117	120	143	126	1384
Toronto, Ont.	T	−3.9	−3.8	0.2	7.0	13.2	19.0	21.9	21.1	16.6	10.6	4.3	−1.8	8.7
	P	67	59	67	66	70	63	74	61	65	60	63	61	776
Montreal, Que.	T	−8.7	−7.8	−2.1	6.2	13.6	18.9	21.6	20.5	15.6	9.4	2.3	−5.9	6.9
	P	87	76	86	83	81	91	102	87	95	83	88	89	1048
Regina, Sask.	T	−16.9	−14.8	−8.1	3.4	11.2	15.3	19.3	17.8	11.9	5.1	−5.4	−12.3	2.2
	P	19	17	21	21	40	83	55	49	34	18	20	17	394
Whitehorse, Yukon	T	−18.1	−14.1	−7.6	−0.2	7.5	12.6	14.2	12.4	7.9	0.7	−8.2	−15.1	−0.7
	P	18	14	15	11	13	27	35	37	25	19	23	20	257

Data sources: National Climatic Center, Asheville, NC, and Atmospheric Environment Service, Canada.

GLOSSARY

absolute humidity The mass of water vapor per unit volume of air containing the water vapor; usually expressed as grams of water vapor per cubic meter of air.

absolute instability Property of an ambient air layer that is unstable for both saturated and unsaturated air parcels.

absolute stability Property of an ambient air layer that is stable for both saturated and unsaturated air parcels.

absolute temperature Temperature scale based on absolute zero and expressed in kelvins (K): 0 °C = 273 K, for example.

absolute zero The theoretical temperature at which a body emits no electromagnetic radiation and all molecular activity ceases (actually some atomic-level activity takes place); 0 K.

absorption The process whereby a portion of the radiation incident on an object is converted to heat.

absorptivity The efficiency of radiation absorption.

acid deposition The combination of acid rain (or snow) plus dry deposition.

acid rain Rain having a pH lower than 5.6, often due to the presence of sulfuric and nitric acids.

adiabatic process Expansional cooling or compressional warming of air parcels in which there is no net heat exchange between the air parcels and the surrounding (ambient) air.

advection Horizontal movement of air from one place to another.

advection fog Ground-level clouds generated by advective cooling of a mild, humid air mass as it travels over a relatively cool surface.

aerosols Tiny liquid or solid particles of various composition that occur suspended in the atmosphere.

AFOS *See* Automation of Field Operations and Service.

agroclimatic compensation Poor growing weather in one area is offset to some extent by better growing weather in other areas.

air density Mass per unit volume of air; about 1.275 kg per cubic meter at 0 °C and 1000 millibars.

air mass A huge volume of air covering thousands of square kilometers that is relatively uniform horizontally in temperature and water vapor concentration.

air mass advection Horizontal movement of air or air masses from one place to another.

air mass climatology Description of climate in terms of frequency of occurrence of various types of air masses.

air mass thunderstorms Thunderstorms that develop almost randomly within a mass of maritime tropical air; usually most frequent during the warmest time of day.

air parcel A unit mass of air—a single gram, for example.

air pollutants Gases and aerosols in air in concentrations that threaten the well-being of living organisms or that disrupt the orderly functioning of the environment.

air pollution episode Time when atmospheric conditions inhibit dilution of air pollutants, and pollutants are thus a hazard to human health.

air pressure The cumulative force exerted on any surface by the molecules composing air; usually expressed as the weight of a column of air.

air pressure gradient Change in air pressure with distance.

air pressure tendency Change in air pressure with time; on a surface weather map, the air pressure change over the prior 3 hours.

albedo The fraction of radiation striking a surface that is reflected.

Aleutian low A semipermanent subpolar cyclone situated over the North Pacific in winter.

altimeter An aneroid barometer calibrated to read altitude or elevation.

altocumulus clouds Middle clouds consisting of patches or puffs of roll-like clouds forming a wavy pattern.

altocumulus lenticularis A lens-shaped altocumulus cloud; a mountain-wave cloud generated by the disturbance of horizontal airflow by a prominent mountain range.

altostratus clouds   Middle layer clouds that are uniformly gray or white.

ambient air   The air surrounding a cloud or rising or sinking air parcels; the outdoor air.

aneroid barometer   A portable instrument that utilizes a flexible metal chamber and spring to measure air pressure; may be used as an altimeter.

anomalies   Departures of temperature, precipitation, or other weather elements from long-term averages.

Antarctic Circle   Latitude 66 degrees 33 minutes S. Poleward of this latitude, there are 24 hours of sunlight at the summer solstice and 24 hours of darkness at the winter solstice.

anticyclone   A dome of air that exerts relatively high pressure compared with the surrounding air; same as a "high." In the Northern Hemisphere, surface winds in an anticyclone blow clockwise and outward.

aphelion   The time of the year when the Earth is farthest from the sun (about 4 July).

arctic (A) air   A very cold and dry air mass that forms primarily in winter over the Arctic Basin, Greenland, and the northern interior of North America.

Arctic Circle   Latitude 66 degrees 33 minutes N. Poleward of this latitude, there are 24 hours of sunlight at the summer solstice and 24 hours of darkness at the winter solstice.

arctic highs   Anticyclones originating in the source regions for cold, dry arctic air.

Arctic sea smoke   Fog that develops when extremely cold, dry air flows over a body of relatively warm open water; a type of steam fog.

atmosphere   A thin envelope of gases (also containing suspended solid and liquid particles and clouds) that encircles the globe.

atmospheric stability   Property of an air layer that imparts a buoyant force to air parcels moving vertically within the air layer; depends on the temperature profile of the air layer.

atmospheric windows   Infrared wavelength bands for which there is little or no absorption by constituent gases of the atmosphere.

aurora australis   Southern Hemisphere equivalent of the aurora borealis.

aurora borealis   Lights, visible at night in the Northern Hemisphere, produced by electrical activity in the ionosphere; northern lights.

auroral zones   Geographical areas where the aurora is visible; in the Northern Hemisphere, centered on northwestern Greenland.

Automation of Field Operations and Service (AFOS)   A computerized National Weather Service communications system that speeds transmission and exchange of weather data.

back (or backing)   Counterclockwise shift in wind direction with time; for example, a wind shift from northeast to north.

barograph   A recording instrument that provides a continuous trace of air pressure variation with time.

barometer   An instrument used to monitor variations in air pressure. *See also* aneroid barometer; mercury barometer.

basic weather network (BWN)   Stations that provide weather data, primarily for aircraft operations, and to supplement other observational data used for weather forecasting.

Beaufort scale   A scale of wind strength based originally on visual assessment of the effects of wind on seas.

bent-back occlusion   Occurs when an occluded front begins to rotate in a counterclockwise sense (in the Northern Hemisphere) about the center of an extratropical cyclone.

Bergeron process   Precipitation formation in cold clouds whereby ice crystals grow at the expense of supercooled water droplets.

blackbody   A perfect radiator or absorber; a material that absorbs 100 percent of the radiation striking it and emits the maximum possible radiation at any given wavelength and temperature.

blizzard warning   Issued when falling or blowing snow is accompanied by winds of over 55 km (35 mi) per hour and reduced visibility.

blocking system   A cutoff cyclone or anticyclone that blocks the usual west-to-east progression of weather systems.

bora   A cold katabatic wind that originates in Yugoslavia and flows onto the coastal plain of the Adriatic Sea.

boundary layer (skin)   A very thin layer of still air in immediate contact with the skin that helps insulate the body from excess heat loss.

Bowen ratio   The ratio of heat energy used for sensible heating to heat energy used for latent heating.

British thermal unit (Btu)   The quantity of heat needed to raise the temperature of one pound of water one Fahrenheit degree.

calorie (cal)   The amount of heat needed to raise the temperature of one gram of water one Celsius degree (from 14.5 to 15.5 °C).

centripetal force   An inward-directed force that confines an object to a curved path; a resultant of other forces.

chinook wind   Air that is adiabatically compressed as it is drawn down the leeward slope of a mountain range. As a consequence, the air is mild and dry.

chromosphere   Portion of the sun above the photosphere; consists of ionized hydrogen and helium at 4000 to 40,000 °C.

circumpolar vortex   Belt of global-scale westerlies that surrounds the Antarctic continent.

cirrocumulus   A high cloud exhibiting a wavelike pattern of small white puffs; composed of ice crystals.

cirrostratus   A high layered cloud, composed of ice crystals, that forms a thin white veil over the sky.

cirrus   High clouds occurring as silky strands and composed of ice crystals.

climate   Weather of some locality averaged over some time period plus extremes in weather behavior observed during the same period.

climatic norm (or normal) Average plus extreme weather at some locality for some period, usually 30 years.

climatic optimum A period about 5000 to 7000 years ago when global temperatures were somewhat warmer than at present.

climatology The study of climate.

cloud condensation nuclei (CCN) Tiny solid and liquid particles on which water vapor condenses.

cloud seeding An attempt to stimulate natural precipitation processes by injecting nucleating agents, such as silver iodide, into clouds.

clouds The visible products (ice crystals and water droplets) of condensation and deposition of water vapor in the atmosphere.

cold air advection Flow of air from relatively cool localities to relatively warm localities.

cold clouds Clouds at temperatures below 0 °C (32 °F); composed of ice crystals or supercooled water droplets or a mixture of both.

cold-core anticyclones (or highs) Shallow high-pressure systems that coincide with domes of relatively cold, dry air.

cold-core lows Cyclones that occupy relatively cold columns of air; migrating midlatitude low-pressure systems whose circulation intensifies with altitude.

cold front A narrow zone of transition between relatively cold, dense air that is advancing and relatively warm, less dense air that is retreating.

cold-type occlusion A front formed when a cold front overtakes a warm front and the air behind the front is colder than the air ahead of the front.

collision–coalescence process The growth of cloud droplets into raindrops within warm clouds; droplets join together upon impact.

comma cloud The pattern of cloudiness associated with a wave cyclone; the head of the comma stretches from the low center to the northwest and its tail follows along the cold front.

compressional warming A temperature rise that accompanies a pressure increase on a parcel of air, as when air parcels descend within the atmosphere.

conceptual model Describes the general functional relationships among components of a system.

condensation Process by which water changes phase from a vapor to a liquid.

conditional stability Property of an ambient air layer that is stable for unsaturated air parcels and unstable for saturated air parcels.

conduction Flow of heat in response to a temperature gradient within an object or between objects that are in physical contact.

confluence zone Zone where air streams having different origins (and often different properties) come together.

continental drift The slow movement of continents as they are carried along on top of gigantic crustal plates.

continental polar (cP) air Relatively dry air mass that develops over the northern interior of North America; very cold in winter and mild in summer.

continental tropical (cT) air Warm, dry air mass that forms over the subtropical deserts of the southwestern United States.

convection Air circulation in the vertical whereby warm air parcels rise and cool air parcels sink.

convective condensation level (CCL) The altitude at which air parcels rising in convection cells reach saturation; coincides with the base of cumuliform clouds.

convergence A wind pattern whereby there is a net inflow of air toward some central point.

cooling degree-day units A measure of the need for air conditioning when average daily temperatures are above 65 °F (18 °C); computed by subtracting 65 °F from the average daily temperature in °F.

Coriolis effect A deflective force arising from the rotation of the Earth on its axis; affects principally synoptic-scale and global-scale winds. Winds are deflected to the right of the initial direction in the Northern Hemisphere and to the left in the Southern Hemisphere.

corona Outermost region of the sun; consists of highly rarefied gases at 1 to 4 million °C.

corona (optical) Colored rings that appear about the moon or sun; due to diffraction of light by spherical cloud droplets.

cumuliform clouds Clouds that exhibit significant vertical development; often produced by updrafts in convection currents.

cumulonimbus clouds Thunderstorm clouds that form as a consequence of deep convection in the atmosphere.

cumulus clouds Clouds that develop as a consequence of the updraft in convection currents; resemble huge puffs of cotton floating in the sky.

cumulus congestus An upward-building convective cloud with vertical development between those of a cumulus cloud and a cumulonimbus cloud.

cumulus stage Initial stage in the life cycle of a thunderstorm cell; consists of towering cumulus clouds with updraft throughout the system.

cup anemometer An instrument used to monitor wind speed. Wind rotation of cups generates an electric current that is calibrated in wind speed.

cutoff high Anticyclonic circulation system that separates from the prevailing westerly airflow and therefore remains stationary.

cutoff low Cyclonic circulation system that separates from the prevailing westerly airflow and therefore remains stationary.

cyclogenesis Birth and development of a cyclone, a low-pressure system.

cyclolysis A process whereby a cyclone weakens, its winds slacken, and its central pressure rises.

cyclone A weather system characterized by relatively low surface air pressure compared with the surrounding air; same as a low. Surface winds blow counterclockwise and inward in the Northern Hemisphere.

Dalton's law The total pressure exerted by a mixture of gases is equal to the sum of the partial pressures of each constituent gas.

dart leaders Surges of negative electrical charge that follow the conductive path formed by the initial stepped leaders and return stroke of a lightning flash.

deepening Process whereby the central pressure of a developing cyclone drops.

degree-day units *See* cooling degree-day units; heating degree-day units.

deposition Process by which water changes phase directly from a vapor into a solid. An example is frost formation.

deposition nuclei Tiny particles on which water vapor is deposited directly as ice.

dew Water droplets formed by condensation of water vapor on a surface.

dew point or dew-point temperature Temperature to which air must be cooled at constant pressure to achieve saturation (if above 0 °C or 32 °F).

diffraction Slight bending of a light wave as it moves along the boundary of an object such as a water droplet.

diffuse insolation Solar radiation that is scattered or reflected by atmospheric components (clouds, for example) to the Earth's surface.

direct insolation Solar radiation that is transmitted directly through the atmosphere to the Earth's surface without interacting with atmospheric components.

dissipating stage (thunderstorm) The final phase in the life cycle of a thunderstorm cell; features downdrafts throughout the system and clouds that gradually vaporize.

divergence A wind pattern whereby there is a net outflow of air from some central point.

Doppler effect A shift in frequency of an electromagnetic or sound wave due to the relative movement of the source or the observer.

Doppler radar Conventional weather radar that has the added capability of determining the detailed motion of target precipitation based on the frequency shift between the outgoing and returning radar beam.

downburst A strong and potentially destructive thunderstorm downdraft; depending on size, classified as either a microburst or a macroburst.

downdraft Downward-moving air, usually associated with a thunderstorm cell.

drainage basin A fixed geographical region from which a river and its tributaries drain water; also called watershed.

drizzle A form of liquid precipitation consisting of water droplets less than 0.5 mm (0.02 in.) in diameter; falls from low stratus clouds.

dropwindsonde A small instrument package equipped with a radio transmitter that is dropped from an aircraft and measures

vertical profiles of temperature, pressure, relative humidity, and wind.

drought An extended period of anomalous moisture deficit.

dry adiabatic lapse rate Rising unsaturated air parcels cool at the rate of about 10 C° per 1000 m of uplift (or 5.5 F° per 1000 ft).

dry deposition Removal of suspended particulates from the air through impaction and gravitational settling; a natural means whereby air is cleansed of pollutants.

dry line Boundary between warm, dry air and warm, humid air in the southeast sector of a mature midlatitude cyclone; likely site for severe thunderstorm development.

dust devil Swirling mass of dust caused by intense solar heating of dry surface areas.

dust dome An accumulation of visibility-restricting aerosols in the air over an urban-industrial area.

dust plume A dust dome that elongates downwind.

Earth-atmosphere system The Earth's surface and atmosphere considered together.

eddy viscosity Frictional resistance arising from eddies (irregular whirls) within a fluid such as air or water.

El Niño An anomalous warming of surface ocean waters off the coasts of Ecuador and Peru and extending westward over the eastern tropical Pacific.

El Niño/Southern Oscillation (ENSO) An episode of anomalously high sea-surface temperatures in the equatorial and tropical eastern Pacific; associated with large-scale swings in surface air pressure between the western and eastern tropical Pacific.

electromagnetic radiation Energy transfer in the form of waves that have both electrical and magnetic properties. Occurs even in a vacuum.

electromagnetic spectrum Range of radiation types arranged by wavelength or by frequency or both.

ENSO *See* El Niño/Southern Oscillation

entrainment Mixing of saturated cloudy air with unsaturated air that surrounds the cloud.

equatorial trough A broad east–west belt of low pressure near the equator.

equinoxes The first days of spring and autumn when day and night are of equal length at all latitudes. The noon sun is directly over the equator.

evaporation Process by which water changes phase from a liquid to a vapor at a temperature below the boiling point of water.

evapotranspiration Vaporization of water through direct evaporation from wet surfaces plus the release of water vapor by vegetation.

expansional cooling A temperature drop that accompanies a pressure reduction on an air parcel, as when air parcels ascend within the atmosphere.

Experiment on Rapidly Intensifying Cyclones over the Atlantic (ERICA) A field study conducted from December 1988 to February 1989 over the North Atlantic with the goal of gathering data on rapidly intensifying cyclones throughout their life cycle.

eye of a hurricane An area of almost cloudless skies, light winds, and gently subsiding air at a hurricane center.

eye wall A circle of cumulonimbus clouds surrounding the eye of a mature hurricane.

F-scale Tornado intensity scale, developed by T. T. Fujita, that rates tornadoes from 0 to 5 on the basis of estimated wind speed.

faculae Small, bright, relatively hot spots on the sun.

fair-weather bias The observation that fair weather days outnumber stormy days almost everywhere.

fetch Distance traveled by wind over a body of water such as a lake or sea.

filling Process whereby a cyclone weakens, its winds slacken, and its central pressure rises.

flash flooding A sudden rise in river or stream levels causing flooding.

foehn European term for chinook wind; warm, dry wind that flows into the Alpine valleys of Austria and Germany.

fog A cloud in contact with the Earth's surface that reduces visibility to less than 1.0 km (0,62 mi).

fog dispersal Clearing of fog either by increasing the air temperature (thereby lowering the relative humidity) or by cloud seeding.

forced convection Convection aided by topographic uplift.

free convection Convection triggered by intense solar heating of the Earth's surface.

freezing nuclei Tiny particles that cause liquid cloud droplets to freeze.

freezing rain Supercooled raindrops that freeze on contact with cold surfaces.

friction The resistance an object encounters as it comes into contact with other objects; in fluids known as viscosity.

friction layer Zone of the atmosphere, between the Earth's surface and an altitude of about 1 km (3,250 ft), where most frictional resistance is confined.

front A narrow zone of transition between air masses of contrasting density, that is, air masses of different temperatures or different water vapor concentrations or both.

frontal fog A cloud in contact with the Earth's surface, formed when precipitation falls from relatively warm air into a wedge of relatively cool air near the surface and raises the vapor pressure in the cool air to saturation; occurs either just ahead of a warm front or just behind a cold front.

frontal thunderstorms Thunderstorms associated with lifting of air along frontal surfaces.

frontal uplift Uplift of air along either a cold or warm frontal surface.

frontogenesis The development or strengthening of a front.

frontolysis The dissipation of a front.

frost Ice crystals formed by deposition of water vapor on a surface.

frost point or frost-point temperature Temperature to which air must be cooled (if at 0 °C or below) at constant pressure to achieve saturation.

frostbite The freezing of body tissue usually on exposure to very cold air.

fumigation Atmospheric stability conditions that favor trapping of air pollutants within an air layer adjacent to the Earth's surface.

funnel cloud A tornadic circulation that does not reach the ground.

gamma radiation Electromagnetic radiation having very short wavelength and great penetrating power.

gas law Relationship among the variables of state; pressure is proportional to the product of density and temperature.

Genesis of Atlantic Lows Experiment (GALE) A field study conducted from January to March 1986 to monitor the initial formation of rapidly intensifying extratropical cyclones off the North Carolina coast.

geologic time A span of millions or billions of years in the past.

geostrophic wind Unaccelerated horizontal wind that flows in straight paths above the friction layer; results from a balance between the horizontal pressure gradient force and the Coriolis effect.

geosynchronous satellite A satellite that orbits Earth at the same rate as the Earth's rotation, so it always scans the same region of the planet.

glacial climate Conditions favorable to the initiation and growth of glacial ice.

glaze A transparent layer of ice caused by slow cooling of supercooled water.

global radiative equilibrium The balance between net incoming solar radiation and the infrared radiation emitted to space by the Earth-atmosphere system.

global-scale circulation The largest spatial scale of weather phenomena; includes global wind belts and semipermanent pressure systems.

glory Concentric rings of color about the shadow of an observer's head that appear on the top of a cloud situated below the observer. A glory is caused by the same optics as a rainbow plus diffraction.

gradient The change in some factor (such as temperature or pressure) with distance.

gradient wind Large-scale, horizontal, and frictionless wind that describes a curved path.

granules A network of huge, irregularly shaped convective cells in the sun's photosphere.

graphical model A compilation or display of data in a form that can be readily useful; a weather map is an example.

graupel Tiny granules of ice or compact snow; a form of precipitation.

gravitation Force of attraction between all objects that depends on the masses of the objects and the distance between them.

gravitational settling Drifting of aerosols downward toward the Earth's surface under the influence of gravity.

gravity The force that holds all objects on the Earth's surface; the net effect of gravitation and the centripetal force due to the Earth's rotation.

greenhouse effect Although nearly transparent to solar radiation, the atmosphere is much less transparent to infrared radiation. Terrestrial infrared radiation is absorbed and reradiated primarily by water vapor and, to a lesser extent, by carbon dioxide, thereby slowing the loss of heat to space from the Earth-atmosphere system.

Greenwich Mean Time (GMT) A worldwide time reference used for synchronizing weather observations; the time at 0 degrees longitude, the prime meridian, which passes through Greenwich, England.

gust front The leading edge of a mass of relatively cool air that flows out of the base of a thunderstorm cloud (downdraft) and spreads along the ground well in advance of the parent thunderstorm cell; a mesoscale cold front.

haboob A dust storm caused by the downdraft of a desert thunderstorm.

Hadley cell Air circulation in tropical and subtropical latitudes of both hemispheres resembling a huge convective cell with rising air over the equator and sinking air in the subtropical anticyclones.

hail or hailstones Precipitation in the form of rounded or jagged chunks of ice, which are often characterized by internal concentric layering. Hail is associated with thunderstorms that have strong updrafts and relatively great moisture content.

hailstreak Accumulation of hail in a long, narrow path along the ground.

hair hygrometer An instrument used to monitor relative humidity by measuring the changes in the length of human hair that accompany humidity variations.

halo Ring of light about the sun (or moon) caused by refraction of sunlight by tiny ice crystals suspended in the upper troposphere.

heat The total kinetic energy of the atoms or molecules composing a substance.

heat equator The latitude of highest mean annual surface air temperature; at about latitude 10 degrees N.

heat lightning Light reflected by clouds from distant thunderstorms occurring beyond the horizon.

heating degree-day units A measure of space heating needs on days when the average air temperature falls below 65 °F (18 °C); computed by subtracting the day's average temperature from 65 °F.

heterosphere The atmosphere above 80 km (50 mi) where gases are stratified, with concentrations of the heavier gases decreasing more rapidly with altitude than concentrations of the lighter gases.

high *See* anticyclone.

hoarfrost Fernlike crystals of ice that form by deposition of water vapor on twigs, tree branches, and other vegetation.

homosphere The atmosphere up to 80 km (50 mi) in which the proportionality of principal gaseous constituents, such as oxygen and nitrogen, is constant.

hook echo A distinctive radar pattern that often indicates the presence of a severe thunderstorm and perhaps tornadic circulation.

horse latitudes Areas of calm winds associated with subtropical anticyclones; near 30 degrees latitude in both hemispheres.

hot-wire anemometer An instrument that measures wind speed based on the rate of heat loss to air flowing by a sensor.

hurricane Intense warm-core oceanic cyclone that originates in tropical latitudes; also called typhoon in the western Pacific Ocean. Winds are in excess of 119 km (74 mi) per hour.

hydrologic cycle Ceaseless flow of water among terrestrial, oceanic, and atmospheric reservoirs.

hydrostatic equilibrium Balance between the vertical air pressure gradient force (directed upward) and the force of gravity (directed downward).

hygrograph An instrument that provides a continuous trace of relative humidity variations with time.

hygroscopic nuclei Tiny particles of matter that have a special chemical affinity for water molecules, so condensation may take place on these nuclei at relative humidities under 100 percent.

ice-forming nuclei (IN) Tiny particles that promote the formation of ice crystals at temperatures well below freezing; include freezing nuclei and deposition nuclei.

ice pellets Frozen raindrops that bounce on impact with the ground; also called sleet.

Icelandic low A subpolar cyclone situated over the North Atlantic and most intense (lowest central air pressure) in winter.

impaction Removal of aerosols from the air through their impact on buildings and other objects at the Earth's surface; a natural cleansing mechanism.

index of continentality Degree of maritime influence on continental air temperatures; usually based on the contrast between mean summer and winter temperatures.

Indian summer A period of mild, sunny weather that occurs in autumn over eastern North America; usually after the first freeze.

infrared radiation Electromagnetic radiation at wavelengths shorter than microwaves and longer than visible red light; emitted by most objects on Earth.

insolation *In*coming *solar* rad*iation. See also* diffuse insolation; direct insolation.

interglacial climate Conditions that favor the melting of glacial ice (if present).

International Meteorological Organization (IMO) Founded in 1878, the predecessor of the World Meteorological Organization.

intertropical convergence zone (ITCZ) Discontinuous belt of thunderstorms paralleling the equator and marking the convergence of the Northern and Southern Hemisphere surface trade winds.

ion An electrically charged atom or molecule.

ionosphere Region of the upper atmosphere from 80 to 900 km (50 to 600 mi) that contains a relatively high concentration of ions (charged particles).

isobars Lines on a map joining localities reporting the same air pressure.

isothermal Constant temperature.

isotherms Lines on a map joining localities reporting the same air temperature.

ITCZ *See* intertropical convergence zone.

January thaw A period of relatively mild weather around 20–23 January that occurs primarily in New England; an example of a singularity in the climatic record.

JAWS *See* Joint Airport Weather Studies.

jet maximum An area of accelerated airflow along a jet stream; same as a jet core.

jet stream Relatively narrow corridor of very strong winds embedded in the planetary winds aloft.

Joint Airport Weather Studies (JAWS) An investigation of microbursts in the vicinity of Denver's Stapleton International Airport conducted during the summer of 1982.

katabatic wind Downslope flow of cold, dense air under the influence of gravity.

kinetic energy Energy possessed by any object in motion.

La Niña A period of strong trade winds and unusually low sea-surface temperatures in the central and eastern tropical Pacific; opposite of El Niño.

lake breeze A relatively cool mesoscale surface wind directed from a lake toward land in response to differential heating between land and lake; develops during the day.

lake-effect snow Snowfall, often highly localized, that occurs along the lee shore of a large lake when cold air flows across a long fetch of relatively mild open water; the cold air becomes less stable and more humid, and uplift along the shoreline triggers cloud development and snowfall.

land breeze A relatively cool mesoscale surface wind directed from land to sea or from land to lake in response to differential cooling between land and water body; develops at night.

latent heat of fusion Heat released when water changes phase from liquid to solid; 80 calories per gram.

latent heat of melting Heat required to change the phase of water from solid to liquid; 80 calories per gram.

latent heat of vaporization Heat required to change the phase of water from liquid to vapor; 540 to 600 calories per gram, depending on the temperature of the water.

latent heating Transport of heat from one place to another within the atmosphere as a consequence of phase changes of water. Heat is supplied for evaporation and sublimation of water at the Earth's surface, and heat is released in condensation and deposition (cloud formation) within the atmosphere.

latitude Distance measured in degrees north or south of the equator. The latitude of the equator is 0 degrees, and the poles are at latitudes 90 degrees N and 90 degrees S.

law of energy conservation Energy is neither created nor destroyed but can change from one form to another.

law of reflection The angle of incident radiation is equal to the angle of reflected radiation.

LCL *See* lifting condensation level.

LDN *See* lightning detection network.

lifting condensation level (LCL) The altitude to which air must be lifted so that expansional cooling leads to condensation (or deposition) and cloud development; corresponds to the base of clouds.

lightning A flash of light produced by an electrical discharge in response to the buildup of an electrical potential between cloud and ground, between clouds, or within a single cloud.

lightning detection network (LDN) System that provides real-time information on the location and severity of lightning strokes.

Little Ice Age A period of relatively cold conditions in many regions of the globe from about 1400 to 1850.

LLWSAS (Low-Level Wind Shear Alert System) An array of ground-level anemometers intended to detect microbursts at airports.

long waves Same as Rossby waves in the westerlies.

longitude The distance measured in degrees east or west of the prime meridian (0 degrees), which passes through Greenwich, England. The maximum longitude is 180 degrees W or 180 degrees E.

low *See* cyclone.

low-level jet stream A surge of maritime tropical air in the lower troposphere northward out of the Gulf of Mexico.

macroburst A downburst that affects a path longer than 4.0 km (2.5 mi).

magnetosphere The region of the upper atmosphere encompassed by the Earth's magnetic field; deflected by the solar wind into a teardrop-shaped cavity.

mammatus clouds Clouds that form on the underside of a thunderstorm anvil and exhibit pouchlike, downward protuberances; may indicate turbulent air.

maritime polar (mP) air Cool, humid air masses that form over the cold ocean waters of the North Pacific and North Atlantic.

maritime tropical (mT) air Warm, humid air masses that form over tropical and subtropical oceans.

mature stage (thunderstorm) The middle and most intense phase of the life cycle of a thunderstorm cell; begins when precipitation reaches the Earth's surface and is characterized by updrafts and downdrafts.

Maunder minimum A 70-year period from 1645 to 1715 when sunspots were rare.

mb   *See* millibar.

MCC   *See* mesoscale convective complex.

McIDAS (Man–computer Interactive Data Access System)   A real-time weather data management system developed at the University of Wisconsin–Madison.

mechanical turbulence   Irregular, random motions of a fluid triggered by obstacles in the path of the fluid.

mercury barometer   A mercury-filled tube used to measure air pressure; the standard barometric instrument, which features great precision.

meridional component   North to south movement parallel to a line of longitude.

meridional flow pattern   Flow of westerlies in a series of deep troughs and sharp ridges; westerlies exhibit considerable amplitude.

mesocyclone   A stage in the development of a tornado; consists of a spinning cylinder of air 10 to 20 km (6.2 to 12.4 mi) in diameter within the updraft of a severe thunderstorm.

mesopause   Narrow zone of transition between the mesosphere below and the thermosphere above; top of the mesosphere.

mesoscale convective complex (MCC)   A nearly circular cluster of many interacting thunderstorms covering an area of many thousands of square kilometers.

mesoscale systems   Weather phenomena operating at the local scale; include thunderstorms and sea breezes, for example.

mesosphere   Thermal subdivision of the upper atmosphere in which air temperature declines with altitude; situated above the stratosphere and below the thermosphere.

meteorology   Scientific study of the atmosphere and atmospheric processes.

METROMEX (Metropolitan Meteorological Experiment) An intensive field investigation of the effects of urban air pollution on cloud and precipitation patterns downwind from St. Louis; the 1971–1975 study concluded that precipitation was enhanced downwind of St. Louis.

Metropolitan Meteorological Experiment (METROMEX) *See* METROMEX.

microburst   A downburst that affects a path that is 4.0 km (2.5 mi.) or shorter.

microscale weather   The smallest spatial subdivision of atmospheric circulation.

microwave   Electromagnetic radiation having wavelengths in the 0.1 to 300 mm range; used in weather radars.

midlatitude westerlies   Global-scale prevailing west-to-east winds in the mid- and upper-troposphere between about 30 and 60 degrees of latitude.

Milankovitch cycles   Systematic changes in three elements of Earth–sun geometry: precession, axial tilt, and orbital eccentricity; may be linked to large-scale fluctuations of the Earth's ice sheets during the Ice Age.

millibar (mb)   The conventional meteorological unit of air pressure. The average sea-level air pressure is 1013.25 millibars.

mist   Very thin fog in which visibility is greater than 1.0 km (0.62 mi).

mistral   A katabatic wind that flows from the Alps down the Rhone River Valley of France to the Mediterranean coast.

mixing depth   Vertical distance between the ground and the altitude to which convection currents reach; the thickness of the mixing layer.

mixing layer   Surface layer of the troposphere in which air is thoroughly mixed by convection. Mixing depth is the thickness of the mixing layer.

mixing ratio   Mass of water vapor per mass of dry air; expressed as grams per kilogram.

mock suns   *See* parhelia.

moist adiabatic lapse rate   A variable rate of cooling applicable to saturated air parcels that are ascending within the atmosphere. This rate is less than the dry adiabatic lapse rate because some of the expansional cooling is compensated for by the release of latent heat that accompanies the phase change of water vapor.

molecular diffusion   Mixing that takes place at the molecular scale; a relatively slow process.

molecular viscosity   Frictional resistance arising from the interactions of molecules composing a fluid such as air or water.

monsoon active phase   A generally cloudy period with frequent deluges of rain.

monsoon circulation   Characterizes regions where seasonal reversals of winds cause wet summers and dry winters.

monsoon dormant phase   A generally sunny and hot period that interrupts rainy monsoon episodes.

mother-of-pearl clouds   *See* nacreous clouds.

mountain breeze   A shallow, gusty downslope flow of cool air that develops at night in some mountain valleys.

mountain wave   A stationary wave situated downwind of a prominent mountain range and caused by the disturbance of the wind by the mountain range.

mountain-wave clouds   Stationary clouds situated downwind of a prominent mountain range and formed as a consequence of the disturbance of the wind by the mountain range.

nacreous clouds   Rarely seen clouds that form in the upper stratosphere; may be composed of ice crystals or supercooled water droplets. Also called mother-of-pearl clouds.

National Climate Data Center (NCDC)   Houses archives of climatic data of the United States; located in Asheville, North Carolina.

National Hurricane Center (NHC)   Responsible for forecasting tropical storms and hurricanes along the East and Gulf coasts and the eastern Pacific Ocean; located in Coral Gables, Florida.

National Oceanic and Atmospheric Administration (NOAA)   Within the U.S. Department of Commerce, the administrative unit that oversees the National Weather Service.

National Severe Storms Forecast Center (NSSFC)   Issues watches for severe thunderstorms and tornadoes; located in Kansas City, Missouri.

National Substation Program   A network of about 11,600 cooperative weather stations in the United States that record weather data for hydrologic, agricultural, and climatic purposes.

National Weather Service (NWS)   The agency of NOAA responsible for weather data acquisition, data analysis, forecast dissemination, and storm watches and warnings.

neutral air layer   An air layer in which an ascending or descending air parcel always has the same temperature as its surroundings.

Newton's first law of motion   An object at rest or in straight-line unaccelerated motion remains that way unless acted upon by a net external force.

Newton's second law of motion   A net force is required to cause a unit mass of a substance to accelerate (or decelerate); force = mass × acceleration.

NEXRAD (Next Generation Weather Radar)   A new national network of Doppler radars scheduled for operation in the early 1990s.

nimbostratus   Low, gray, layered clouds that resemble stratus clouds but are thicker and yield more substantial precipitation.

NIMROD (Northern Illinois Meteorological Research on Downbursts)   A study conducted in the western suburbs of Chicago in 1978 to gather data on the frequency and conditions of downburst formation.

noctilucent clouds   Clouds that occur in the upper mesosphere, are rarely seen, and probably are composed of meteoric dust.

nocturnal radiational cooling   Emission of infrared radiation by the Earth-atmosphere system to space at night; most extreme when skies are clear, humidity is low, and surface winds are light.

Norwegian cyclone model   The original description of the structure and life cycle of a midlatitude low-pressure system, first proposed during World War I by researchers at the Norwegian School of Meteorology at Bergen.

nuclei   Tiny solid or liquid particles of matter on which condensation or deposition of water vapor can take place.

numerical model   One or more mathematical expressions that approximate the behavior of a system.

occluded front   A front formed when a cold front overtakes a warm front; represents the final stage in the life cycle of a midlatitude cyclone. *Also see* cold-type occlusion and warm-type occlusion.

occlusion   The final stage in the life cycle of a midlatitude cyclone. Occlusion occurs when the cold front overtakes the warm front.

occlusion stage   The final phase in the life cycle of an extratropical cyclone; the cold front catches up with the warm front to form an occluded front.

orographic lifting   Rising motion of air induced by topography.

orographic precipitation   Rainfall or snowfall from clouds induced by topographic uplift.

overrunning   The process whereby less dense air displaces more dense air by flowing up and over the denser air; characterizes a warm front.

ozone hole   A thinning of stratospheric ozone over Antarctica during the Southern Hemisphere spring; recent deepening of the ozone hole may be linked to air pollution.

ozone shield   Ozone (O_3) within the stratosphere that filters out potentially lethal intensities of ultraviolet radiation from the sun.

Pacific air   A North American air mass originating over the Pacific Ocean that travels through the western mountains and emerges on the Great Plains warmer and drier than in its source region.

parhelia   Two bright spots of light appearing on either side of the sun; each is separated from the sun by an angle of 22 degrees. Parhelia are caused by refraction of sunlight by ice crystals; also called mock suns and sundogs.

penumbra   Outer, lighter area of a sunspot.

perihelion   The time of year when the Earth's orbital path brings it closest to the sun (about 3 January).

persistence   Tendency for weather episodes to continue for some period of time.

pH scale   A system used to specify the range of acidity and alkalinity of different substances. A pH of 7 is neutral; acids have pH values less than 7, and alkaline substances have pH values greater than 7.

photodissociation   The process by which radiation breaks down molecules into their smaller components.

photosphere   The visible surface of the sun.

photosynthesis   The process whereby plants use sunlight, water, and carbon dioxide to manufacture their food.

physical model   A miniaturized version of a real system.

planetary albedo   The fraction of solar radiation that is scattered and reflected back into space by the Earth-atmosphere system.

point of occlusion   Where the cold, warm, and occluded fronts all come together within an extratropical cyclone late in its life cycle; location favorable for development of a new cyclone.

polar front   Transition zone between cold polar easterlies and mild midlatitude westerlies.

polar front jet stream   A jet stream situated in the upper troposphere between the midlatitude tropopause and the polar tropopause and directly over the polar front.

polar highs   Anticyclones originating in the source regions for continental polar air.

polar-orbiting satellite   Satellite in relatively low orbit that travels over or near the geographical poles on meridional trajectories.

poleward heat transport   Flow of heat from tropical to middle and high latitudes in response to latitudinal imbalances in radiational heating and cooling. Poleward heat transport is accomplished by air mass exchange, storms, and surface ocean currents.

precipitable water The depth of water produced when all the water vapor in a column of air is condensed; usually the column of air reaches from the Earth's surface to the "top" of the atmosphere.

precipitation Water in solid or liquid form that falls to the Earth's surface from clouds.

precipitation scavenging Washing of pollutants from air by rainfall and snowfall.

pressure gradient force A force operating in the atmosphere that causes air parcels to accelerate from regions of high pressure toward regions of low pressure in response to an air pressure gradient.

primary air pollutants Substances that are pollutants immediately upon entry into the atmosphere.

Project STORMFURY A project aimed at weakening hurricanes by cloud seeding.

Project Whitetop A cloud-seeding project, carried out in Missouri during the 1950s, which may have reduced precipitation because of overseeding.

psychrometer An instrument used to determine relative humidity. The psychrometer depends on the difference in readings between a dry-bulb thermometer and a wet-bulb thermometer.

pyranometer The standard instrument for measuring solar radiation incident on a horizontal surface; calibrates the temperature response of a special sensor in units of radiation flux.

radar (radio detection and ranging) An instrument that sends and receives microwaves for the purpose of determining the location and movement of areas of precipitation.

radar echo Microwaves scattered by distant rain or snow back to a receiver where they are displayed as bright spots on a cathode ray tube.

radiation Energy transport via electromagnetic waves capable of traveling through a vacuum.

radiation fog A ground-level cloud formed by the nocturnal radiational cooling of a humid air layer so that its relative humidity approaches 100 percent.

radiational cooling A temperature drop, most pronounced at night, that is induced by emission of infrared radiation.

radiational temperature inversion Cooling of a surface air layer by loss of infrared radiation, so the coldest air is adjacent to the Earth's surface and the air temperature increases with altitude.

radio waves Long-wavelength, low-frequency electromagnetic waves.

radiosonde A small balloon-borne instrument package equipped with a radio transmitter that measures vertical profiles of temperature, pressure, and relative humidity in the atmosphere.

rain A form of precipitation consisting of liquid water drops having diameters between 0.5 and 5.0 mm (0.02 and 0.2 in.).

rainbow An arc of concentric colored bands formed by refrac-

tion and internal reflection of sunlight by raindrops. Observer must be looking at a distant rain shower with the sun at his/her back.

rain gauge A device—usually a cylindrical container—for collecting and measuring rainfall.

rain shadow A region situated downwind of a high mountain barrier and characterized by descending air and, as a consequence, a relatively dry climate.

rawinsonde A radiosonde tracked from the ground by a direction-finding antenna to measure variations in horizontal wind direction and wind speed with altitude.

reduction to sea level An adjustment applied to surface air pressure readings in order to eliminate the influence of station elevation.

reflection The process whereby a portion of the radiation that is incident on a surface is returned by that surface.

refraction The bending of a light ray as it passes from one transparent medium to another (from air to water, for example). The bending is due to the differing speeds of light in the two different media.

relative humidity A measure of how close an air sample is to saturation at a specific temperature; expressed as a percentage.

return stroke A positively charged electrical flow that emanates from the ground and meets a downward-moving stepped leader; part of a lightning stroke.

rime An opaque, granular layer of ice formed by the rapid freezing of supercooled water.

roll cloud A low, cylindrically shaped and elongated cloud occurring behind a gust front; associated with but detached from a cumulonimbus cloud.

Rossby waves Series of long-wavelength troughs and ridges that characterize the westerlies as they encircle the globe; also called long waves.

Saffir–Simpson Hurricane Intensity Scale Hurricane intensity scale based on central pressure, wind speed, and storm surge and damage potential; 1 is minimal, 5 is most intense.

Santa Ana wind A hot, dry (chinook-type) wind that blows from the desert plateaus of Utah and Nevada toward coastal southern California.

saturation mixing ratio Maximum concentration of water vapor in a given volume of air at a specific temperature.

saturation vapor pressure The maximum possible vapor pressure in a sample of air at a specific temperature.

scattering The process whereby small particles disperse radiation in all directions.

scientific method A systematic form of inquiry that involves observation, speculation, and reasoning.

scientific model An approximation or simulation of a real system; omits all but the most essential variables of the system.

sea breeze A relatively cool mesoscale surface wind directed from sea toward land in response to differential heating between land and sea; develops during the day.

secondary air pollutants   Pollutants generated by chemical reactions occurring within the atmosphere.

semipermanent pressure systems   Persistent cyclones and anticyclones that are components of the global-scale circulation. These pressure cells exhibit some seasonal changes in location and in surface pressures.

sensible heating   The transport of heat from one location or object to another via conduction, convection, or both.

severe thunderstorms   Thunderstorms accompanied by locally damaging surface winds, frequent lightning, or large hail.

sheet lightning   Bright flashes across the sky due to cloud-to-cloud electrical discharges.

shelf cloud   A low, wedge-shaped, and elongated cloud that occurs along a gust front; associated with and attached to a cumulonimbus cloud.

short waves   Relatively small ripples (troughs and ridges) superimposed on long waves in the planetary westerlies.

Sierra Cooperative Pilot Project (SCPP)   A long-term experimental seeding of winter orographic clouds on the windward western slopes of California's Sierra Nevada.

single-station forecasts   Weather forecasts based on observations at one location.

singularity   A weather event that occurs on or near a certain date with unusual regularity; the January thaw is an example.

sleet   *See* ice pellets.

snow   A type of precipitation consisting of an assemblage of ice crystals in the form of plates, columns, or flakes.

snowbelts   Areas to the lee of large lakes such as the Great Lakes that are subject to frequent lake-effect snows.

snowburst   A particularly intense fall of snow that leads to significant accumulations in a short period of time; usually applied to lake-effect snows.

snow grains   Frozen form of precipitation consisting of particles of white ice having diameters less than 1 mm; originates in the same way as drizzle.

snow pellets   Frozen form of precipitation consisting of soft conical or spherical white ice particles having diameters of 1 to 5 mm.

solar altitude   The angle of the sun above the horizon.

solar constant   The flux of solar radiational energy falling on a surface positioned at the top of the atmosphere and oriented perpendicular to the solar beam when Earth is at its average distance from the sun; about 2.00 calories per square centimeter per minute.

solar flare   Gigantic disturbance on the sun that emits to space high-velocity streams of electrically charged subatomic particles.

solar irradiance   The rate of total radiational energy output of the sun.

solar wind   A stream of charged subatomic particles (mainly protons and electrons) flowing out into space from the sun.

solstices   When the sun is at its maximum poleward locations relative to the Earth (23 degrees 30 minutes North and South); first days of summer and winter.

soundings   Continuous altitude measurements that provide profiles of such variables as temperature, humidity, and wind speed.

southern oscillation   Opposing swings of surface air pressure between the eastern and western tropical Pacific Ocean.

specific heat   Amount of heat required to raise the temperature of 1 gram of a substance by 1 C°.

specific humidity   The mass of water vapor per mass of air containing the water vapor; usually expressed as grams of water vapor per kilogram of air.

split flow pattern   Westerlies to the north have a wave configuration that differs from that of westerlies to the south.

Spörer minimum   A period of reduced sunspot activity from 1450 to 1550.

squall line   A line of intense thunderstorms occurring parallel to and ahead of a fast-moving, well-defined cold front.

stable air layer   Air layer characterized by a vertical temperature profile such that air parcels return to their original altitudes following upward or downward displacements.

standard atmosphere   The mean vertical profiles of temperature, pressure, and density within the atmosphere.

standing wave   A stationary undulation of the horizontal airflow that develops downwind of a large obstacle; a mountain wave is an example.

station model   A conventional representation on a weather map, using standard symbols, of weather conditions at some locality.

stationary front   A nearly stationary narrow zone of transition between contrasting air masses; winds blow parallel to the front but in opposite directions on either side of the front.

steam fog   The general name for fog produced when cold air comes in contact with relatively warm water; has the appearance of rising streamers.

Stefan–Boltzmann law   The total energy radiated by a blackbody at all wavelengths is directly proportional to the fourth power of the absolute temperature (in kelvins) of the body.

stepped leaders   The initial electrical discharges in a lightning flash; consist of negative electrical charge that travels from a cloud base to within 100 m (300 ft) of the ground.

storm surge   A hurricane-induced rise in sea level that reaches the shoreline ahead of the storm; caused by strong winds and low air pressure.

stratiform clouds   Layered clouds, such as altostratus, often produced by air mass overrunning.

stratocumulus clouds   Low clouds consisting of large, irregular puffs or rolls arranged in a layer.

stratopause   Transition zone between the stratosphere and the mesosphere.

stratosphere   The atmosphere's thermal subdivision situated between the troposphere and mesosphere; primary site of ozone formation. Within the stratosphere, air temperature in the lower part is constant with altitude, and then temperature increases with altitude.

stratus   Low clouds that occur as a uniform gray layer stretching from horizon to horizon. They may produce drizzle, and where they intersect the ground, are classified as fog.

streamlines   The flow pattern of air moving horizontally; on a map, lines that are drawn everywhere parallel to wind direction.

sublimation   Process by which water changes from a solid into a vapor without passing through the liquid phase.

sublimation nuclei   *See* deposition nuclei.

subpolar lows   High-latitude, semipermanent cyclones marking the convergence of global-scale surface southwesterlies of midlatitudes with surface northeasterlies of polar latitudes; Icelandic flow and Aleutian low are examples.

subsidence temperature inversion   A temperature inversion that develops aloft as a result of air gradually sinking over a wide area and warmed by adiabatic compression; occurs on the eastern flanks of the subtropical anticyclones.

subtropical anticyclones   Semipermanent warm-core, high-pressure systems centered over subtropical latitudes of the Atlantic, Pacific, and Indian oceans.

subtropical jet stream   A zone of unusually strong winds situated between the tropical tropopause and the midlatitude tropopause.

sundogs   *See* parhelia.

sunspots   Relatively large, dark blotches that appear on the face of the sun.

supercells   Severe thunderstorm cells.

supercooled water droplets   Cloud droplets that remain liquid even though the temperature is below the freezing point.

supersaturation   Property of an air sample having a relative humidity greater than 100 percent.

synoptic-scale weather   Weather phenomena operating at the continental or oceanic spatial scale; includes migrating high-pressure and low-pressure systems, air masses, and fronts.

synoptic weather network (SWN)   Weather observing stations that provide data primarily for weather map preparation and weather forecasting.

teleconnection   A linkage between weather changes occurring in widely separated regions of the globe.

temperature   A measure of the average kinetic energy of the individual atoms or molecules composing a substance.

temperature gradient   Temperature change with distance.

temperature inversion   An extremely stable air layer in which temperature increases with altitude, the inverse of the usual temperature profile in the troposphere.

terminal velocity   Constant downward-directed speed of a particle within a fluid due to a balance between gravity and fluid resistance.

thermal conductivity   Ability of a material to conduct heat.

thermal low   A low pressure system that develops as a consequence of intense solar heating of the ground, which in turn heats the air, thereby lowering its density; same as a warm-core low.

thermal stability   Resistance to a change in temperature.

thermal turbulence   Irregular, random motions of a fluid triggered by unequal heating and cooling of the fluid.

thermograph   A recording instrument that gives a continuous trace of the variation of temperature with time.

thermometer   An instrument used to measure temperature.

thermosphere   Outermost thermal subdivision of the atmosphere in which temperatures increase with altitude.

thunder   Sound accompanying lightning; produced by violent expansion of air due to intense heating by a lightning discharge.

thunderstorm   A mesoscale weather system produced by strong convection currents that reach to great altitudes within the troposphere. Consists of cumulonimbus clouds accompanied by lightning and thunder and, often, locally heavy rainfall (or snowfall) and gusty surface winds.

tilted updraft   An updraft that is inclined so that precipitation is deflected away from the main updraft. The tilted updraft therefore persists, and the thunderstorm builds to great altitudes.

tipping-bucket rain gauge   A device that collects rainfall in increments of 0.01 in. by containers that alternately fill and empty (tip).

tornado   A small mass of air that whirls rapidly about an almost vertical axis. The tornado is made visible by clouds, and by dust and debris sucked into the system.

tornado alley   Region of maximum tornado frequency in North America; a corridor stretching from central Texas northward into Oklahoma, Kansas, and Nebraska, and eastward into central Illinois and Indiana.

trade winds   Prevailing global-scale surface winds in tropical latitudes; blow from the northeast in the Northern Hemisphere and from the southeast in the Southern Hemisphere.

transmissivity   The fraction of radiation incident on an object that passes through that object.

transpiration   Process by which water vapor escapes from plants through leaf pores.

triple point   The point of occlusion where the cold, warm, and occluded fronts all come together.

Tropic of Cancer   Latitude 23 degrees 27 minutes N; a solstice position of the sun.

Tropic of Capricorn   Latitude 23 degrees 27 minutes S; a solstice position of the sun.

tropical depression   An early stage in the development of a hurricane; winds are at least 37 km (23 mi) per hour but less than 63 km (39 mi) per hour.

tropical disturbance   A region of convective activity over tropical seas with a a detectable center of low pressure; the initial stage in the development of a hurricane.

tropical storm   A tropical cyclone having wind speeds of 63 to 119 km (39 to 74 mi) per hour prehurricane stage.

tropopause   Zone of transition between the troposphere below and the stratosphere above; top of the troposphere.

troposphere   Lowest thermal subdivision of the atmosphere in which air temperature normally drops with altitude; site of most weather.

turbidity   Dustiness of the atmosphere.

turbulence   Irregular, random motions of a fluid such as air or water.

typhoons   Hurricanes that form in the western tropical Pacific Ocean.

ultraviolet radiation (UV)   Short-wave, energetic electromagnetic radiation that is emitted by the sun; much of the solar ultraviolet radiation is absorbed in the stratosphere where it is involved in the formation of ozone.

umbra   Central dark area of a sunspot.

unstable air layer   An air layer characterized by a vertical temperature profile such that air parcels accelerate upward or downward and away from their original altitudes.

updraft   Upward-moving air, usually in a convective or thunderstorm cell.

upslope fog   Ground-level cloud formed as a consequence of the expansional cooling of humid air that is forced to ascend a mountain slope.

upwelling   The upward circulation of cold, nutrient-rich bottom water toward the ocean surface.

urban heat island   The relative warmth of a city compared with surrounding areas.

UV   *See* ultraviolet radiation.

UVB   The biologically effective portion of the solar ultraviolet radiation, in the 0.28 to 0.32 micrometer wavelength range, that reaches the Earth's surface; can cause sunburning and skin cancer.

valley breeze   A shallow, upslope flow of air that develops during the day within mountain valleys.

vapor pressure   That portion of the total air pressure exerted by the water vapor in a sample of air.

variables of state   Temperature, pressure, and density of air.

veer (or veering)   Clockwise shift in wind direction with time. For example, a wind shift from south to southwest.

vertical pressure gradient   Decrease of air pressure with altitude.

virga   A shaft of rain or snow falling from a distant cloud that vaporizes before reaching the ground.

viscosity   Friction within fluids such as air and water.

visible light   Electromagnetic radiation having wavelengths in the range of about 0.40 (violet) to 0.70 (red) micrometers.

wake   A region of turbulent (irregular) airflow that develops to the lee (downwind) of an obstacle such as a building.

warm air advection   Flow of air from a relatively warm locality to a relatively cool locality.

warm clouds   Clouds at temperatures above 0 °C (32 °F); composed of liquid water droplets only.

warm-core anticyclones   High-pressure systems occupying a thick column of subsiding warm, dry air. Subtropical anticyclones are examples.

warm-core low   A surface, synoptic-scale stationary cyclone that develops as a consequence of intense solar heating of a large, relatively dry geographical area; same as a thermal low.

warm front   A narrow zone of transition between relatively warm air that is advancing and relatively cool air that is retreating.

warm-type occlusion   A front formed when a cold front overtakes a warm front and the air behind the front is warmer than the air ahead of the front.

water budget   Balance sheet for the inputs and outputs of water to and from the various global water reservoirs.

watershed   *See* drainage basin.

waterspout   A tornadolike disturbance that forms over a large body of water; usually much weaker than a tornado but associated with a cumulonimbus cloud.

wave cyclone   A low-pressure system that develops along the polar front; cyclonic circulation causes a wave to form along the front.

wave frequency   Number of crests or troughs of a wave that pass a given point in a specified period of time, usually 1 second.

wavelength   Distance between successive crests or successive troughs of a wave.

weather   The state of the atmosphere at some place and time in terms of such variables as temperature, cloudiness, precipitation, and radiation.

weather modification   Any change in weather that is induced by human activity, either intentionally or unintentionally. Cloud seeding and fog dispersal are examples of intentional weather modification, while enhancement of rainfall by air pollution is an example of unintentional weather modification.

weather warning   Issued when hazardous weather is observed.

weather watch   Issued when hazardous weather is considered possible based on current or anticipated atmospheric conditions.

weighing-bucket rain gauge   A device that is calibrated so that the weight of rainfall is recorded directly in terms of rainfall in millimeters or inches.

wet-bulb depression   On a psychrometer, the difference in readings between the wet-bulb thermometer and the dry-bulb thermometer; used to determine relative humidity.

Wien's displacement law   The higher the temperature of a radiating object, the shorter is the wavelength of maximum radiation intensity; applies to blackbodies.

wind pressure   The force per unit area exerted by the wind on some surface.

wind shear   An abrupt change in wind speed or direction with distance.

wind sock   A large, conical, open bag designed to indicate wind direction and relative speed; usually used at small airports.

wind vane   An instrument used to monitor wind direction by always pointing into the wind.

windchill equivalent temperature (WET)   An air temperature index that attempts to gauge the sensible heat loss from exposed skin resulting from the combined effect of low air temperature and wind.

World Meteorological Organization (WMO)   Coordinates weather data collection and analysis by more than 145 member nations; based in Geneva, Switzerland.

World Weather Watch (WWW)   International weather-monitoring network coordinated by the World Meteorological Organization.

X rays   Highly energetic, short-wavelength electromagnetic radiation.

zonal component   East–west motion of air.

zonal flow pattern   Flow of the westerlies almost directly from west to east; westerlies exhibit little amplitude.

zonda   A foehn-type wind that flows into the Andes valleys of Argentina.

Index

Psychrometric Tables

As explained in Chapter 6, a psychrometer is a standard instrument for measuring how close the air is to saturation. It consists of two thermometers mounted side by side: a dry-bulb thermometer and a wet-bulb thermometer. The dry bulb gives the actual air temperature while the wet bulb gives the temperature produced by evaporative cooling. The difference between the dry-bulb and wet-bulb readings is the wet-bulb depression.

The relative humidity and the dew point can be obtained from the dry-bulb temperature and the wet-bulb depression. Use Table A to obtain the relative humidity: find the dry-bulb temperature in the left column and then read across to the relative humidity that corresponds to the wet-bulb depression. Follow the same procedure to determine the dew point in Table B.

TABLE A
Relative Humidity (Percent)

Dry-bulb temp. (°C)	Wet-bulb depression (°C)															
	0.5	1.0	1.5	2.0	2.5	3.0	3.5	4.0	4.5	5.0	7.5	10.0	12.5	15.0	17.5
−10.0	85	69	54	39	24	10	—	—	—	—	—	—	—	—	—
− 7.5	87	73	60	48	35	22	10	—	—	—	—	—	—	—	—
− 5.0	88	77	66	54	43	32	21	11	0	—	—	—	—	—	—
− 2.5	90	80	70	60	50	41	31	22	12	3	—	—	—	—	—
0.0	91	82	73	65	56	47	39	31	23	15	—	—	—	—	—
2.5	92	84	76	68	61	53	46	38	31	24	—	—	—	—	—
5.0	93	86	78	71	65	58	51	45	38	32	1	—	—	—	—
7.5	93	87	80	74	68	62	56	50	44	38	11	—	—	—	—
10.0	94	88	82	76	71	65	60	54	49	44	19	—	—	—	—
12.5	94	89	84	78	73	68	63	58	53	48	25	4	—	—	—
15.0	95	90	85	80	75	70	66	61	57	52	31	12	—	—	—
17.5	95	90	86	81	77	72	68	64	60	55	36	18	2	—	—
20.0	95	91	87	82	78	74	70	66	62	58	40	24	8	—	—
22.5	96	92	87	83	80	76	72	68	64	61	44	28	14	1	—
25.0	96	92	88	84	81	77	73	70	66	63	47	32	19	7	—
27.5	96	92	89	85	82	78	75	71	68	65	50	36	23	12	1
30.0	96	93	89	86	82	79	76	73	70	67	52	39	27	16	6
32.5	97	93	90	86	83	80	77	74	71	68	54	42	30	20	11
35.0	97	93	90	87	84	81	78	75	72	69	56	44	33	23	14
37.5	97	94	91	87	85	82	79	76	73	70	58	46	36	26	18
40.0	97	94	91	88	85	82	79	77	74	72	59	48	38	29	21